Lecture Notes in Computer Science   14422

Founding Editors

Gerhard Goos
Juris Hartmanis

The series Lecture Notes in Computer Science (LNCS), including its subseries Lecture Notes in Artificial Intelligence (LNAI) and Lecture Notes in Bioinformatics (LNBI), has established itself as a medium for the publication of new developments in computer science and information technology research, teaching, and education.

LNCS enjoys close cooperation with the computer science R & D community, the series counts many renowned academics among its volume editors and paper authors, and collaborates with prestigious societies. Its mission is to serve this international community by providing an invaluable service, mainly focused on the publication of conference and workshop proceedings and postproceedings. LNCS commenced publication in 1973.

Weili Wu · Guangmo Tong

Editors

Computing and Combinatorics

29th International Conference, COCOON 2023
Hawaii, HI, USA, December 15–17, 2023
Proceedings, Part I

 Springer

Editors
Weili Wu (iD)
University of Texas at Dallas
Richardson, TX, USA

Guangmo Tong (iD)
University of Delaware
Newark, DE, USA

ISSN 0302-9743 ISSN 1611-3349 (electronic)
Lecture Notes in Computer Science
ISBN 978-3-031-49189-4 ISBN 978-3-031-49190-0 (eBook)
https://doi.org/10.1007/978-3-031-49190-0

This Springer imprint is published by the registered company Springer Nature Switzerland AG
The registered company address is: Gewerbestrasse 11, 6330 Cham, Switzerland

Paper in this product is recyclable.

Preface

The papers in these proceedings, which consist of two volumes, were presented at the 29th International Computing and Combinatorics Conference (COCOON 2023), on December 15–17, 2023, in Honolulu, Hawaii, USA. The topics cover most aspects of theoretical computer science and combinatorics pertaining to computing.

In total 60 papers were selected from 146 submissions by an international program committee consisting of a large number of scholars from various countries and regions, distributed all over the world, including Asia, North America, Europe, and Australia. Each paper was evaluated by at least three reviewers. The decision was made based on those evaluations through a process containing a discussion period.

Authors of selected papers come from the following countries and regions: Australia, Canada, China (including Hong Kong, Macau, and Taiwan), Czechia, France, Germany, India, Israel, Japan, Sweden, and the USA. Many of these papers represent reports of continuing research, and it is expected that most of them will appear in a more polished and complete form in scientific journals.

We wish to thank all who have made this meeting possible and successful, the authors for submitting papers, the program committee members for their excellent work in reviewing papers, the sponsors, the local organizers, and Springer for their support and assistance. We are especially grateful to Lian Li and Xiaoming Sun, who lead the Steering committee, for making the ranking of COCOON go up significantly in recent years, and to Yi Zhu and Xiao Li, who made tremendous efforts on local arrangements and set-up.

December 2023

Weili Wu
Guangmo Tong

Preface

Organization

General Co-chairs

Peter Varman   Rice University, USA
Ding-Zhu Du   University of Texas at Dallas, USA

PC Co-chairs

Weili Wu   University of Texas at Dallas, USA
Guangmo Tong   University of Delaware, USA

Web Co-chairs

Xiao Li   University of Texas at Dallas, USA
Ke Su   University of Texas at Dallas, USA

Finance Co-chair

Jing Yuan   University of Texas at Dallas, USA

Registration Chair

Xiao Li   University of Texas at Dallas, USA

Local Chair

Yi Zhu   Hawaii Pacific University, USA

Program Committee Members

An Zhang	Hangzhou Dianzi University, China
Bhaskar Dasgupta	University of Illinois at Chicago, USA
Bo Li	Hong Kong Polytechnic University, China
Boting Yang	University of Regina, Canada
C. Pandu Rangan	Indian Institute of Technology Madras, India
Chee Yap	New York University, USA
Chia-Wei Lee	National Tatung University, Taiwan
Christos Zaroliagis	University of Patras, Greece
Chung-Shou Liao	National Tsing Hua University, Taiwan
Deshi Ye	Zhejiang University, China
Dominik Köppl	Tokyo Medical and Dental University, Japan
Eddie Cheng	Oakland University, USA
Gruia Calinescu	Illinois Institute of Technology, USA
Guohui Lin	University of Alberta, Canada
Haitao Wang	University of Utah, USA
Hans-Joachim Boeckenhauer	ETH Zurich, Switzerland
Ho-Lin Chen	National Taiwan University, Taiwan
Hsiang-Hsuan Liu	Utrecht University, The Netherlands
Jiangxiong Guo	Beijing Normal University at Zhuhai, China
Joong-Lyul Lee	University of North Carolina at Pembroke, USA
Jou-Ming Chang	National Taipei University of Business, Taiwan
Kai Jin	Sun Yat-sen University, China
Kunihiko Sadakane	University of Tokyo, Japan
Ling-Ju Hung	National Taipei University of Business, Taiwan
M. Sohel Rahman	Bangladesh University of Engineering and Technology, Bangladesh
Manki Min	Louisiana Tech University, USA
Micheal Khachay	Ural Federal University, Russia
Ovidiu Daescu	University of Texas at Dallas, USA
Pavel Skums	Georgia State University, USA
Peng Li	Chongqing University of Technology, China
Peng Zhang	Shandong University, China
Peter Rossmanith	RWTH Aachen University, Germany
Prudence Wong	University of Liverpool, UK
Qilong Feng	Central South University, China
Qiufen Ni	Guangdong University of Technology, China
Raffaele Giancarlo	University of Palermo, Italy
Ralf Klasing	CNRS and University of Bordeaux, France
Ryuhei Uehara	Japan Advanced Institute of Science and Technology, Japan

Sharma V. Thankachan	North Carolina State University, USA
Shengxin Liu	Harbin Institute of Technology at Shenzhen, China
Sun-Yuan Hsieh	National Cheng Kung University, Taiwan
Takeshi Tokuyama	Tohoku University, Japan
Thomas Erlebach	Durham University, UK
Travis Gagie	Dalhousie University, Canada
Van Bang Le	University of Rostock, Germany
Vassilis Zissimopoulos	National and Kapodistrian University of Athens, Greece
Vincent Chau	Southeast University, China
Wenguo Yang	University of Chinese Academy of Sciences, China
Wing-Kai Hon	National Tsing Hua University, Taiwan
Wolfgang Bein	University of Nevada, USA
Xianyue Li	Lanzhou University, China
Xiaowei Wu	University of Macau, China
Xinjian Ding	Beijing University of Technology, China
Xujin Chen	University of Chinese Academy of Sciences, China
Yifei Zou	Shandong University, China
Yitong Yin	Nanjing University, China
Yixin Cao	Hong Kong Polytechnic University, China
Yong Chen	Hangzhou Dianzi University, China
Yuqing Zhu	California State University, Los Angeles, USA
Zhao Zhang	Zhejiang Normal University, China
Zhipeng Cai	Georgia State University, USA

Contents – Part I

Contents – Part II

Combinatorics and Algorithms

Algorithmic Solution in Applications

Algorithm in Networks

Complexity and Approximation

Complexity and Enumeration in Models
of Genome Rearrangement

Lora Bailey[1], Heather Smith Blake[2], Garner Cochran[3], Nathan Fox[4],
Michael Levet[5(✉)], Reem Mahmoud[6], Elizabeth Bailey Matson[7],
Inne Singgih[8], Grace Stadnyk[9], Xinyi Wang[10], and Alexander Wiedemann[11]

[1] Department of Mathematics, Grand Valley State University, Allendale, MI, USA
baileylo@gvsu.edu
[2] Department of Mathematics and Computer Science, Davidson College,
Davidson, NC, USA
hsblake@davidson.edu
[3] Department of Mathematics and Computer Science, Berry College,
Mount Berry, GA, USA
gcochran@berry.edu
[4] Department of Quantitative Sciences, Canisius University, Buffalo, NY, USA
fox42@canisius.edu
[5] Department of Computer Science, College of Charleston, Charleston, SC, USA
levetm@cofc.edu
[6] Department of Computer Science, Virginia Commonwealth University,
Richmond, VA, USA
mahmoudr@vcu.edu
[7] The Division of Mathematics and Computer Science, Alfred University,
Alfred, NY, USA
matson@alfred.edu
[8] Department of Mathematical Sciences, University of Cincinnati,
Cincinnati, OH, USA
inne.singgih@uc.edu
[9] Department of Mathematics, Furman University, Greenville, SC, USA
grace.stadnyk@furman.edu
[10] Department of Computational Mathematics, Science, and Engineering,
Michigan State University, East Lansing, MI, USA
wangx249@msu.edu
[11] Department of Mathematics, Randolph–Macon College, Ashland, VA, USA
alexanderwiedemann@rmc.edu

Abstract. In this paper, we examine the computational complexity of
enumeration in certain genome rearrangement models. We first show
that the PAIRWISE REARRANGEMENT problem in the Single Cut-and-
Join model (Bergeron, Medvedev, & Stoye, *J. Comput. Biol.* 2010) is

We wish to thank the American Mathematical Society for organizing the Mathematics Research Community workshop where this work began. This material is based
upon work supported by the National Science Foundation under Grant Number DMS
1916439. ML was partially supported by J. A. Grochow's NSF award CISE-2047756
and the University of Colorado Boulder, Department of Computer Science Summer
Research Fellowship. ML thanks J.A. Grochow for helpful discussions.

W. Wu and G. Tong (Eds.): COCOON 2023, LNCS 14422, pp. 3–14, 2024.
https://doi.org/10.1007/978-3-031-49190-0_1

#P-complete under polynomial-time Turing reductions. Next, we show that in the Single Cut or Join model (Feijao & Meidanis, *IEEE ACM Trans. Comp. Biol. Bioinf.* 2011), the problem of enumerating all medians (#MEDIAN) is logspace-computable (FL), improving upon the previous polynomial-time (FP) bound of Miklós & Smith (RECOMB 2015).

Keywords: Genome Rearrangement · Phylogenetics · Single Cut-and-Join · Single Cut or Join · Computational Complexity

1 Introduction

With the natural occurrence of mutations in genomes and the wide range of effects this can incite, scientists seek to understand the evolutionary relationship between species. Several discrete mathematical models have been proposed (which we discuss later) to model these mutations based on biological observations. Genome rearrangement models consider situations in which large scale mutations alter the order of the genes within the genome. Sturtevant [26,27] observed the biological phenomenon of genome rearrangement in the study of strains of *Drosophila* (fruit flies), only a few years after he produced the first genetic map [25]. Palmer & Herbon [21] observed similar phenomenon in plants. McClintock [15] also found experimental evidence of genes rearranging themselves, or "transposing" themselves, within chromosomes. Subsequent to his work on *Drosophila*, Sturtevant together with Novitski [28] introduced one of the first genome rearrangement problems, seeking a minimum length sequence of operations (in particular, so-called *reversals* [12]) that would transform one genome into another.

In this paper, we consider genome rearrangement models where each genome consists of directed edges, representing genes. Each directed edge receives a unique label, and each vertex has degree 1 or 2 (where we take the sum of both the in-degree and out-degree). There are no isolated vertices. Notably, each component in the associated undirected graph is either a path or a cycle. Biologically, each component in the graph represents a chromosome. Paths correspond to linear chromosomes, such as in eukaryotes, and cycles correspond to circular chromosomes, which play a role in tumor growth [22].

A genome model specifies the number of connected components (chromosomes), the types of components (linear, circular, or a mix of the two), and the permissible operations. The models we will consider allow for removing (cutting) and creating (joining) instances where two edges (genes) are incident, with certain models allowing for multiple cuts or joins to occur as part of a single operation. The *reversal* model [28], for example, takes as input a genome consisting precisely of a single linear chromosome. In a now classical paper, Hannenhalli & Pevzner [12] exhibited a polynomial-time algorithm for computing the distance between two genomes in the reversal model. Later, those same authors generalized the reversal model to allow for multiple chromosomes and additional operations [11]. There are also several models that permit genomes which consist of both linear and circular chromosomes, including, for instance, the *Single Cut or Join (SCoJ)* [10], *Single Cut-and-Join (SCaJ)* [3], and *Double Cut-and-Join (DCJ)* [30] models (see Sect. 2.1 for a precise formulation). When choosing an appropriate model, it is important to balance biological relevance with computational tractibility. This motivates the study of the computational complexity for genome rearrangement problems.

There are several natural genome rearrangement problems. We have already mentioned the DISTANCE problem, which asks for the minimum number of operations needed to transform one genome into another. Other natural problems include PAIRWISE REARRANGEMENT (see Definition 3) and MEDIAN (Definition 4). We summarize the known complexity-theoretic results in Table 1.

Table 1. Theorems are denoted by T, conjectures by C, and problems with unknown complexity by U (those problems without enough evidence for a conjecture). The entry "not in FP" is under the assumption that P ≠ NP. The entry "not in FPRAS" is under the assumption that RP ≠ NP. Those marked with † and ‡ follow from the fact that the corresponding decision problem is NP-hard, [7] and [29] respectively. Those marked with * indicate results in this paper.

	Reversal	SCoJ	SCaJ	DCJ
DISTANCE	T : in FP [12]	T : in FP [3]	T : in FP [10]	T : in FP [4]
PAIRWISE REARRANGEMENT	C: #P-complete C: In FPRAS	T: in FP [16]	T: #P-complete* U: in/not in FPRAS	C: #P-complete T: in FPRAS [20]
MEDIAN	T: not in FP† T: not in FPRAS†	T: in FP [17] T: in FL*	U: FP/NP-hard U: in/not in FPRAS	T: not in FP‡ T: not in FPRAS‡

Main Results. Our first main result concerns the computational complexity of the PAIRWISE REARRANGEMENT problem in the Single Cut-and-Join model:

Theorem 1. *In the Single Cut-and-Join model, the* PAIRWISE REARRANGEMENT *problem is #P-complete under polynomial-time Turing reductions.*

Remark 1. We establish Theorem 1 in the special case when the adjacency graph (see Definition 5) is a disjoint union of cycles. A related question that remains open is whether this #P-completeness holds when the adjacency graph consists of only paths.

We also improve the known computational complexity of the #MEDIAN problem in the Single Cut or Join model. Miklós & Smith [17] previously showed that counting the number of medians—the #MEDIAN problem—belongs to FP. We carefully analyze their work to obtain the following improved complexity-theoretic upper bound:

Theorem 2. *In the Single Cut or Join model, the #MEDIAN problem belongs to FL.*

A complete proof of Theorem 2 will appear in the full version; we briefly sketch the ideas here. We may, in FL, identify the connected components [23] of an auxiliary graph that Miklós & Smith refer to as the *conflict graph*. We may then evaluate the relevant formulas of [17] in TC⁰, and therefore in FL. As we rely crucially on Reingold's FL connectivity algorithm [23], we conjecture that in the Single Cut or Join model, #MEDIAN is FL-complete.

The full version of our work is available on arXiv [6], and we refer the reader there for the full details of our work.

Further Related Work. There has been significant work on efficient computational approaches, such as sampling and approximation (see, for instance, [8,9,14,16,17,19]), to cope with the intractability of enumeration. In addition to the problems in Table 1, we run into issues of combinatorial explosion when examining statistics such as the breakpoint reuse [2,5] and the size/positions of reversals [1,8]. Developing an efficient uniform or near-uniform sampler would allow for obtaining a statistically significant sample for hypothesis testing. Such samples are needed, for instance, to test the Random Breakpoint Model [2,5] and check if there is natural selection for maintaining balanced replicators [8]. Jerrum, Valiant, & Vazirani [13] showed that finding a near-uniform sampler has the same complexity as enumerating the size of the space. Thus, approximate counting and sampling are closely related. Past work on these samplers has often utilized a rapidly mixing Markov chain on the full evolutionary history space [18].

2   Preliminaries

2.1   Genome Rearrangement

Definition 1. *A* genome *is an edge-labeled directed graph in which each label is unique and the total degree of each vertex is 1 or 2 (in-degree and out-degree combined). In particular, a genome consists of disjoint paths and cycles. The components of a genome we call* chromosomes. *Each edge begins at its* tail *and ends at its* head, *collectively referred to as its* extremities. *Degree 2 vertices are called* adjacencies, *and degree 1 vertices are called* telomeres.

Adjacencies can be viewed as unordered sets of two extremities, and telomeres as sets containing exactly one extremity. For simplicity, we write adjacency $\{a, b\}$ as ab and telomere $\{c\}$ as c. Each genome is then uniquely defined by its set of adjacencies and telomeres. Consider the following operations on a given genome:

(i) *Cut*: an adjacency ab is separated into two telomeres, a and b,
(ii) *Join*: two telomeres a and b become one adjacency, ab,
(iii) *Cut-join*: replace adjacency ab and telomere c with adjacency ac and telomere b, and

(iv) *Double-cut-join*: replace adjacencies ab and cd with adjacencies ac and bd.

Note that a cut-join operation combines a single cut and a single join into one operation, and a double-cut-join operation performs two cuts and two joins in one operation.

Several key models are based on these operations. The *Double Cut-And-Join (DCJ)* model was initially introduced by Yancopoulos, Attie, & Friedberg [30] and permits all four operations. Later, Feijao & Meidanis [10] introduced the *Single Cut Or Join (SCoJ)* model, which only allows operations (i) and (ii). Alternatively, the *Single Cut-And-Join (SCaJ)* model [3] allows operations (i)-(iii), but not operation (iv). In this paper, we consider the Single Cut-And-Join and Single Cut Or Join models.

Definition 2. *For any genome rearrangement model J, it is always possible to perform a sequence of operations from J that transforms genome G_1 into G_2 if they share the same set of edge labels. Such a sequence is called a* scenario. *The minimum length of such a scenario is called the* distance *and is denoted $d^J(G_1, G_2)$. When J is understood, we simply write $d(G_1, G_2)$. An operation on a genome G_1 that (strictly) decreases the distance to genome G_2 is called a* sorting operation *for G_1 and G_2. A scenario requiring $d(G_1, G_2)$ operations to transform G_1 into G_2 is called a* most parsimonious scenario *or* sorting scenario. *When G_2 is understood, we refer to the action of transforming G_1 into G_2 using the minimum number of operations as* sorting G_1. *The number of most parsimonious scenarios transforming G_1 into G_2 is denoted $\#MPS(G_1, G_2)$.*

We now turn to defining the key algorithmic problems that we will consider in this paper.

Definition 3. *Let J be a model of genome rearrangement, and let G_1 and G_2 be genomes. The* Distance *problem asks to compute $d(G_1, G_2)$. The* Pairwise Rearrangement *problem asks to compute $\#MPS(G_1, G_2)$.*

Definition 4. *Let J be a model of genome rearrangement, and let $\mathcal{G}$ be a collection of $k \geq 3$ genomes. A* median *for $\mathcal{G}$ is a genome G that minimizes*

$$\sum_{G_i \in \mathcal{G}} d(G_i, G).$$

The Median *problem asks for one median for $\mathcal{G}$. The $\#$Median *problem asks for the number of medians for $\mathcal{G}$.*

To investigate these computational problems, we begin by introducing the adjacency graph.

Definition 5. *Given two genomes G_1 and G_2 with the same set of edge labels, the* adjacency graph *$A(G_1, G_2)$ is a bipartite multigraph $(V_1 \cup V_2, E)$ where each vertex in V_i corresponds to a unique adjacency or telomere in G_i and the number of edges between two vertices is the size of the intersection of the corresponding adjacency or telomere sets.*

Note that each vertex in an adjacency graph $A(G_1, G_2)$ must have either degree 1 or 2 (corresponding, respectively, to telomeres and adjacencies in the original genome), and so $A(G_1, G_2)$ is composed entirely of disjoint cycles and paths. Note also that every operation on G_1 corresponds to an operation on V_1 in $A(G_1, G_2)$. Whether or not an operation on $A(G_1, G_2)$ corresponds to a sorting operation on G_1—that is, whether it decreases the distance to G_2 or not—depends highly on the structure of the components acted on. To better describe such sorting operations, we adopt the following classification:

Definition 6. *Components of $A(G_1, G_2)$ are classified as follows, where the size of a component B is defined to be $\lfloor |E(B)|/2 \rfloor$:*

- *A W-shaped component is an even path with its two endpoints in V_1.*
- *An M-shaped component is an even path with its two endpoints in V_2.*
- *An N-shaped component is an odd path, further called a trivial path if it is size 0 (a single edge).*
- *A crown is an even cycle, further called a trivial crown if it is size 1 (a 2-cycle).*

The language "trivial" is motivated by the fact that such components indicate where G_1 and G_2 already agree, and hence no sorting operations are required on vertices belonging to trivial components. Indeed, a sorting scenario can be viewed as a minimal length sequence of operations which produces an adjacency graph consisting of only trivial components.

Observation 3. *In the SCaJ model, a case analysis yields precisely these sorting operations on $A(G_1, G_2)$:*

(a) A cut-join operation on a non-trivial N-shaped component, producing an N-shaped component and a trivial crown

(b) A cut-join operation on a W-shaped component of size at least 2, producing a trivial crown and a W-shaped component

(c) A join operation on a W-shaped component of size 1, producing a trivial crown

(d) A cut operation on an M-shaped component, producing two N-shaped components

(e) A cut operation on a non-trivial crown, producing a W-shaped component

(f) A cut-join operation on an M-shaped component and a W-shaped component, where an adjacency in the M-shaped component is cut and joined to a telomere in the W-shaped component, producing two N-shaped components

(g) A cut-join operation on a non-trivial crown and an N-shaped component, where an adjacency in the crown is cut and joined to the telomere in the N-shaped component, producing an N-shaped component

(h) A cut-join operation on a non-trivial crown and a W-shaped component, where an adjacency from the crown is cut and joined with a telomere from the W-shaped component, producing a W-shaped component

Note that (a)–(e) are sorting operations on G_1 that operate on only one component in the adjacency graph, though they may produce two different components. On the other hand, (f)–(h) are sorting operations on G_1 that operate on two separate components in the adjacency graph.

Using these sorting operations on $A(G_1, G_2)$, the distance between two genomes G_1 and G_2 for the SCaJ model is given by

$$d(G_1, G_2) = n - \frac{\#N}{2} - \#T + \#C \tag{1}$$

where n is the number of genes in G_1 (equivalently, one half of the number of edges in $A(G_1, G_2)$), $\#N$ is the number of N-shaped components, $\#T$ is the number of trivial crowns, and $\#C$ is the number of non-trivial crowns [3].

Let $\mathcal{B}$ be the set of all components of $A(G_1, G_2)$ and let $\mathcal{B}'$ be a subset of $\mathcal{B}$. Define

$$d(\mathcal{B}') := \sum_{B \in \mathcal{B}'} \text{size}(B) - \#T + \#C.$$

Note that $d(\mathcal{B}) = d(G_1, G_2)$. In general, $d(\mathcal{B}')$ is the minimum number of operations needed to transform all components of $\mathcal{B}'$ into trivial components, with no operation acting on a component not belonging to $\mathcal{B}'$.

Definition 7. *Let A and B be components of an adjacency graph, and consider a particular sorting scenario. We say $A \sim B$ if either $A = B$ or there is a cut-join operation in the scenario where an extremity a from A and an extremity b from B are joined into an adjacency. The transitive closure of $\sim$, which we call $\equiv$, is an equivalence relation which we call* sort together. *We will be particularly interested in subsets of the equivalence classes of $\equiv$. We abuse terminology by referring to such a subset as a set that* sorts together.

Note that if two components in $A(G_1, G_2)$ sort together, the cut-join witness of this does not need to occur immediately. For example, two non-trivial crowns C_1 and C_2 can sort together by first cutting C_1 to produce a W-shaped component, then operation (b) can be applied multiple times before operation (h) sorts C_2 and the remaining W-shaped component together.

We will now introduce additional notation that we will use in this paper. Let $\mathcal{B}$ be the collection of all components of a given adjacency graph $A(G_1, G_2)$. Let $\Pi(\mathcal{B})$ denote the set of all partitions of $\mathcal{B}$. Define $\#\text{MPS}(\mathcal{B})$ to be the number of most parsimonious scenarios transforming G_1 into G_2. For a partition $\pi \in \Pi(\mathcal{B})$, define $\#\text{MPS}(\mathcal{B}, \pi)$ to be the number of most parsimonious scenarios transforming G_1 into G_2, where two components A and B belong to the same part of π if and only if A and B sort together. For a subset $\mathcal{B}'$ of $\mathcal{B}$, let $\#\text{ST}(\mathcal{B}')$ denote the number of sequences with $d(\mathcal{B}')$ operations in which the components of $\mathcal{B}'$ sort together and are transformed into trivial components with no operation acting on a component not belonging to $\mathcal{B}'$.

We conclude by restricting our attention to a single component of an adjacency graph and determine the number of most parsimonious scenario which sort that component, independent of the other components. We will later use these counts as building blocks to enumerate the most parsimonious scenarios for multiple components in the adjacency graph.

Lemma 1. *Let $A(G_1, G_2)$ be an adjacency graph with component B.*

(a) If B is an N-shaped component, then $\#ST(\{B\}) = 1$.
(b) If B is a W-shaped component of size w, then $\#ST(\{B\}) = 2^{w-1}$.
(c) If B is a M-shaped component of size m, then $\#ST(\{B\}) = 2^{m-1}$.
(d) If B is a non-trivial crown of size c, $\#ST(\{B\}) = c \cdot 2^{c-1}$.

Proof. To appear in the full version.

3   Enumerating All-Crowns Sorting Scenarios

Let $\mathcal{C}$ be a collection of crowns. Recall that $\Pi(\mathcal{C})$ is the set of all partitions of $\mathcal{C}$. For any partition $\pi = (\pi_1, \ldots, \pi_k)$ in $\Pi(\mathcal{C})$, let g_i be the sum of the sizes of the crowns in π_i and p_i be the number of crowns in π_i.

Theorem 4. *Consider an adjacency graph consisting entirely of a collection $\mathcal{C} = \{C_1, C_2, \ldots, C_q\}$ of q crowns where C_i has size c_i. Then*

$$\#MPS(\mathcal{C}) = \sum_{\pi \in \Pi(\mathcal{C})} \binom{d}{g_1 + p_1, \ldots, g_k + p_k} 4^{q-k} 2^{d-q-k} \prod_{i=1}^{q} c_i \left(\prod_{j=1}^{k} \prod_{\ell=0}^{p_j-2} (g_j + \ell) \right) \tag{2}$$

where $d := d(\mathcal{C})$ and, for each $\pi \in \Pi(\mathcal{C})$, $k = k(\pi)$ is the number of parts in π.

Proof. To appear in the full version.

We will now discuss some consequences that will be useful in Sect. 4.

Corollary 1. *Let $\mathcal{C} = \{C_1, C_2, \ldots, C_{2n}\}$ be a collection of $2n$ crowns for some positive integer n. Denote by c_i the size of crown C_i. Suppose that $\sum_{i=1}^{2n} c_i = 2p - 2n$ for some odd prime p. The number of most parsimonious scenarios where all of the crowns sort together is*

$$2^{2p+2n-3} \prod_{i=1}^{2n} c_i \prod_{\ell=0}^{2n-2} (2p - 2n + \ell).$$

Proof. This is Eq. (2) with the powers of 4 and 2 combined, setting $q = 2n$ and $d = 2p$.

Corollary 2. *Let $\mathcal{C} = \{C_1, C_2, \ldots, C_{2n}\}$ be a collection of $2n$ crowns for some positive integer n. Denote by c_i the size of crown C_i. Suppose that $\sum_{i=1}^{2n} c_i = 2p - 2n$ for some odd prime p. Suppose $\mathcal{C}$ is partitioned by (π_1, π_2) with $|\pi_1| = |\pi_2|$ such that the crowns in each π_i sort together and the sum of the sizes of the crowns in π_i is $p - n$. Then*

$$\#MPS(\mathcal{C}, (\pi_1, \pi_2)) = \binom{2p}{p} 2^{2p+2n-6} \prod_{i=1}^{2n} c_i \prod_{\ell=0}^{n-2} (p - n + \ell)^2.$$

Proof. The desired count is one term in Eq. (2) arising from a partition with $k = 2$ parts. Since both parts have sum $p - n$, we have $g_1 = g_2 = p - n$. Hence, the multinomial coefficient becomes $\binom{2p}{p}$. The result follows from combining the powers of 4 and 2 along with the recognition that $k = 2$, $q = 2n$ and $d = 2p$.

Lemma 2. *Let $A = \{a_1, a_2, \ldots, a_n\}$ be a multiset whose elements are each at least 3 and sum to kp, with p an odd prime and k a positive integer. Let $C = \{C_1, C_2, \ldots, C_n\}$ be a collection of n crowns where C_i has size $a_i - 1$. Let π be a partition of C. If $\#MPS(C, \pi)$ is not divisible by p then, in the corresponding partition π' of A, (i) each part $\pi'_i = \{a_{i_1}, \ldots, a_{i_q}\}$ has $q \leq p$ elements, and (ii) $p \mid \sum_{j=1}^q a_{i_j}$.*

Proof. To appear in the full version.

4   PAIRWISE REARRANGEMENT is #P-Complete

In this section, we establish Theorem 1. Due to space constraints, we will outline the construction. A full proof of correctness will appear in the full version.

Our starting point is the MULTISET-PARTITION problem, which is known to be #P-complete under parsimonious reductions [24]. Precisely, the MULTISET-PARTITION problem takes as input a multiset $A = \{a_1, \ldots, a_n\}$ with each $a_i > 0$ is an integer. The goal is to count the number of ways of partitioning A into two multisets $B = \{b_1, \ldots, b_j\}$ and $C = \{c_1, \ldots, c_k\}$ with equal sum, i.e. such that

$$\sum_{i=1}^{j} b_i = \sum_{i=1}^{k} c_i.$$

We say that B and C have equal size if $j = k$. The MULTISET-EQUAL-PARTITION is a variant of MULTISET-PARTITION in which we wish to count partitions of A into two multisets in which both the sizes and sums are equal. We first establish the following:

Proposition 1. MULTISET-EQUAL-PARTITION   *is*   *#P-complete*   *under polynomial-time Turing reductions.*

Proof. To appear in the full version.

In order to establish Theorem 1, we will establish a polynomial-time Turing reduction from MULTISET-EQUAL-PARTITION to PAIRWISE REARRANGEMENT in the SCaJ model. Let A be an instance of MULTISET-EQUAL-PARTITION. Our first step will involve a series of transformations from A to a new multiset A'', where (i) every equal-sum partition of A'' has equal size, and (ii) the number of equal-size, equal-sum partitions of A'' is twice that of A. Then starting from A'', we will construct a polynomial-time Turing reduction to PAIRWISE REARRANGEMENT in the SCaJ model.

Without loss of generality, we may assume that $|A| = 2n$ for some integer n, and that $\sum_{x \in A} x$ is even. Otherwise, A has no equal-size, equal-sum partitions.

Let $a := 1 + \sum_{x \in A} x$. Define A' to be the multiset obtained by adding a to every element of A. We will later construct a multiset A'' from A'. We begin by establishing the following properties about A':

Claim. A' has the same number of equal-size, equal-sum partitions as A.

Claim. Every partition of A' with equal sum must also have equal size.

Given A', we now construct a second multiset A''. Let

$$b := \sum_{x \in A'} x = (2n+1)a - 1.$$

As a is odd, we have that b is even. Choose an integer $c > b$ such that $b+2c = 2p$, for some prime $p > \binom{2n+2}{n+1}$. (We will later construct a set $\mathcal{C}$ of crowns such that the pair $(p, \mathcal{C})$ satisfy the hypotheses of Corollary 1.) Observe that $p \nmid c$. Define $A'' := A' \cup \{c, c\}$. Note that $\sum_{x \in A''} x = 2p$.

Claim. Every partition of A'' with equal sum must also have equal size. In particular, the number of equal-size, equal-sum partitions of A'' is twice the number of equal-size, equal-sum partitions of A'.

We now construct a set of crowns $\mathcal{C}$ from A'' as follows. Let $\mathcal{C} = \{C_1, \ldots, C_{2n+2}\}$ be a set of $2n+2$ crowns, one with size $m-1$ for each $m \in A''$. Observe that the (multipartite) partitions of A'' are in bijection with the (multipartite) partitions of $\mathcal{C}$ (where we stress that each crown in its entirety belongs to a single part; that is, we are partitioning the set $\mathcal{C}$ of crowns and not the underlying genes or telomeres).

Denote c_i to be the size of C_i. Let N be the number of sorting scenarios for $\mathcal{C}$. Let

$$N' = N - 2^{2p+2n-1} \prod_{i=1}^{2n+2} c_i \prod_{\ell=0}^{2n} (2p - 2n - 2 + \ell).$$

By Corollary 1, N' equals the number of most parsimonious scenarios for the $2n+2$ crowns in $\mathcal{C}$ where the crowns do not all sort together. Let $w \equiv N'$ (mod p) with $0 \leq w < p$. We will show later that $p \nmid N'$ (see Claim 4). Now let

$$M = \binom{2p}{p} 2^{2p+2n-4} \prod_{i=1}^{2n+2} c_i \prod_{\ell=0}^{n-1} (p - n - 1 + \ell)^2.$$

By Corollary 2, M equals the number of most parsimonious scenarios for any partition of the crowns into two parts of equal size, where the sum of the sizes of the crowns in each part is the same and the crowns within the same part sort together. Let $u \equiv M$ (mod p) with $0 \leq u < p$. We will now show that $u > 0$.

Claim. M is not divisible by p. Consequently, $u > 0$.

In order to complete the reduction and the proof of Theorem 1, we will recover the number of equal-size, equal-sum partitions of A'' in polynomial-time, relative to the PAIRWISE REARRANGEMENT oracle.

Claim. The number of equal-size, equal-sum partitions of A'' is the least positive integer m satisfying $m \equiv u^{-1}w$ (mod p).

5   Conclusion

We investigated the computational complexity of genome rearrangement problems in the Single Cut-and-Join and Single Cut or Join models. In particular, we showed that the PAIRWISE REARRANGEMENT problem in the Single Cut-and-Join model is #P-complete under polynomial-time Turing reductions (Theorem 1) and in the Single Cut or Join model, #MEDIAN ∈ FL (Theorem 2).

Natural next steps for the Single Cut-and-Join model include investigating the complexity of MEDIAN– it remains open whether this problem is NP-hard. It would also be interesting to investigate whether there is particular combinatorial structure in the Single Cut-and-Join model that forces the PAIRWISE REARRANGEMENT problem to be #P-complete. It is open whether PAIRWISE REARRANGEMENT remains #P-hard when the adjacency graph does not contain any crowns. In light of our result that PAIRWISE REARRANGEMENT is #P-complete, we seek efficient algorithmic approaches to cope with this intractability. One natural approach would be to investigate if we can efficiently sample sorting scenarios; that is, whether PAIRWISE REARRANGEMENT belongs to FPRAS.

References

1. Ajana, Y., Jean-François, L., Tillier, E.R.M., El-Mabrouk, N.: Exploring the set of all minimal sequences of reversals—an application to test the replication-directed reversal hypothesis. In: Guigó, R., Gusfield, D. (eds.) WABI 2002. LNCS, vol. 2452, pp. 300–315. Springer, Heidelberg (2002). https://doi.org/10.1007/3-540-45784-4_23
2. Alekseyev, M.A., Pevzner, P.A.: Comparative genomics reveals birth and death of fragile regions in mammalian evolution. Genome Biol. 11(11), R117 (2010)
3. Bergeron, A., Medvedev, P., Stoye, J.: Rearrangement models and single-cut operations. J. Comput. Biol. J. Comput. Mol. Cell Biol. 17, 1213–1225 (2010)
4. Bergeron, A., Mixtacki, J., Stoye, J.: A unifying view of genome rearrangements. In: Bücher, P., Moret, B.M.E. (eds.) WABI 2006. LNCS, vol. 4175, pp. 163–173. Springer, Heidelberg (2006). https://doi.org/10.1007/11851561_16
5. Bergeron, A., Mixtacki, J., Stoye, J.: On computing the breakpoint reuse rate in rearrangement scenarios. In: Nelson, C.E., Vialette, S. (eds.) RECOMB-CG 2008. LNCS, vol. 5267, pp. 226–240. Springer, Heidelberg (2008). https://doi.org/10.1007/978-3-540-87989-3_17
6. Bailey, L., et al.: Complexity and Enumeration in Models of Genome Rearrangement. arXiv:2305.01851 (2023)
7. Caprara, A.: Formulations and hardness of multiple sorting by reversals. In: Proceedings of the Third Annual International Conference on Computational Molecular Biology, RECOMB 1999, pp. 84–93. Association for Computing Machinery, New York, NY, USA (1999)
8. Darling, A., Miklós, I., Ragan, M.: Dynamics of genome rearrangement in bacterial populations. PLoS Genet. 4, e1000128 (2008)
9. Durrett, R., Nielsen, R., York, T.: Bayesian estimation of genomic distance. Genetics 166, 621–629 (2004)

10. Feijão, P., Meidanis, J.: SCJ: a breakpoint-like distance that simplifies several rearrangement problems. IEEE/ACM Trans. Comput. Biol. Bioinf. **8**, 1318–1329 (2011)
11. Hannenhalli, S., Pevzner, P.A.: Transforming men into mice (polynomial algorithm for genomic distance problem). In: Proceedings of IEEE 36th Annual Foundations of Computer Science, pp. 581–592 (1995)
12. Hannenhalli, S., Pevzner, P.A.: Transforming cabbage into turnip: polynomial algorithm for sorting signed permutations by reversals. J. ACM **46**(1), 1–27 (1999)
13. Jerrum, M.R., Valiant, L.G., Vazirani, V.V.: Random generation of combinatorial structures from a uniform distribution. Theoret. Comput. Sci. **43**, 169–188 (1986)
14. Larget, B., Simon, D.L., Kadane, J.B., Sweet, D.: A Bayesian analysis of metazoan mitochondrial genome arrangements. Mol. Biol. Evol. **22**(3), 486–495 (2004)
15. McClintock, B.: Chromosome organization and genic expression. In: Cold Spring Harbor Symposia on Quantitative Biology, vol. 16, pp. 13–47. Cold Spring Harbor Laboratory Press (1951)
16. Miklós, I., Kiss, S.Z., Tannier, E.: Counting and sampling SCJ small parsimony solutions. Theor. Comput. Sci. **552**, 83–98 (2014)
17. Miklós, I., Smith, H.: Sampling and counting genome rearrangement scenarios. BMC Bioinf. **16**, S6 (2015)
18. Miklós, I., Smith, H.: The computational complexity of calculating partition functions of optimal medians with hamming distance. Adv. Appl. Math. **102**, 18–82 (2019)
19. Miklós, I., Tannier, E.: Bayesian sampling of genomic rearrangement scenarios via double cut and join. Bioinformatics **26**(24), 3012–3019 (2010)
20. Miklós, I., Tannier, E.: Approximating the number of double cut-and-join scenarios. Theoret. Comput. Sci. **439**, 30–40 (2012)
21. Palmer, J.D., Herbon, L.A.: Plant mitochondrial DNA evolves rapidly in structure, but slowly in sequence. J. Mol. Evol. **28**, 87–97 (1988)
22. Raphael, B., Pevzner, P.: Reconstructing tumor amplisomes. Bioinformatics (Oxford, England) **20**(Suppl 1), i265–73 (2004)
23. Reingold, O.: Undirected connectivity in log-space. J. ACM **55**(4) (2008)
24. Simon, J.: On the difference between one and many. In: Salomaa, A., Steinby, M. (eds.) ICALP 1977. LNCS, vol. 52, pp. 480–491. Springer, Heidelberg (1977). https://doi.org/10.1007/3-540-08342-1_37
25. Sturtevant, A.H.: The linear arrangement of six sex-linked factors in drosophila, as shown by their mode of association. J. Exp. Zool. **14**(1), 43–59 (1913)
26. Sturtevant, A.H.: Genetic factors affecting the strength of linkage in drosophila. Proc. Natl. Acad. Sci. U.S.A. **3**(9), 555–558 (1917)
27. Sturtevant, A.H.: Known and probably inverted sections of the autosomes of Drosophila melanogaster. Carnegie Inst. Washington Publisher **421**, 1–27 (1931)
28. Sturtevant, A.H., Novitski, E.: The homologies of the chromosome elements in the genus drosophila. Genetics **26**(5), 517–541 (1941)
29. Tannier, C.Z., Sankoff, D.: Multichromosomal median and halving problems under different genomic distances. BMC Bioinf. **10**, 120 (2009)
30. Yancopoulos, S., Attie, O., Friedberg, R.: Efficient sorting of genomic permutations by translocation, inversion and block interchange. Bioinformatics (Oxford, England) **21**, 3340–3346 (2005)

Conditional Automatic Complexity and Its Metrics

Bjørn Kjos-Hanssen$^{(\boxtimes)}$ ⓘD

University of Hawai'i at Mānoa, Honolulu, HI 96822, USA
bjoernkh@hawaii.edu
https://math.hawaii.edu/wordpress/bjoern/

Abstract. Li, Chen, Li, Ma, and Vitányi (2004) introduced a similarity metric based on Kolmogorov complexity. It followed work by Shannon in the 1950s on a metric based on entropy. We define two computable similarity metrics, analogous to the Jaccard distance and Normalized Information Distance, based on conditional automatic complexity and show that they satisfy all axioms of metric spaces.

Keywords: Automatic complexity · Kolmogorov complexity · Jaccard distance

1 Introduction

In this article we show that metrics analogous to the Jaccard distance and the Normalized Information Distance can be defined based on conditional nondeterministic automatic complexity A_N. Our work continues the path of Shannon (1950) on entropy metrics and Gács (1974) on symmetry of information among others.

Shallit and Wang (2001) defined the automatic complexity of a word w as, somewhat roughly speaking, the minimum number of states of a finite automaton that accepts w and no other word of length $|w|$. This definition may sound a bit artificial, as it is not clear the length of w is involved in defining the complexity of w. In this article we shall see how *conditional* automatic complexity neatly resolves this issue.

Definition 1 ([4,12]). *Let $L(M)$ be the language recognized by the automaton M. Let x be a sequence of finite length n. The (unique-acceptance) nondeterministic automatic complexity $A_N(w) = A_{Nu}(w)$ of a word w is the minimum number of states of an NFA M such that M accepts w and the number of walks along which M accepts words of length $|w|$ is 1.*

The exact-acceptance nondeterministic automatic complexity $A_{Ne}(w)$ of a word w is the minimum number of states of an NFA M such that M accepts w and $L(M) \cap \Sigma^{|w|} = \{w\}$.

This work was partially supported by a grant from the Simons Foundation (#704836 to Bjørn Kjos-Hanssen).

The (deterministic) automatic complexity $A(w)$ is the minimum number of states of a DFA M with $L(M) \cap \Sigma^{|w|} = \{w\}$.

Finally, $A^-(w)$ is the minimum number of states of a deterministic but not necessarily complete (total) NFA with $L(M) \cap \Sigma^{|w|} = \{w\}$.

Remark 1. $A^-(w)$ is so named because it satisfies $A^-(w) \leq A(w) \leq A^-(w) + 1$.

Lemma 1. *For each word w, we have $A_{Ne}(w) \leq A_{Nu}(w)$.*

Proof. We simply note that if M uniquely accepts w, then M exactly accepts w.

Remark 2. Let $w = 00$ and let M be the following NFA:

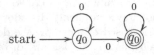

We see that M accepts w on two distinct walks, (q_0, q_0, q_1) and (q_0, q_1, q_1). Hence M exactly accepts w, but does not uniquely accept w. This example is contrived in the sense that we could remove an edge without changing $L(M)$, but it is not clear that this will always be possible for other NFAs. Thus the question whether $A_{Nu} = A_{Ne}$, considered in [6], remains open.

2   Conditional Complexity

Definition 2 (Track of two words). *Let Γ and Δ be alphabets. Let $n \in \mathbb{N}$, $x \in \Gamma^n$ and $y \in \Delta^n$. When no confusion with binomial coefficients is likely, we let $\binom{a}{b} = (a, b) \in \Sigma \times \Delta$. The track of x and y, $x \# y \in (\Gamma \times \Delta)^n$, is defined to be the word*

$$\binom{x_0}{y_0} \binom{x_1}{y_1} \cdots \binom{x_{n-1}}{y_{n-1}},$$

which we may also denote as $\binom{x}{y}$.

Definition 3 (Projections of a word). *Let Γ and Δ be alphabets. Let $n \in \mathbb{N}$, $x \in \Gamma^n$ and $y \in \Delta^n$. The projections π_1 and π_2 are defined by $\pi_1(x \# y) = x$, $\pi_2(x \# y) = y$.*

$x \# y$ can be thoughts of as a parametrized curve $i \mapsto (x_i, y_i)$. The symbol $\#$ reminds us of the two "tracks" corresponding to x and y. This use of the word "track" can be found in [11].

Theorem 1. *There exist words x, y with $A_N(x \# y) \not\leq A_N(x) + A_N(y)$.*[1]

Proof. Let $x = (010)^4$ and $y = (01)^6$. Then we check that $A_N(x \# y) = 6$, $A_N(x) = 3$, and $A_N(y) = 2$.

---

[1] Recall that $A_N := A_{Nu}$.

Definition 4. *A permutation word is a word that does not contain two occurrences of the same symbol.*

Theorem 2. *For all words x, y, we have $\max\{A_N(x), A_N(y)\} \leq A_N(x\#y)$. There exist words x, y with $\max\{A^-(x), A^-(y)\} \not\leq A^-(x\#y)$.*

Proof. Let x be a word of some length n with $A^-(x) > n/2 + 1$. An example can be found among the maximum-length sequences for linear feedback shift registers as observed in [5]. Let y be a permutation word (Definition 4) of the same length. Whenever y is a permutation word, so is $x\#y$. Therefore $A^-(x\#y) \leq n/2 + 1 < A^-(x)$.

Definition 5. *Let Γ and Δ be alphabets. Let $n \in \mathbb{N}$ and $x \in \Gamma^n, y \in \Delta^n$. The conditional (nondeterministic) automatic complexity of x given y, $A_N(x \mid y)$, is the minimum number of states of an NFA over $\Gamma \times \Delta$ such that Item i and Item ii hold.*

(i) Let m be the number of accepting walks of length $n = |x| = |y|$ for which the word w read on the walk satisfies $\pi_1(w) = y$. Then $m = 1$.
(ii) Let w be the word in Item i. Then $\pi_2(w) = x$.

An example of Definition 5 is given in Remark 3.

The conditional complexity $A(x \mid y)$ must be defined in terms of a unique sequence of edges rather than a unique sequence of states, since we cannot assume that there is only one edge from a given state q to given state q'.

Theorem 3. $A_N(x\#y) \leq A_N(x \mid y) \cdot A_N(y)$. *In relativized form,* $A_N(x \mid z) \leq A_N(y \mid z) \cdot A_N(x \mid y\#z)$.

Proof. We describe the unrelativized form only. We use a certain product of NFAs.[2] Let two NFAs

$$M_1 = (Q_1, \Gamma \times \Delta, \delta_1, q_{0,1}, F_1), \quad M_2 = (Q_2, \Gamma, \delta_2, q_{0,2}, F_2)$$

be given. The product is $M_1 \times_1 M_2 = (Q_1 \times Q_2, \Delta, \delta, (q_{0,1}, q_{0,2}), F_1 \times F_2)$ where $\delta((q, q'), a) \ni (r, r')$ if $\delta_1(q, (b, a)) \ni r$ and $\delta_2(q', b) \ni r'$ for some b.

(We can also form $M_1 \times_2 M_2$ where $\delta((q, q'), (b, a)) \ni (r, r')$ if $\delta_1(q, (b, a)) \ni r$ and $\delta_2(q', b) \ni r'$.).

For a walk w, let word(w) be the word read on the labels of w. Consider an accepting walk w from (q_1, q_2) to (r_1, r_2). By definition of the start and final states of $M_1 \times_1 M_2$, the projection $\pi_1(w)$ is also accepting. Hence by *the $A_N(y)$ witness assumption* $\pi_1(w)$ is the only accepting walk of its length and word$(\pi_1(w)) = y$. Since word$(\pi_1(w)) = y$, by *the $A_M(x \mid y)$ witness assumption* w is the unique walk with word$(\pi_1(w)) = y$, and word$(\pi_2(w)) = x$. Thus the accepted word is $x\#y$ and w is the unique accepting walk of its length.

Theorem 4. *There exist x and y with $A_N(x\#y) \neq A_N(x \mid y) \cdot A_N(y)$.*

---

[2] This construction may well have appeared elsewhere but we are not aware of it.

Proof. Let $y = 0001$ and $x = 0123$. It is enough to note that $A_N(x \# y) = 3$, $A_N(y) = 2$, and $A_N(x \mid y)$ is an integer.

Remark 3. Theorem 3 is not optimal in the sense of Theorem 4. On the other hand, Theorem 3 is optimal in the sense that there is a class of word pairs for which it cannot be improved: let $y = (012345)^k$ for some large k, and let $x = (0123)^l$ where $4l = 6k$, so that $|x| = |y|$. We have $A_N(x \# y) = \mathrm{lcm}(4, 6) = 12$, $A_N(y) = 6$, and $A_N(x \mid y) = 2$ as witnessed by the NFA in (1).

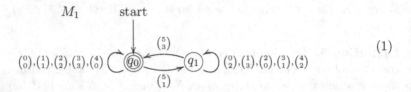

$$(1)$$

The product of this and a cyclic M_2 automaton for y is another cyclic automaton, shown in (2).

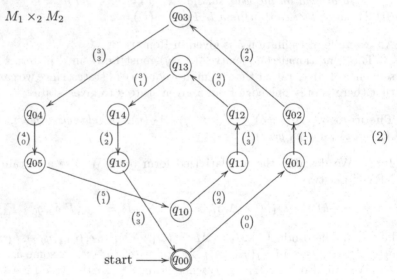

$$(2)$$

Definition 6. *A word x induces an equivalence relation $i \sim_x j \iff x_i = x_j$.*

Theorem 5. *If $\sim_x$ refines $\sim_y$ then $A_N(y \mid x) = 1$.*
   If y is a constant word then $A_N(x \mid y) = A_N(x)$.

An equivalent characterization is that $A(x \mid y)$ is the minimum number of states of an NFA that accepts y on only one walk (but may accept other words of the same length), such that the equivalence relation induced by the sequence of labeled edges used refines $\sim_x$.

3 Bearing on the Unique vs. Exact Problem

A central problem in automatic complexity is whether $A_{Ne} = A_{Nu}$ [6]. Moving to conditional complexity sheds new light.

Definition 7. *A* sparse witness *for $A_{Ne}(x \mid y)$ is an NFA M that witnesses the value of $A_{Ne}(x \mid y)$, with the additional properties that*

(i) if any edge is removed from M, then it is no longer a witness of $A_{Ne}(x \mid y)$; and

(ii) M has fewer or equal number of edges as some witness M_1 of $A_{Nu}(x \mid y)$.

Note that if $A_{Ne}(x \mid y) = A_{Nu}(x \mid y)$ then any witness for $A_{Nu}(x \mid y)$ is a sparse witness for $A_{Ne}(x \mid y)$. The converse fails:

Theorem 6. *There exist binary words x, y such that there is a $A_{Ne}(x \mid y)$ sparse witness that is not an $A_{Nu}(x \mid y)$-witness.*

Proof. We will display slightly more than promised: an $A_{Ne}(x \mid y)$ witness that has *strictly fewer* edges than some $A_{Nu}(x \mid y)$ witness for the same x, y. Consider $x = 0000110$, $y = 0010100$, and the state sequences 00111200 and 01111200. They both generate the same NFA:

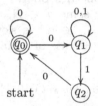

They are sparse witnesses, and have only 6 edges, whereas an A_{Nu} witness with 7 edges is the state sequence 01200210.

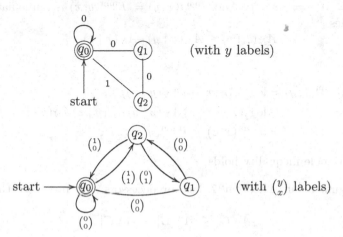

Here, the arrowless edges represent two edges with the same label in opposite directions.

We do not know if sparse witnesses can ever be found in the unconditional case $A_{Ne}(x)$. For example, for $x = 01110$ the state sequence is 011220 is a non-sparse witness; and there are no sparse witnesses for any $|x| \leq 8$.

4   A Jaccard Distance Metric

For binary words x, y let

$$J^{\text{num}}(x, y) = \log(A_N(x \mid y) A_N(y \mid x)). \tag{3}$$

The base of logarithm chosen is not important, but it is sometimes convenient to let $\log = \log_2$.

Let us briefly recall the definitions of metric and pseudometric spaces.

Definition 8. *Let X be a set. A function $d : X \times X \to \mathbb{R}_{\geq 0}$ is a pseudometric if it is commutative, satisfies the triangle inequality, and satisfies $d(x, x) = 0$ for all $x \in X$. If in addition $d(x, y) = 0 \implies x = y$ then d is a metric.*

The underlying space for our metrics will be $\alpha^n / \text{Sym}(\alpha)$ where α is an alphabet, $n \in \mathbb{N}$, and $\text{Sym}(\alpha)$ is the symmetric group of all permutations of α. Our metrics arise from pseudometrics on α^n. If α is finite we can assume $\alpha = [a] = \{0, 1, \ldots, a - 1\}$ for some $a \in \mathbb{N}$, and choose *slow* sequences as representatives. Here, a sequence $s \in \alpha^n$ is slow if $s(0) = 0$ and for each i, $s(i) \leq \max\{s(j) \mid j < i\} + 1$. In the case of a binary alphabet $\{0, 1\}$, this set of representatives is simply $0\{0, 1\}^{n-1}$.

Theorem 7. *J^{num} is a metric on the set $0\{0, 1\}^{n-1}$ for any $n \geq 1$.*

Proof. We have $J^{\text{num}}(x, y) = 0$ iff x and y are isomorphic under some permutation of Σ. If we restrict attention to binary words of the form $0z$ we get $J^{\text{num}}(x, y) = 0 \iff x = y$. And $J^{\text{num}}(x, y) = J^{\text{num}}(y, x)$ is immediate. Theorem 3 implies

$$A_N(x \mid y) \leq A_N(z \mid y) \cdot A_N(x \mid z),$$

hence

$$
\begin{aligned}
J^{\text{num}}(x, y) &= \log_2(A_N(x \mid y) A_N(y \mid x)) \\
&\leq \log_2(A_N(x \mid z) A_N(z \mid y) A_N(y \mid z) A_N(z \mid x)) \\
&= J^{\text{num}}(x, z) + J^{\text{num}}(y, z)
\end{aligned}
$$

hence the triangle inequality holds.

The argument in Theorem 7 has predecessors: for instance, the simple inequality

$$|A \setminus C| \leq |A \setminus B| + |B \setminus C|$$

was used in the analysis of the Jaccard distance (see Kjos-Hanssen 2022 [8]). Horibe (1973) [3] gives the following argument which is credited to a conference talk by Shannon (1950) later published in 1953 [13] (Shannon writes that it is "readily shown" but does not give the argument). Let $H(x \mid y)$ denote the entropy of x given y. Then

$$H(x \mid z) \leq H(x, y \mid z) = H(x \mid y, z) + H(y \mid z)$$
$$\leq H(x \mid y) + H(y \mid z).$$

Similarly, Gács, Tromp, and Vitányi [1] show

$$K(x \mid y^*) \leq^+ K(x, z \mid y^*) \leq^+ K(z \mid y^*) + K(x \mid z^*).$$

Here, K is the prefix-free Kolmogorov complexity, $K(x \mid y)$ its conditional version; y^* is a shortest program for y, so $K(y) = |y^*|$; and $a \leq^+ b$ means $a \leq b + O(1)$. The heavy lifting was already done in 1974 by Gács [2] who showed "symmetry of information", $K(x, y) =^+ K(x) + K(y \mid x^*)$. Symmetry of information does not exactly hold in our setting:

Theorem 8. *There exist words x, y with $A_N(x \mid y)A_N(y) \neq A_N(y \mid x)A_N(x)$.*

Proof. Let $x = 0001$, $y = 0011$, then it would imply $2 \cdot 3 = 2 \cdot 2$.

Remark 4. Let a, b, c, a', b', c' be natural numbers. If $a \leq bc$ and $a' \leq b'c'$ then $\max\{a, a'\} \leq \max\{b, b'\} \max\{c, c'\}$,

Theorem 9. *The following function $J_{\max}^{\mathrm{num}}$ is a metric on $0\{0, 1\}^{n-1}$.*

$$J_{\max}^{\mathrm{num}}(x, y) = \log_2 \max\{A_N(x \mid y), A_N(y \mid x)\}$$
$$= \max\{\log_2 A_N(x \mid y), \log_2 A_N(y \mid x)\}.$$

Proof. To show the triangle inequality, applying Remark 4,

$$\max\{A(x \mid y), A_N(y \mid x)\} \leq \max\{A_N(x \mid z), A_N(z \mid x)\} \cdot \max\{A_N(y \mid z), A_N(z \mid y)\}.$$

Lemma 2 ([8, **Lemma 5**]). *Let $d(x, y)$ be a metric and let $a(x, y)$ be a non-negative symmetric function. If $a(x, z) \leq a(x, y) + d(y, z)$ for all x, y, z, then $d'(x, y) = \frac{d(x,y)}{a(x,y)+d(x,y)}$, with $d'(x, y) = 0$ if $d(x, y) = 0$, is a metric.*

Theorem 10. *The following Jaccard distance type function is a metric on $0\{0, 1\}^n$ (with the convention $0/0 = 0$):*

$$J(x, y) = \frac{\log(A_N(x \mid y)A_N(y \mid x))}{\log(A_N(x \mid y)A_N(y \mid x)A_N(x)A_N(y)) - \log(A_N(x\#y))}$$

Proof. It suffices to prove the triangle inequality. We apply Lemma 2. Namely, let $d(x, y) = \log_2(A_N(x \mid y)A_N(y \mid x))$ and let $a(x, y) = \log_2\left(\frac{A_N(x)A_N(y)}{A_N(x\#y)}\right)$. Then we must show $a(x, z) \leq a(x, y) + d(y, z)$ which is equivalent to (writing $A = A_N$ temporarily)

$$A(x\#y)A(x)A(z) \leq A(x)A(y)A(y \mid z)A(z \mid y)A(x\#z)$$

$$A(x\#y)A(z) \leq A(y)A(y \mid z)A(z \mid y)A(x\#z)$$

Since $A(z) \leq A(z \mid y)A(y)$, it suffices to show

$$A(x\#y) \leq A(y \mid z)A(x\#z)$$

This is shown as follows:

$$A(x\#y) \leq A(y\#(x\#z)) \leq A(y \mid x\#z)A(x\#z) \leq A(y \mid z)A(x\#z).$$

One advantage of J is that it does not depend on the base of the logarithm chosen.

Lemma 3. *For the special case where $y = 0^n$, a constant word, and $x \neq y$, we have $J(x, y) = 1$.*

Proof. We compute

$$J(x, y) = \log(A_N(x))/\log(A_N(x)A_N(x)/A_N(x)) = 1.$$

If this metric is very close to the discrete metric, it is of course not very interesting. This is fortunately not the case:

Theorem 11. *There exist words $x, y \in 0\{0, 1\}^{n-1}$ with $0 < J(x, y) < 1/2$, i.e.,*

$$A_N(x)A_N(y) \not\leq A_N(x \mid y)A_N(y \mid x)A_N(x\#y).$$

Proof. Let $x = u0$, $y = u1$, where $u = 0000100$. Then $J(u0, u1) = 0.46$.

Calculating $A_N(x \mid y)$ for independent random x, y of lengths n up to 20 we find that the mode of the distribution is around $n/4$ (see Appendix).

We do not know whether the problem "$J(x, y) = 1$?" is decidable in polynomial time.

Theorem 12 (Hyde [4]). *For all words x, $A_N(x) \leq |x|/2 + 1$.*

Theorem 13. *Let $x_1, x_2 \in 0\{0, 1\}^{n-1}$, $n \leq 12$, with $x_1 \neq x_2$. Let $S = \{x_1, x_2\}$ and let $A = \{001, 010, 011\}$. Then the following are equivalent:*

1. *$J(x_1, x_2) = 1$;*
2. *either*
   (a) $0^n \in S$, or
   (b) $n \geq 10$, and $S = \{(01)^{n/2}, \alpha^{n/3}\}$ for some $\alpha \in A$.

Proof. This is done by computerized search, so we merely give some remarks on the simplifications making the computation feasible. If there is an example of length 9 or more then by Theorem 12, the equation $A_N(x\#y) = A_N(x)A_N(y)$ must be of the form $a = a \cdot 1$, $a \leq 5$ (already ruled out) or $4 = 2 \cdot 2$. So it is enough to check words of complexity 2. At length 10 we also have to include the possible equation $6 = 2 \cdot 3$, so we consider all words of complexity 3 or less.

Lemma 4. *If α is a permutation word and $k \in \mathbb{N}$ then α^k is $(k+1)$-powerfree.*

Proof. If $w = \alpha^k$ contains a $(k+1)$-power u then let a be the first symbol in u. Then a appears at least $k+1$ times in w. However, there are $|\alpha|$ distinct symbols in w and they each appear k times.

Theorem 14 ([4]). *Let $k \in \mathbb{N}$. If an NFA M uniquely accepts w of length n, and visits a state p at least $k+1$ times during its computation on input w, then w contains a kth power.*

Theorem 15. *Let $k \in \mathbb{N}$, $k \geq 1$. If a word w is kth-powerfree, then $A_N(w) \geq \frac{|w|+1}{k}$.*

Proof. Let k and w be given. Let $q = A_N(w)$ and let M witness that $A_N(w) \leq q$. For a contrapositive proof, assume $q < \frac{|w|+1}{k}$. Thus $k < \frac{|w|+1}{q}$.

Let p be a most-visited state in M during its computation on input w. Then p is visited at least $(|w|+1)/q > k$ times, hence at least $k+1$ times. By Theorem 14, w contains an kth power.

Proposition 1. *For each nonempty permutation word α and each $k \in \mathbb{N}$ with $k \geq |\alpha| - 1$, we have $A_N(\alpha^k) = |\alpha|$.*

Proof. Let $a = |\alpha| \geq 1$, let $k \geq a - 1$, and $w = \alpha^k$. Then by Lemma 4 and Theorem 15,

$$A_N(w) \geq \frac{|w|+1}{k+1} = \frac{ka+1}{k+1} = a - \frac{a-1}{k+1} \geq a - \frac{a-1}{a} > a - 1.$$

Since $A_N(w)$ is an integer, $A_N(w) \geq a$. The other direction just uses a single cycle.

From Proposition 1 we have an infinite family of examples with $J(x,y) = 1$ as in Theorem 13. Namely, x and y can be powers of permutation words of relatively prime length.

Lemma 5. *If α and β are permutation words of lengths a and b, then*

$$(\alpha^{\text{lcm}(a,b)/a}) \# (\beta^{\text{lcm}(a,b)/b})$$

is a permutation word.

Proof. Let this word $w = w_1 \ldots w_n$, $w_i \in \Sigma$. If $w_i = w_j$ then the first and second coordinates of w_i and w_j are equal, so i is congruent to j mod a and mod b. Hence i is congruent to j mod $\text{lcm}(a,b)$, so $i = j$.

The case $k = 2$, $\alpha = 01$, $\beta = 012$ exemplifies Theorem 16. There $a = 2, b = 3$, $u = 010101$, $v = 012012$, $x = 010101010101$, and $y = 012012012012$.

Theorem 16. *Let α and β be permutation words of relatively prime lengths $a = |\alpha|$ and $b = |\beta|$. Let $k \in \mathbb{N}$, $k \geq 2$. Let $x = u^k$, $y = v^k$, where $u = \alpha^b$ and $v = \beta^a$. Then $J(x,y) = 1$.*

Proof. It suffices to show $A_N(x\#y) = A_N(x)A_N(y)$. By Lemma 5, $u\#v$ is a permutation word, and $x\#y = (u\#v)^k$, and so by Proposition 2,

$$A_N(x\#y) = |u\#v| = |u| = ab = A_N(x)A_N(y).$$

In Proposition 2, the condition $k \geq |\alpha| - 1$ in Proposition 1 is strengthened to simply $k \geq 2$ (but of course not to $k \geq 1$).

Proposition 1 still gives nonredundant information in the case where $2 > |\alpha| - 1$, i.e., $\alpha \in \{1,2\}$ and $k \in \{0,1\}$.

Theorem 17 ([7]). *Let $q \geq 1$ and let x be a word such that $A_N(x) \leq q$. Then x contains a set of powers $x_i^{\alpha_i}$, $\alpha_i \geq 2$, $1 \leq i \leq m$ such that all the $|x_i|, 1 \leq i \leq m$ are distinct and nonzero, and satisfying (4).*

$$n + 1 - m - \sum_{i=1}^{m}(\alpha_i - 2)|x_i| \leq 2q. \tag{4}$$

Proposition 2. *For each nonempty permutation word α, and $k \geq 2$, we have $A_N(\alpha^k) = |\alpha|$.*

Proof. First assume $k = 2$. Theorem 17 implies that if a square of a permutation word has complexity at most q, then $n/2 \leq q$.

Now let $k > 2$. Since α^2 is a prefix of α^k, we have $A_N(\alpha^k) \geq A_N(\alpha^2) = |\alpha|$.

We may wonder whether as long as α is primitive, or at least if α has maximal A_N-complexity (achieving Hyde's bound in Theorem 12), without necessarily being a permutation word, Proposition 2 still holds, i.e., $A_N(\alpha^2) = |\alpha|$. But this fails:

Definition 9. *A word w has* emergent simplicity *if $A_N(w)$ is maximal, but $A_N(w^2) < |w|$.*

Proposition 3. *There exists a word having emergent simplicity. The minimal length of such a word in $\{0,1\}^*$ is 7.*

Proof. Let $w = 0001000$. Then w is maximally complex, but $A_N(w^2) = 6$. The only other example of length 7 is 0010100.

Hence not all maximal-complexity words are like 001, 010, 011 in Theorem 13. That is, in general, maximal-complexity words cannot be used to get instances of $J(x,y) = 1$. We next show that this emergent simplicity can only go so far, in Theorem 19.

Definition 10. *A word $\tilde{w}$ is a* cyclic shift *(or a* conjugate*) of another word w if there are words a, b with $w = ab$, $\tilde{w} = ba$.*

Lemma 6 ([11, **Theorem 2.4.2**]). *A cyclic shift of a primitive word is primitive.*

Theorem 18 (Lyndon and Schützenberger [10]; see [11, Theorem 2.3.3]).
Let $x, y \in \Sigma^+$. Then the following four conditions are equivalent:

1. $xy = yx$.
2. *There exists $z \in \Sigma$ and integers $k, l > 0$ such that $x = z^k$ and $y = z^l$.*
3. *There exist integers $i, j > 0$ such that $x^i = y^j$.*

Theorem 19. *If $A_N(w^{|w|}) < |w|$ then w is not primitive (and hence does not have maximal A_N-complexity).*

Proof. If $A_N(w^{|w|}) < |w|$ then $A_N(w^{|w|}) \leq |w| - 1$ and hence, since $|w^{|w|}| = |w|^2$ and $k - 1 \not\geq (k^2 + 1)/(k + 1)$ for all k, $A_N(w^{|w|}) \not\geq \frac{|w|^2+1}{|w|+1}$. Therefore, by Theorem 15 $w^{|w|}$ contains an $(|w| + 1)$th power $u^{|w|+1}$. Thus $(|w| + 1)|u| \leq |w|^2$, so $|u| < |w|$ (*). Since $w^{|w|}$ also contains $u^{|w|}$, there is a cyclic shift $\tilde{w}$ of w such that $u^{|w|} = \tilde{w}^{|u|}$. Then letting $g = \gcd(|w|, |u|)$, we have $u^{|w|/g} = \tilde{w}^{|u|/g}$ as well, and now the exponents are relatively prime. By Theorem 18, there is a word α with $u = \alpha^{|u|/g}$ and $\tilde{w} = \alpha^{|w|/g}$. If $|w|/g > 1$ then $\tilde{w}$ is nonprimitive and hence so is w by Lemma 6. If $|w|/g = 1$ then $|w|$ divides $|u|$ which is a contradiction to (*). $\square$

Theorem 20. *Let $x, y \in 0\{0, 1\}^{n-1}$. Then $A_N(x \mid y) = 1$ iff $x = y$ or $x = 0^n$.*

Proof. Suppose $x \neq 0^n$ and M is a 1-state NFA such that there is only one labeled walk on which there exists z with M accepting $y\#z$ (which we can also write $\binom{y}{z}$); and on that one walk, z is x. M has at most four edges, with labels from among $\binom{0}{0}, \binom{0}{1}, \binom{1}{0}, \binom{1}{1}$.

Since $x = 0u$ and $y = 0v$ both start with 0, the walk must start with an edge labeled $\binom{0}{0}$. Then there is no edge labeled $\binom{0}{1}$, or else M would accept both $y\#x$ and $y\#(1u)$.

Since $x \neq 0^n$, x contains a 1, so there is an edge labeled $\binom{0}{1}$ or $\binom{1}{1}$. So there is an edge labeled $\binom{1}{1}$.

Then there is no edge labeled $\binom{1}{0}$, or else two different z's would occur. (Let $x = 0^k 1w$, then M would accept both $y\#x$ and $y\#(0^k 0w)$.). So M is just the 1-state NFA with two edges labeled $\binom{0}{0}$ and $\binom{1}{1}$ only. Consequently, $x = y$. $\square$

This J deems x and y to be "disjoint" when $A_N(x\#y) = A_N(x)A_N(y)$, which for example happens when x and y are high powers of short words of relatively prime length.

5   A Normalized Information Distance Metric

The following metric $J_{\max}$, more analogous to the Normalized Information Distance [9] than the Jaccard distance, seems better than J above in the following way: If $A_N(x \mid y) = A_N(x)$ and $A_N(y \mid x) = A_N(y)$, then x and y are of no help in compressing each other. This should mean that their distance is maximal ($J_{\max}(x, y) = 1$). This argument can be compared to that made by Li et al. [9]. The condition we get from J, that $J(x, y) = 1$ when $A(x\#y) = A_N(x)A_N(y)$, seems relatively unmotivated in comparison.

Definition 11. *Let x, y be words of length $n \in \mathbb{N}$. We define*

$$J_{\max}(x, y) = \frac{\log \max\{A(x \mid y), A(y \mid x)\}}{\log \max\{A(x), A(y)\}}$$

Theorem 21. $J_{\max}$ *is a metric on the set* $0\{0, 1\}^{n-1}$.

The proof is similar to that of Theorem 10 and is given in the Appendix. For this metric the case $J(x, y) = 1$ is perhaps not much easier to understand. We already have examples where $A(y \mid x) = A(y)$ even though x and y are both nontrivial, such as $x, y \in \{001, 010, 011\}$. We can ask whether this metric embeds in Euclidean space but it fails already at $n = 4$.

Conclusion and Future Work. We have seen that while automatic complexity A_N is quite different from Kolmogorov complexity K, surprisingly we can obtain metrics similar to those for K using $\log A_N$ and a conditional version of A_N. In fact, these metrics are genuine metrics (not just up to some accuracy) and are computable. We also saw that the conditional version of A_N sheds more light upon the problem of distinguishing $A_N = A_{Nu}$ and its word-counting version A_{Ne}. In the future, we hope to characterize when $J = 1$ and $J_{\max} = 1$, and determine whether A_N, J or $J_{\max}$ are efficiently computable.

Appendix

The distributions of $q = A_{Nu}(x \mid y)$ for $n \leq 10$ are as follows (modes indicated in brackets):

$n \backslash q$	1	2	3	4	5	6
0	[1]					
1	[1]					
2	[3]	1				
3	7	[9]				
4	15	[45]	4			
5	31	[197]	28			
6	63	[755]	191	15		
7	127	[2299]	1561	109		
8	255	5905	[9604]	571	49	
9	511	14005	[47416]	3205	399	
10	1023	31439	[206342]	21066	2102	172

For example, the table entry for $(n, q) = (2, 2)$ is 1 since the only instance of $x, y \in 0\{0, 1\}$ with $A_N(x \mid y) = 2$ is $A_N(01 \mid 00)$.

Proof of Theorem 21.

Proof. The nontrivial part is the triangle inequality. Let

$$a_{xy} = \log\max\{A(x), A(y)\} - \log\max\{A(x \mid y), A(y \mid x)\};$$

then by Lemma 2 it suffices to show that $a_{xz} \leq a_{xy} + d_{yz}$. In other words:

$$\log\max\{A(x), A(z)\} - \log\max\{A(x \mid z), A(z \mid x)\}$$
$$\leq \log\max\{A(x), A(y)\} - \log\max\{A(x \mid y), A(y \mid x)\} + \log\max\{A(y), A(z)\}.$$

Equivalently, we must show that

$$\max\{A(x \mid y), A(y \mid x)\}\max\{A(x), A(z)\}$$
$$\leq \max\{A(x \mid z), A(z \mid x)\}\max\{A(y), A(z)\}\max\{A(x), A(y)\}.$$

There are now two cases.
Case 1: $A(x) \leq A(z)$. Then in fact

$$\max\{A(x \mid y), A(y \mid x)\}\max\{A(x), A(z)\}$$
$$\leq \max\{A(y), A(z)\}\max\{A(x), A(y)\}.$$

Case 2: $A(z) < A(x)$. Then we use $A(x \mid y) \leq A(x) \leq A(x \mid z)A(z)$ and $A(y \mid x) \leq A(y)$.

References

1. Gács, P., Tromp, J.T., Vitányi, P.M.B.: Algorithmic statistics. IEEE Trans. Inform. Theory **47**(6), 2443–2463 (2001). https://doi.org/10.1109/18.945257
2. Gač, P.: The symmetry of algorithmic information. Dokl. Akad. Nauk SSSR **218**, 1265–1267 (1974)
3. Horibe, Y.: A note on entropy metrics. Inf. Control **22**, 403–404 (1973)
4. Hyde, K.K., Kjos-Hanssen, B.: Nondeterministic automatic complexity of overlap-free and almost square-free words. Electron. J. Combin. **22**(3), 18 (2015)
5. Kjos-Hanssen, B.: Automatic complexity of shift register sequences. Discrete Math. **341**(9), 2409–2417 (2018). https://doi.org/10.1016/j.disc.2018.05.015
6. Kjos-Hanssen, B.: Few paths, fewer words: model selection with automatic structure functions. Exp. Math. **28**(1), 121–127 (2019). https://doi.org/10.1080/10586458.2017.1368048
7. Kjos-Hanssen, B.: An incompressibility theorem for automatic complexity. Forum Math. Sigma **9**, e62 (2021). https://doi.org/10.1017/fms.2021.58
8. Kjos-Hanssen, B.: Interpolating between the Jaccard distance and an analogue of the normalized information distance. J. Logic Comput. **32**(8), 1611–1623 (2022). https://doi.org/10.1093/logcom/exac069. https://doi-org.eres.library.manoa.hawaii.edu/10.1093/logcom/exac069
9. Li, M., Chen, X., Li, X., Ma, B., Vitányi, P.M.B.: The similarity metric. IEEE Trans. Inform. Theory **50**(12), 3250–3264 (2004). https://doi.org/10.1109/TIT.2004.838101
10. Lyndon, R.C., Schützenberger, M.P.: The equation $a^M = b^N c^P$ in a free group. Michigan Math. J. **9**, 289–298 (1962)

11. Shallit, J.: A Second Course in Formal Languages and Automata Theory, 1st edn. Cambridge University Press, New York, NY, USA (2008)
12. Shallit, J., Wang, M.W.: Automatic complexity of strings. J. Autom. Lang. Comb. **6**(4), 537–554 (2001). 2nd Workshop on Descriptional Complexity of Automata, Grammars and Related Structures, London, ON (2000)
13. Shannon, C.: The lattice theory of information. Trans. IRE Prof. Group Inf. Theory **1**(1), 105–107 (1953). https://doi.org/10.1109/TIT.1953.1188572

Streaming and Query Once Space Complexity of Longest Increasing Subsequence

Xin Li[1] and Yu Zheng[2(✉)]

[1] Johns Hopkins University, Baltimore, MD 21218, USA
`lixints@cs.jhu.edu`
[2] Meta Platforms Inc., Menlo Park, CA 94025, USA
`hizzy1027@gmail.com`

Abstract. Longest Increasing Subsequence (LIS) is a fundamental problem in combinatorics and computer science. Previously, there have been numerous works on both upper bounds and lower bounds of the time complexity of computing and approximating LIS, yet only a few on the equally important space complexity.

In this paper, we further study the space complexity of computing and approximating LIS in various models. Specifically, we prove non-trivial space lower bounds in the following two models: (1) the adaptive query-once model or read-once branching programs, and (2) the streaming model where the order of streaming is different from the natural order.

As far as we know, there are no previous works on the space complexity of LIS in these models. Besides the bounds, our work also leaves many intriguing open problems.

Keywords: longest increasing subsequence · streaming algorithm · approximation algorithm · branching program

1 Introduction

Longest Increasing Subsequence (LIS) is a natural measure of a sequence where the alphabet has a total order, and finding (the length of) LIS in a sequence is a fundamental problem in both combinatorics and computer science, which has been studied for decades. For example, the well known Erdös-Szekeres theorem in combinatorics states that, given any natural numbers r and s, any sequence of at least $(r-1)(s-1)+1$ objects with a total order contains a monotonically increasing subsequence of length r or a monotonically decreasing subsequence of length s. Besides being interesting in its own right, LIS is also closely related to other important string problems. For example, it is a special case of the problem of finding the longest common subsequence (LCS) between two strings, where one string is arranged in the increasing order. As such, algorithms for LIS are often used as subroutines for LCS, which in turn has wide applications in bioinformatics due to its connections to gene sequences.

W. Wu and G. Tong (Eds.): COCOON 2023, LNCS 14422, pp. 29–60, 2024.
https://doi.org/10.1007/978-3-031-49190-0_3

In terms of computing LIS, the classical Patience Sorting algorithm [Ham72, Mal62] can find an LIS of a sequence of length n over an alphabet Σ in time $O(n \log n)$ and space $O(n \log n)$, while the work of [KOO+20] generalizes this by providing a trade-off between the time and space used. Specifically, for any $s \in \mathbb{N}$ with $\sqrt{n} \leq s \leq n$, [KOO+20] gives an algorithm that uses $O(s \log n)$ space and $O(\frac{n^2}{s} \log n)$ time for computing LIS-length, and $O(\frac{n^2}{s} \log^2 n)$ time for finding an actual subsequence. However, no algorithm is known to achieve a better trade-off. We remark that the decision version of LIS is in the class NL (non-deterministic logspace), and thus by Savitch's theorem [Sav70], it can be solved in time $n^{O(\log n)}$ and space $O(\log^2 n)$.

Several works studied the problem of *approximating* LIS, with the goal of achieving either better time complexity or better space complexity. For time complexity, one aims to obtain a good approximation of LIS using *sublinear time*. The known results in this category depend heavily on the length of LIS. For example, the work of [SS17] provides an $(1 + \varepsilon)$ approximation of LIS-length in truly sublinear time if the length is at least $(1-\lambda)n$ for $\lambda = \Omega(\log \log n / \log n)$. When the length of LIS is at least λn for an arbitrary $\lambda < 1$, [RSSS19] gave an algorithm that provides an $O(1/\lambda^3)$ approximation in $\tilde{O}(\sqrt{n}/\lambda^7)$ time. A subsequent work of [MS21] improved the approximation to $O(1/\lambda^\varepsilon)$ and the time complexity to $O(n^{1-\Omega(\varepsilon)}(\log n/\lambda)^{O(1/\varepsilon)})$. The work of [NV21] introduced an $O(1/\lambda)$ approximation algorithm with non-adaptive query complexity $\tilde{O}(\sqrt{r} \operatorname{poly}(1/\lambda))$ assuming there are r distinct values in the sequence. Finally, a recent work of [ANSS22] achieves a $1/\lambda^{o(1)}$ approximation in $O(n^{o(1)}/\lambda)$ time for any $\lambda = o(1)$. We note that all these algorithms are randomized algorithms.

The situation becomes better for space complexity. Here, the goal is to obtain a good approximation of LIS using *sublinear space*, while still maintaining polynomial time. In this context, the work of [Sah17] provides an algorithm for LIS-length that achieves an εn additive approximation using space $O(\frac{\log n}{\varepsilon})$, while a recent work [CFH+21] achieves a $1 + \varepsilon$ approximation using $\tilde{O}_{\varepsilon,\delta}(n^\delta)$ space and $\tilde{O}_{\varepsilon,\delta}(n^{2-2\delta})$ time, for any constants $\delta \in (0, \frac{1}{2})$ such that $\frac{1}{\delta}$ is an integer, and $\varepsilon \in (0, 1)$. [CFH+21] further provides an algorithm that achieves a $1 + O(\frac{1}{\log \log n})$ approximation using $O(\frac{\log^4 n}{\log \log n})$ space and $n^{5+o(1)}$ time.

In addition, due to applications on large data sets, LIS is also well studied in the *streaming model*, where the sequence is accessed from a data stream with one or a small number of passes, rather than random access. In this model, the work of [GJKK07] provides a one-pass streaming algorithm that achieves a $1 + \varepsilon$ approximation of LIS-length, using $O(n \log n)$ time and $O(\sqrt{n/\varepsilon} \log n)$ space, where the space complexity is known to be tight due to the lower bounds given in [GG10, EJ08]. Interestingly, all these algorithms are deterministic, and it is an intriguing question to see if randomized algorithms can achieve better space complexity.

In this work, we further study the lower bounds of space complexity for computing and approximating LIS in various models. Recall that the decision version of LIS is in the class NL, and the question of whether NL = L (L stands for deterministic logspace) is still a major open problem in theoretical computer science.

Hence, to get any non-trivial space lower bound, it is natural and necessary to restrict the models. Here, we study two different models: the query-once model and the streaming model.

Query-Once Model. In this model, we allow the algorithm to have random access to the input sequence, but impose the restriction that the algorithm can only query each element of the sequence at most once, using any adaptive strategy. It is thus a natural and strict generalization of the one-pass streaming model.

In fact, we study the slightly more general model of *read-once branching program*, which can be used to represent the deterministic query-once algorithm. Informally, a read-once branching program models the query-once model as a directed graph, where at each node of the branching program, the program queries one position of the input sequence. Depending on the queried input symbol, the program jumps to another node and continues the process. The read-once property ensures that in any computation path, any input element is queried at most once. The size of the branching program is defined as the number of nodes, which roughly corresponds to $2^{O(s)}$ for a space s computation. A formal description is given in Appendix A. We note that the model of read-once branching program is a non-uniform model, hence is more general than the uniform query-once algorithm model.

Size lower bounds of read-once branching programs for explicit functions have also been the subject of extensive study. Following a long line of research [Weg88, Zák84, Dun85, Juk88, KMW91, SS92, Pon98, Gál97, BW98, ABCR99, Kab03, Li23], the current best lower bound for a function in P is $2^{n-O(\log n)}$ [Li23], which is optimal up to the constant in $O(\cdot)$. There are even functions in uniform-AC^0 that give strong size lower bounds for read-once branching programs [Juk88, KMW91, Gál97, BW98, LZ23], where the current best lower bound is $2^{(1-\delta)n}$ for any constant $\delta > 0$ [LZ23].

However, many of the above lower bounds are achieved by somewhat contrived functions, thus it is also important and interesting to study the size of read-once branching programs computing natural functions. Notable examples include the integer multiplication function [Pon98, BW01, AK03], where a lower bound of $\Omega(2^{n/4})$ [BW01] is known for deterministic read-once branching programs and a lower bound of $2^{\Omega(n/\log n)}$ [AK03] is known for randomized read-once branching programs; and the clique-only function [Zák84, BRS93], where a lower bound of $2^{\Omega(\sqrt{n})}$ is known even for non-deterministic read-once branching programs.

Following this direction, in this paper we study the size of deterministic read-once branching programs for LIS-length.

Streaming Model with Arbitrary Order. In this model, the input sequence is again given to the algorithm in a data stream. However, unlike the standard streaming model for LIS where the sequence is given from the first element to the last element, here we study the streaming model where the elements are given in some arbitrary order, according to a permutation of $[n]$. Indeed, in practice the data stream containing the input sequence may not be exactly in the natural order of the elements. For example, consider the situation of an asynchronous network, where a client sends a long sequence to a server for processing. Even if the client sends the sequence from the first element to the last element, the elements received by the server may be in a different order due to transmission delays in the network. Hence, this model is also a natural and practical generalization of the standard streaming model for LIS. In this paper we study two types of streaming orders, which we define later. The motivation of these orders comes from the fact that a random order falls into either type with high probability. Therefore, these orders actually capture most of the streaming orders.

We study LIS space lower bounds for both deterministic and randomized algorithms in these models. Previously, the only known non-trivial space lower bounds for approximating LIS are in the streaming model with standard order, where the aforementioned works [GG10, EJ08] give a lower bound of $\Omega\left(\frac{1}{R}\sqrt{\frac{n}{\varepsilon}}\log\left(\frac{|\Sigma|}{\varepsilon n}\right)\right)$ for any R pass deterministic streaming algorithm achieving a $1+\varepsilon$ approximation of LIS-length when the alphabet size $|\Sigma| \geq n$, and the subsequent work of [LZ21] which extends this bound to $\Omega(\min(\sqrt{n}, |\Sigma|)/R)$ for any constant $\varepsilon > 0$ with any alphabet size. In contrast, there is no known non-trivial space lower bound for randomized algorithms that $1+\varepsilon$ approximate LIS-length, for any constant $\varepsilon > 0$ in the streaming model.

For exact computation of the LIS-length, [SW07] establishes a space lower bound of $\Omega(n)$ for any $O(1)$-pass randomized streaming algorithm, as long as $|\Sigma| \geq n$, again in the standard order.

In summary, we stress that the only known lower bounds for either approximating or exact computation of the LIS-length in the streaming model are in the *standard order*, as far as we know. Furthermore, there is no known space lower bound (even for deterministic computation of LIS-length) in the query-once model.

1.1   Our Results

In this paper we establish new space lower bounds for computing and approximating LIS-length in the query-once model and the streaming model. We start by stating our results in the query-once model. Below, for a sequence $x \in \Sigma^n$, we use LIS(x) to stand for the length of a longest increasing subsequence in x. In this paper we always assume that the alphabet size $|\Sigma| > n$.

Query-Once Model

Theorem 1. *Given input sequences $x \in \Sigma^n$, any read-once branching program that computes LIS(x) has size $2^{\Omega(n)}$.*

Remark 1. Since our alphabet size $|\Sigma| > n$, our input size is actually $n' = O(n\log n)$ and hence in terms of n', the lower bound is $2^{\Omega(n'/\log n')}$. Also, in our model the read-once branching program reads a symbol in Σ each time, instead of just one bit. It is an interesting open question to see if one can get better lower bounds or in the model where the read-once branching program reads an input bit each time.

This gives the following corollaries.

Corollary 1. *Given input sequences* $x \in \Sigma^n$, *any deterministic algorithm that computes* $\mathsf{LIS}(x)$ *and queries each symbol of* x *at most once (the queries can be adaptive) needs to use* $\Omega(n)$ *space.*

Corollary 2. *Given input sequences* $x \in \Sigma^n$, *any one-pass deterministic streaming algorithm computing* $\mathsf{LIS}(x)$ *needs to use* $\Omega(n)$ *space, regardless of the order of the elements in the stream.*

It is clear that these lower bounds are almost tight, up to a $\log|\Sigma|$ factor.

Remark 2. We note that [SW07] establishes a space lower bound of $\Omega(n)$ for any $O(1)$-pass randomized streaming algorithm that computes $\mathsf{LIS}(x)$, as long as $|\Sigma| \geq n$. However, we stress that their result does NOT supersede ours, since their result only applies to the standard streaming order, where the input sequence is read from left to right. On the other hand, our Corollary 2 applies to any arbitrary streaming order. Therefore, these two results are incomparable. Furthermore, our Theorem 1 and Corollary 1 give lower bounds in the read-once branching program model and query-once model, which are strictly stronger than the streaming model.

Streaming Model in Special Orders. Given an alphabet Σ and an input sequence $x = x_1x_2\cdots x_n \in \Sigma^n$, we represent the order of streaming as a permutation $\pi : [n] \rightarrow [n]$ and write $\pi = \pi_1\pi_2\cdots\pi_n$, where each $\pi_i = \pi(i) \in [n]$. The streaming algorithm has access to x in the order of π, i.e., it sees x_{π_1}, then x_{π_2} and so on. In other words, the index i refers to the i-th symbol in the stream, while the index π_i refers to the π_i-th symbol in the original input sequence x, so the i-th symbol in the stream corresponds to the π_i-th symbol in the original input sequence x.

We prove space lower bounds for two types of orders.

Definition 1 (Type 1 order). π *is a type 1 order with parameter* m *if there are two sets of indices* $I = \{i_1, i_2, \ldots, i_m\}$ *and* $J = \{j_1, j_2, \ldots, j_m\}$, *with* $|I| = |J| = m$ *such that* $\max(I) < \min(J)$ *and* $1 \leq \pi_{i_1} < \pi_{j_1} < \pi_{i_2} < \pi_{j_2} < \cdots < \pi_{i_m} < \pi_{j_m} < n$.

Notice that $\max(I) < \min(J)$ guarantees that for any $i \in I$ and $j \in J$, x_{π_i} appears before x_{π_j} in order π. The constraint $1 \leq \pi_{i_1} < \pi_{j_1} < \pi_{i_2} < \pi_{j_2} < \cdots < \pi_{i_m} < \pi_{j_m} < n$ says that the original indices of the symbols corresponding to I and J in x are interleaved.

For example, any order that first reveals all symbols in odd positions and then all symbols in even positions is an order of Type 1 with parameter $n/2$ since we can let $I = \{1, 2, \ldots, n/2\}$ and $J = \{n/2 + 1, n/2 + 2, \ldots, n\}$.

We have the following lower bounds for deterministic approximation algorithms in Type 1 order.

Theorem 2. *Given input sequences $x \in \Sigma^n$ in any* **type 1** *order with parameter m, any R-pass deterministic algorithm that achieves a $1 + 1/32$ approximation of $\mathsf{LIS}(x)$ needs to use $\Omega(m/R)$ space.*

We believe that type 1 order is interesting, since one can show that, with high probability, a random streaming order is a type 1 order with parameter $\Omega(n)$. Hence type 1 order actually captures most of the streaming orders. In turn, this gives the following corollary.

Corollary 3. *Given input sequences $x \in \Sigma^n$ in a random order sampled uniformly from all permutations on $[n]$, with probability $1 - 2^{-\Omega(n)}$, any R-pass deterministic algorithm that achieves a $1 + 1/32$ approximation of $\mathsf{LIS}(x)$ needs to use $\Omega(n/R)$ space.*

For randomized algorithms, we show a simple lower bound for exact computation.

Theorem 3. *Given input sequences $x \in \Sigma^n$ in any* **type 1** *order with parameter m, any R-pass randomized algorithm that computes $\mathsf{LIS}(x)$ correctly with probability at least $2/3$ needs to use $\Omega(m/R)$ space.*

The second type of orders generalizes type 1 orders, and corresponds to orders with interleaving blocks.

Definition 2 (Type 2 order). *π is a type 2 order with parameters r and s if the following holds. There are r disjoint sets of indices $B_1, B_2, \ldots, B_r$ each of size s such that*

$$\max \left(\bigcup_{l \text{ is odd}} B_l \right) < \min \left(\bigcup_{l \text{ is even}} B_l \right),$$

and for any $1 \le l < r$, we have $\max_{i \in B_l}(\pi_i) < \min_{i \in B_{l+1}}(\pi_i)$.

For example, if we divide $[n]$ evenly into $\sqrt{n}$ blocks each of size n, then any order that first reveals symbols in the odd blocks and then symbols in the even blocks is a Type 2 order with $r = s = \sqrt{n}$. This is because we can pick $r = s = \sqrt{n}$ such that $B_1, B_2, \ldots, B_{r/2}$ are the odd blocks and $B_{r/2+1}, B_{r/2+2}, \ldots, B_r$ are the even blocks.

Remark. Type 1 order with parameter m can be viewed as a special case of type 2 order as it is essentially type 2 order with parameter $r = 2m$ and $s = 1$.

We have the following lower bound for deterministic approximation algorithms in type 2 order.

Theorem 4. *Given input sequences $x \in \Sigma^n$ in any* **type 2** *order with parameters r and s, any R-pass deterministic algorithm that gives a $1 + 1/400$ approximation of $\mathsf{LIS}(x)$ needs to use $\Omega(r \cdot s/R)$ space.*

Intuitively, the streaming order that is most friendly to computing or approximating LIS-length is the natural order (or the reverse order). Indeed, in these models the algorithm given in [GJKK07] achieves a one-pass $1+\varepsilon$ approximation of LIS-length, using $O(n \log n)$ time and $O(\sqrt{n/\varepsilon} \log n)$ space. We thus conjecture that this is the best one can do, and $1 + \varepsilon$ approximation of LIS-length in any streaming order requires $\Omega(\sqrt{n})$ space. Specifically, we have the following conjecture, which seems quite natural but we haven't been able to prove.

Conjecture 1. Given input sequences $x \in \Sigma^n$, for any one-pass streaming order, any deterministic algorithm that achieves $1 + \varepsilon$ approximation of $\mathsf{LIS}(x)$ needs to use $\Omega_\varepsilon(\sqrt{n})$ space.

In particular, it is not clear if there is any streaming order where one can get a constant factor deterministic approximation algorithm for LIS-length that uses $o(\sqrt{n})$ space.

1.2   Technique Overview

We now give an overview of the techniques used in our paper. Full proofs are deferred to the appendix.

Query-Once Model and Read-Once Branching Programs. Our first lower bound is for any read-once branching program. Specifically, we prove that any read-once branching program computing LIS exactly must have a large size, which in turn implies a space lower bound. Assume the alphabet size is m and the input length is n, and let x and y be two different input strings. Let $I_j(x)$ denote the set of positions the branching program queries at the j-th level on input x. Note that $I_j(x)$ may not be equal to $I_j(y)$, due to the adaptivity of the algorithm.

However, we show that if the computation paths of two different inputs x and y go through the same node at the j-th level, then we must have $I_j(x) = I_j(y)$. This is because if not, we can find two new sequences x' and y' such that x' (resp. y') follows the same computation path as x (resp. y) until the j-th level, and at the same time x' and y' follow the same computation path after the j-th level. Since the branching program can query each position of an input at most once, the computation path of x' and y' after the j-th level can not query any position in $I_j(x) \cup I_j(y)$. That means the branching program will output the same result for x' and y', and at the same time, there must be at least one position in x' (and y') not queried by the branching program. We can now change the symbol in that unqueried position of x' to get another sequence with a different LIS-length, but since this position is not queried the branching program will still give the same output. This is a contradiction.

With the above observation, we show that there must be a level in the branching program with $2^{\Omega(n)}$ nodes. The proof is based on the following two claims.

First, for any two different increasing sequences $x, y \in \Sigma^n$ and a subset S of $[n]$ with size $n/5$, if x and y are not equal when restricted to S (i.e. $x|_S \neq y|_S$), then we can find two new sequences x' and y' with unequal LIS-length such that for positions in S, $x' = x$ and $y' = y$ (i.e. $x'|_S = x|_S$, $y'|_S = y'|_S$) and for positions not in S, $x' = y'$ (i.e. $x'|_{[n]\setminus S} = y'|_{[n]\setminus S}$) (Claim 1 in the appendix). The construction of x' and y' is given in the proof of Claim 1.

Second, there exists a set of $2^{\Omega(n)}$ increasing sequences such that for any $S \subseteq [n]$ with size $n/5$ and any two sequences in the set, they are not equal when restricted to S (Claim 2 in the appendix). The proof is based on a probabilistic argument, where we show that by independently randomly choosing $2^{\Omega(n)}$ increasing sequences, there is a non-zero probability that they satisfy the claim.

Now, consider the set in Claim 2 and the $n/5$-th level of the branching program, we argue that any two sequences in the set cannot go through the same node at that level. This is because if they do, then by Claim 1, we can build two new sequences with different LIS-length but the branching program will give the same output. Thus, the $n/5$-th level must have $2^{\Omega(n)}$ nodes. This yields our $2^{\Omega(n)}$ size lower bound for read-once branching programs and $\Omega(n)$ space lower bound any adaptive query-once algorithm.

Streaming in Special Orders. As many other streaming space lower bounds, our proof is based on reductions from communication complexity problems. Specifically, we consider a 2-party communication problem where Alice and Bob each holds a different part of the input sequence based on the streaming model. In addition, our proof uses error-correcting codes to create gaps that are necessary for our approximation lower bounds. A more detailed description is given below.

Type 1 Order. In type 1 order with parameter m, there are $2m$ positions and the streaming first reveals the odd positions and then the even positions. To make the presentation easier, let's consider the special case where we first see all the odd positions of the input sequence, and then all the even positions. We can translate this streaming order into a 2-party communication problem, where Alice holds all the odd positions of the input sequence and Bob holds all the even positions. Their goal is to approximate the LIS-length. The space required by any streaming algorithm is at least the communication complexity between Alice and Bob.

The communication complexity lower bound is established by constructing a large fooling set. Specifically, our fooling set is obtained from a simple binary asymptotically good error-correcting code C with codeword length $n/4$. The size of the code C (number of codewords) is $2^{\Omega(n)}$ and for any two different codewords c_1 and c_2, their Hamming distance is $\Omega(n)$.

Assume Alice holds a codeword $u \in \{0,1\}^{n/4}$ and Bob holds a codeword $v \in \{0,1\}^{n/4}$. Our proof gives a construction that transforms u and v into a sequence such that the odd positions only depend on u and the even positions only depend on v. Denote this new sequence by $z = z(u,v)$. We divide z into $n/4$ small blocks of size 4, such that, for the i-th block, if $u_i = v_i$, or $u_i = 0$ and $v_i = 1$, then the LIS-length of this block is 1; on the other hand, if $u_i = 1$ and $v_i = 0$, then the LIS-length of this block is 0. We design the blocks so that the total LIS-length of z is the summation of the LIS-lengths of all blocks. Thus, when $u = v$, the LIS of z is always a fixed number. But when $u \neq v$, since u

and v are both codewords of C, there is a constant fraction of positions such that $u_i \neq v_i$. Then the LIS-length of one of $z(u,v)$ and $z(v,u)$ is smaller by a constant factor. This makes C a fooling set and gives the $\Omega(n)$ space lower bound. Generalizing this to any parameter $m < n$, we get the lower bound of $\Omega(m)$ for type 1 order with parameter m.

We also show that any randomized algorithm computing LIS-length exactly in this order must use $\Omega(m)$ space. The proof is based on a reduction from the set-disjointness problem to computing LIS exactly.

Type 2 Order. In type 2 order, we assume Alice and Bob each holds r interleaved blocks of size s. Since type 1 order is a special case of type 2 order with $r = m$ and $s = 1$, naturally, we want to extend our previous techniques to type 2 order and construct another fooling set.

Recall that for type 1 order, our fooling set is obtained from an asymptotically good error-correcting code, where Alice and bob each holds a codeword, u and v. The input sequence constructed from these codewords is divided into m small blocks, where the i-th block is determined by u_i and v_i. Whenever $u_i \neq v_i$, we can potentially reduce the LIS-length by 1. Since the code C has $\Omega(m)$ distance, this gives a lower bound for constant factor approximation algorithms.

In the case of type 2 order, each block has size s. If we can only create a gap of 1 in LIS-length for each pair of blocks depending on whether the corresponding bits of u and v are equal, then we can only get a lower bound for $1 + o(1)$ approximation. To amplify the gap, we use another asymptotically good error-correcting code.

Here we present a simplified version of our construction to illustrate the high-level idea, while the actual construction is slightly more complicated. We use two asymptotically good error-correcting codes, $C^{(1)}$ and $C^{(2)}$. $C^{(1)} \subseteq \{0,1\}^s$ is a binary code, which has codeword length s and distance $s/4$. Note that $C^{(1)}$ is a subset of $\{0,1\}^s$. $C^{(2)} \subseteq (C^{(1)})^r$ uses $C^{(1)}$ as its alphabet, and it has codeword length r and distance $r/2$. Since both $C^{(1)}$ and $C^{(2)}$ are asymptotically good, we have $|C^{(1)}| = 2^{\Omega(s)}$ and $|C^{(2)}| = 2^{\Omega(r \cdot s)}$.

We assume Alice and Bob each holds a codeword of $C^{(2)}$, denoted by A and B. The first step is to construct a $s \times 2r$ weighted directed grid graph using A and B. The grid has $2r$ columns and each column has s nodes. In this graph, each node has two outgoing edges, one connecting to the node on its right, and the other connecting to the node below. The graph has the following properties. All edges going downward has weight 1, and all edges going from an even column to an odd column has weight 1. For edges going from an odd column to an even column, the weights depend on A and B. Specifically, for an edge going from the j-th node in the $(2i-1)$-th column to the j-th node in the $2i$-th column, it has weight 1 if $(A_i)_j \neq (B_i)_j$ and 0 otherwise. Here, A_i is the i-th symbol of A, which is a codeword of $C^{(1)}$; and $(A_i)_j$ the j-th symbol of A_i, which is in $\{0,1\}$. The same notation applies to $(B_i)_j$.

If $A = B$, then any edge going from an odd column to its right have weight 0. Otherwise, by the distance property of our codes, there are at least $r/2$ odd columns such that for each column, at least $1/4$ of the edges going out of this column to its right must have weight 1. This is because the code $C^{(2)}$ has distance $r/2$ and $C^{(1)}$ has distance $s/4$.

We now transform the graph into a 2-party communication problem where Alice holds the odd columns and Bob holds the even columns. Their goal is to approximate the largest weight of any path going from the top-left node to the bottom-right node. We can use a combinatorial argument based on the Erdös-Szekeres theorem to show that if $A = B$, the largest weight of any path is a fixed number; however, if $A \neq B$, the largest weight of any path increases by a constant factor. Thus, the code $C^{(2)}$ gives a fooling set for this problem and any deterministic algorithm that solves the problem needs communication complexity at least $\log(|C^{(2)}|) = \Omega(r \cdot s)$.

Finally, we reduce this problem to approximating LIS-length in type 2 order, by assigning appropriate values to each node of the grid graph and reordering the nodes into a sequence. The high-level idea is that for each path in the grid graph, we can find a sequence whose LIS-length is equal to the weight of the path. This gives our space lower bound in type 2 order.

1.3   Related Work

There have also been several works studying the "complement" problem of LIS, namely approximating the distance to monotonicity, i.e., $d_m(x) = n - \mathsf{LIS}(x)$, in both the sublinear-time and the streaming settings (note that computing $d_m(x)$ exactly is equivalent to computing $\mathsf{LIS}(x)$). For time complexity, this was first studied by [ACCL07], and a subsequent work of [SS17] gave a $(1 + \varepsilon)$ approximation algorithm in time $\mathsf{poly}(1/d_m(x), \log n)$ for any constant $\varepsilon > 0$.

In the streaming setting, [GJKK07, EJ08] gave algorithms that achieve $O(1)$ approximation of $d_m(x)$ using $\mathsf{polylog}(n)$ space. This was later improved by [SS13] to achieve a randomized $(1+\varepsilon)$-approximation algorithm using $\mathsf{polylog}(n)$ space, and further by [NS14] to achieve a deterministic $(1 + \varepsilon)$-approximation algorithm using $\mathsf{polylog}(n)$ space. [NS14] also proved streaming space lower bound of $\Omega(\log^2 n/\varepsilon)$ and $\Omega(\log^2 n/(\varepsilon \log \log n))$ for deterministic and randomized $(1 + \varepsilon)$-approximation of $d_m(x)$, respectively.

LIS has also been recently studied in the Massively Parallel Computation (MPC) model and the fully dynamic model. In the former, [IMS17] gave a $O(1/\varepsilon^2)$-round algorithm that achieves $(1 + \varepsilon)$-approximation of LIS-length, as long as each machine uses space $n^{3/4+\Omega(1)}$. In the latter, a sequence of works [MS20, GJ21, KS21] resulted in an exact algorithm [KS21] with sublinear update time, together with a deterministic $1 + o(1)$-algorithm algorithm with update time $n^{o(1)}$.

2   Computing LIS in the Query-Once Model

In this section, we consider the *query-once model* where algorithms only allowed to access each symbol of the input sequence once. We show that in this model, any algorithm that computes LIS exactly must use $\Omega(n)$ space.

Theorem 1. *Given input sequences $x \in \Sigma^n$, any read-once branching program that computes $\mathsf{LIS}(x)$ has size $2^{\Omega(n)}$.*

We introduce some notations. For any set of indices $S \subseteq [n]$, the function f_S takes two sequences $x, y \in \Sigma^n$ as input and outputs a sequence σ such that $\sigma|_S = x|_S$ and $\sigma|_{[n]\setminus S} = y|_{[n]\setminus S}$. So if $\sigma = f_S(x, y)$, then for any $i \in [n]$,

$$\sigma_i = \begin{cases} x_i, & \text{if } i \in S, \\ y_i, & \text{if } i \notin S. \end{cases}$$

Our proof is based on the following 2 claims.

Claim 1. *Assume the alphabet size $m > c \cdot n$ for some large enough constant c, and $x, y \in [m]^n$ are two increasing sequences. Then $\forall\, S \subset [n]$ with $|S| = n/5$ such that $x|_S \neq y|_S$, there is a sequence $z \in [0, m+1]^n$ such that*

$$\mathsf{LIS}(f_S(x, z)) \neq \mathsf{LIS}(f_S(y, z)).$$

Claim 2. *Assume the alphabet size $m > c \cdot n$ for some large enough constant c, then there exists a set T of increasing subsequences in $[m]^n$ with size $2^{\Omega(n)}$, such that $\forall\, S \subset [n]$ with $|S| = n/5$ and $\forall\, x, y \in T$, we have $x|_S \neq y|_S$.*

Due to the page limit, the proof of Claim 1 and Claim 2 are deferred to Appendix B.

Proof (Proof of Theorem 1). Assume there is a levelled branching program that computes LIS exactly. Let $t = n/5$, we show that the number of nodes at the t-th level is $2^{\Omega(n)}$.

We consider the computation path of different input sequences. Given an input sequence x, we let $I_j(x)$ denote the set of positions that have been queried in the first j levels by the branching program. Meanwhile, we say a sequence is even if all its symbols are even numbers. We have the following observation.

Consider two sequences x and y that takes only even values, if their computation path both go through the same node in j-th level, then we must have $I_j(x) = I_j(y)$. To see this, assume $I_j(x) \neq I_j(y)$, let $I_j = I_j(x) \cup I_j(y)$. If $j = n$, $I_j(x) \neq I_j(y)$ means some positions in x or y are not queried. Say there is a position l that is not queried in x, then fix any longest increasing subsequence of x, we can change x_l to an odd value and increase the LIS by 1. This will not influence the computation path since x_l is not queried. If $1 \leq j < n$, we can build two new sequences $\tilde{x}$ and $\tilde{y}$ by letting $\tilde{x}|_{I_j} = x|_{I_j}$, $\tilde{y}|_{I_j} = y|_{I_j}$, and fill any positions in $[n] \setminus I_j$ with symbol 1. Then after j-th level, $\tilde{x}$ and $\tilde{y}$ will follow the same computation path since after j-th level, the branching program will not query any position in I_j (or it will query some position in x or y twice). We must have $\mathsf{LIS}(\tilde{x}) = \mathsf{LIS}(\tilde{y})$. Since the LIS is only determined by positions in I_j, we have $x|_{I_j} = y|_{I_j}$. By the same argument as the case of $j = n$, we can get a contradiction.

Also notice that, for two even increasing sequences x and y, if their computation path both go through the same node in $t = n/5$-th level, let $S = I_t(x)$, then we must have $x|_S = y|_S$. This follows from Claim 1. To see this, if $x|_S \neq y|_S$, then according to Claim 1, there exists $\tilde{x}$ and $\tilde{y}$ such that $\tilde{x}|_S = x|_S$, $\tilde{y}|_S = y|_S$, $\tilde{x}|_{[n]/S} = \tilde{y}_{[n]/S}$ and $\mathsf{LIS}(\tilde{x}) \neq \mathsf{LIS}(\tilde{y})$. If we use $\tilde{x}$ as input to the branching program, in the first t levels, it will follow the exact same computation path as x since $\tilde{x}|_S = x|_S$. $\tilde{y}$ will also follow the same computation path as y. Thus, the computation path of $\tilde{x}$ and $\tilde{y}$ will collide at t-th level. Notice that $\tilde{x}|_{[n]/S} = \tilde{y}|_{[n]/S}$, after t-th level, $\tilde{x}$ and $\tilde{y}$ will follow the same computation path, which means our branching program will output the same result for $\tilde{x}$ and $\tilde{y}$. However, this is contradictory to $\mathsf{LIS}(\tilde{x}) \neq \mathsf{LIS}(\tilde{y})$.

Let T be a set of increasing sequences guaranteed by Claim 2. We can turn it into a set of even increasing subsequences by multiplying each symbols by a factor of two. This will not affect the properties guaranteed by Claim 2 and will only increase the alphabet size by a factor of 2, which is still $O(n)$. We denote this new set by T'. By the above observation, the computation paths of any two sequences in T' must go through different nodes in t-th level. Thus, the t-th level must have $2^{\Omega(n)}$ nodes. The space used by the branching program is $\Omega(n)$.

Corollary 4. *Assume the input sequence x is given as a stream, then no matter what is the order of the stream, any 1-pass deterministic algorithm computing $\mathsf{LIS}(x)$ takes $\Omega(n)$ space.*

Proof. The streaming model can be viewed as a restricted version of the query-once model where the algorithm can only access input sequence in a specific order. Thus, the space lower bound for query-once model also holds for the streaming model.

3   Lower Bounds for Streaming LIS in Different Orders

Due to page limit, we defer this section to Appendix C.

4   Open Problems

Our work leaves several natural open problems. First, can one get better lower bounds for the query-once model/read-once branching programs? Can one generalize the bounds to query-k times/read-k times branching programs? Can one get any lower bounds for randomized or non-deterministic branching programs? How about lower bounds for approximation?

Second, can one get any lower bounds for randomized algorithms approximating LIS-length in the streaming model? Finally, is our conjecture in the streaming model true, or is there any streaming order where one can get a constant factor deterministic approximation of LIS-length that uses $o(\sqrt{n})$ space?

Acknowledgments. The first author is supported by NSF CAREER Award CCF-1845349 and NSF Award CCF-2127575. The second author was partially supported by NSF CAREER Award CCF-1845349, with work mostly done while being a graduate student at Johns Hopkins University.

A Preliminaries

Notations. We use $[n]$ to denote the set of all positive integers that are at most n, and use $[a, b]$ to denote the set of all integers that are at least a and at most b.

For any set of indices $I \subseteq [n]$, we use $x|_I$ to denote the subsequence of x restricted to the indices in I. In other words, assume $I = \{i_1, i_2, \ldots, i_t\}$ and $1 \leq i_1 < i_2 < \cdots < i_t \leq n$, then $x|_I = x_{i_1} x_{i_2} \cdots x_{i_t}$.

Branching Program. The following definition is from [Bea89]. An R-way branching program consists of a directed acyclic rooted graph of out-degree $R = R(n)$ with each non-sink node labelled by an index from $\{1, 2, \ldots, n\}$ and with R out-edges of each node labelled $1, \ldots, R$. Edges of the branching program may also be labelled by a sequence of values from some output domain. The *size* of a branching program is the number of nodes it has.

Let $x = (x_1, x_2, \ldots, x_n)$ be an n-tuple of integers chosen from the range $[R]$. An R-way branching program computes a function of input x as follows. The computation starts at the root of the branching program. At each non-sink node v encountered, the computation follows the out edge labelled with the value of x_i where i is the index that labels node v (i.e. variable x_i is queried at v). The computation terminates when it reaches a sink node. The sequence of nodes and edges encountered is the *computation path* followed by x. The output of the branching program on input x is determined by the sink node the computation path ends in.

The *time* used by a branching program is the length of the longest computation path followed by any input. The *space* used by a branching program is the logarithm base 2 of its size.

An R-way branching program is *levelled* if the nodes of the underlying graph are assigned levels so that the root has level 0 and the out edges of a node at level l only go to nodes at level $l+1$. It is known that any branching program can be levelled without changing its time and with at most squaring its size [Pip79]. This will not change the time used and will increase the space complexity by a factor of 2 at most. Thus we can assume R-way branching programs are levelled without loss of generality. If the branching program has the additional property that in any computation path, any x_i is queried at most once, then it is called a read-once branching program.

The Log Sum Inequality. In Appendix C.2, we will use the following inequality.

Lemma 1 (The log sum inequality). *Given $2n$ positive numbers $a_1, a_2, \ldots, a_n$ and $b_1, b_2, \ldots, b_n$, let $a = \sum_{i=1}^{n} a_i$ and $b = \sum_{i=1}^{n} b_i$, then we have*

$$\sum_{i=1}^{n} a_i \log \frac{a_i}{b_i} \geq a \log \frac{a}{b}.$$

Communication Complexity and Fooling Set. We will consider the 2-party communication model where 2 players Alice and Bob each holds input $x \in X$, $y \in Y$ respectively. The goal is to compute a function $f : X \times Y \to \{0,1\}$. We define the *deterministic communication complexity* of f in this model as the minimum number of bits required to be sent by the players in every deterministic communication protocol that always outputs a correct answer. Correspondingly, the *randomized communication complexity* (denoted by $R_\epsilon(f)$) is the minimum number of bits required to be sent by the players in every randomized communication protocol that can output a correct answer with probability at least $1 - \epsilon$.

Some of our proofs use the classical fooling set argument. Fooling set is defined as following.

Definition 3 (Fooling set). *Let $f : X \times Y \to \{0,1\}$. A subset $S \subseteq X \times Y$ is a fooling set for f if there exists $z \in \{0,1\}$ such that $\forall (x,y) \in S$, $f(x,y) = z$ and for any two distinct $(x_1,y_1),(x_2,y_2) \in S$, either $f(x_1,y_2) \neq z$ or $f(x_2,y_1) \neq z$.*

We have the following Lemma.

Lemma 2 (Corollary 4.7 of [Rou15]). *If f has a fooling set S of size t, then the deterministic communication complexity of f is at least $\log_2 t$.*

Set-Disjointness Problem. We will use some classical results about the communication complexity of 2 party set-disjointness problem. 2 party set-disjointness problem is defined as following. Assuming there are two parties, Alice and Bob, each of them holds a k-subset of $[n]$, i.e. Alice holds a set $A \subseteq [n]$ with $|A| = k$ and Bob holds another set $B \subseteq [n]$ with $|B| = k$. Alice and Bob wants to compute whether A and B are disjoint ($\mathsf{DISJ}_{n,k}$). Here,

$$\mathsf{DISJ}_{n,k}(A,B) = \begin{cases} 1, & \text{if } A \cap B = \emptyset, \\ 0, & \text{if } A \cap B \neq \emptyset. \end{cases}$$

We have the following result about the randomized communication complexity of set-disjointness.

Theorem 5 (Theorem 1.2 of [HW07]). *For any $c < 1/2$, the randomized communication complexity $R_{1/3}(\mathsf{DISJ}_{n,k}) = \Omega(k)$ for every $k \leq cn$.*

Erdös-Szekeres Theorem.

Theorem 6 (Erdös-Szekeres theorem). *Given two integers $r > 0$ and $s > 0$, any sequence of distinct real numbers with length at least $(r-1)(s-1)+1$ contains a monotonically increasing subsequence of length r or a monotonically decreasing subsequence of length s.*

B Proofs Omitted in Section 2

In this section, we provide formal proofs omitted in Sect. 2.

B.1 Proof of Claim 1 in Section 2

Proof. In other words, the claim says if two increasing sequences x and y are not equal on a subset of indices S with $|S| = n/5$, then there is a way to build two sequences $\tilde{x}, \tilde{y} \in [0, m+1]^n$ with $\tilde{x}|_S = x|_S$, $\tilde{y}|_S = y|_S$ and $\tilde{x}|_{[n]/S} = \tilde{y}|_{[n]/S}$ such that $\mathsf{LIS}(\tilde{x}) \neq \mathsf{LIS}(\tilde{y})$.

Let us fix index set $S \in [n]$ and increasing sequences x, y such that $x|_S \neq y|_S$. Our goal is two build $\tilde{x}$ and $\tilde{y}$ satisfying the above properties. We first fix $\tilde{x}|_S = x|_S$, $\tilde{y}|_S = y|_S$ and say that the positions not in S are unfixed.

Let $l \in S$ be the first index that $x_l \neq y_l$ and r be the last position that $x_r \neq y_r$. There are two cases. First, there are at least $2n/5$ unfixed positions after l (with index larger than l). Second, there are less than $2n/5$ unfixed positions after l.

For the first case, without loss of generality, we assume $x_l < y_l$. Let E be the set of symbols that appeared in $x|_S$ or $y|_S$ with value larger than x_l, i.e.

$$E = \{a \mid a > x_l \text{ and } a = x_i \text{ or } a = y_i \text{ for some } i \in S\}$$

Let u_E be the sequence of all elements in E concatenated in the increasing order. Since $|S| = n/5$, we have $|E| = |u_E| \leq 2n/5 - 1$. We can build $\tilde{x}$ and $\tilde{y}$ as follows. First, we put 0 into all unfixed positions before l (if any). Then, for unfixed positions after l, we first put u_E into the unfixed positions (notice that we assume there are at least $2n/5$ unfixed positions after l and $|u_E| \leq 2n/5 - 1$) and for the remaining unfixed positions, we put $m+1$ into them.

We argue that $\mathsf{LIS}(\tilde{x}) \neq \mathsf{LIS}(\tilde{y})$. To see this, notice that $\tilde{x}$ and $\tilde{y}$ are equal before l-th position and all symbols before l-th position are smaller than x_l. Also notice that, in both $\tilde{x}|_{[l+1:n]}$ and $\tilde{y}|_{[l+1:n]}$, there are exactly $|E| + 1$ distinct symbols (symbols in E plus the symbol $m+1$) and we can find an increasing subsequence of length $|E| + 1$ (u_E plus the symbol $m+1$). Since all symbols after l-th position are larger than x_l (in both $\tilde{x}$ and $\tilde{y}$). For $\tilde{x}$, the longest increasing subsequence is the longest increasing subsequence before l-th position plus the symbol x_l, and plus the longest increasing subsequence after l-th position. For $\tilde{y}$, the longest increasing subsequence we can find is similar except we can not include the symbol x_l (which does not appear in $\tilde{y}$). Thus, we have $\mathsf{LIS}(\tilde{x}) = \mathsf{LIS}(\tilde{y}) + 1$.

For the second case, the argument is symmetric. There are $4n/5$ unfixed positions. If there are less than $2n/5$ unfixed positions after l. There are at least $2n/5$ unfixed positions before r. Without loss of generality, we assume $x_l > y_l$ and let E be the set of symbols that appeared in $x|_S$ or $y|_S$ with value smaller than x_r. Similarly, let u_E be the sequence of all elements in E concatenated in the increasing order.

We can build $\tilde{x}$ and $\tilde{y}$ as follows. First, we put $m+1$ into all unfixed positions after r (if any). Then, for unfixed positions before r, we first put a 0 in the first unfixed position and then put u_E into the unfixed positions. If there are any remaining unfixed positions, we put 0 into them. By a similar analysis in the first case, we can show that $\mathsf{LIS}(\tilde{x}) = \mathsf{LIS}(\tilde{y}) + 1$. This finishes the proof.

B.2   Proof of Claim 2 in Section 2

Proof. We prove the claim by a probabilistic argument. Let A be the set of all increasing sequence in $[m]^n$ ($|A| = \binom{m}{n}$) and x, y are two sequences sampled i.i.d. from A uniformly. Then, consider a fixed set $S \subset [n]$ with $|S| = n/5$. We now give an estimation of the probability of $x|_S = y|_S$. Before the computation, we introduce some notations. For simplicity, let $t = n/5$. Let $S = \{i_1, i_2, \ldots, i_t\}$ such that $i_1 < i_2 < \cdots < i_t$. In addition, we let $i_0 = 0$ and $i_{t+1} = n + 1$. For $i \in [t+1]$, let

$$d_l = i_l - i_{l-1} - 1.$$

And for any fixed sequence $z \in A$, we let $z_{t+1} = m + 1$ and

$$a_l(z) = z_{i_l} - z_{i_{l-1}} - 1.$$

Clearly, $a_l(z) \geq d_l$ since z is an increasing sequence. Also, we have $\sum_{l=1}^{t+1} d_l = n - t$, and $\sum_{l=1}^{t+1} a_l(z) = m - t$.

We can write $\Pr[x|_S = y|_S] = \frac{1}{|A|} \sum_{z \in A} \Pr[x|_S = z|_S]$. For a fixed z, $\Pr[x|_S = z|_S]$ is exactly the number of sequences in A satisfies $x|_S = z|_S$ divided by the size of A. Notice that $x|_S = z|_S$ fixed t positions in x. For any $l \in [t]$, there are $\binom{a_l(z)}{d_l}$ ways to pick symbols between i_l-th position and i_{l-1}-th position. Here, we allow $i_l = i_{l-1} + 1$ since in this case $\binom{a_l(z)}{d_l} = 1$. For simplicity, we write $a_l = a_l(z)$ for simplicity. Thus, for fixed $z \in A$,

$$\Pr[x|_S = z|_S] = \frac{1}{|A|} \cdot \prod_{l=1}^{t+1} \binom{a_l}{d_l}$$

$$\leq \frac{1}{|A|} \cdot \prod_{l=1}^{t+1} \left(\frac{e \cdot a_l}{d_l}\right)^{d_l} \tag{1}$$

$$\leq \frac{1}{|A|} \cdot e^{n-t} \cdot \prod_{l=1}^{t+1} \left(\frac{a_l}{d_l}\right)^{d_l}$$

By the log sum inequality (Lemma 1) and the fact that $\sum_{l=1}^{t+1} d_l = n - t$, and $\sum_{l=1}^{t+1} a_l(z) = m - t$, we have

$$\log\left(\prod_{l=1}^{t+1} \left(\frac{a_l}{d_l}\right)^{d_l}\right) = \sum_{l=1}^{t+1} d_l \log \frac{a_l}{d_l}$$

$$= -\sum_{l=1}^{t+1} d_l \log \frac{d_l}{a_l} \tag{2}$$

$$\leq -(n-t) \cdot \log \frac{n-t}{m-t}$$

$$= (n-t) \cdot \log \frac{m-t}{n-t}$$

Combine Eq. (1), Eq. (2) and $|A| = \binom{m}{n}$, we have

$$\Pr[x|_S = z|_S] \leq \frac{1}{|A|} \cdot e^{n-t} \cdot \prod_{l=1}^{t+1} \left(\frac{a_l}{d_l}\right)^{d_l}$$

$$\leq \frac{1}{|A|} \cdot e^{n-t} \cdot \left(\frac{m-t}{n-t}\right)^{n-t} \tag{3}$$

$$\leq \left(\frac{n}{m}\right)^n \cdot \left(e \cdot \frac{m-t}{n-t}\right)^{n-t}$$

Plug in $t = n/5$, we have $\Pr[x|_S = z|_S] \leq \left(\frac{n}{m}\right)^n \cdot \left(e \cdot \frac{m-t}{n-t}\right)^{n-t} \leq \left(\frac{n}{m}\right)^{n/5} \cdot \left(\frac{5e}{4}\right)^{4n/5}$. Thus, for any two x, y sampled i.i.d. from A, we have

$$\Pr[x|_S = y|_S] \leq \left(\frac{n}{m}\right)^{n/5} \cdot \left(\frac{5e}{4}\right)^{4n/5}.$$

Let $I = \{S \subset [n]$ such that $|S| = n/5\}$. We have

$$\Pr[\exists\, S \in I, x|_S = y|_S] \leq \sum_{S \in I} \Pr[x|_S = y|_S]$$

$$\leq \binom{n}{n/5} \cdot \left(\frac{n}{m}\right)^{n/5} \cdot \left(\frac{5e}{4}\right)^{4n/5} \tag{4}$$

$$\leq \left(5e \cdot \frac{n}{m} \cdot \left(\frac{5e}{4}\right)^4\right)^{n/5}.$$

Taking $m = \Omega(n)$ to be large enough, we have $\Pr[\exists\, S \in I, x|_S = y|_S] \leq 2^{-n/5}$.

If we take a random subset $T \subseteq A$ with size $2^{n/20}$. By a union bound,

$$\Pr[\forall\, x, y \in T \text{ and } S \in I, x|_S \neq y|_S] \geq 1 - \binom{|T|}{2} \cdot 2^{-n/5} > 0. \tag{5}$$

Thus, there exists a set $T \subset [m]^n$ of increasing sequences with size $2^{n/20}$ such that for any two sequences $x, y \in T$ and any $S \in I$, we have $x|_S \neq y|_S$. This proves the claim.

C   Lower Bounds for Streaming LIS in Different Orders

In this section, we consider the problem of computing/approximating LIS in the streaming model but with different orders.

We restate the definition of the orders we studied here.

Definition 1 (Type 1 order). π *is a type 1 order with parameter m if there are two sets of indices $I = \{i_1, i_2, \ldots, i_m\}$ and $J = \{j_1, j_2, \ldots, j_m\}$, with $|I| = |J| = m$ such that $\max(I) < \min(J)$ and $1 \leq \pi_{i_1} < \pi_{j_1} < \pi_{i_2} < \pi_{j_2} < \cdots < \pi_{i_m} < \pi_{j_m} < n$.*

Definition 2 (Type 2 order). π *is a type 2 order with parameters* r *and* s *if the following holds. There are* r *disjoint sets of indices* $B_1, B_2, \ldots, B_r$ *each of size* s *such that*

$$\max \Big(\bigcup_{l \text{ is odd}} B_l \Big) < \min \Big(\bigcup_{l \text{ is even}} B_l \Big),$$

and for any $1 \leq l < r$, *we have* $\max_{i \in B_l}(\pi_i) < \min_{i \in B_{l+1}}(\pi_i)$.

C.1   Lower Bounds for Type 1 Orders

Lower Bounds for Deterministic Algorithms

Theorem 2. *Given input sequences* $x \in \Sigma^n$ *in any* **type 1** *order with parameter* m, *any* R-*pass deterministic algorithm that achieves a* $1 + 1/32$ *approximation of* $\mathsf{LIS}(x)$ *needs to use* $\Omega(m/R)$ *space.*

Proof. Without loss of generality, we assume n is an even number. Consider the order $\pi = 1, 3, \ldots, n-1, 2, 4, \ldots n$, i.e. in order π, we first see symbols in odd positions in natural order and then symbols in even positions in natural order. We first show an $\Omega(n)$ lower bound for this order to illustrate our idea. Theorem 2 then follows from the proof by a simple observation.

The idea is to build a large fooling set. Let $C \subseteq \{0,1\}^{n/4}$ be an asymptotically good error-correcting code with constant rate and distance $n/16$. We consider the following two party communication scenario: Alice holds a codeword $u \in C$ and Bob holds a codeword $v \in C$.

Consider the following construction. Alice first turns $u \in \{0,1\}^{n/4}$ into $u' \in \mathbb{N}^{n/2}$ by the following transformation. Let u' be an empty sequence at first. Then for each $i \in [n/4]$, let $u_i \in 0, 1$ be the i-th symbol of u. If $u_i = 0$, we attach $0, 2i - 1$ to the end of u' and if $u_i = 1$, we attach $2i - 1, 0$ to the end of u'. Bob does a similar transformation to get $v' \in \mathbb{N}^{n/2}$. The only difference is that if $v_i = 0$, we attach $0, 2i$ and if $v_i = 1$, we attach $2i, 0$ to the end of v'. We note both u' and v' has exactly $n/4$ symbols that are 0 and $n/4$ that are nonzero.

We can get a sequence $z \in \mathbb{N}^n$ such that

$$z = u'_1, v'_1, u'_2, v'_2, \ldots, u'_{n/2}, v'_{n/2}$$

.

In other words, z is equal to u' if we only look at the odd positions of z and is equal to v' if we only look at the even positions. Notice that the sequence z is different if Alice holds v and Bob holds u. Thus, we use $z_{u,v}$ to denote the sequence we get when Alice hols u and Bobs holds v and $z_{v,u}$ vice versa.

We now show that if $u = v$, $\mathsf{LIS}(z_{u,v}) \geq n/2$. Otherwise,

$$\min\{\mathsf{LIS}\,(z_{u,v}), \mathsf{LIS}\,(z_{v,u})\} \leq \frac{15n}{32} + 1.$$

To see this, we can divide $z = z_{u,v}$ into $n/4$ blocks each with 4 symbols. Let z^i be the i-th block. By our construction of z, $z^i = u'_{2i-1}, v'_{2i-1}, u'_{2i}, v'_{2i}$. There are four cases:

- Case 1: $u_i = v_i = 0$, $z^i = 0, 0, 2i - 1, 2i$.
- Case 2: $u_i = 1, v_i = 0$, $z^i = 2i - 1, 0, 0, 2i$.
- Case 3: $u_i = 0, v_i = 1$, $z^i = 0, 2i, 2i - 1, 0$.
- Case 4: $u_i = v_i = 1$, $z^i = 2i - 1, 2i, 0, 0$.

We note that in all cases except case 3, the two non-zero symbols $2i - 1, 2i$ form an increasing sequence. Thus if $u = v$, then each block z^i is either in case 1 or case 4. Thus, $1, 2, 3, \ldots, n/2 - 1, n/2$ is a subsequence of z since $2i - 1, 2i$ is a subsequence of block z^i. We have $\mathsf{LIS}(z) \geq n/2$.

If $u \neq v$, since u, v are both codewords from C, by our assumption, the Hamming distance between u and v is at least $n/16$. Thus, there are at least $n/16$ positions $i \in [n/4]$ such that z^i is in case 2 or case 3. Let a be the number of blocks z^i that are of case 2 and b be the number of blocks z^i that are in case 3. Thus $a + b > n/16$ and one of a or b is at least $n/32$. Without loss of generality, we can assume that $b \geq n/32$ since if not, we can look at the sequence $z_{v,u}$. This is because if $z_{u,v}^i$ is in case 2, then $z_{v,u}^i$ is in case 3 and vice versa.

Notice that for any $1 \leq i < j \leq n/4$, the nonzero symbols in z^i is always strictly smaller than the nonzero symbols in z^j. The best strategy to pick a longest increasing subsequence is to pick the longest increasing subsequence with nonzero symbols in each block z^i and then combine them (with an additional 0 symbol from the first block in some cases). In each block z^i, the length of longest increasing subsequence with nonzero symbols is 2 if the block is in case 1, 2, and 4. And it is 1 if the block is in cases 3. By our assumptions that the number of blocks in cases 3 is at least $n/32$. We know $\mathsf{LIS}(z) \leq \frac{15n}{32} + 1$.

We can define a function $f : C \times C \to \{0, 1\}$ such that

$$f(u, v) = \begin{cases} 1, & \text{if } \mathsf{LIS}(z_{u,v}) \geq n/2 \\ 0, & \text{if } \mathsf{LIS}(z_{u,v}) \leq \frac{15n}{32} + 1 \end{cases}$$

We have shown that C is a fooling set for function f. The deterministic communication complexity of f is $\log |C| = \Omega(n)$. Also notice that if we can get a $1 + 1/32$ approximation of LIS in the order that first reveals odd symbols and then even symbols, we can use it as a protocol to compute f exactly. Thus, any R pass streaming algorithm for this order that gives a $1 + 1/32$ approximation must use a $\Omega(n/R)$ space.

Finally, we generalize this result to any order of type 1 with parameter m. This is because we can use an error correcting code with codeword length $m/2$ as a fooling set. The above construction gives us a sequence $z \in \mathbb{N}^{2m}$. We can then obtain another sequence z' such that when z' is restricted to indices in $I \cup J$, it is equal to z and all other symbols are fixed to 0. In the two party communication scenario, Alice holds symbols with indices in I and Bob holds symbols with indices in J. Their goal is to compute $\mathsf{LIS}(z')$. By the same fooling set argument, we get the lower bound.

We now show that, with high probability, deterministically approximating LIS of a sequence given in a random streaming order requires $\Omega(n)$ space.

Corollary 3. *Given input sequences $x \in \Sigma^n$ in a random order sampled uniformly from all permutations on $[n]$, with probability $1 - 2^{-\Omega(n)}$, any R-pass deterministic algorithm that achieves a $1 + 1/32$ approximation of $\mathsf{LIS}(x)$ needs to use $\Omega(n/R)$ space.*

Before the proof of Corollary 3, we first show the following claim.

Claim 3. *Assume n is a multiple of 32, and a streaming order is uniformly randomly sampled from all permutations of $[n]$, with probability at least $1 - 2^{-\Omega(n)}$, it is a type 1 order with parameter $n/32$.*

Proof. Let π be a streaming order of n elements, which is also a permutation of $[n]$. Denote the set of all permutations over n by S_n. Let A_π be the first $n/2$ elements revealed by the order π.

First, for a subset $A \subseteq [n]$, we consider A can be divided into how many intervals of consecutive integers. In other words, each subset A can be represented as the union of multiple intervals of consecutive integers. Let $g(A)$ be the minimal number of intervals that A can be represented as, i.e.

$$g(A) = \min\{k \mid A = \bigcup_{t=1}^{k} [i_t, j_t) \text{ for some } i_1 < j_1 < i_2 < j_2 < \cdots < i_k < j_k\}$$

Here, $[i, j)$ is the set of all integers at least i and smaller then j. For example, if $A = 1, 2, \ldots, n/2$, then $g(A) = 1$. If A is the subset of all even integers in $[n]$, then $g(A) = n/2$.

For any permutation π, if $g(A_\pi) = m + 1$ distinct gaps, then π must a type 1 order with parameter m. This is because we can represent A_π as $\bigcup_{t=1}^{m+1} [i_t, j_t)$ for some $i_1 < j_1 < i_2 < j_2 < \cdots < i_{m+1} < j_{m+1}$. By the definition, we have $j_i \notin A_\pi$. We can pick $I = \{i_1, i_2, \ldots, i_m\}$ and $J = \{j_1, j_2, \ldots, j_m\}$ (j_{m+1} may be equal to $n + 1$). Since $I \subseteq A_\pi$ and $J \cap A_\pi = \emptyset$, Any elements in I will be revealed before J in the order of π. Thus, π is a type 1 order with parameter m.

Then, we show that if A is uniformly sampled from the set of all subsets with size $n/2$, then $g(A) \geq n/2$ with probability at least $1 - 2 - \Omega(n)$.

We compute how many subsets A has $g(A) \leq n/32$. Given an integer k, consider the how many set A has $g(A) = k$. A can be divided into k intervals. There are $\binom{n/2-1}{k-1}$ different ways to divide $n/2$ consecutive integers into k consecutive intervals.

For elements not in A, depending on whether $i_0 = 0$ or $j_k = n + 1$, there are $k - 1$, k or $k + 1$ consecutive intervals. For simplicity, we add 2 dummy elements 0 and $n + 1$, and consider how many ways we can divide $\{0, 1, \ldots, n/2 + 1\}$ into $k + 1$ consecutive intervals. Similarly to elements in A, there are $\binom{n/2+1}{k}$ choice of the size of intervals.

Thus, the number of subsets A with $g(A) = k$ and $|A| = n/2$ is $\binom{n/2-1}{k-1} \cdot \binom{n/2}{k}$. We can bound the number of sets A with $g(A) \leq n/32$ as following

$$\left|\{A \subset [n] \mid |A| = n/2 \text{ and } g(A) \le n/32\}\right| = \sum_{i=1}^{n/32} \binom{n/2-1}{k-1} \cdot \binom{n/2}{k}$$

$$= \sum_{i=1}^{n/32} \frac{(k-1)!(n/2-k)!}{(n/2-1)!} \cdot \frac{k!(n/2-k)!}{(n/2!)}$$

$$\le \frac{n}{2} \cdot \sum_{i=1}^{n/32} \binom{n/2}{k}^2$$

$$\le \frac{n}{2} \cdot \left(\sum_{i=1}^{n/32} \binom{n/2}{k}\right)^2. \tag{6}$$

Notice that

$$\sum_{i=1}^{n/32} \binom{n/2}{k} 2^{-n/2} \le \binom{n/2}{n/32} 2^{-15n/32}. \tag{7}$$

This is because we can view $\sum_{i=1}^{n/32} \binom{n/2}{k} 2^{-n/2}$ as the probability of doing $n/2$ independent coin flip and there are at most $n/32$ coin flips with head up. This happens if and only if there is a set of $(n/2 - n/32) = 15n/32$ coin flips ends with tail up. There are $\binom{n/2}{15n/32} = \binom{n/2}{n/32}$ choices if subsets with size $15n/32$, and for each, the probability of all coin flips is tail up is $2^{-15n/32}$. Taking a union bound, we get Eq. (7). Combining with Eq. (6), we have the following bound

$$\left|\{A \subset [n] \mid |A| = n/2 \text{ and } g(A) \le n/32\}\right| \le \frac{n}{2} \cdot \left(\sum_{i=1}^{n/32} \binom{n/2}{k}\right)^2$$

$$\le \frac{n}{2} \cdot \left(\binom{n/2}{n/32} \cdot 2^{n/32}\right)^2$$

$$\le \frac{n}{2} \cdot \left((16e)^{n/32} \cdot 2^{n/32}\right)^2$$

$$\le \frac{n}{2} \cdot (32e)^{n/16}$$

$$\le \frac{n}{2} \cdot 2^{7n/16}$$

There are $\binom{n}{n/2}$ subsets of $[n]$ with size exactly $n/2$. When n is a large enough integer, if we sample a subset A from all subsets of $[n]$ with size $n/2$, we have

$$\Pr[g(A) \le n/32] = \frac{\left|\{A \subset [n] \mid |A| = n/2 \text{ and } g(A) \le n/32\}\right|}{\binom{n}{n/2}}$$

$$\le \frac{\frac{n}{2} \cdot 2^{7n/16}}{2^{n/2}}$$

$$\le 2^{-n/32}.$$

Finally, we notice that if π is uniformly sampled from S_n, then A_π is uniformly distributed over all subset of $[n]$ with size $n/2$. It is because for any two subsets $A, B \subset [n]$, both with size $n/2$, we have

$$\left| \{\pi \in S_n | A_\pi = A\} \right| = \left| \{\pi \in S_n | B_\pi = B\} \right| = \left((\frac{n}{2})! \right)^2.$$

Thus, with probability at lease $1 - 2^{-n/32}$, the permutation π has $g(A_\pi) > n/32$ and is a type 1 order with parameter $n/32$.

Proof (Proof of Corollary 3). Since with probability at least $1 - 2^{-\Omega(n)}$, a random permutation (or streaming order) is a type 1 order with parameter $n/32$. By Theorem 2, with probability at least $1 - 2^{-\Omega(n)}$, a $1 + 1/32$ deterministic approximation of $\mathsf{LIS}(x)$ needs $\Omega(n)$ space.

Lower Bounds for Randomized Algorithms. For randomized algorithms, we show an $\Omega(m)$ lower bound for exact computation when the input is given in **type 1** order with parameter m. We prove the following.

Theorem 3. *Given input sequences $x \in \Sigma^n$ in any **type 1** order with parameter m, any R-pass randomized algorithm that computes $\mathsf{LIS}(x)$ correctly with probability at least $2/3$ needs to use $\Omega(m/R)$ space.*

Proof. The idea is to reduce set-disjointness problem to computing LIS exactly.

Consider a **type 1** order with parameter m and let I and J be the subsets in the definition. We only focus on the subsequence $\tilde{x}$ limited to the indices in $I \cup J$, denoted by $\tilde{x} = \tilde{x}_1 \tilde{x}_2 \cdots \tilde{x}_{2m}$. The streaming order, when restricted to $\tilde{x}$, is $\tilde{x}_1, \tilde{x}_3, \cdots, \tilde{x}_{2m-1}, \tilde{x}_2, \tilde{x}_4, \cdots, \tilde{x}_{2m}$. In the 2 party communication model, Alice holds $\tilde{x}_1, \tilde{x}_3, \cdots, \tilde{x}_{2m-1}$ and Bob holds $\tilde{x}_2, \tilde{x}_4, \cdots, \tilde{x}_{2m}$.

Consider an instance A, B of $\mathsf{DISJ}_{2m,k}$, where A and B are both k-subset of $[2m]$. For simplicity, represent A and B as characteristic vectors in $\{0,1\}^m$, i.e. A and B are both 0,1 vector with exactly k 1 s.

We construct the sequence $\tilde{x}$ such that $\tilde{x}_1, \tilde{x}_3, \cdots, \tilde{x}_{2m-1}$ is determined by A and $\tilde{x}_2, \tilde{x}_4, \cdots, \tilde{x}_{2m}$ is determined by B. The construction is straightforward. For every $i \in [m]$, let $\tilde{x}_{2i-1} = A_i \cdot 2i$ and $\tilde{x}_{2i} = B_i \cdot (2i - 1)$.

For simplicity, we attach a dummy 0 character in front of $\tilde{x}$. This won't affect our asymptotic lower bound.

If $\mathsf{DISJ}_{2m,k}(A, B) = 1$, for any $i \in [m]$, A_i and B_i must not both be 1. There are $2k$ positions that one of $\tilde{x}_{2i-1}$ and $\tilde{x}_{2i}$ is non-zero. Since these non-zero positions are increasing, $\mathsf{LIS}(\tilde{x}) = 2k + 1$ (plus the 0 at the first position).

If $\mathsf{DISJ}_{2m,k}(A, B) = 0$, there exists $i \in [m]$ such that $A_i = B_i = 1$. We have $\tilde{x}_{2i-1} = 2i$ and $\tilde{x}_{2i} = 2i - 1$. Since there are only $2k$ non-zero positions, we have $\mathsf{LIS}(\tilde{x}) \le 2k$ (plus the 0 at the first position).

Thus, if Alice and Bob can compute $\mathsf{LIS}(\tilde{x})$, they can also determine $\mathsf{DISJ}_{2m,k}(A, B)$. The communication complexity lower bound given in Theorem 5 yields our space lower bound.

C.2 Lower Bounds for Type 2 Orders

Our communication complexity based argument relies on the analysis of a carefully constructed matrix. In the following, we represent any position in a matrix with a pair of indices (i, j) where i is the row number and j is the column number. For any two distinct positions (i_1, j_1) and (i_2, j_2), we say $(i_1, j_1) < (i_2, j_2)$ if $i_1 < i_2$ and $j_1 < j_2$.

We first show the following lemma, which is a result of Erdös-Szekeres theorem.

Lemma 3. *Given a matrix $M \in \{0, 1\}^{s \times r}$ with s rows and r columns, if each column of M has exactly $\frac{s}{4}$ 1's, then there exists $t = \lfloor \frac{r \cdot s}{8(r+s)} \rfloor$ positions (denoted by $(i_1, j_1), (i_2, j_2), \ldots, (i_t, j_t))$ in the matrix with value 1, such that $(i_1, j_1) < (i_2, j_2) < \cdots < (i_t, j_t)$.*

Proof. Consider the function $f : [s] \times [r] \to [rs]$ such that $f(i, j) = r \cdot i - j + 1$. f transforms any position (i, j) into an integer in $[rs]$ and the integers corresponding to different positions are distinct.

For each column, there are exactly $s/4$ positions that have value 1. We transform them into integers in $[rs]$ with function f and arrange them in a decreasing order to get a sequence of integers. We denote the sequence corresponding to j-th column by $\sigma^{(j)}$. For example, assume $(i_1, 1), \ldots, (i_{s/4}, 1)$ are the positions with value 1 in the first column with $i_1 < \cdots < i_{s/4}$, then $\sigma^{(1)} = r \cdot i_{s/4}, \ldots, r \cdot i_1$.

Let $\sigma = \sigma^{(1)} \circ \cdots \circ \sigma^{(r)}$. So σ is the concatenation from $\sigma^{(1)}$ to $\sigma^{(r)}$. We show that there is an increasing subsequence of σ with length $t = \lfloor \frac{r \cdot s}{8(r+s)} \rfloor$.

To see this, we first show that the longest decreasing subsequence of σ has length at most $r + s$. Consider any decreasing subsequence of σ and the positions corresponding to this decreasing subsequence in the matrix. Assume the first element of this decreasing subsequence is in the range $[r \cdot i + 1, r \cdot (i + 1)]$ for some integer i, for the next element, there are two cases. First, the next element is in the range $[r \cdot i + 1, r \cdot (i + 1)]$, then it means the element is in the same row but in a column to the right of the first element. Second, the next element is in the same column as the first element, then it must be strictly smaller than $r \cdot i + 1$, which means it is in a row below the first element. This is true for any two consecutive elements in the decreasing subsequence. Notice that σ is concatenated by columns from left to right and there are r columns and s rows, the decreasing subsequence has length at most $r + s$.

The next step is followed directly from Erdös-Szekeres theorem (Theorem 6). σ has exactly $r \cdot s/4$ elements. Let $t = \lfloor \frac{r \cdot s}{8(r+s)} \rfloor$ and $l = r + s + 1$. Then

$$
t \cdot l - t - l + 1 = \lfloor \frac{r \cdot s}{8(r+s)} \rfloor \cdot (r+s+1) \quad t - r - s \leq \frac{r \cdot s}{4}.
$$

Since there are no decreasing subsequence in σ with length $l = r + s + 1$, by Erdös-Szekeres theorem, there must exists an increasing subsequence with length t.

Given an increasing subsequence of length t, consider the positions of any two consecutive elements in this increasing subsequence, say (i_1, j_1) and (i_2, j_2). By the construction of σ, we know $j_1 \leq j_2$. If $j_1 = j_2$, then by $f(i_1, j_2) < f(i_2, j_2)$, we must have $i_1 < i_2$. But this is impossible since elements in the same column are arranged in decreasing order. Thus, the only possible case is $(i_1, j_1) < (i_2, j_2)$. The increasing subsequence in σ corresponding to a sequence of increasing positions $(i_1, j_1) < (i_2, j_2) < \cdots < (i_t, j_t)$.

We restate the main result here.

Theorem 4. *Given input sequences $x \in \Sigma^n$ in any* **type 2** *order with parameters r and s, any R-pass deterministic algorithm that gives a $1 + 1/400$ approximation of $\mathsf{LIS}(x)$ needs to use $\Omega(r \cdot s/R)$ space.*

In the following, we sometimes view sequences as vectors. For example, a sequence $\{0, 1\}^n$ can also be viewed as a vector of dimension n.

Proof. We first assume $r \cdot s = n$. Our argument can be extended to the case where $r \cdot s < n$. This is because our lower bound is only related to r and s. Given a lower bound for sequence of length $r \cdot s$, we can put dummy elements in the last $n - r \cdot s$ positions and our lower bound also holds for the sequence of length n.

Without loss of generality, we assume s can be divided by 36 and r is an even number. This is because our lower bound is asymptotic, if s can not be divided by 36, we can consider the largest multiple of 36 smaller than s and the similar for r.

In the following, let $p = s/36$ and $q = r/2$.

Our proof needs to use two asymptotically good error-correcting codes, $C^{(1)}$ and $C^{(2)}$. Here, $C^{(1)} \subseteq \{0, 1\}^p$ is over binary alphabet. It has codeword length p and distance $p/4$. $C^{(1)}$ can be viewed as a subset of $\{0, 1\}^p$. $C^{(2)} \subseteq (C^{(1)})^q$ uses $C^{(1)}$ as its alphabet set. It has codeword length q and distance $q/2$. Since both $C^{(1)}$ and $C^{(2)}$ are asymptotically good, $|C^{(1)}| = 2^{\Omega(s)}$ and $|C^{(2)}| = 2^{\Omega(r \cdot s)}$.

Assume there are two parties, Alice and Bob, each holds a codeword of $C^{(2)}$, say Alice holds u and Bob holds v. We now show how to use $C^{(2)}$ as a fooling set to obtain our lower bound. The high level idea is, we build a sequence $\sigma(u, v) \in \mathbb{Z}^n$ depending on u and v, such that σ can be divided evenly into r blocks each of length s. Those odd blocks are only depending on u (thus only known to Alice) and those even blocks are only depending on v (thus only known to Bob).

Our construction guarantees that, if $u = v$, then $\mathsf{LIS}(\sigma) = s/9 + 3r + \min(p, q)$, and if $u \neq v$, $\mathsf{LIS}(\sigma) = s/9 + 3r + \min(p, q) + \frac{pq}{8(p+q)}$. We first give the construction of σ.

Construction of σ. We first build a 01 matrix $M \in \{0, 1\}^{\frac{s}{4} \times 4r}$ using u and v. M has $s/4$ rows and $4r$ columns. We can group the elements of the matrix into r blocks such that, for $1 \leq i \leq r$, the i-th block are elements from the 4 columns with number $4i - 3, 4i - 2, 4i - 1, 4i$. In the matrix M, the odd blocks are only determined by u (the codeword known to Alice) and the even blocks are only determined by v (the codeword known to Bob). A pictorial representation is given in Fig. 1.

Fig. 1. pictorial representation of M. Each small square represents an element of M. We partition elements of M into r blocks where each block contains 4 consecutive columns. Odd blocks (elements in red) are determined by Alice's codeword u and even blocks (elements in blue) are determined by Bob's codeword, v. (Color figure online)

We turn this 01 matrix M into a sequence of integers $\sigma(M)$ as following. We first build a new matrix $M' \in \mathbb{Z}^{\frac{s}{4} \times 4r}$, such that, for each position $(i, j) \in [\frac{s}{4}] \times [4r]$, we set

$$M'_{i,j} = \begin{cases} 0, & \text{if } M_{i,j} = 0, \\ 4r \cdot (i-1) + j, & \text{if } M_{i,j} = 1. \end{cases} \tag{8}$$

Then, σ is the concatenation of elements in M' column by column. Specifically, let $\sigma^{(j)}$ be the concatenation of all symbols in the j-th column of M', i.e. $\sigma^{(j)} = M'_{1,j} \circ M'_{2,j} \circ \cdots \circ M'_{s/4,j}$. The final sequence σ we get is

$$\sigma(M) = \sigma^{(1)} \circ \cdots \circ \sigma^{(r)}.$$

View Matrix M as a Grid Graph. We can view the matrix M as a **directed** grid graph, such that for each position (i, j), there are two out going edges from this it connecting to $(i+1, j)$ and $(i, j+1)$ respectively (assuming they exist). A **path** in the matrix M is a path in the corresponding graph that goes from $(1, 1)$ to $(\frac{s}{4}, 4r)$ (from left-top to right bottom). We define the **weight** of a path is the number of 1 node (position with value 1) the path covers. We say a node is **covered** by a path if the path goes through that node.

Claim 4. $LIS(\sigma(M))$ *is equal to the largest weight of any path in M.*

Proof. First, for any path in the grid graph, non-zero nodes it covers are in increasing order. This is because any path can only go down or right. By our assignment of the values in M' (Eq. (8)), non-zero elements in any row or column are in strict increasing order. Also, since σ is the concatenation of all columns of M' from left to right. For any path, nodes it covers is a subsequence of σ. Thus, we can say for any path with weight w in the grid graph, there is an increasing subsequence of σ with length w.

On the other hand, we show that for any increasing subsequence, there is a path covering all the corresponding nodes in the grid graph. To see this, we show that for any two consecutive nodes (i, j), (i', j') in the increasing subsequence, we have $i \leq i'$ and $j \leq j'$. $j \leq j'$ is by definition since we concatenate columns from left to right. Since it is an increasing subsequence, we must have $4r \cdot (i-1) + j < 4r \cdot (i'-1) + j'$. By the fact that there are only $4r$ columns, we have $i \leq i'$. Thus, (i', j') must appear on the down-right side of (i, j). We can find a path in the grid graph connecting them. This shows for any increasing subsequence of σ with length w, there is a path in the grid graph with weight w.

Constructing M with u and v. In the following, we show how to build the matrix M with u, v (codewords hold by Alice and Bob) and then prove Claim 5 and Claim 6, which yields Theorem 4.

We start with Alice's part. Alice holds $u \in C^{(2)}$, or it can be viewed as q codewords of $C^{(1)}$. We denote them by $u^{(1)}, u^{(2)}, \ldots, u^{(q)}$. Alice has control over all $q = r/2$ odd blocks. In our construction, the $(2i-1)$-th block (or the i-th block hold by Alice) is determined by $u^{(i)}$. The construction is the same for all i.

We take $2i-1$-th block for an example. There are 4 columns in this block, i.e. columns with number $8i-7$, $8i-6$, $8i-5$ and $8i-4$. Given $u^{(i)} \in C^{(1)} \subseteq \{0,1\}^p$, we turn it into $\bar{u}^{(i)} \in \{0,1\}^{9p}$ such that each 1 in $u^{(i)}$ is replaced by $(1,1,0,0,0,0,1,1,0)$ and each 0 is replaced by $(0,0,1,1,1,1,0,0,0)$. This is the $(8i-4)$-th column of M. For the other 3 columns, they all have 1 at rows $9j$ for $j \geq 1$ and 0 everywhere else.

For Bob, the construction is similar. We take $2i$-th block for an example. Given each $v^{(i)} \in C^{(1)} \subseteq \{0,1\}^p$, we turn it into $\bar{v}^{(i)} \subseteq \{0,1\}^{9p}$ such that each 1 in $v^{(i)}$ is replaced by $(1,1,0,0,0,0,1,1,0)$ and each 0 is replaced by $(0,0,1,1,1,1,0,0,0)$, which is the same as Alice's transform. Then we use $\bar{v}^{(i)}$ as the first column of the $2i$-th block (Note: for Alice's construction, we use it as the last column of the corresponding block). For the other 3 columns, they all have 1 at rows $9j$ for $j \geq 1$ and 0 everywhere else.

Let us see an example. Say $p = q = 2$, $u^{(1)} = (0,1)$, $u^{(2)} = (1,1)$, $v^{(1)} = (0,0)$ and $v^{(2)} = (1,0)$, then the matrix M is given in Fig. 2. Notice that M is a matrix with $s/4 = 9p = 18$ rows and $4r = 8q = 16$ columns. In Fig. 2, only elements in red are determined by u and v. The rest of the matrix is fixed for any u and v. In other words, u and v determined the content of $p \cdot q$ sub-matrices in M, each of size 8×2.

$$
M = \begin{pmatrix}
0 & 0 & 0 & 0 & 0 & 0 & 0 & 0 & 0 & 0 & 0 & 1 & 1 & 0 & 0 & 0 \\
0 & 0 & 0 & 0 & 0 & 0 & 0 & 0 & 0 & 0 & 0 & 1 & 1 & 0 & 0 & 0 \\
0 & 0 & 0 & 1 & 1 & 0 & 0 & 0 & 0 & 0 & 0 & 0 & 0 & 0 & 0 & 0 \\
0 & 0 & 0 & 1 & 1 & 0 & 0 & 0 & 0 & 0 & 0 & 0 & 0 & 0 & 0 & 0 \\
0 & 0 & 0 & 1 & 1 & 0 & 0 & 0 & 0 & 0 & 0 & 0 & 0 & 0 & 0 & 0 \\
0 & 0 & 0 & 1 & 1 & 0 & 0 & 0 & 0 & 0 & 0 & 0 & 0 & 0 & 0 & 0 \\
0 & 0 & 0 & 0 & 0 & 0 & 0 & 0 & 0 & 0 & 0 & 1 & 1 & 0 & 0 & 0 \\
0 & 0 & 0 & 0 & 0 & 0 & 0 & 0 & 0 & 0 & 0 & 1 & 1 & 0 & 0 & 0 \\
1 & 1 & 1 & 0 & 0 & 1 & 1 & 1 & 1 & 1 & 1 & 0 & 0 & 1 & 1 & 1 \\
0 & 0 & 0 & 1 & 0 & 0 & 0 & 0 & 0 & 0 & 0 & 1 & 0 & 0 & 0 & 0 \\
0 & 0 & 0 & 1 & 0 & 0 & 0 & 0 & 0 & 0 & 0 & 1 & 0 & 0 & 0 & 0 \\
0 & 0 & 0 & 0 & 1 & 0 & 0 & 0 & 0 & 0 & 0 & 0 & 1 & 0 & 0 & 0 \\
0 & 0 & 0 & 0 & 1 & 0 & 0 & 0 & 0 & 0 & 0 & 0 & 1 & 0 & 0 & 0 \\
0 & 0 & 0 & 0 & 1 & 0 & 0 & 0 & 0 & 0 & 0 & 0 & 1 & 0 & 0 & 0 \\
0 & 0 & 0 & 0 & 1 & 0 & 0 & 0 & 0 & 0 & 0 & 0 & 1 & 0 & 0 & 0 \\
0 & 0 & 0 & 1 & 0 & 0 & 0 & 0 & 0 & 0 & 0 & 1 & 0 & 0 & 0 & 0 \\
0 & 0 & 0 & 1 & 0 & 0 & 0 & 0 & 0 & 0 & 0 & 1 & 0 & 0 & 0 & 0 \\
1 & 1 & 1 & 0 & 0 & 1 & 1 & 1 & 1 & 1 & 1 & 0 & 0 & 1 & 1 & 1
\end{pmatrix}
$$

Fig. 2. matrix M when $u^{(1)} = (0,1)$, $u^{(2)} = (1,1)$, $v^{(1)} = (0,0)$ and $v^{(2)} = (1,0)$.

We look at the sub-matrices determined by u and v. For each pair of indices $(i,j) \in [p] \times [q]$, the sub-matrix determined by the value of $u_i^{(j)}$ and $v_i^{(j)}$ has 4 cases:

1. $u_i^{(j)} = 1$ and $v_i^{(j)} = 0$.
2. $u_i^{(j)} = v_i^{(j)} = 1$.
3. $u_i^{(j)} = v_i^{(j)} = 0$.
4. $u_i^{(j)} = 0$ and $v_i^{(j)} = 1$.

A pictorial depiction of these 4 cases (from left to right) is given in Fig. 3. As shown in Fig. 3, if $u_i^{(j)} = v_i^{(j)}$, any path in the sub-matrix has weight at most 5 (marked in red). However, if $u_i^{(j)} \neq v_i^{(j)}$, there exists a path in the sub-matrix with weight 6 (marked in green).

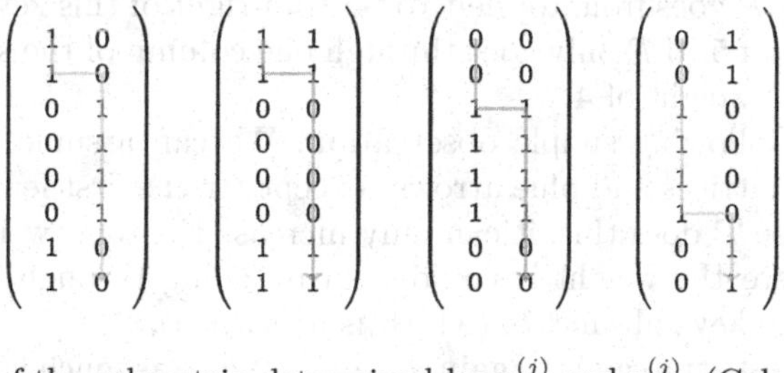

$$
\begin{pmatrix} 1 & 0 \\ 1 & 0 \\ 0 & 1 \\ 0 & 1 \\ 0 & 1 \\ 0 & 1 \\ 1 & 0 \\ 1 & 0 \end{pmatrix}
\quad
\begin{pmatrix} 1 & 1 \\ 1 & 1 \\ 0 & 0 \\ 0 & 0 \\ 0 & 0 \\ 0 & 0 \\ 1 & 1 \\ 1 & 1 \end{pmatrix}
\quad
\begin{pmatrix} 0 & 0 \\ 0 & 0 \\ 1 & 1 \\ 1 & 1 \\ 1 & 1 \\ 1 & 1 \\ 0 & 0 \\ 0 & 0 \end{pmatrix}
\quad
\begin{pmatrix} 0 & 1 \\ 0 & 1 \\ 1 & 0 \\ 1 & 0 \\ 1 & 0 \\ 1 & 0 \\ 0 & 1 \\ 0 & 1 \end{pmatrix}
$$

Fig. 3. 4 cases of the submatrix determined by $u_i^{(j)}$ and $v_i^{(j)}$. (Color figure online)

To simplify the argument a bit, we call the red sub-matrices in Fig. 2 *key sub-matrices* since everything else are invariant and only these sub-matrices are determined by u and v. There are $p \cdot q$ key sub-matrices. We label them by (i, j) such that the key sub-matrix (i, j) is determined by $u_i^{(j)}$ and $v_i^{(j)}$. Here, $1 \leq i \leq p$ and $1 \leq j \leq q$.

A path in M can go through those key sub-matrices to gain higher weight. It can also cover the 1's outside these key sub-matrices. An abstraction of M is given in Fig. 4. In Fig. 4, key sub-matrices are represented as red rectangles and the blue arrow are the 1's that paths can take. They are all on rows with row-number that is a multiple of 9. 0's are not shown in the figure since paths going through these nodes won't gain any weight. We are only interested in paths with largest weight.

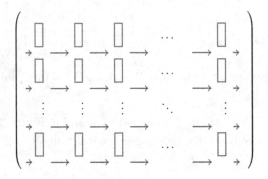

Fig. 4. abstraction of M

We show that the largest weight of any path in M can vary depending on whether $u = v$. We have the following two claims.

Claim 5. *If $u = v$, then* $\mathsf{LIS}(\sigma(M)) \leq 4p + 6q + \min(p, q)$.

Proof. By Claim 4, only need to show that any path in M has weight at most $4p + 6q + \min(p, q)$.

Let $\mathcal{P}$ be one of the path in M that has largest weight. Since $u = v$, for each key sub-matrix, if $\mathcal{P}$ goes from top-left to bottom-right of this key sub-matrix, it will gain a weight of 5. If $\mathcal{P}$ only goes through one column of the key sub-matrix, it will gain a total weight of 4.

We have the following simple observation. We can assume $\mathcal{P}$ does not go beyond key sub-matrices and blue arrows (except for the first few steps). This is because every time $\mathcal{P}$ does this, it can only increase the total weight by at most 1, but it will waste the weight it could gain by going through the blue arrow (which is 6) or the key sub-matrix (which is at least 4).

If $p \geq q$, the best strategy is to gain a weight of 5 in as much key sub-matrices as possible. But we can only do this for q times since every time it gain a total weight of 5 in one key sub-matrix, it has to go to a new column. The total weight it can gain in blue arrows is at most $6q$ since we assume it will not go to the zero region. We can gain an additional weight of $4(p - q)$ in key sub-matrices (in these sub-matrices, $\mathcal{P}$ only uses one column). Thus, the total weight in this case is at most $5q + 6q + 4(p - q) = 4p + 6q + q$.

If $p < q$, the best strategy is again to gain a weight of 5 in as much key sub-matrices as possible. We can do this for p times. The total weight it can gain in blue arrows is at most $6q$. Thus, the total weight in this is at most $5p + 6q$.

Combine these two cases, we conclude the largest weight of any path in M is at most $4p + 6q + \min(p, q)$.

Claim 6. *If $u \neq v$, then $\mathsf{LIS}(\sigma(M)) \geq 4p + 6q - 3 + \min(p, q) + \frac{pq}{16(p+q)}$.*

Proof. By Claim 4, only need to show there exists a path in M with weight at least $4p + 6q - 3 + \min(p, q) + \frac{pq}{16(p+q)}$.

Since u and v are chosen from $C^{(2)}$, which is an error-correcting code with distance $q/2$. If $u \neq v$, there are $q/2$ positions what u and v that are not equal. Let j be on of them ($u^{(j)} \neq v^{(j)}$). Since $u^{(j)}$ and $v^{(j)}$ are both codewords of $C^{(1)} \subseteq \{0, 1\}^p$. There are $p/4$ positions that $u^{(j)}$ and $v^{(j)}$ does not equal.

Let A be a $p \times q$ matrix such that $A_{i,j} = 1$ if $u^{(j)} \neq v^{(j)}$ and $A_{i,j} = 0$ if $u^{(j)} = v^{(j)}$. Then in matrix A there are at least $q/2$ columns with at least $p/4$ 1's. Then, if we only look at those columns with at least $p/4$ 1's, by Lemma 3, there are $t = \lfloor \frac{r \cdot s}{16(r+s)} \rfloor$ positions in the matrix (denoted $(i_1, j_1), (i_2, j_2), \ldots, (i_t, j_t)$) with value 1, such that $(i_1, j_1) < (i_2, j_2) < \cdots < (i_t, j_t)$.

The best strategy for a path to gain largest weight is to go through key sub-matrices that can provide a total weight of 6 as much as possible ($u^{(j)} \neq v^{(j)}$). Lemma 3 guarantees there are at least $t = \lfloor \frac{r \cdot s}{16(r+s)} \rfloor$ that a path can go through. For the remaining, it can go from top-left to bottom-right of $\min p, q - t$ key sub-matrices. Each will give a weight of at least 5. If $p > q$, it can gain additional $4(p - q)$ by going through $p - q$ key sub-matrices but using only one column. Along the blue arrows, a path can again at least $6q - 3$ (minus the 3 it misses on the first row). Thus, we can find a path with weight at least $4p + 6q - 3 + \min(p, q) + \frac{pq}{16(p+q)}$.

By Claim 5 and Claim 6, we can conclude that $C^{(2)}$ is a fooling set for **type 2**. Even if Alice and Bob can approximate LIS to within a factor of $1 + 1/400$. When p, q are large enough, we have

$$(1 + 1/400) \cdot \left(4p + 6q + \min(p, q)\right) \leq 4p + 6q + \min(p, q) + \frac{11 \max(p, q)}{400}$$

$$< 4p + 6q + \min(p, q) + \frac{22 \cdot p \cdot q}{400 \cdot (p + q)}$$

$$\leq 4p + 6q + \min(p, q) + \frac{pq}{16(p + q)} - 3.$$

In this case, they can still not determine whether $u = v$ or not. Since both $C^{(1)}$ and $C^{(2)}$ are asymptotically good, $C^{(2)}$ has size $2^{\Omega(rs)}$. To deterministically approximate LIS to within a $1 + 1/400$ factor (know whether $u = v$ or not), any R-round streaming algorithm would need a space of at least $\Omega(rs/R)$.

References

[ABCR99] Andreev, A.E., Baskakov, J.L., Clementi, A.E.F., Rolim, J.D.P.: Small pseudo-random sets yield hard functions: new tight explicit lower bounds for branching programs. In: Wiedermann, J., van Emde Boas, P., Nielsen, M. (eds.) ICALP 1999. LNCS, vol. 1644, pp. 179–189. Springer, Heidelberg (1999). https://doi.org/10.1007/3-540-48523-6_15

[ACCL07] Ailon, N., Chazelle, B., Comandur, S., Liu, D.: Estimating the distance to a monotone function. In: Jansen, K., Khanna, S., Rolim, J.D.P., Ron, D. (eds.) APPROX/RANDOM -2004. LNCS, vol. 3122, pp. 229–236. Springer, Heidelberg (2004). https://doi.org/10.1007/978-3-540-27821-4_21

[AK03] Ablayev, F., Karpinski, M.: A lower bound for integer multiplication on randomized ordered read-once branching programs. Inf. Comput. $186(1)$, 78–89 (2003)

[ANSS22] Andoni, A., Nosatzki, N.S., Sinha, S., Stein, C.: Estimating the longest increasing subsequence in nearly optimal time. In: FOCS 2022, pp. 708–719 (2022)

[Bea89] Beame, P.: A general sequential time-space tradeoff for finding unique elements. In: STOC, pp. 197–203 (1989)

[BRS93] Borodin, A., Razborov, A., Smolensky, R.: On lower bounds for read-k-times branching programs. Comput. Complex. $3(1)$, 1–18 (1993)

[BW98] Bollig, B., Wegener, I.: A very simple function that requires exponential size read-once branching programs. Inf. Process. Lett. (1998)

[BW01] Bollig, B., Woelfel, P.: A read-once branching program lower bound of $\Omega(2^{n/4})$ for integer multiplication using universal. In: STOC (2001)

[CFH+21] Cheng, K.: Streaming and small space approximation algorithms for edit distance and longest common subsequence. In: ICALP (2021)

[Dun85] Dunne, P.E.: Lower bounds on the complexity of 1-time only branching programs (preliminary version). In: Budach, L. (ed.) FCT 1985. LNCS, vol. 199, pp. 90–99. Springer, Heidelberg (1985). https://doi.org/10.1007/BFb0028795

[EJ08] Ergun, F., Jowhari, H.: On distance to monotonicity and longest increasing subsequence of a data stream. In: Proceedings of the Nineteenth Annual ACM-SIAM Symposium on Discrete Algorithms, pp. 730–736 (2008)

[Gál97] Gál, A.: A simple function that requires exponential size read-once branching programs. Inf. Process. Lett. $62(1)$, 13–16 (1997)

[GG10] Gál, A., Gopalan, P.: Lower bounds on streaming algorithms for approximating the length of the longest increasing subsequence. SIAM J. Comput. $39(8)$, 3463–3479 (2010)

[GJ21] Gawrychowski, P., Janczewski, W.: Fully dynamic approximation of LIS in polylogarithmic time. In: STOC 2021. ACM (2021)

[GJKK07] Gopalan, P., Jayram, T.S., Krauthgamer, R., Kumar, R.: Estimating the sortedness of a data stream. In: SODA (2007)

[Ham72] Hammersley, J.: A few seedlings of research. In: Proceedings of the Sixth Berkeley Symposium on Mathematical Statistics and Probability, pp. 345–394 (1972)

[HW07] Håstad, J., Wigderson, A.: The randomized communication complexity of set disjointness. Theory Comput. $3(1)$, 211–219 (2007)

[IMS17] Im, S., Moseley, B., Sun, X.: Efficient massively parallel methods for dynamic programming. In: STOC, pp. 798–811. ACM (2017)

[Juk88] Jukna, S.: Entropy of contact circuits and lower bounds on their complexity. Theor. Comput. Sci. **57**, 113–129 (1988)

[Kab03] Kabanets, V.: Almost k-wise independence and hard Boolean functions. Theor. Comput. Sci. **297**(1–3), 281–295 (2003)

[KMW91] Krause, M., Meinel, C., Waack, S.: Separating the eraser turing machine classes L_e, NL_e, $coNL_e$ and P_e. Theor. Comput. Sci. (1991)

[KOO+20] Kiyomi, M., Ono, H., Otachi, Y., Schweitzer, P., Tarui, J.: Space-efficient algorithms for longest increasing subsequence. Theory Comput. Syst. **64**(3), 522–541 (2020)

[KS21] Kociumaka, T., Seddighin, S.: Improved dynamic algorithms for longest increasing subsequence. In: STOC, pp. 640–653. ACM (2021)

[Li23] Li, X.: Two source extractors for asymptotically optimal entropy, and (many) more. arXiv:2303.06802 (2023)

[LZ21] Li, X., Zheng, Y.: Lower bounds and improved algorithms for asymmetric streaming edit distance and longest common subsequence. In: FSTTCS 2021. Schloss Dagstuhl-Leibniz-Zentrum für Informatik (2021)

[LZ23] Li, X., Zhong, Y.: Explicit directional affine extractors and improved hardness for linear branching programs. arXiv:2304.11495 (2023)

[Mal62] Mallows, C.L.: Patience sorting. SIAM Rev. **4**(2), 148–149 (1962)

[MS20] Mitzenmacher, M., Seddighin, S.: Dynamic algorithms for LIS and distance to monotonicity. In: STOC, pp. 671–684. ACM (2020)

[MS21] Mitzenmacher, M., Seddighin, S.: Improved sublinear time algorithm for longest increasing subsequence. In: SODA (2021)

[NS14] Naumovitz, T., Saks, M.: A polylogarithmic space deterministic streaming algorithm for approximating distance to monotonicity. In: SODA, pp. 1252–1262. SIAM (2014)

[NV21] Newman, I., Varma, N.: New sublinear algorithms and lower bounds for LIS estimation. In: 48th International Colloquium on Automata, Languages, and Programming (ICALP 2021) (2021)

[Pip79] Pippenger, N.: On simultaneous resource bounds. In: SFCS 1979 (1979)

[Pon98] Ponzio, S.: A lower bound for integer multiplication with read-once branching programs. SIAM J. Comput. **28**(3), 798–815 (1998)

[Rou15] Roughgarden, T.: Communication complexity (for algorithm designers). CoRR, abs/1509.06257 (2015)

[RSSS19] Rubinstein, A., Seddighin, S., Song, Z., Sun, X.: Approximation algorithms for LCS and LIS with truly improved running times. In: FOCS 2019, Baltimore, Maryland, USA, pp. 1121–1145. IEEE (2019)

[Sah17] Saha, B.: Fast space-efficient approximations of language edit distance and RNA folding: an amnesic dynamic programming approach. In: FOCS (2017)

[Sav70] Savitch, W.J.: Relationships between nondeterministic and deterministic tape complexities. J. Comput. Syst. Sci. (1970)

[SS92] Simon, J., Szegedy, M.: A new lower bound theorem for read-only-once branching programs and its applications. In: Advances in Computational Complexity Theory (1992)

[SS13] Saks, M., Seshadhri, C.: Space efficient streaming algorithms for the distance to monotonicity and asymmetric edit distance. In: SODA (2013)

[SS17] Saks, M., Seshadhri, C.: Estimating the longest increasing sequence in polylogarithmic time. SIAM J. Comput. **46**(2), 774–823 (2017)

[SW07] Sun, X., Woodruff, D.P.: The communication and streaming complexity of computing the longest common and increasing subsequences. In: SODA, pp. 336–345 (2007)

[Weg88] Wegener, I.: On the complexity of branching programs and decision trees for clique functions. J. ACM **35**(2), 461–471 (1988)

[Zák84] Žák, S.: An exponential lower bound for one-time-only branching programs. In: Chytil, M.P., Koubek, V. (eds.) MFCS 1984. LNCS, vol. 176, pp. 562–566. Springer, Heidelberg (1984). https://doi.org/10.1007/BFb0030340

Approximating Decision Trees
with Priority Hypotheses

Jing Yuan[1] and Shaojie Tang[2]($\boxtimes$)

[1] Department of Computer Science and Engineering, University of North Texas,
Denton, USA
jing.yuan@unt.edu

[2] Naveen Jindal School of Management, University of Texas at Dallas, Richardson,
USA
shaojie.tang@utdallas.edu

Abstract. This paper addresses the problem of creating decision trees for identifying hypotheses, also known as entities, in a setting where the cost of an action is dependent on the true hypothesis. Specifically, we consider the scenario where n hypotheses are divided into m groups based on their priority levels. Taking an action on a higher priority hypothesis incurs a higher cost. This is relevant to many real-world applications where cost-sensitive decisions need to be made. For example, in a medical diagnosis task, the goal is to take a series of actions (such as medical tests) to identify a cause. Each action in this process requires conducting a test on the patient and observing the outcome, which can take anywhere from a few minutes to several weeks depending on the test. In this case, the cost (the result of waiting for the outcome) is higher if the true hypothesis is more time-sensitive. For example, if the true hypothesis is toxic chemical exposure (as opposed to a chronic disease such as diabetes), a delay of a few minutes could significantly increase the patient's risk of mortality. We propose a group greedy algorithm to solve this problem. We demonstrate that under worst-case scenarios, our algorithm has an approximation ratio of $O(m \log n)$. Importantly, when $m = 1$, meaning there is only one group of hypotheses, our result is consistent with the logarithmic approximation bound for the traditional optimal decision tree problem.

1 Introduction

The goal of the traditional optimal decision tree problem is to identify an unknown hypothesis through a series of actions, also known as tests or queries, while minimizing the cost. Many applications, such as active learning and disease/fault diagnosis, can be formulated as an optimal decision tree problem; see [1,3]. For instance, in the problem of disease diagnosis, a physician may perform a series of medical tests on a patient. By ruling out diseases that are inconsistent with the reported outcome, the physician can identify the cause at the lowest cost. In this context, it is important for the physician to strategically and adaptively choose a sequence of tests while taking into account the cost associated with each action.

Previous studies on the optimal decision tree (or active learning) problem [2,4,6,9] typically assume that the cost of taking an action is known in advance, so the cost of taking any sequence of tests is also known in advance. There are various ways to define the incurred cost of an action. In the context of disease diagnosis, the cost of a medical test is often defined as the time it takes to perform and observe the outcome from that medical test. For example, when an imaging test is performed, it may take several days for the results. In this setting, the incurred cost of an action is not dependent on the true disease. However, in practice, the consequences of taking a test can vary depending on the (unknown) true disease. For example, if the true hypothesis is toxic chemical exposure (versus a chronic disease such as diabetes), a delay of a few minutes could significantly increase the patient's risk of mortality. Another example would be detecting fraudulent credit transactions, where the cost of investigating an honest transaction might be higher than that of investigating a fraud one. Given the above discussions, it is crucial to develop a new model in which the costs are dependent on the true hypothesis.

To achieve this, we divide the n hypotheses into m groups based on their priority levels. We assume that taking an action on a higher priority hypothesis incurs a higher cost (or penalty). In the example of disease diagnosis, we can prioritize diseases based on their emergency levels, such that the cost of long wait times is greater for emergent diseases. It's important to note that the algorithm may not know the cost of an action during the learning process until it observes the true hypothesis. Our objective is to design a decision tree that minimizes the total cost. To solve this problem, we propose a group greedy algorithm. In summary, our algorithm prioritizes identifying hypotheses from high priority groups. We prove that our algorithm has an approximation ratio of $O(m \log n)$ under worst-case scenarios. Notably, when $m = 1$, meaning there is only one group of hypotheses, our result is consistent with the logarithmic approximation bound for the traditional optimal decision tree problem.

Additional Related Work. Our work is closely related to response-dependent active learning [5,7,8] where they assume that the cost of an action is decided by the outcome from that action. Moreover, [5,8] assume there are only two possible outcomes of every action. Our hypothesis-dependent cost model is more general: our cost function allows an action to have different incurred costs on two hypotheses even if they cause the same outcome from that action. However, our current results rely on the assumption that the incurred cost of an action is always higher on a higher priority hypothesis. In the future, we would like to generalize this work by relaxing this assumption.

2   Preliminaries

An instance of our problem, known as Optimal Decision Trees with Priority Hypotheses (ODTH), is defined as $\mathcal{I} = (H, G, A, \mathbf{c})$; here H is a set of n hypotheses; G is a partition of H into m *priority* groups; A is a set of actions; $\mathbf{c}$ is a cost function that assigns a cost $c(a, g)$ for each pair of action $a \in A$ and group $g \in G$ such that $c(a, g)$ is the cost of taking action a given that the true (unknown) hypothesis belongs to group g. We extend the notation of g by letting $g(h) \in G$ denote the group that contains hypothesis $h \in H$. Hence, selecting action $a \in A$ on hypothesis $h \in H$ incurs a cost of $c(a, g(h))$ and returns an outcome $a(h)$ from a set of possible outcomes O. We assume that A is complete such that for every distinct $h \in H$ and $h' \in H$, there exists an action $a \in A$ such that $a(h) \neq a(h')$. Moreover, we assume that $c(a, g') \geq c(a, g)$ for any action $a \in A$ and any pair of groups g and g' such that g' has a higher priority than g. That is, higher priority groups are assumed to be more sensitive to the overhead such as wait times of an action. In the motivating example of disease diagnosis, we may prioritize diseases according to their emergency levels such that the consequences of long wait times are greater for emergent diseases. Note that the actual cost of an action is unknown during the learning process before the true hypothesis is observed.

Given an instance of ODTH, our goal is to select a sequence of actions to determine the true *unknown* hypothesis. Formally, any adaptive policy can be represented as a decision tree where each internal node represents an action and each edge represents an outcome from an action. We say that a tree D covers $\mathcal{I}$ if each hypothesis from H is associated with an unique leaf in D. Given instance I and decision tree D, the cost of D on a hypothesis $h \in H$ is

$$\text{cost}(D, h) = \sum_{a \in \text{path}(D, h)} c(a, g(h)) \tag{1}$$

where $\text{path}(D, h)$ denotes the set of all nodes that appear in the root-to-leaf path in D traced by h. We define the total cost as

$$\text{cost}(D) = \max_{h \in H} \text{cost}(D, h).$$

The objective of ODTH is to identify an optimal feasible decision tree $D^*_{\mathcal{I}}$ such that the worst-case cost is minimized, i.e.,

$$D^*_{\mathcal{I}} \in \underset{D \in \mathcal{D}(\mathcal{I})}{\text{argmin}}\, \text{cost}(D), \tag{2}$$

where $\mathcal{D}(\mathcal{I})$ denotes the set of all decision trees that cover $\mathcal{I}$.

3   Group Greedy Algorithm and Analysis

Before presenting our algorithm, we first introduce a well studied problem called Discrete Function Evaluation Problem (DFEP) [2]. We show that DFEP is closely related to our problem. A solution to DFEP is required by our final algorithm as a subroutine.

3.1 Discrete Function Evaluation Problem

Given an instance of ODTH $\mathcal{I} = (H, G, A, \mathbf{c})$, assuming group $g_1 \in G$ has the highest priority, we define the corresponding DFEP instance of $\mathcal{I}$ as $\mathcal{P}(\mathcal{I}) = (H, C, A, \mathbf{c}')$; here C is a partition of H into $|g_1|+1$ classes, where each hypothesis from g_1 constitutes a class and the rest of the hypotheses constitute another class; $\mathbf{c}'$ is a cost function that assigns a cost $c'(a) = c(a, g_1)$ for each action $a \in A$, that is, $c'(a)$ is the highest possible cost of a w.r.t. $\mathcal{I}$. Selecting action $a \in A$ on hypothesis $h \in H$ incurs a cost of $c'(a)$ w.r.t. $\mathcal{P}(\mathcal{I})$ and returns an outcome $a(h)$ from a set of possible outcomes O. Unlike the cost function defined in ODTH, $c'(a)$ is hypothesis independent. Given an instance of DFEP, our goal is to select a sequence of actions to determine the class of an *unknown* true hypothesis. Every adaptive policy w.r.t. $\mathcal{P}(\mathcal{I})$ can be represented as a decision tree where each internal node is associated with an action and each edge is associated with an outcome from an action. We say that a tree D covers $\mathcal{P}(\mathcal{I})$ if each leaf of D is associated with a set of hypotheses that belong to the same class. Given instance $\mathcal{P}(\mathcal{I})$ and decision tree D, the testing cost of $h \in H$ w.r.t. $\mathcal{P}(\mathcal{I})$ is

$$\text{cost}'(D, h) = \sum_{a \in \text{path}(D,h)} c'(a) \tag{3}$$

where $\text{path}(D, h)$ denotes the set of all nodes that appear in the root-to-leaf path from the root of D to the leaf associated with h. We define the total cost of D w.r.t. $\mathcal{P}(\mathcal{I})$ as

$$\text{cost}'(D) = \max_{h \in H} \text{cost}'(D, h).$$

The goal of DFEP is to identify a decision tree $D^*_{\mathcal{P}(\mathcal{I})}$ such that the worst-case cost is minimized, i.e.,

$$D^*_{\mathcal{P}(\mathcal{I})} \in \underset{D \in \mathcal{D}(\mathcal{P}(\mathcal{I}))}{\text{argmin}} \ \text{cost}'(D), \tag{4}$$

where $\mathcal{D}(\mathcal{P}(\mathcal{I}))$ denotes the set of all decision trees that cover $\mathcal{P}(\mathcal{I})$.

Greedy Algorithm for DFEP. To solve DFEP, [2] developed a greedy algorithm that attains an $O(\log n)$ approximation ratio. Before introducing this greedy algorithm, we introduce some additional notations. Given an instance of DFEP $\mathcal{P}(\mathcal{I})$, two hypotheses are defined as a *pair* if they belong to different classes. Let $F(S)$ denote the total number of pairs in $S \subseteq H$, i.e.,

$$F(S) = \sum_{t=1}^{k-1} \sum_{t'=t+1}^{k} z_{c_t} z_{c_{t'}}, \tag{5}$$

where z_c is the number of hypotheses in S that belongs to class $c \in C$, i.e., $z_c = |S \cap c|$.

The greedy algorithm (labeled as GreedyTree) greedily chooses action with the largest worst-case ratio of the reduction in F and cost until it covers $\mathcal{P}(\mathcal{I})$. It is easy to verify that upon the termination of GreedyTree, $F(S)$ equals zero for all $S \subseteq H$ that is associated with a leaf node of the returned tree. We next explain GreedyTree (listed in Algorithm 1) in details. In the first round of GreedyTree, it takes $V = H$ as input and computes $R(a) := \min_{o \in O(a,V)} \frac{F(V) - F(V_a^o)}{c'(a)}$ for each action $a \in A$, where $O(a, V)$ is the set of possible outcomes for action a assuming the true hypothesis belongs to V, i.e., $O(a, V) = \cup_{h \in V} a(h)$; V_a^o is the set of hypotheses in V that has outcome o for action a, i.e., $V_a^o = \{h \in V \mid a(h) = o\}$. Then it selects action with the largest R value (labeled as $\hat{a}$) as the root. For each outcome $o \in O(\hat{a}, V)$ of $\hat{a}$, it applies the same procedure on $V = H_{\hat{a}}^o$ recursively, where $H_{\hat{a}}^o$ is the set of hypotheses in H that has outcome o for action $\hat{a}$, i.e., $H_{\hat{a}}^o = \{h \in V \mid \hat{a}(h) = o\}$.

It has been shown that GreedyTree attains an $O(\log n)$ approximation ratio w.r.t. $\mathcal{P}(\mathcal{I})$.

Lemma 1 *[2]. Let $D_{\mathcal{P}(\mathcal{I})}^*$ denote the optimal decision tree w.r.t. $\mathcal{P}(\mathcal{I})$ and $D_{\mathcal{P}(\mathcal{I})}^{greedy}$ denote the decision tree returned from GreedyTree($\mathcal{P}(\mathcal{I}), H$), we have*

$$cost'(D_{\mathcal{P}(\mathcal{I})}^{greedy}) \leq O(\log n) cost'(D_{\mathcal{P}(\mathcal{I})}^*). \qquad (6)$$

Algorithm 1. GreedyTree($\mathcal{P}(\mathcal{I}), V$)

1: **if** $F(V) = 0$ **then**
2:    return
3: **for** $a \in A$ **do**
4:    compute $R(a) := \min_{o \in O(a,V)} \frac{F(V) - F(V_a^o)}{c'(a)}$, where $O(a, V)$ is the set of possible outcomes for action a assuming the true hypothesis belongs to V and V_a^o is the set of hypotheses in V that has outcome o for action a.
5: $\hat{a} \leftarrow \text{argmax}_a R(a)$
6: **for** each outcome $o \in O(\hat{a}, V)$ of $\hat{a}$ **do**
7:    GreedyTree($\mathcal{P}(\mathcal{I}), V_{\hat{a}}^o$)

Connecting DFEP to ODTH. We next establish a connection between the optimal decision tree w.r.t. $\mathcal{P}(\mathcal{I})$ and the optimal decision tree w.r.t. $\mathcal{I}$.

Lemma 2. *Let $D_{\mathcal{I}}^*$ denote the optimal decision tree w.r.t. $\mathcal{I}$ and $D_{\mathcal{P}(\mathcal{I})}^*$ denote the optimal decision tree w.r.t. $\mathcal{P}(\mathcal{I})$, we have*

$$cost'(D_{\mathcal{P}(\mathcal{I})}^*) \leq cost(D_{\mathcal{I}}^*). \qquad (7)$$

Proof: The proof of this lemma relies on the observation that given an optimal decision tree $D_{\mathcal{I}}^*$ for our original problem $\mathcal{I}$, we can build a decision tree for its corresponding instance of DFEP $\mathcal{P}(\mathcal{I})$ from $D_{\mathcal{I}}^*$. Given a decision tree $D_{\mathcal{I}}^*$, rooted at r, we next show how to identify the class of an unknown hypothesis

$h^* \in H$. First, we select action $r \in A$ and observe the outcome of $a(h^*)$; then we follow the branch associated with $a(h^*)$ to reach a child $r' \in A$ of r; we apply the same procedure recursively for the decision tree rooted at r'. This procedure continues until it reaches some node which determines the class of h^*. Let $\hat{D}_{\mathcal{P}(\mathcal{I})}$ denote this decision tree. Consider the parent node $a \in A$ of an arbitrary leaf node in $\hat{D}_{\mathcal{P}(\mathcal{I})}$, it is easy to verify that the decision tree rooted at a contains at least one leaf that is associated with some hypothesis from g_1. Hence, there exists at least one "most expensive" (w.r.t. the cost function cost') root-to-leaf path in $\hat{D}_{\mathcal{P}(\mathcal{I})}$ such that its leaf is associated with some hypothesis $\hat{h} \in g_1$, i.e.,

$$\mathsf{cost}'(\hat{D}_{\mathcal{P}(\mathcal{I})}) = \mathsf{cost}'(\hat{D}_{\mathcal{P}(\mathcal{I})}, \hat{h}). \tag{8}$$

By the definition of $\mathsf{cost}'(\hat{D}_{\mathcal{P}(\mathcal{I})}, \hat{h})$, we have

$$\mathsf{cost}'(\hat{D}_{\mathcal{P}(\mathcal{I})}, \hat{h}) = \sum_{a \in \mathsf{path}(\hat{D}_{\mathcal{P}(\mathcal{I})}, \hat{h})} c'(a) = \sum_{a \in \mathsf{path}(\hat{D}_{\mathcal{P}(\mathcal{I})}, \hat{h})} c(a, g_1), \tag{9}$$

where the second equality is by the definition of $c'(a)$. According to the construction of $\hat{D}_{\mathcal{P}(\mathcal{I})}$, we have $\mathsf{path}(\hat{D}_{\mathcal{P}(\mathcal{I})}, \hat{h})$ is identical to $\mathsf{path}(D_{\mathcal{I}}^*, \hat{h})$. Hence,

$$\sum_{a \in \mathsf{path}(\hat{D}_{\mathcal{P}(\mathcal{I})}, \hat{h})} c(a, g_1) = \sum_{a \in \mathsf{path}(D_{\mathcal{I}}^*, \hat{h})} c(a, g_1) \leq \mathsf{cost}(D_{\mathcal{I}}^*), \tag{10}$$

where the inequality is by the definition of $\mathsf{cost}(D_{\mathcal{I}}^*)$.

Inequalities (8) (9) (10) together imply that

$$\mathsf{cost}'(D_{\mathcal{P}(\mathcal{I})}^*) \leq \mathsf{cost}(D_{\mathcal{I}}^*). \tag{11}$$

$\square$

The following corollary follows from Lemma 1 and Lemma 2.

Corollary 1. *Let $D_{\mathcal{I}}^*$ denote the optimal decision tree w.r.t. our original problem $\mathcal{I}$ and $D_{\mathcal{P}(\mathcal{I})}^{\mathsf{greedy}}$ denote the decision tree returned from* GreedyTree($\mathcal{P}(\mathcal{I}), H$), *we have*

$$\mathsf{cost}'(D_{\mathcal{P}(\mathcal{I})}^{\mathsf{greedy}}) \leq O(\log n)\mathsf{cost}(D_{\mathcal{I}}^*). \tag{12}$$

3.2   Design of Group Greedy Decision Tree

Now we are in position to present our final algorithm (labeled as Group-GreedyTree). At a high level, GroupGreedyTree processes hypotheses in accordance to their priorities. That is, GroupGreedyTree starts with identifying the true hypothesis from the highest priority group g_1, if the true hypothesis is not from g_1, then GroupGreedyTree proceeds to identifying the true hypothesis from the second highest priority group, and this process continues until the true hypothesis has been identified. We next describe GroupGreedyTree in details. We first introduce some notations. Given a ODTH instance $\mathcal{I}$ and any subset of hypotheses $S \subseteq H$, define a new ODTH instance $\mathcal{I}_{\cap S} = (S, G_{\cap S}, A, \mathbf{c})$; here $G_{\cap S} = \{g \in G \mid g \cap S \neq \emptyset\}$ is the set of groups from G that contains at least one item from S. GroupGreedyTree works as follows:

1. We first build a decision tree $D_{\mathcal{P}(\mathcal{I})}^{\text{greedy}}$ from GreedyTree($\mathcal{P}(\mathcal{I}), H$). Define a leaf r from $D_{\mathcal{P}(\mathcal{I})}^{\text{greedy}}$ as *incomplete* if it is associated with more than one hypotheses, i.e., $|H(r)| > 1$ where $H(r)$ denotes the set of possible hypotheses associated with r. Note that if such r exists, then it must be the case that $H(r) \subseteq H \setminus g_1$, this is because all hypotheses from the highest priority group g_1 must be uniquely identified by GreedyTree($\mathcal{P}(\mathcal{I}), H$). We traverse $D_{\mathcal{P}(\mathcal{I})}^{\text{greedy}}$ to find the set U of all incomplete leaf nodes.

2. For each incomplete leaf node $r \in U$, define a new instance $\mathcal{I}_{\cap H(r)}$. Replace r using a decision tree $D_{\mathcal{P}(\mathcal{I}_{\cap H(r)})}^{\text{greedy}}$ returned from GreedyTree($\mathcal{P}(\mathcal{I}_{\cap H(r)}), H(r)$).

3. Apply the same procedures recursively for $D_{\mathcal{P}(\mathcal{I}_{\cap H(r)})}^{\text{greedy}}$ for each $r \in U$ to further expand the current tree. This process continues until each leaf node in the current tree is associated with a single hypothesis.

Algorithm 2. GroupGreedyTree($\mathcal{P}(\mathcal{I}), V$)

1: **if** $F(V) = 0$ **then**
2:    **if** $|V| = 1$ **then**
3:       return
4:    **else**
5:       GroupGreedyTree($\mathcal{P}(\mathcal{I}_{\cap V}), V$)
6: **for** $a \in A$ **do**
7:    compute $R(a) := \min_{o \in O(a, V)} \frac{F(V) - F(V_a^o)}{c'(a)}$.
8: $\hat{a} \leftarrow \text{argmax}_a R(a)$
9: **for** each outcome $o \in O(\hat{a}, V)$ of $\hat{a}$ **do**
10:   GroupGreedyTree($\mathcal{P}(\mathcal{I}_{\cap V_{\hat{a}}^o}), V_{\hat{a}}^o$)

The pseudocode of GroupGreedyTree is listed in Algorithm 2. We next present the main theorem of this paper.

Theorem 1. *Let $D_{\mathcal{I}}^*$ denote the optimal decision tree w.r.t. our original problem $\mathcal{I}$ and $D_{\mathcal{P}(\mathcal{I})}^{g\text{-}greedy}$ denote the decision tree returned from GroupGreedyTree($\mathcal{P}(\mathcal{I}), H$), we have*

$$cost(D_{\mathcal{P}(\mathcal{I})}^{g\text{-}greedy}) \leq O(m \log n) cost(D_{\mathcal{I}}^*). \tag{13}$$

Proof: We prove this theorem through induction on the number of groups m. The base case when $m = 1$ is trivial. If $m = 1$, i.e., all hypotheses have the same priority, then ODTH is reduced to the classical optimal decision tree problem, which is a special case of DFEP, and the group greedy algorithm Group-GreedyTree$(\mathcal{P}(\mathcal{I}), H)$ is reduced to GreedyTree$(\mathcal{P}(\mathcal{I}), H)$. Lemma 1 indicates that the greedy decision tree returned from GroupGreedyTree$(\mathcal{P}(\mathcal{I}), H)$ achieves a $O(\log n)$ approximation ratio. I.e.,

$$\text{cost}(D_{\mathcal{P}(\mathcal{I})}^{\text{g-greedy}}) \leq O(\log n)\text{cost}(D_{\mathcal{I}}^*). \tag{14}$$

Assume this theorem holds for $m \leq k$, i.e., $\text{cost}(D_{\mathcal{P}(\mathcal{I})}^{\text{g-greedy}}) \leq O(k\log n)$ $\text{cost}(D_{\mathcal{I}}^*)$, we next prove the case when $m = k + 1$. Recall that in Corollary 1, we show that

$$\text{cost}'(D_{\mathcal{P}(\mathcal{I})}^{\text{greedy}}) \leq O(\log n)\text{cost}(D_{\mathcal{I}}^*). \tag{15}$$

Observe that for every hypothesis h from g_1, we have

$$\text{cost}'(D_{\mathcal{P}(\mathcal{I})}^{\text{greedy}}, h) = \text{cost}(D_{\mathcal{P}(\mathcal{I})}^{\text{greedy}}, h). \tag{16}$$

This is because by the definition of $\mathbf{c}'$, we have $c'(a) = c(a, g_1)$ for all $a \in A$. Moreover, for every hypothesis h that does not belong to g_1, we have

$$\text{cost}'(D_{\mathcal{P}(\mathcal{I})}^{\text{greedy}}, h) \geq \text{cost}(D_{\mathcal{P}(\mathcal{I})}^{\text{greedy}}, h). \tag{17}$$

This is because $c'(a) = c(a, g_1) \geq c(a, g_i)$ for all $a \in A$ and $i \neq 1$ where the inequality is by the fact that g_1 has the highest priority. Inequalities (16) and (17) together imply that for all hypotheses $h \in H$, we have

$$\text{cost}'(D_{\mathcal{P}(\mathcal{I})}^{\text{greedy}}, h) \geq \text{cost}(D_{\mathcal{P}(\mathcal{I})}^{\text{greedy}}, h). \tag{18}$$

This together with (15) implies that for all hypotheses $h \in H$, we have

$$\text{cost}(D_{\mathcal{P}(\mathcal{I})}^{\text{greedy}}, h) \leq \text{cost}'(D_{\mathcal{P}(\mathcal{I})}^{\text{greedy}}, h) \leq \text{cost}'(D_{\mathcal{P}(\mathcal{I})}^{\text{greedy}}) \leq O(\log n)\text{cost}(D_{\mathcal{I}}^*), \tag{19}$$

where the second inequality is by the definition of $\text{cost}'(D_{\mathcal{P}(\mathcal{I})}^{\text{greedy}})$, that is,

$$\text{cost}'(D_{\mathcal{P}(\mathcal{I})}^{\text{greedy}}) = \max_{h \in H} \text{cost}'(D_{\mathcal{P}(\mathcal{I})}^{\text{greedy}}, h).$$

Now we are in position to prove this theorem. To prove this theorem, it suffices to show that for all hypotheses $h \in H$, we have

$$\text{cost}(D_{\mathcal{P}(\mathcal{I})}^{\text{g-greedy}}, h) \leq O((k+1)\log n)\text{cost}(D_{\mathcal{I}}^*). \tag{20}$$

The proof of the case when $h \in g_1$ is trivial. If $h \in g_1$, then h must be uniquely identified by both $D_{\mathcal{P}(\mathcal{I})}^{\text{g-greedy}}$ and $D_{\mathcal{P}(\mathcal{I})}^{\text{greedy}}$. Hence h is a leaf node in both $D_{\mathcal{P}(\mathcal{I})}^{\text{g-greedy}}$ and $D_{\mathcal{P}(\mathcal{I})}^{\text{greedy}}$. It follows that $\text{cost}(D_{\mathcal{P}(\mathcal{I})}^{\text{g-greedy}}, h) = \text{cost}(D_{\mathcal{P}(\mathcal{I})}^{\text{greedy}}, h)$. This together with (18) implies that $\text{cost}(D_{\mathcal{P}(\mathcal{I})}^{\text{g-greedy}}, h) \leq O(\log n)\text{cost}(D_{\mathcal{I}}^*) \leq O((k+1)\log n)\text{cost}(D_{\mathcal{I}}^*)$.

We next prove the case when $h \notin g_1$. By abuse of notation, define $H(h)$ as the set of hypotheses that can not be distinguished from h in $D_{\mathcal{P}(\mathcal{I})}^{\text{greedy}}$. Observe that for all hypotheses $h \notin g_1$, we have $\text{cost}(D_{\mathcal{P}(\mathcal{I})}^{\text{g-greedy}}, h) = \text{cost}(D_{\mathcal{P}(\mathcal{I})}^{\text{greedy}}, h) + \text{cost}(D_{\mathcal{P}(\mathcal{I}_{\cap H(h)})}^{\text{g-greedy}})$ by the design of $D_{\mathcal{P}(\mathcal{I})}^{\text{g-greedy}}$. Hence,

$$\text{cost}(D_{\mathcal{P}(\mathcal{I})}^{\text{g-greedy}}, h) = \text{cost}(D_{\mathcal{P}(\mathcal{I})}^{\text{greedy}}, h) + \text{cost}(D_{\mathcal{P}(\mathcal{I}_{\cap H(h)})}^{\text{g-greedy}}) \tag{21}$$

$$\leq O(\log n)\text{cost}(D_{\mathcal{I}}^*) + \text{cost}(D_{\mathcal{P}(\mathcal{I}_{\cap H(h)})}^{\text{g-greedy}}) \tag{22}$$

$$\leq O(\log n)\text{cost}(D_{\mathcal{I}}^*) + O(k \log n)\text{cost}(D_{\mathcal{I}_{\cap H(h)}}^*) \tag{23}$$

$$\leq O(\log n)\text{cost}(D_{\mathcal{I}}^*) + O(k \log n)\text{cost}(D_{\mathcal{I}}^*) \tag{24}$$

$$= O((k+1) \log n)\text{cost}(D_{\mathcal{I}}^*), \tag{25}$$

where the first inequality is by (19), the second inequality is by the inductive assumption, the third inequality is by the observation that any decision tree that covers $\mathcal{I}$ is also a decision tree that covers $\mathcal{I}_{\cap H(h)}$, hence, $\text{cost}(D_{\mathcal{I}_{\cap H(h)}}^*) \leq \text{cost}(D_{\mathcal{I}}^*)$ where $D_{\mathcal{I}_{\cap H(h)}}^*$ denotes the optimal decision tree that covers instance $\mathcal{I}_{\cap H(h)}$. This finishes the proof of the case when $h \notin g_1$. $\square$

4   Conclusion

In conclusion, we have addressed the limitations of previous studies on the optimal decision tree problem by considering costs that are dependent on the unknown true hypothesis. By dividing the hypotheses into priority groups and introducing a group greedy algorithm, we have developed a model that strategically selects actions based on their priority level, minimizing the total cost in scenarios where the consequences of actions vary with the true hypothesis.

Our proposed algorithm achieves an approximation ratio of $O(m \log n)$ in worst-case scenarios, providing an efficient solution for optimizing decision trees with cost-dependent actions. Furthermore, our result aligns with the logarithmic approximation bound for the traditional optimal decision tree problem when there is only one group of hypotheses.

For future work, we can explore extensions of our model to handle more complex cost dependencies and investigate algorithms with improved approximation ratios. Additionally, empirical evaluations on real-world datasets and applications can further validate the effectiveness and practicality of our approach. By advancing the understanding of optimal decision trees with cost-dependent actions, we can contribute to various domains such as active learning, disease diagnosis, and fraud detection, enhancing decision-making processes and minimizing costs in practical scenarios.

References

1. Chakaravarthy, V.T., Pandit, V., Roy, S., Awasthi, P., Mohania, M.: Decision trees for entity identification: approximation algorithms and hardness results. In: Proceedings of the Twenty-Sixth ACM SIGMOD-SIGACT-SIGART Symposium on Principles of Database Systems, pp. 53–62 (2007)
2. Cicalese, F., Laber, E., Saettler, A.M.: Diagnosis determination: decision trees optimizing simultaneously worst and expected testing cost. In: International Conference on Machine Learning, pp. 414–422. PMLR (2014)
3. Dasgupta, S.: Analysis of a greedy active learning strategy. In: Advances in Neural Information Processing Systems, vol. 17 (2004)
4. Golovin, D., Krause, A.: Adaptive submodularity: theory and applications in active learning and stochastic optimization. J. Artif. Intell. Res. **42**, 427–486 (2011)
5. Kapoor, A., Horvitz, E., Basu, S.: Selective supervision: guiding supervised learning with decision-theoretic active learning. In: IJCAI, vol. 7, pp. 877–882 (2007)
6. Margineantu, D.D.: Active cost-sensitive learning. In: IJCAI, vol. 5, pp. 1622–1623 (2005)
7. Sabato, S.: Submodular learning and covering with response-dependent costs. Theoret. Comput. Sci. **742**, 98–113 (2018)
8. Saettler, A., Laber, E., Cicalese, F.: Approximating decision trees with value dependent testing costs. Inf. Process. Lett. **115**(6–8), 594–599 (2015)
9. Yuan, J., Tang, S.: Worst-case adaptive submodular cover. In: Proceedings of the 2023 International Conference on Autonomous Agents and Multiagent Systems, pp. 1915–1922 (2023)

Approximating the λ-low-density Value

Joachim Gudmundsson⬛, Zijin Huang(✉)⬛, and Sampson Wong⬛

University of Sydney, Darlington, NSW 2008, Australia
zijin.huang@sydney.edu.au

Abstract. The use of realistic input models has gained popularity in the theory community. Assuming a realistic input model often precludes complicated hypothetical inputs, and the analysis yields bounds that better reflect the behaviour of algorithms in practice.

One of the most popular models for polygonal curves and objects is λ-low-density. To select the most efficient algorithm for a certain input, one often needs to approximate the λ-low-density value, or density for short. In this paper, we show that given a set of n objects in $\mathbb{R}^2$, one can $(2 + \varepsilon)$-approximate the density value in $O(n \log n + \lambda n/\varepsilon^4)$ time.

Finally, we argue that some real-world trajectory data sets have small density values, warranting the recent development of specialised algorithms. This is done by computing approximate density values for 12 real-world trajectory data sets.

Keywords: realistic input models · c-packedness · low-density · computational geometry

1 Introduction

Theoretical algorithmic analysis is an essential tool for understanding the complexity of a problem. It allows researchers to establish upper and lower bounds by carefully constructing worst-case scenarios, giving us insight into the difficulty of a problem. However, the lower bounds established on these worst-case scenarios may not accurately reflect the difficulty of a problem in real-world situations.

Realistic input models, such as fatness [3], low-density [3], uncluttered [3], and c-packedness [6] describe real-world patterns. These models rule out unlikely scenarios by placing constraints on the input. For example, a set of segment S is c-packed if for any ball B of radius r, $\|B \cap S\| \le c \cdot r$ — S has low congestion. The packedness value c is the maximum value for which S is c-packed, and c is the respective realistic input model parameter. There are extensive studies on realistic input models [3,6,8], and many problems have more efficient solutions when the input data has these constraints [4–7, 11–13].

In this paper, we study the notion of low-density. Measuring the size of an object by the radius of its smallest enclosing ball, we say a set $S = \{P_1, ..., P_n\}$

J. Gudmundsson—Funded by the Australian Government through the Australian Research Council DP180102870.

© The Author(s), under exclusive license to Springer Nature Switzerland AG 2024
W. Wu and G. Tong (Eds.): COCOON 2023, LNCS 14422, pp. 71–82, 2024.
https://doi.org/10.1007/978-3-031-49190-0_5

of objects with sizes $\{\rho_1, ..., \rho_n\}$ is λ-low-density if for any ball B of radius r, there are at most λ objects with $\rho_i \geq r$ intersecting B. We say S is low-density if λ is a small constant. Van der Stappen and Overmars [12] originally proposed low-density as a realistic constraint for the environments of robotic navigation. More recent studies of low-density focus on its application on road networks and GPS trajectories analysis.

There are many examples where there are much more efficient algorithm if the data set is low-density. Approximating the Frechet distance [6] and map-matching [4,5] problems have much more efficient solutions if the input is low-density.

Before employing these algorithms, it is often important to be confident that λ is sufficiently small. As an example, Driemel et al. [6] proposed an algorithm to approximate the Fréchet distance between a pair of d-dimensional low-density curves. The running time of the algorithm is $O((\lambda n)^{2(d-1)/d} \log n)^1$. However, for any pair of general curves, there is an $O(dn^2 \log n)$ time algorithm [2]. Therefore, the specialised algorithm of [6] is more efficient than the general algorithm of [2] if and only if $\lambda = o(n^{1/(d-1)})$.

Another example is Buchin et al.'s [4] data structure for map matching queries on λ-low density road networks. For road network of complexity n, constant spanning ratio, and constant query complexity, their queries run in $O(\lambda^6 \sqrt{n} \text{ polylog}(n))$ time. However, a general algorithm solves the map matching problem in $O(n \log n)$ time [1], if the trajectory has constant complexity. Therefore, the specialised algorithm of [4] is more efficient than the general algorithm of [1] if and only if $\lambda = o(n^{1/12})$. In both examples, one would need to either compute or approximate the value of λ to decide whether the specialised algorithm or the general algorithm would be more efficient.

Our Contribution. Given a set S of n objects in $\mathbb{R}^2$, we propose an algorithm that $(2 + \varepsilon)$-approximates the density value of a set of objects in $O(n \log n + \lambda n/\varepsilon^4)$ time. Our algorithm is the first to approximate the density value of intersecting objects, and the first to do so in a dynamic setting. Using this algorithm, we also approximate the low-density values of 12 real-world trajectory data sets in $\mathbb{R}^2$, and show that about half of the data sets we tested have low density values.

In many applications, it suffices to decide whether λ is small. It is straightforward to modify our algorithm from a minimisation to a decision problem. Given a parameter $\lambda \in \mathbb{R}$ and a set S of n objects in $\mathbb{R}^2$, our algorithm can decide in $O(n \log n + \lambda n/\varepsilon^4)$ time whether the density value is at least λ, or at most $(2 + \varepsilon)\lambda$. For example, it is sufficient to run the decision version on $\lambda = o(n^{1/(d-1)})$ to choose between the algorithms [2,6], and on $\lambda = o(n^{1/12})$ to choose between data structures [1,4].

Related Work. De Berg et al. [3] proposed an $O(n \log^3 n + \lambda n \log^2 n + \lambda^2 n)$ algorithm to compute the density of planar scenes (environments with non-intersecting objects). However, in data sets such as vehicle trajectories, city road

---

[1] The dependence on λ is based on our own calculation, as it is not stated in [6].

networks, and virtual environments, objects do intersect, so the algorithm of [3] does not apply. Our algorithm works on intersecting objects.

Another related work is by Har-Peled and Zhou [10], in which they presented a randomised algorithm that, with high probability, computes a $(288+\varepsilon)$-approximation of the c-packedness of a set of segments in $O(n \log^2 n)$ time. They used canonical squares of similar size to cover a segment, and stored these canonical squares in a compressed quadtree. Once the quadtree is built, they use it to compute the number of long segments intersecting a canonical square to approximate the so-called α-long congestion. Their insights may be extended to obtain an $O(1)$-approximation of the density value. However, it is not discussed in [10]. Our approach is similar in storing objects in a quadtree, and computing the maximum number of intersecting objects of a canonical square.

Organisation. In Sect. 2, we formally define low-density, and discuss a property that will be used throughout our study. In Sect. 3, we will discuss how to efficiently store a set of λ-low-density objects in a compressed quadtree by storing at most $O(\lambda)$ objects in every square. In Sect. 4, we will discuss how to approximate the density value by carefully placing balls in the plane such that a small number of them can cover an optimal ball (a ball that intersects the most number of objects of size greater than its radius).

2   Overview

In this section, we formally introduce low-density and its basic properties. We will first give a formal definition of low-density. De Berg et al. [3] defined low-density for non-intersecting objects, and we modified their definition to remove this restriction. In our study, we will measure the size of an object P_i by the radius ρ_i of P_i's minimum enclosing ball.

Definition 1. *Let $S = \{P_1, ..., P_n\}$ be a set of d-dimensional objects, and let $\lambda \geq 1$ be a parameter. We say S is λ-low-density if for any ball B, the number of objects $P_i \in S$ with $\rho_i \geq radius(B)$ that intersect B is at most λ. The density of S is defined to be the smallest λ for which S is λ-low-density.*

In addition, a ball B is density optimal, or optimal for short, with respect to a set S if S is λ-low-density and B intersects exactly λ objects P_i with $\rho_i \geq radius(B)$.

2.1   A Basic Property

De Berg et al. [3] made the useful observation that we can restrict our attention to balls with radius ρ_i for some $P_i \in S$, as shown in the following.

Observation 1. *Let $S = \{P_1, ..., P_n\}$ be a set of d-dimensional objects that is λ-low-density, and let P_i be the smallest object that intersects an optimal ball. Then there exists an optimal ball B with radius ρ_i.*

As a result, if P_i is the smallest object that intersects an optimal ball B, we can increase the radius of B to ρ_i, and B will still be optimal.

3   Augmenting the Compressed Quadtree

In this section, we will discuss how to build an augmented compressed quadtree to store objects in $\mathbb{R}^2$. The main result of this section is a compressed quadtree storing a set of λ-low-density objects that supports fuzzy range queries around objects in S efficiently (Theorem 3). We will start with some preliminaries on general compressed quadtree.

3.1   Preliminaries

We assume that, without loss of generality, all objects lie in a unit square. In a quadtree, squares are recursively partitioned into four equal squares. A node v in a quadtree corresponds to a square $\square_v$. If $\square_v$ is partitioned into four equal squares $\{\square_a, \square_b, \square_c, \square_d\}$, then node a, b, c, d are the children of node v. We say a square $\square$ is a *canonical square* if and only if $\square$ is a node in the quadtree, where the unit square is the root (see [9] for a formal definition). In the rest of the paper, we blur the difference between a canonical square and its corresponding node in the quadtree.

From the construction of a quadtree, one can define the grid and grid cell. We will use the definition in [9]. For any $i, j \in \mathbb{Z}$, the intersection of the halfplanes $x \geq ai, x < a(i+1), y \geq aj$ and $y < a(j+1)$ is a *grid cell* of grid G_a. Since each canonical square is recursively partitioned into four canonical squares, a canonical square must be a cell in $G_{1/2^i}$ (see Fig. 1).

Observation 2. ([9]) *A square $\square$ is a canonical square if and only if $\square$ is a cell in $G_{1/2^i}$, where i is a non-negative integer.*

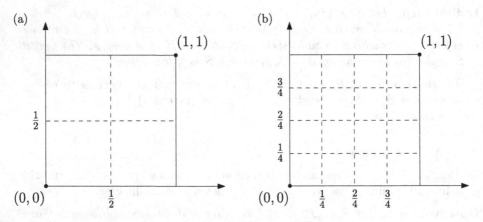

Fig. 1. The grid described by the squares of size $1/2^i$ is denoted $G_{1/2^i}$. (a) Showing $G_{1/2}$, and (b) $G_{1/4}$.

3.2 Covering a Minimal Enclosing Box with Canonical Squares

With unit square and canonical square defined, we will show that we can cover the minimal enclosing box Q of P with a set $\mathcal{C}_P$ of at most four large canonical squares. Then we will partition $\mathcal{C}_P$ into a set of small canonical squares, and delete those that do not intersect P. In particular, we highlight that the minimal enclosing balls or boxes of a set S of n objects can be computed in linear time with respect to the size of the input (the total number of edges in S). Therefore we assume that the minimal enclosing boxes of S can be computed in $O(n)$ time.

In order to cover Q with canonical squares, we first need to define the *most significant separator* of Q. Consider the vertical line at $x = \frac{k}{2^i}$, $i, k \in \mathbb{Z}$, where k is odd. We say it is a vertical *separator* of Q if it intersects Q. We say $x = \frac{k}{2^i}$ is more significant than $x = \frac{k'}{2^j}$ if $i < j$. We define a horizontal separator analogously. Notice that each Q can intersect exactly one most significant vertical and one most significant horizontal separator. Indeed, if Q intersects two most significant vertical separators $x = \frac{k}{2^i}$ and $x = \frac{k'}{2^i}$, $k \neq k'$, there exists some $k'' \in (k, k')$ such that k'' is even, since both k and k' are odd. Then Q also intersects $x = \frac{k''}{2^i} = \frac{k''/2}{2^{i-1}}$, which means $x = \frac{k}{2^i}$ cannot be the most significant separator.

In order to find the most significant separators of Q, we need to assume the unit RAM model. In particular, we assume that $\log_2(\cdot)$ and XOR operations of arbitrarily large numbers take constant time.

Lemma 1. *Finding the most significant separators of Q takes $O(1)$ time under the unit RAM model.*

Proof. We use the *bit twiddling* method in [9]. We assume that $\alpha, \beta \in [0, 1)$ are written in base two as $\alpha = 0.\alpha_1\alpha_2...$, and $\beta = 0.\beta_1\beta_2...$ Let $\mathrm{bit}_\triangle(\alpha, \beta)$ be the index of the first bit after the period in which they differ.

Let α, β be the x-coordinate of the top-left and top-right corner of Q, respectively. Let $i = \mathrm{bit}_\triangle(\alpha, \beta)$. Notice that if the first $i - 1$ bits of α and β are the same, they must reside in a canonical square which is a cell of $G_{1/2^{i-1}}$. For example, if $\alpha_1 = \beta_1 = 1$, then $\alpha, \beta \in [1/2, 1)$. Furthermore, if $\alpha_2 = \beta_2 = 0$, then $\alpha, \beta \in [2/4, 3/4)$. Therefore, the most significant vertical separator must be either $x = \alpha$, $x = \beta$ or the immediate next separator to the right side of α which takes the form $x = k/2^i$, where k is odd.

We can then find the most significant separator by adding $1/2^i$ to α, and set the bits after ith bit to 0. Let $l = 0.\alpha_1\alpha_2...\alpha_{2^i-1}1$. We report the more significant separator among α, β, and l as the most significant vertical separator of Q. And since it takes $O(1)$ time to calculate $\mathrm{bit}_\triangle(\alpha, \beta)$, calculating l and comparing l, α, and β also takes $O(1)$ time.

Har-Peled [9] justified why it is reasonable to assume that $\mathrm{bit}_\triangle(\cdot, \cdot)$ can be computed in constant time. Exponents and mantissas represent modern float numbers. If two numbers have different exponents, we can compute $\mathrm{bit}_\triangle(\cdot, \cdot)$ by giving the larger component. Otherwise, we can XOR the mantissas, and $\log_2(\cdot)$ the result.

The most significant separators of Q partitions Q into four quadrants. We can expand each quadrant into a canonical square $\square$, and partition $\square$ into a constant number of smaller canonical squares in Lemma 2.

Lemma 2. *Let P be an object in $\mathbb{R}^2$. One can construct $O(1)$ canonical squares to cover the minimal enclosing box Q of P such that the size of each canonical square is at most ρ, the radius of P's minimum enclosing ball.*

3.3   Merging Canonical Squares

We have shown that we can compute a set C_i of canonical squares to cover P_i's minimal enclosing box. Next, we will show that we can efficiently merge these canonical squares to remove duplicates, while making sure that each square stores its intersecting objects in decreasing order of size. We say that the canonical squares in C_i are the *associated squares* of P_i, and analogously, P_i is their associated object. We will merge the canonical squares in the below step.

1. For each object P_i, perform a linear scan of the canonical squares in C_i, and delete squares that do not intersect P_i.
2. Let $C = \cup_{P_i \in S} C_i$ and sort all the canonical squares in C based on their sizes, then by coordinates of the bottom-left corners, and, finally, by the size of the object they intersect in decreasing order.
3. Perform a linear scan of the sorted squares in the list, and merge adjacent squares if they are identical. While merging two squares, also merge the set of objects they intersect. The objects that intersect the same square are ordered in decreasing order of size.

At the end of the preprocessing step, we have $O(n)$ canonical squares. Each canonical square stores its intersecting object(s) in decreasing order of size. It takes linear time with respect to the input size to compute the minimum bounding boxes of S. Scaling them to fit into the unit square takes $O(n)$ time. Generating these squares takes $O(n)$ time by Lemmas 1 and 2. Sorting C takes $O(n \log n)$ time. Merging two adjacent squares requires inserting the larger object to the start of the intersecting object list, which takes $O(n)$ time in total. As a result, we get the below lemma.

Lemma 3. *One can generates a set of $O(n)$ canonical squares to cover a set $S = \{P_1, ..., P_n\}$ of objects in $\mathbb{R}^2$ in $O(n \log n)$ time.*

3.4   An Efficient Construction Using Compressed Quadtrees

We will construct the compressed quadtree from the $O(n)$ canonical squares constructed in the previous section. Let $\square_v$ denote the square associated with a node v. We use the below lemma from [9].

Lemma 4 (Lemma 2.11 in [9]). *Given a list C of n canonical squares, all lying inside the unit square, one can construct a (minimal) compressed quadtree T such that for any square $c \in C$, there exists a node $v \in T$, such that $\square_v = c$. The construction time is $O(n \log n)$.*

We now apply Lemma 4, with the $O(n)$ canonical squares constructed in Sect. 3.3 as input, we obtain a compressed quadtree, where each canonical square corresponds to an internal node.

Once we complete the construction of the compressed quadtree $\mathcal{T}$, we start the *push-down step* as follows. First, sort the objects in increasing order of size. Then, for each object $P_i \in S$, and for each descendant $\square_v$ of P_i's associated canonical squares, if $\square_v$ intersects P_i, insert P_i at the beginning of the list of intersecting objects of v.

An important property of the canonical squares in $\mathcal{T}$ is that each $\square \in \mathcal{T}$ stores at most $O(\lambda)$ objects in decreasing order of size. For an object P_i, the sizes of its associated canonical squares are at most ρ_i by Lemma 2. During the push-down step, an object P_i is only stored in canonical squares with sizes at most ρ_i.

Lemma 5. *Each canonical square $\square$ in the compressed quadtree $\mathcal{T}$ stores all $O(\lambda)$ intersecting objects $P_i \in S$ with $\rho_i \geq size(\square)$ in decreasing order of size.*

According to Lemma 3 and Lemma 4, one can construct a compressed quadtree with $O(n)$ canonical squares in $O(n \log n)$ time. It remains only to analyse the running time of the push-down step. Sorting the objects by sizes takes $O(n \log n)$ time. By Lemma 5, each canonical square stores $O(\lambda)$ intersecting objects. Charging $O(\lambda)$ to each of $O(n)$ objects, the push-down step takes $O(\lambda n)$ time, and the compressed quadtree uses $O(\lambda n)$ space in total.

Lemma 6. *One can construct a compressed quadtree of a set S of n objects in $\mathbb{R}^2$ in $O(n \log n + \lambda n)$ time using $O(\lambda n)$ space. A canonical square in $\mathcal{T}$ stores $O(\lambda)$ intersecting objects in decreasing order of size.*

3.5   Query Equal or Larger Nearby Objects

To use our data structure to approximate the density, we augment the quadtree further to query nearby objects. Given a set of λ-low-density objects $S = \{P_1, ..., P_n\}$, and a set of parameters $\{d_1, ..., d_n\}$, where $d_i \geq 0$, we will show that we can construct our quadtree to answer the following query efficiently: given an object P_i stored in $\mathcal{T}$, report all objects P_j with $\rho_j \geq \rho_i$ that are at most d_i apart from P_i. We will call this query the *near-by query*.

Recall that during the construction of the quadtree, we compute a set of canonical squares to cover an object P_i. The main modification to support the near-by query is to compute a set $\mathcal{C}_i$ of canonical squares to cover P_i, as well as the area within distance d_i from P_i, i.e., $\mathcal{C}_i$ covers $M_i = P_i \oplus B(0, d_i/2)$. Naturally, the number of canonical squares required to cover M_i depends on d_i. With an increasing number of canonical squares, the size of the quadtree increases. To perform the near-by query, one needs to visit each $\square \in \mathcal{C}_i$, and report intersecting objects of $\square$. We summarise our results in the below Theorem 3.

Theorem 3. *Let $S = \{P_1, ..., P_n\}$ be a set of λ-low-density objects in $\mathbb{R}^2$, and let $\{d_1, ..., d_n\}$ be a set of parameters where $d_i \geq 0$. If one can generate n_i canonical* squares of size at most ρ_i to cover $P_i \oplus B(0, d_i/2)$, then one can compute a compressed quadtree $\mathcal{T}$ to store S in $O(N \log N + \lambda N)$ time, where $N = \sum_i n_i$, using $O(\lambda N)$ space. Each canonical square in $\mathcal{T}$ stores $O(\lambda)$ objects, and it takes $O(n_i \lambda)$ time to report all objects P_j with $\rho_j \geq \rho_i$ that are within distance d_i from P_i.

4   Approximating the Low-Density Value

In this section, we use a simple algorithm to approximate the density value of a set of objects $S = \{P_1, ..., P_n\}$. Each $P_i \in S$ is a possible smallest object intersecting an optimal ball, and by Observation 1, we can focus our attention on balls of size ρ_i. As such, we are interested in objects P_j with $\rho_j \geq \rho_i$ that are within distance ρ_i from P_i, since a ball of size ρ_i can intersect both P_i and P_j.

To query the nearby objects of P_i, we will use the augmented compressed quadtree in Sect. 3. With a slight abuse of notation, we redefine $M_i = P_i \oplus B(0, \rho_i)$. As the diameter of M_i is at most a constant number times ρ_i, we can compute a constant number of canonical squares with sizes at most ρ_i to cover M_i by Lemma 2; combining with Theorem 3, this leads to the following corollary.

Corollary 1. *Let $S = \{P_1, ..., P_n\}$ be a set of λ-low-density objects with sizes $\{\rho_1, ..., \rho_n\}$ in $\mathbb{R}^2$. One can cover $M_i = P_i \oplus B(0, \rho_i)$ with $O(1)$ canonical squares C_i of size at most ρ_i to cover M_i. One can construct a compressed quadtree $\mathcal{T}$ using $\cup_{P_i \in S} C_i$ in $O(n \log n + \lambda n)$ time. Using $\mathcal{T}$, one can query all objects P_j intersecting M_i with $\rho_j \geq \rho_i$ in $O(\lambda)$ time.*

We now assume that P_i is the smallest object intersecting an optimal ball B^*. These optimal balls must lie in M_i, and we can query the intersecting objects of M_i efficiently using the above corollary. Our approximation algorithms relies on a simple technique: covering potential optimal balls with equal-size balls. Let us assume that we know an optimal ball B^*, and we can cover B^* with c balls $\{O_1, ..., O_c\}$ of equal sizes. If O_i intersects the most number of objects, O_i intersect at least λ/c objects, and every O_i intersects at most λ objects. Reporting c times the number of intersecting objects of O_i yield a c-approximation of the density value.

Observation 4. *Let $S = \{P_1, ..., P_n\}$ be a set of λ-low-density objects. If we can cover an optimal ball B^* with c balls of radius at most $radius(B^*)$, we can c-approximate the density value of S.*

We present our $(2 + \varepsilon)$-approximation in the following section.

4.1   Improving the Approximation Factor to $2 + \varepsilon$

We will first show that for any optimal ball B^* and any object P intersecting B^*, we can move B^* in a circular sector of B^* by a certain distance such that B^* still intersects P. A circular sector $\triangledown$ is a closed portion of a disk enclosed by

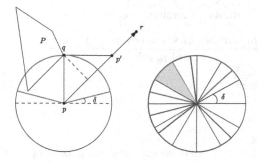

Fig. 2. Left: a circular sector (red) defined by a furthest point q of an object P. Moving the center of the ball from p to p', it still intersects q. Right: a slice (orange) covered by two circular sectors (red and blue). (Color figure online)

two radii and an arc (see Fig. 2, left). To simplify the description, in this section, we will assume, without loss of generality, that the smallest object intersecting an optimal ball B^* has size 1, and B^* has radius 1.

Lemma 7. *Let $B^* = B(p, 1)$ be an optimal ball, and let P intersect B^*. There exists a circular sector $\triangledown$ in B^* with angle $\pi - 2\delta$ such that we can move the center of B^* in any direction inside $\triangledown$ by $2\cos(\pi/2 - \delta)$ and B^* still intersects P.*

Proof. Say we move B^* by distance d in a direction $\overrightarrow{pr}$ in $\triangledown$ (see Fig. 2, left). Clearly, $0 \leq \angle qpr \leq \pi/2 - \delta$. We can move B^* by a distance $d = 2 \cdot \cos(\angle qpr) \geq 2 \cdot \cos(\frac{\pi}{2} - \delta)$, and B^* still intersects q.

Next, we will show that there is a small circular sector $\triangledown$ with angle δ, which we will call a *slice*, such that if we move the center of B^* in the direction of $\triangledown$ by $2\cos(\pi/2 - \delta)$, B^* intersects roughly $\lambda/2$ objects. The intuition is that if we construct all the circular sectors within B^* for all objects intersecting B^* based on the method in Lemma 7. Then on average, roughly half of the circular sectors, say c of them, will cover a slice (see Fig. 2, right). Move the center of B^* within a circular sector defined by the object P, and B^* still intersects P. Therefore if we move the center of B^* in a slice covered by most number of circular sectors, which we will call an *optimal slice*, then B^* intersects at least c objects. We formalise the above insight in the below Lemma 8. For simplicity, we assume that a ball can be partitioned into slices of angle δ.

Lemma 8. *Let $B^* = B(p, 1)$ be an optimal ball. If we partition $B(p, 2\cos(\pi/2 - \delta))$ into slices with angles δ, there exists a slice $\triangledown$ such that any ball $B(q, 1)$ with its centre q residing in $\triangledown$ intersects at least $(1/2 - 2\delta/\pi) \cdot \lambda$ objects.*

Proof. The average number of circular sectors that cover a slice is

$$\frac{\text{minimum number of slices covered by one circular sector}}{\text{total number of slices}}$$

$$= \frac{\frac{\pi - 2\delta}{\delta} - 2}{2\pi/\delta}$$

$$= \frac{1}{2} - \frac{2\delta}{\pi}$$

On average, $1/2 - 2\delta/\pi$ circular sectors cover a slice. Let $\triangledown$ be the slice that is covered by the most number of circular sectors. Note that for any $q \in \triangledown$, $B(q,1)$ completely covers $\triangledown$. Therefore, $B(q,1)$ intersects at least $(1/2 - 2\delta/\pi)\lambda$ objects, since B^* intersects λ objects.

Any ball $B(\triangledown,1)$ intersects at least $(1/2 - 2\delta/\pi)\lambda$ objects.

Based on the above, if the center of a ball B lies in an optimal slice, then B intersects at least $\lambda(\frac{1}{2} - \frac{2\delta}{\pi})$ segments. Observe that one can construct an incircle for a smaller triangle that lies within a slice. If our grid is so refined that the distance from the center of an incircle to the nearest grid point is less than its radius, then one of the grid points must lie in an optimal slice. Recall that if P is the smallest object intersecting an optimal ball B^*, B^* must exists in $M = P \oplus B(0,\rho)$. We formalise the above arguments in Lemma 9.

Lemma 9. *One can cover $M = P \oplus B(0,\rho)$ with $O(\cos^4(\pi/2 - \delta))$ grid points such that at least one grid point lies inside an optimal slice.*

Proof. Consider a smaller triangle $\triangle$ that lies in an optimal slice with two long sides with lengths $x = 2\cos(\pi/2 - \delta)$ and one short side with length x^2. Let the lengths of the three sides of $\triangle$ be a, b, and c. The radius of $\triangle$'s incircle is[2]:

$$r = \frac{1}{2}\sqrt{\frac{(b+c-a)(c+a-b)(a+b-c)}{a+b+c}}$$

$$= \frac{1}{2}\sqrt{\frac{2x^2 \cdot 2x^2 \cdot (2x - 2x^2)}{2x + 2x^2}} \geq \frac{x^4}{\sqrt{2}} \qquad \text{where } 0 < \delta < \frac{\pi}{2} \text{ and } 0 < x < \cos(\frac{\pi}{4}).$$

Therefore, if we cover M with a square grid such that the distance between two adjacent grid points is $x^4/2$. Then the incircle of any optimal slice contains at least one such grid point since $\cos(\pi/2 - \delta/2) < \cos(\pi/2 - \delta)$ when $0 < \delta < \pi/2$.

Based on Lemma 9, one can place $O(1/\cos^4(\pi/2 - \delta))$ balls to cover each M_i, and one such ball B must have its centre lying in an optimal slice. Based on Lemma 8, B must intersect at least $(1/2 - 2\delta/\pi) \cdot \lambda$ objects, assuming that P_i is the smallest object that intersects an optimal ball. Using the quadtree to query objects intersecting M_i, and covering M_i with $O(1/\varepsilon^4)$ balls, we obtain the following theorem.

---

[2] Weisstein, Eric W. "Incircle." From MathWorld–A Wolfram Web Resource. https://mathworld.wolfram.com/Incircle.html.

Theorem 5. *A* $(2 + \varepsilon)$*-approximation of the* λ*-low-density value of a set* S *of objects in* $\mathbb{R}^2$ *can be computed in* $O(n \log n + \lambda n/\varepsilon^4)$ *time, where* $0 < \varepsilon < 1$.

Proof. Given the approximation error ε, we first compute δ, the angle of a slice.

$$\frac{1}{\frac{1}{2} - \frac{2\delta}{\pi}} = 2 + \varepsilon \implies (2 + \varepsilon)(\frac{1}{2} - \frac{2\delta}{\pi}) = 1 \implies \delta = \frac{\frac{1}{2}\varepsilon\pi}{4 + 2\varepsilon} \in O(\varepsilon)$$

Using Corollary 1, we can construct a compressed quadtree in $O(n \log n + \lambda n)$ time. Querying $O(\lambda)$ intersecting objects of $M_i = P_i \oplus B(0, \rho_i)$ takes $O(\lambda)$ time. Combining Lemma 7, 8, and 9, we can cover each M_i with $1/\varepsilon^4$ balls of size ρ_i, and compute the number of their intersecting objects in $O(\lambda n/\varepsilon^4)$ time. In total, it takes $O(n \log n + \lambda n/\varepsilon^4)$ time to $(2 + \varepsilon)$-approximate the density of S.

5   Experiments

We implemented a simple algorithm to obtain the 4-approximate density values on several trajectory data sets. Among the twelve 2D trajectory data sets, we observe that the density of six are much smaller than the size of the trajectories.

6   Concluding Remarks

In this paper we considered the problem of approximating the λ-low-density value of a set of objects. Our main results is a $(2 + \varepsilon)$-approximation algorithm running in $O(n \log n + \lambda n/\varepsilon^4)$ time. Previously, only an $O(n \log^3 n + \lambda n \log^2 n + \lambda^2 n)$ time algorithms was known for the special case when the objects are disjoint.

We also implemented a simple 4-approximation algorithm to estimate the density values of twelve real-world trajectory data sets. We observed that the estimated densities for most of the data sets are small constants. We also observed that the trajectories in half of the data sets have low density-to-size ratios, which indicates that low density is a practical, and realistic input model for trajectories.

References

1. Alt, H., Efrat, A., Rote, G., Wenk, C.: Matching planar maps. J. Algorithms. **49**(2), 262–283 (2003). https://doi.org/10.1016/S0196-6774(03)00085-3, https://www.sciencedirect.com/science/article/pii/S0196677403000853
2. Alt, H., Godau, M.: Computing the Fréchet distance between two polygonal curves. Int. J. Comput. Geom. Appl. **05**, 75–91 (1995). https://doi.org/10.1142/S0218195995000064, https://www.worldscientific.com/doi/abs/10.1142/S0218195995000064
3. de Berg, M., Katz, M., van der Stappen, A.F., Vleugels, J.: Realistic input models for geometric algorithms. In: Proceedings of the thirteenth annual symposium on Computational geometry, pp. 294–303. SCG 1997, Association for Computing Machinery, New York, NY, USA (1997). https://doi.org/10.1145/262839.262986

4. Buchin, K., Buchin, M., Gudmundsson, J., Popov, A., Wong, S.: Map matching queries under Fréchet distance on low-density spanners. In: European Workshop on Computational Geometry (2023)
5. Chen, D., Driemel, A., Guibas, L.J., Nguyen, A., Wenk, C.: Approximate map matching with respect to the Fréchet distance. In: 2011 Proceedings of the Workshop on Algorithm Engineering and Experiments (ALENEX), pp. 75–83. Proceedings, Society for Industrial and Applied Mathematics, January 2011. https://doi.org/10.1137/1.9781611972917.8, https://epubs.siam.org/doi/abs/10.1137/1.9781611972917.8
6. Driemel, A., Har-Peled, S., Wenk, C.: Approximating the Fréchet distance for realistic curves in near linear time. Discr. Comput. Geom. **48**(1), 94–127 (2012). https://doi.org/10.1007/s00454-012-9402-z
7. Gudmundsson, J., Seybold, M.P., Wong, S.: Map matching queries on realistic input graphs under the Fréchet distance. In: Proceedings of the 2023 Annual ACM-SIAM Symposium on Discrete Algorithms, pp. 1464–1492. Society for Industrial and Applied Mathematics, January 2023. https://doi.org/10.1137/1.9781611977554.ch53
8. Gudmundsson, J., Sha, Y., Wong, S.: Approximating the Packedness of polygonal curves. Comput. Geom. **108**, 101920 (2023). https://doi.org/10.1016/j.comgeo.2022.101920, https://www.sciencedirect.com/science/article/pii/S0925772122000633
9. Har-Peled, S.: Geometric Approximation Algorithms. American Mathematical Society, USA (2011)
10. Har-Peled, S., Zhou, T.: How Packed Is It, Really? May 2021. https://arxiv.org/abs/2105.10776v1
11. Schwarzkopf, O., Vleugels, J.: Range searching in low-density environments. Inf. Process. Lett. **60**(3), 121–127 (1996). https://doi.org/10.1016/S0020-0190(96)00154-8
12. van der Stappen, A.F., Overmars, M.H.: Motion planning amidst fat obstacles. In: Proceedings of the Tenth Annual Symposium on Computational Geometry, pp. 31–40. SCG 1994, Association for Computing Machinery, New York, NY, USA, June 1994. https://doi.org/10.1145/177424.177453
13. Van Der Stappen, A.F.: The complexity of the free space for motion planning amidst fat obstacles. J. Intell. Robot. Syst. **11**(1), 21–44 (1994). https://doi.org/10.1007/BF01258292

Exponential Time Complexity
of the Complex Weighted Boolean #CSP

Ying Liu[(✉)]

State Key Laboratory of Computer Science, Institute of Software, Chinese Academy
of Sciences, University of Chinese Academy of Sciences, Beijing, China
liuy@ios.ac.cn

Abstract. Cai, Lu, and Xia [8] proved a dichotomy for complex
weighted Boolean #CSP. If the parameter set of Boolean constraint func-
tions $\mathcal{F}$ is a subset of either of the affine-type function set $\mathcal{A}$ and the
product type function set $\mathcal{P}$, then $\#\mathrm{CSP}(\mathcal{F})$ is polynomial-time solvable; otherwise, $\#\mathrm{CSP}(\mathcal{F})$ is #P-hard. Furthermore, the result holds for
$\#R_3\text{-}\mathrm{CSP}(\mathcal{F})$, which additionally restricts every variable constrained by
no more than 3 constraint functions (not necessarily distinct).

We strengthen the #P-hardness to the tight sub-exponential time
lower bound under the counting *Exponential Time Hypothesis* (#ETH).
We demonstrate that, if #ETH holds, then $\#\mathrm{CSP}(\mathcal{F})$ with $\mathcal{F} \not\subseteq \mathcal{A}$ and
$\mathcal{F} \not\subseteq \mathcal{P}$ has no sub-exponential time algorithm. The result also holds for
$\#R_D\text{-}\mathrm{CSP}(\mathcal{F})$ with some integer $D > 0$, even $D = 3$.

Additionally, we demonstrate that a vital tool *pinning*, forcing some
variables to be 0 or 1, is still available in the context of $\#R_D\text{-}\mathrm{CSP}$ when
proving the sub-exponential time lower bound.

Keywords: Block interpolation · Computational complexity ·
Dichotomy · Pinning · Counting exponential time hypothesis · #CSP

1 Introduction

This paper focuses on the complexity of a general class of counting problems:
counting constraint satisfaction problems (#CSP). A #CSP problem is parameterized by a set $\mathcal{F}$ of local constraint functions. Suppose $\mathcal{F}$ is defined over a
domain $[q] = \{1, 2, 3, ..., q\}$ for some positive integer q. The problem $\#\mathrm{CSP}(\mathcal{F})$
accepts an input tuple $I = (X, C)$ and outputs

$$Z(I) = \sum_{x_1, x_2, ..., x_n \subset [q]} \prod_{(F, x_{i_1}, x_{i_2}, ..., x_{i_k}) \in C} F(x_{i_1}, x_{i_2}, ..., x_{i_k}),$$

where X denotes a finite set of variables $\{x_1, x_2, ..., x_n\}$ and C denotes a finite
set of *clauses*. Each clause is a constraint $F \in \mathcal{F}$ of some arity k depending on F

Supported by NSFC61932002 and NSFC 62272448.

together with a sequence of k variables (not necessarily distinct) $x_{i_1}, x_{i_2}, ..., x_{i_k} \in \{x_1, x_2, ..., x_n\}$. If $q = 2$, $[q]$ denotes the Boolean domain, and $\#CSP(\mathcal{F})$ belongs to the class Boolean $\#CSP$. Studying the computational complexity of $\#CSP$ problems, is a significant and fantastic sub-field of complexity theory.

Beyond the complexity of individual $\#CSP$ problems [19–21], people are interested in the dichotomy theorems for $\#CSP$ problems. Such a dichotomy states that any $\#CSP$ problem is either solvable in polynomial time (P) or $\#P$-hard [21]. A series of dichotomy theorems [2,8,10,14] for Boolean $\#CSP$ problems were developed in the past years. An overview is available in [4]. Based on these results, a series of dichotomies for $\#CSP$ problems with parameterized sets over a general finite domain [3,5,6,15] were also developed and reached a peak [5]. So a new type of dichotomy theorems has grabbed researchers' attention.

The new type is called the *fine-grained* dichotomy, which states that a problem in a class is either in P or unsolvable in sub-exponential time, if the counting *Exponential Time Hypothesis* ($\#ETH$) holds. $\#ETH$ [13] states that counting the number of satisfying assignments to a given Boolean formula ($\#SAT$) can not be computed in sub-exponential time. Dell et al. developed fine-grained dichotomies for Tutte polynomial [1,11,13], Chen et al. developed a fine-grained dichotomy for unweighted $\#GH$ [9], and Curticapean demonstrated a fine-grained dichotomy for immanant families [12]. In the class $\#CSP$, only a fine-grained dichotomy existed for the unweighted Boolean $\#CSP$ [1].

1.1 Main Results and Proof Outlines

We develop the fine-grained dichotomy for complex weighted Boolean $\#CSP$.

After introducing the necessary preliminary knowledge in Sect. 2, we prepare a powerful instrument *pinning*, which uses two special unary functions $\delta_0 = [1, 0]$ or $\delta_1 = [0, 1]$ to fix some variables to the constant 0 or 1 respectively. The pinning lemma [14] shows that δ_0 and δ_1 is available in the context of $\#CSP(\mathcal{F})$ for any set $\mathcal{F}$. We further demonstrate that they are also available in the context of $\#R_D\text{-}CSP(\mathcal{F})$, where each instance has every variable constrained by no more than D constraints (not necessarily distinct) for some constant $D > 0$.

Theorem 1. *Let $\mathcal{F}$ be a set of functions. For any constant $D' > 0$, there exists some constant $D > 0$ such that*

$$\#R_{D'}\text{-}CSP(\mathcal{F} \cup \{\delta_0, \delta_1\}) \quad \leq_{\text{serf}} \quad \#R_D\text{-}CSP(\mathcal{F}).$$

The symbol $A \leq_{\text{serf}} B$ denotes the existence of a sub-exponential time reduction algorithm T for A, with oracle access for problem B.

We utilize some special *black-box* unary functions to help build a divide and conquer reduction algorithm T, showed in Sect. 3. Actually, such a sub-exponential time algorithm can also be established by the idea of *block interpolation* [11], with classification discussion on the characters of functions in $\mathcal{F}$.

Section 4 presents the main theorem: a fine-grained dichotomy for complex weighted Boolean $\#CSP$.

Theorem 2. *Let $\mathcal{F}$ be a set of functions mapping Boolean inputs to complex numbers. If $\mathcal{F}$ is a subset of either $\mathcal{A}$ or $\mathcal{P}$, $\#CSP(\mathcal{F})$ can be computed in polynomial time. Otherwise, if #ETH holds, then there exists a constant $\varepsilon > 0$ such that $\#CSP(\mathcal{F})$ has no $O(2^{\varepsilon N})$ time deterministic algorithm, where N denotes the number of variables in the input instance.*
   The result still holds for $\#R_3\text{-}CSP(\mathcal{F})$.

Cai, Lu, and Xia [8] presented the polynomial algorithms for $\#CSP(\mathcal{A})$ and $\#CSP(\mathcal{P})$, so we focus on the hardness.

In Sect. 4.1, we prove the lower bound for $\#R_D\text{-}CSP(\{H\})$ with some constant $D > 0$ and a binary function $H \notin \mathcal{A} \cup \mathcal{P}$, by classified discussion on the first element of H. If $H(0,0) = 0$, we prove that $\#R_D\text{-}CSP(\{[0,1,1]\}) \leq_{\text{serf}} \#R_{D'}\text{-}CSP(\{H\})$ for some constants D, D', where $\#R_D\text{-}CSP(\{[0,1,1]\})$ has subexponential time lower bound [11, Theorem 1.2]. Else if $H(0,0) \neq 0$, we prove that we can simulate a binary function $H' \notin \mathcal{A} \cup \mathcal{P}$ with $H'(0,0) = 0$ in the context of $\#R_D\text{-}CSP(\{H\})$, when extra two unary functions U_x, U_y are available. And we prove that we can also simulate any unary function in the context of $\#R_D\text{-}CSP(\{H\})$, with the help of the pinning lemma (Theorem 1).

In Sect. 4.2, we follow the original proofs in [8] to prove that, for $\#CSP(\mathcal{F})$ with $\mathcal{F} \not\subseteq \mathcal{A}$ and $\mathcal{F} \not\subseteq \mathcal{P}$, $\#R_D\text{-}CSP(\{H\}) \leq_{\text{serf}} \#R_{D'}\text{-}CSP(\mathcal{F}) \leq_{\text{serf}} \#R_3\text{-}CSP(\mathcal{F})$ for some binary function $H \notin \mathcal{A} \cup \mathcal{P}$, where D' is some constant.

2 Preliminaries

2.1 Definitions and Notations

Let $\mathbb{N}$, $\mathbb{Z}$, and $\mathbb{C}$ denote the set of natural numbers, integers, and complex numbers, respectively. Let $[q]$ denote a finite domain $\{1, 2, ..., q\}$ for a positive integer q. If $q = 2$, $[q]$ denotes the Boolean domain, in which an element is 0 or 1. A complex valued Boolean function F maps $\{0,1\}^k$ to $\mathbb{C}$ for some integer k, and k is called the *arity* of F. A unary Boolean function U is written as $[U(0), U(1)]$, and a binary Boolean function F can be expressed as a 2×2 matrix $\begin{pmatrix} F(0,0) & F(0,1) \\ F(1,0) & F(1,1) \end{pmatrix}$.

A function F is a *symmetric function* if its value is invariant under the permutations of its variables. A Boolean symmetric function F of some arity k can be written as $[f_0, f_1, ..., f_k]$, where f_j is the value of F when j, $0 \leq j \leq k$, variables are assigned the value 1. For example, an equality function $(=_k)$ of arity $k \geq 3$ and the binary disequality function $(\neq_2)$ are symmetric functions that can be written as $[1, 0, ..., 0, 1]$ and $[0, 1, 0]$ respectively. δ_0 and δ_1 denotes two special unary functions $[1, 0]$ and $[0, 1]$, respectively. A function is also called a *constraint* or a *signature*.

Let $\otimes$ denote the tensor product (i.e., the Kronecker product), that is, for two matrices $X = X_{a \times b}$ and $Y = Y_{c \times d}$, $X \otimes Y$ is an $ac \times bd$ matrix with entry $X_{i,j} Y_{k,l}$ at $(i, k) \in [a] \times [c]$ row and $(j, l) \in [b] \times [d]$ column. Tensor product is defined recursively $X^{\otimes k} = X^{\otimes(k-1)} \otimes X$.

Definition 1. $\mathcal{D} = \{[a_1, b_1] \otimes [a_2, b_2] \otimes \cdots \otimes [a_k, b_k] \mid a_i, b_i \in \mathbb{C} \text{ for any } i \in [k]$ *and* $k = 1, 2, ...\}$.

Definition 2. *A function is of* product type *if it can be written as a product of unary functions, binary equality functions* $(=_2)$ *and binary dis-equality functions* $(\neq_2)$, *on not necessarily disjoint subsets of variables.* $\mathcal{P}$ *denotes the set of all functions of product type.*

A function F is *degenerate* if $F \in \mathcal{D}$. A binary function in $\mathcal{P}$ is either degenerate or of the form $\begin{pmatrix} a & 0 \\ 0 & b \end{pmatrix}$ or $\begin{pmatrix} 0 & a \\ b & 0 \end{pmatrix}$ with $a, b \in \mathbb{C}$.

Let F be a function on variables $x_1, x_2, ..., x_k \in \{0, 1\}$. The *support* of F is defined as $R_F = \{(x_1, x_2, ..., x_k) | F(x_1, x_2, ..., x_k) \neq 0\}$. R_F is *affine* if, for any $\alpha, \beta, \gamma \in R_F$, $\alpha \oplus \beta \oplus \gamma \in R_F$, i.e., $A(x_1, x_2, ..., x_k, 1)^T = 0$ for some matrix $A \in GL_{k+1}(\mathbb{Z})$ and any $(x_1, x_2, ..., x_k) \in R_F$. $\oplus$ denotes the operator: pairwise XOR. Define $X = (x_1, x_2, ..., x_k, 1)^T$ and an indicator function χ_{AX}: if $AX = 0$ then $\chi_{AX} = 1$; else $\chi_{AX} = 0$.

Definition 3. *A function is of* affine type *if it can be written as* $\lambda \chi_{AX} i^{P(X)}$, *where* $\lambda \in \mathbb{C}$, $i = \sqrt{-1}$, *and* $P(X) \in \mathbb{Z}(X)$ *is a homogeneous quadratic polynomial with every cross term* $x_i x_j$ $(i \neq j)$ *having an even coefficient.* $\mathcal{A}$ *denotes the class of all functions of affine type.*

$P(X) = P(x_1, x_2, ..., x_k, 1) = a_0 + \sum_{i=1}^{k} a_i x_i^2 + \sum_{1 \leq i < j \leq k} 2b_{ij} x_i x_j$ with $a_i, b_{ij} \in \mathbb{Z}$. Considering the form of a binary function $F \in \mathcal{A}$, the four elements of F do not have exactly three non-zero values because F has affine support. If $F = \lambda \begin{pmatrix} i^{p_{00}} & i^{p_{01}} \\ i^{p_{10}} & i^{p_{11}} \end{pmatrix}$ for some constant λ and $p_{00}, p_{01}, p_{10}, p_{11} \in \{0, 1, 2, 3\}$, then $p_{00} + p_{01} + p_{10} + p_{11} \equiv 0 \pmod 2$ [7]. For example, a binary symmetric function in $(\mathcal{A} - \mathcal{P})^1$ has the form $\lambda[1, \pm i, 1]$ or $\lambda[1, \pm 1, -1]$. More information about $\mathcal{A}$ can be seen in [4,8].

Definition 4. *Given a set* $\mathcal{F}$ *of functions over the Boolean domain, we define the counting Boolean constraint satisfaction problem* #CSP$(\mathcal{F})$ *as*
    Input: A tuple $I = (X, \mathcal{C})$
    Output: $Z(I) = \sum_{x_1, x_2, ..., x_n \in \{0, 1\}} \prod_{(F, x_{i_1}, x_{i_2}, ..., x_{i_k}) \in \mathcal{C}} F(x_{i_1}, x_{i_2}, ..., x_{i_k})$,
where X *denotes a set of variables* $x_1, x_2, ..., x_n$ *and* $\mathcal{C}$ *is a finite set of clauses. Each clause in* $\mathcal{C}$ *is a constraint* $F \in \mathcal{F}$ *of some arity* k *depending on* F *together with a sequence of* k *variables (not necessarily distinct)* $x_{i_1}, x_{i_2}, ..., x_{i_k} \in \{x_1, x_2, ..., x_n\}$. *A satisfying assignment is an assignment to* X *such that the product of functions is not zero.*

In the context of a #CSP problem, F and λF denote the same function because the non-zero scalar λ only introduces a multiplication factor into the final value. So we always *normalize* a function by multiplying it with some non-zero constants.

---

[1] Given two sets A and B, $(A - B)$ denotes the set $\{x | x \in A \text{ and } x \notin B\}$.

Let $G(V, E)$ denote a graph, where V is the set of vertices and E is the set of edges. A vertex cover of graph G is a vertex set $S \subseteq V$ such that $v \in S$ or $u \in S$ for every edge $e = (v, u) \in E$. #Vertex Cover (#VC) denotes the problem of counting the vertex covers of a given graph.

Any instance $I = (X, \mathcal{C})$ of a #CSP has a graphical representation $G(V, E)$.

If every constraint is binary, every vertex $v \in V$ represents a variable in X, and every edge $e = (u, v) \in E$ represents a function F with $(F, u, v) \in \mathcal{C}$. For example, an instance of #CSP($\{[0, 1, 1]\}$ is a graph $G(V, E)$ and every edge $e = (u, v)$ represents a function $OR_2 = [0, 1, 1]$ on two variables u, v. The set of variables assigned 1 in a satisfying assignment corresponds to a vertex cover of G; therefore, $Z(I) = \#VC(G)$. So #VC can be expressed as #CSP($\{[0, 1, 1]\}$).

If there are some constraints of arity $k > 2$, then an instance $I = (X, \mathcal{C})$ can be expressed as a bipartite graph $G(V_L \cup V_R, E)$. Every vertex $v \in V_L$ represents a variable, whereas every vertex $u \in V_R$ is assigned a function F of arity k. Edges $(u, v_1), ..., (u, v_k)$ in E represents the constraint relationship $(F, v_1, v_2, ..., v_k) \in \mathcal{C}$. A graphic representation G is an undirected multigraph that allows parallel edges and self-loops, and G is typically used to represent the input tuple I. Correspondingly, $Z(I)$ is also written as $Z(G)$.

$\#R_D$-CSP($\mathcal{F}$) of some integer $D > 0$ denotes the restriction to #CSP($\mathcal{F}$) that each instance has every variable constrained by no more than D constraints (not necessarily distinct). Each instance of $\#R_D$-CSP($\mathcal{F}$) is bounded degree.

2.2   Counting Exponential Time Hypothesis

The counting *Exponential Time Hypothesis* (#ETH) is a relaxed version of ETH which is introduced by Impagliazzo et al. in [16, 17].

Conjecture 1 (#ETH [13]). There exists a constant $\varepsilon > 0$ such that no deterministic algorithm can compute #3SAT in $O(2^{\varepsilon \cdot n})$ time, where n is the number of variables in the input formula.

A class of Turing reductions preserving the sub-exponential time lower bound is called *sub-exponential time reduction families* (SERF) [11, 17]. We restrict the definition of SERF on problems with graphs as inputs.

Definition 5 ([11, 17]). *Let A and B be two problems on graphs. A sub-exponential time reduction family from A to B is an algorithm T with oracle access for B. T accepts a tuple (G, ε) as input, where G is an input graph of A and $\varepsilon > 0$ is a running time parameter of T, and*

(1) computes $A(G)$ in time $f(\varepsilon)2^{(\varepsilon|V(G)|)}$ with
(2) only invoking the oracle of B on graphs G' with no more than $g(\varepsilon)(|V(G)| + |E(G)|)$ vertices and edges,

where $f(\varepsilon)$ and $g(\varepsilon)$ are computable functions.

If such an algorithm exists, We say that A is SERF-reducible to B, written as $A \leq_{\mathrm{serf}} B$.

SERF reductions are known to be transitive [17, Section 1.1.4], and they preserve the sub-exponential time lower bound.

Lemma 1 ([11]). *Let A and B be problems that satisfy $A \leq_{serf} B$. If there exist constants $\varepsilon, D > 0$ such that A has no $O(2^{\varepsilon n})$ time algorithm on graphs with n vertices and no more than Dn edges, then there must exist constants $\varepsilon', D' > 0$ such that B has no $O(2^{\varepsilon' n'})$ time algorithm on graphs with n' vertices and no more than $D'n'$ edges.*

2.3 Gadget Construction

A *gadget* is a signature grid $G(V, E \cup X)$ (π is omitted), where E is the set of edges with two endpoints in V, and X is the set of dangling edges with one endpoint in V and another end dangling. G defines a signature F_G of arity $k = |X|$ which takes the value $F_G(a_1, a_2, ..., a_k) = \sum_{\sigma: E \to \{0,1\}} \prod_{v \in V} F_v(\hat{\sigma}|_{N(v)})$, where $\hat{\sigma}$ is the extension of σ by the assignment $(a_1, a_2, ..., a_k) \in \{0,1\}^k$ on the dangling edges X. G is called an $\mathcal{F}$-*gate* with signature F_G, if $F_v \in \mathcal{F}$ for every $v \in V$. Therefore, we can replace each occurrence of F_G by the gadget G. That is, $\#\mathrm{CSP}(\{F_G\} \cup \mathcal{F})$ can be reduced to $\#\mathrm{CSP}(\mathcal{F})$. Besides, this reduction is a SERF reduction.

Lemma 2. *Let $\mathcal{F}$ be a finite set of functions. If we can construct an $\mathcal{F}$-gate with the signature F, then $\#R_D$-$\mathrm{CSP}(\mathcal{F} \cup \{F\}) \leq_{serf} \#R_{D'}$-$\mathrm{CSP}(\mathcal{F})$ for some constants $D, D' > 0$.*

2.4 Block Interpolation

Polynomial interpolation is widely applied to building reductions between counting problems [19–21]. However, it can not always build the SERF reductions since the generated instance maybe with non-linearly size growth and not sparse. Curticapean [11] introduces the idea of *Block interpolation*, which builds the SERF reduction. We prove the following lemma to show how it works.

Lemma 3. *Let $x \in \mathbb{C}$ be a nonzero constant. $\#R_D$-$\mathrm{CSP}(\{[0, 1, 1]\})$ is SERF-reducible to $\#R_{D'}$-$\mathrm{CSP}(\{[0, 1, x]\})$ for some constants $D, D' > 0$.*

Proof. Let $G(V, E)$ with $|V| = n$ and $|E| = m \leq Dn$ be an instance of $\#R_D$-$\mathrm{CSP}(\{[0, 1, 1]\})$. If $x^k = 1$ for some integer $k > 0$, then we replace each edge with k parallel edges with signature $[0, 1, x]$. The new instance G' has $Z(G') = Z(G)$, and we finish the proof. And $D' \leq kD$ in this case.

Otherwise, x is not a root of unity. Given an integer $d > 0$, we can assume m is divisible by d. Otherwise, we can add some isolated edges assigned $[0, 1, 1]$ to fit the assumption. The addition only bring a multiplication constant to $Z(G)$. **Step 1: (Set up block interpolation)** We divide the E to disjoint blocks $E_1, E_2, .., E_{\frac{m}{d}}$ such that $|E_i| = d$ for any $i \in [\frac{m}{d}]$. For every satisfying assignment to vertices of G, we label it a type $\boldsymbol{t} = (t_1, t_2, ..., t_{\frac{m}{d}}) \in \{0, 1, ..., d\}^{\frac{m}{d}}$ where t_i

records the number of edges $e = (u, v) \in E_i$ with $u = v = 1$. And $(d - t_i)$ is the number of edges $e = (u, v) \in E_i$ with $u \neq v$. Then

$$Z(G) = \sum_t \rho_t \prod_{i=1}^{\frac{m}{d}} (1)^{d - t_i} (1)^{t_i}$$

where ρ_t is the number of satisfying assignments with type t. We define a multivariate polynomial $\mu(y_1, y_2, ..., y_{\frac{m}{d}}) = \sum_t \rho_t \prod_{i=1}^{\frac{m}{d}} y_i^{t_i}$ on indeterminates $y_1, y_2, ..., y_{\frac{m}{d}}$, and $Z(G) = \mu(1, 1, ..., 1)$.

Step 2: (Recover coefficients) Given $l = (l_1, l_2, ..., l_{\frac{m}{d}}) \in \mathbb{N}^{\frac{m}{d}}$, we construct a graph G_l from G by replacing every edge in E_i by l_i parallel edges with signature $[0, 1, x]$.

$$Z(G_l) = \sum_t \rho_t \prod_{i=1}^{\frac{m}{d}} (1^{l_i})^{d - t_i} (x^{l_i})^{t_i} = \sum_t \rho_t \prod_{i=1}^{\frac{m}{d}} (x^{l_i})^{t_i} = \mu(x^{l_1}, x^{l_2}, ..., x^{l \frac{m}{d}}).$$

Since x is not a root of unity, we can construct a series of graphs G_l with $l \in [d+1]^{\frac{m}{d}}$ to obtain the values of μ at $(d+1)^{\frac{m}{d}}$ distinct points. Then we can recover all ρ_t in $\mathrm{poly}((d+1)^{\frac{m}{d}})$ time by Lagrange interpolation.

Each instance G_l, with n vertices and no more than $(d+1)m$ edges, is an instance of #$R_{D'}$-CSP($\{[0, 1, x]\}$) where $D' = (d+1)D$. The time to generate a new instance is $\mathrm{poly}(m)$ time. Therefore, the total time to compute $Z(G)$ is $(d+1)^{\frac{m}{d}} \mathrm{poly}(m) + \mathrm{poly}((d+1)^{\frac{m}{d}}) \leq c \cdot 2^{\frac{\log(d+1)}{d} \cdot Dn}$ for some constant $c > 0$, with the oracle access for #$R_{D'}$-CSP($\{[0, 1, x]\}$). Given a running time parameter ε, we choose d such that $D \frac{\log(d+1)}{d} \leq \varepsilon$. The above algorithm is a SERF.

Block interpolation has been widely used to prove the sub-exponential time lower bounds for individual counting problems [1,9,11].

Theorem 3 (Theorem 1.2 in [11]). *If #ETH holds, then there exist constants $\varepsilon, D > 0$ such that there is no $O(2^{\varepsilon n})$ time deterministic algorithm for #R_D-CSP($\{[0, 1, 1]\}$), where n is the number of vertices in the input graph.*

3   Pinning

Pinning is a vital tool to reduce the arity of a function. Inspired by the proof of the signature decomposition theorem [18, Lemma 3.1], we prove that the two special unary functions $[1, 0]$ and $[0, 1]$ is available in the context of #R_D-CSP($\mathcal{F}$) for any nonempty set $\mathcal{F}$, where $D > 0$ is some constant.

Proof (of Theorem 1). Let $G(V_L \cup V_R, E)$ with n variables and m clauses be an instance of #$R_{D'}$-CSP($\mathcal{F} \cup \{\delta_0\}$). Each vertex in V_L is a variable, each vertex in V_R is a constraint function in $\mathcal{F} \cup \{\delta_0\}$ and the edges in E represent the constraint relationships between variables and constraints. $m \leq |E| \leq D'n$. Suppose $X \subseteq V_L$ is the set of variables that are constrained by δ_0, i.e., all variables in X are forced to be assigned 0. So the variables in X can be replaced by other variables assigned 0 without changing the value $Z(G)$.

We can divide X to $q = \frac{|X|}{d}$ disjoint blocks $X_1, X_2, ..., X_q$ for some positive integer d, such that $|X_i| = d$ (w.l.o.g, $|X|$ is divisible by d). We merge all variables in X_i into one variable x_i and the corresponding clauses (δ_0, v) with $v \in X_i$ are merged into one (δ_0, x_i). The new bipartite graph $G_q(V'_L \cup V'_R, E')$ with $V'_L = \{x_1, ..., x_q\} \cup (V_L - X)$ has exactly $q = \frac{|X|}{d}$ variables constrained by δ_0, and $Z(G_q) = Z(G)$. $|V'_L| \leq n$, $|V'_R| \leq m$, $|E'| \leq |E|$, and the max-degree of vertices in V'_L is no more than dD'. The process is shown in Fig. 1.

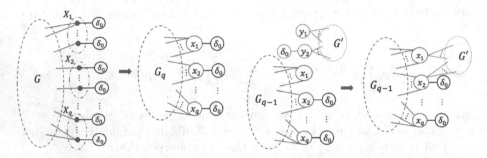

Fig. 1. Constructing G_q from G.

Fig. 2. Constructing G'_{q-1} from G_{q-1} and G', where the blue part is graph G_{q-1} and the orange part denotes graph G'. (Color figure online)

We present a divide and conquer algorithm to compute $Z(G_q)$. We construct a new instance G_{q-1} from G_q by deleting the clause (δ_0, x_1). We treat G_{q-1} as a signature on variable x_1. $G_{q-1}(0) = Z(G_{q-1}|x_1 = 0)$ and $G_{q-1}(1) = Z(G_{q-1}|x_1 = 1)$. Therefore, $Z(G_{q-1}) = G_{q-1}(0) + G_{q-1}(1)$ and $Z(G_q) = G_{q-1}(0)$.

(1). If $G_{q-1}(x_1) = [c, c]$ with some $c \in \mathbb{C}$, no matter what inner structure G_{q-1} has, then $Z(G_q) = \frac{1}{2}Z(G_{q-1})$.

(2). There exists some graph G_{q-1} such that $G_{q-1}(x_1) = [a, b]$ with $a \neq b$. Rename such a graph G_{q-1} as G' with $G'(x_1) = [a, b]$. The vertex x_1 in G' is also dubbed y_1, and other vertices adjacent to δ_0 are merged into one vertex y_2. Then we add the constraint $G'(x_1)$ to G_{q-1}, to construct a new graph G'_{q-1}. We actually construct G'_{q-1} by copying G_{q-1} and G', merging x_1, y_1 into x_1, merging x_2, y_2 into x_2 and merging the corresponding clauses $(\delta_0, x_2), (\delta_0, y_2)$ into one (δ_0, x_2), showed in Fig. 2. $Z(G_q) = \frac{1}{a-b}(Z(G'_{q-1}) - bZ(G_{q-1}))$.

Since G' is a given constraint function, of which the size is considered as a constant, then G'_{q-1} with $O(n)$ vertices has maximum degree $O(dD')$.

We can compute the values $Z(G_{q-1})$ and $Z(G'_{q-1})$ in the same way. Let $T(l; |V(G_l)|)$ to denote the computation time of $Z(G_l)$ where G_l has exactly l

variables constrained by δ_0. Then the time recurrence inequality is

$$T(l; |V(G_l)|) \leq \mathrm{poly}(|V(G_l)|) + T(l-1; |V(G_{l-1})|) + T(l-1; |V(G'_{l-1})|) + O(1)$$

for $0 \leq l \leq q$, with $|V(G_{l-1})| = O(|V(G_l)|)$, $|V(G'_{l-1})| = O(|V(G_l)|)$ and $T(0; *) = O(1)$. Besides, the maximum degree of each generated graph is $O(dD')$. So the total time $T(q; |V(G_q)|) \leq 2^{q+1}\mathrm{poly}(|V(G_q)|) = 2^{\frac{|X|}{d}+1}\mathrm{poly}(n+m)$. So we can compute $Z(G) = Z(G_q)$ in $O(2^{\frac{n}{d}})$ time, since $|X| \leq n$ and $m \leq D'n$.

By the above algorithm, for any constant $D' > 0$, there exists some constant $D = O(dD')$ such that $\#R_{D'}\text{-CSP}(\mathcal{F} \cup \{\delta_0\}) \leq_{\mathrm{serf}} R_D\text{-CSP}(\mathcal{F})$. The proof of $\#R_{D'}\text{-CSP}(\mathcal{F} \cup \{\delta_1\}) \leq_{\mathrm{serf}} \#R_D\text{-CSP}(\mathcal{F})$ is similar.

4   The Fine-Grained Dichotomy

Cai, Lu, and Xia [8] presented the polynomial time algorithm for $\#\mathrm{CSP}(\mathcal{F})$ where $\mathcal{F}$ is a subset of either $\mathcal{A}$ or $\mathcal{P}$. We focus on the hardness.

4.1   One Binary Function

We firstly consider the sub-exponential lower bound for $\#\mathrm{CSP}(\{H\})$ with the binary function $H \notin \mathcal{A} \cup \mathcal{P}$. Suppose $H = \begin{pmatrix} a & b \\ c & d \end{pmatrix}$ with $a, b, c, d \in \mathbb{C}$. If $a = 0$, then $d \neq 0$ and $bc \neq 0$ since $H \notin \mathcal{A} \cup \mathcal{P}$.

Lemma 4. *Let H be a binary function $\begin{pmatrix} 0 & b \\ c & d \end{pmatrix}$ with $bcd \neq 0$. If $\#ETH$ holds, then there exist constants $\varepsilon, D > 0$ such that $\#R_D\text{-CSP}(\{H\})$ can not be computed in $O(2^{\varepsilon n})$ time, where n is the number of variables in the input instance.*

Proof. We normalize H to $\begin{pmatrix} 0 & 1 \\ c & d \end{pmatrix}$. We realize $H' = [0, c, d^2]$ by $H'(x_1, x_2) = H(x_1, x_2)H(x_2, x_1)$. Normalize H' to $[0, 1, \frac{d^2}{c}]$.

According to Lemma 2 and Lemma 3, we build the reduction chain $\#R_D\text{-CSP}(\{[0, 1, 1]\}) \leq_{\mathrm{serf}} \#R_{D'}\text{-CSP}(\{[0, 1, \frac{d^2}{c}]\}) \leq_{\mathrm{serf}} \#R_{2D'}\text{-CSP}(\{H\})$, and conclude this lemma by Lemma 1 and Theorem 3.

Lemma 5. *Let H be a binary function $\begin{pmatrix} 1 & b \\ c & d \end{pmatrix}$ with $bc \neq d$ and $bcd \neq 0$. If $\#ETH$ holds, there exist constants $\varepsilon, D > 0$ such that $\#R_D\text{-CSP}(\{H, [1, x], [1, y]\})$ can not be computed in $O(2^{\varepsilon n})$ time, where n is the number of variables in the input instance, and $[1, x], [1, y]$ are some unary functions.*

Proof. Let $U_x = [1, x]$ and $U_y = [1, y]$. If $d \neq -bc$, then we realize $H' = \begin{pmatrix} 0 & \frac{bc-d}{b} \\ \frac{bc-d}{b} & \frac{(bc)^{\frac{2}{2}}-d^2}{bc} \end{pmatrix}$ by $H'(x_1, x_2) = \sum_{x_3 \in \{0,1\}} H(x_1, x_3)H(x_3, x_2)U_x(x_3)$, where

$x = -\frac{1}{bc}$. Otherwise, we choose $x = -\frac{2}{bc}$ and $H' = \begin{pmatrix} -1 & 3b \\ 3c & -bc \end{pmatrix}$. Then we real-

ize $H'' = \begin{pmatrix} 0 & -\frac{8}{3}b \\ -\frac{8}{3}c & \frac{80}{9}bc \end{pmatrix}$ by $H''(x_1, x_2) = \sum_{x_3 \in \{0,1\}} H'(x_1, x_3) H'(x_3, x_2) U_y(x_3)$,

where $y = -\frac{1}{9bc}$.

By the above, $\#R_{D'}\text{-CSP}(\{H', H''\}) \leq_{\text{serf}} \#R_D\text{-CSP}(\{H, U_x, U_y\})$ for some constants $D, D' > 0$. So this lemma is true by Lemma 4 and Lemma 1.

Lemma 6. *Let H be a binary function $\begin{pmatrix} a & b \\ c & d \end{pmatrix} \notin \mathcal{A} \cup \mathcal{P}$ with $a, b, c, d \in \mathbb{C}$. If $\#ETH$ holds, there exist constants $\varepsilon, D > 0$ such that $\#R_D\text{-CSP}(\{H\})$ can not be computed in $O(2^{\varepsilon n})$ time, where n is the number of variables in the input instance.*

Proof. We discuss the values of a, b, c, d. If $a = 0$ (the case $d = 0$ is symmetric), then we finish by Lemma 4. Suppose $a, d \neq 0$. H is normalized to $\begin{pmatrix} 1 & b \\ c & d \end{pmatrix}$. Since $H \notin \mathcal{P}$, then $bc \neq d$ and at most one of b, c is 0. W.l.o.g, we assume $c \neq 0$.

1. $b \neq 0$.
   (1) At least one of b, c, d is not a root of unity. We realize $U_1 = [1, b]$, $U_2 = [1, c]$, $U_3 = [1, d]$ from H by $U_1(x_1) = H(x_1, x_2)\delta_0(x_2)$, $U_2(x_2) = H(x_1, x_2)\delta_0(x_1)$, $U_3(x_1) = H(x_1, x_1)$, respectively. δ_0 and δ_1 are available according to Theorem 1. One of U_1, U_2, U_3 can interpolate $[1, y]$ for any $y \in \mathbb{C}$, similar as the way in the proof of Lemma 3. So we build the SERF reduction from $\#R_{D'}\text{-CSP}(\{H, [1, x], [1, y]\})$ to $\#R_D\text{-CSP}(\{H\})$ for some constants x, y, D, D', and proof this lemma by Lemma 4 and Lemma 1.
   (2) b, c, d are roots of unity. Suppose $b^k = c^t = 1$ for some integers $k, t > 0$. We normalize $U_1 = [b, 1]$ and $U_2 = [c, 1]$, and realize $H' = bc[1, 1, \frac{d}{bc}] \notin \mathcal{P}$ by $H'(x_1, x_2) = H(x_1, x_2)U_1(x_2)U_2(x_1)$. If $\frac{d}{bc} \neq -1$, then $H' \notin \mathcal{A}$. We realize $H'' = 2[1, \frac{bc+d}{bc}]$ by $H''(x) = \sum_{y \in \{0,1\}} H'(x, y)$, and $\frac{bc+d}{bc}$ is not a root of unity since $\frac{d}{bc} \neq 1$. Then we do as the case (1). Otherwise, $\frac{d}{bc} = -1$ and $H' = [1, 1, -1]$. If $U_1, U_2 \in \mathcal{A}$, then $b, c \in \{\pm 1, \pm i\}$ and $H = \begin{pmatrix} 1 & b \\ c & -bc \end{pmatrix} \in \mathcal{A}$, which is a contradiction. So at least one of U_1, U_2 is not in $\mathcal{A}$. W.l.o.g, we assume $U_1 = [1, b] \notin \mathcal{A}$, so $b \notin \{\pm 1, \pm i\}$. We realize $H'' = 2[1, b, -b^2] \notin \mathcal{A} \cup \mathcal{P}$ by $H''(x_1, x_2) = H'(x_1, x_2)U_1(x_1)U_1(x_2)$ and $H''' = 2(1 + b)[1, \frac{b-b^2}{1+b}]$ by $H'''(x) = \sum_{y \in \{0,1\}} H''(x, y)$. The element $\frac{b-b^2}{1+b}$ of H''' is not a root of unity. Then we turn to the case (1).

2. $b = 0$ and $H = \begin{pmatrix} 1 & 0 \\ c & d \end{pmatrix}$. We realize $H' = [1, c, c^2 + d^2] \notin \mathcal{P}$ by $H'(x_1, x_2) = \sum_{x_3 \in \{0,1\}} H(x_1, x_3)\overline{H}(x_2, x_3)$. If $H' \notin \mathcal{A}$, then we turn to case 1. Otherwise, H' has the form $[1, \pm i, 1]$ or $[1, \pm 1, -1]$. So H is either $\begin{pmatrix} 1 & 0 \\ \pm i & \sqrt{2} \end{pmatrix}$ or $\begin{pmatrix} 1 & 0 \\ \pm 1 & \sqrt{2}i \end{pmatrix}$. Then $U_3 = [1, \sqrt{2}]$ or $[1, \sqrt{2}i]$ can interpolate $[1, y]$ for any $y \in \mathbb{C}$.

We interpolate the function $U(x) = [1,2]$ to realize $H'' = [1,\pm i, 3]$ or $[1, \pm 1, -3]$ by $H''(x_1, x_2) = \sum_{x_3 \in \{0,1\}} H(x_1, x_3) H(x_2, x_3) U(x_3)$. Since $H'' \notin \mathcal{A} \cup \mathcal{P}$, we turn to the case 1-(1).

4.2 Proof of Theorem 2

Polynomial time algorithms for #CSP($\mathcal{A}$) and #CSP($\mathcal{P}$) are presented in [8]. Then we consider the sub-exponential time lower bound of $\#R_D$-CSP($\mathcal{F}$) with $\mathcal{F} \not\subseteq \mathcal{A}$ and $\mathcal{F} \not\subseteq \mathcal{P}$, where $D \geq 3$ is some constant.

In [8], the original proofs of Lemma 5.6, Lemma 5.8, Lemma 5.9 and the original proof in Sect. 5.4 built the SERF reductions from #CSP(H) of some binary function $H \notin \mathcal{A} \cup \mathcal{P}$ to #CSP($\mathcal{F}$) with a set $\mathcal{F} \not\subseteq \mathcal{A}$ and $\mathcal{F} \not\subseteq \mathcal{P}$. With the help of our pinning lemma (Theorem 1), these original proofs can be used to demonstrate $\#R_{D'}$-CSP(H) $\leq_{\text{serf}} \#R_D$-CSP($\mathcal{F}$) for some constants $D, D' > 0$.

The original proofs in [8, Section 6] proved $\#R_D$-CSP($\mathcal{F}$) $\leq_{\text{serf}} \#R_3$-CSP($\mathcal{F} \cup \{=_2\}$) and $\#R_3$-CSP($\mathcal{F} \cup \{Q\}$) $\leq_{\text{serf}} \#R_3$-CSP($\mathcal{F}$), where Q is some non-degenerate binary function. The original proof of [8, Lemma 6.1] used polynomial interpolation to reduce $\#R_3$-CSP($\mathcal{F} \cup \{=_2\}$) to $\#R_3$-CSP($\mathcal{F} \cup \{Q\}$). So the reduction is not a SERF reduction. We transform the polynomial interpolation to block interpolation and prove that $\#R_3$-CSP($\mathcal{F} \cup \{=_2\}$) $\leq_{\text{serf}} \#R_3$-CSP($\mathcal{F} \cup \{Q\}$). Then we conclude the sub-exponential time lower bound for $\#R_D$-CSP($\mathcal{F}$) by Lemma 6 and Lemma 5, even $D = 3$.

Lemma 7. *Let Q be a non-degenerate binary function and $\mathcal{F}$ be a finite set of functions.*

$$\#R_3 - \text{CSP}(\mathcal{F} \cup \{=_2\}) \leq_{\text{serf}} \#R_3 - \text{CSP}(\mathcal{F} \cup \{Q\}).$$

Proof. The Jordan normal of Q is either $\Lambda = \begin{pmatrix} \lambda_1 & 0 \\ 0 & \lambda_2 \end{pmatrix}$ or $\Lambda = \begin{pmatrix} \lambda & 1 \\ 0 & \lambda \end{pmatrix}$ such that $Q = T\Lambda T^{-1}$ for some invertible matrix $T \in GL_2(\mathbb{C})$, where $\lambda_1, \lambda_2, \lambda \neq 0$.

Let $G(V, E)$ be an instance of $\#R_3$-CSP($\mathcal{F} \cup \{=_2\}$) and $S \subseteq V$ is the set of vertices assigned signature $(=_2)$. We replace each vertex in S by a chain $T, (=_2), T^{-1}$, that is, we replace each function $(=_2)(x, w)$ by $T(x, y)(=_2)(y, z)T^{-1}(z, w)$, where $x, y, z, w \in \{0, 1\}$. This defines a new instance $G'(V', E')$ with $|V'| \leq 3|V|$ and $|E'| \leq 2|E|$. Let $S' \subseteq V'$ be the set of vertices assigned signature $(=_2)$. There is a one-to-one mapping between the vertices of S and S'. Because $T \begin{pmatrix} 1 & 0 \\ 0 & 1 \end{pmatrix} T^{-1} = \begin{pmatrix} 1 & 0 \\ 0 & 1 \end{pmatrix}$, $Z(G) = Z(G')$. We consider interpolating $(=_2)$ by Q.

1. $\Lambda = \begin{pmatrix} \lambda_1 & 0 \\ 0 & \lambda_2 \end{pmatrix}$. If $(\frac{\lambda_2}{\lambda_1})^k = 1$ for some positive integer k, we replace each vertex in S by a k-length chain $Q, Q, ..., Q$. The chain realizes the signature $(T\Lambda T^{-1})^k = T\Lambda^k T^{-1} = (\lambda_1)^k (T(=_2)T^{-1}) = (\lambda_1)^k (=_2)$. This defines a new instance G'', and $Z(G'') = (\lambda_1)^{k|S|} Z(G)$.

Otherwise, we divide S' to $r = \frac{|S'|}{d}$ disjoint blocks $S_1', S_2', ..., S_r'$ such that each block $|S_i'| = d$ for some integer $d > 0$ (w.l.o.g, $|S'|$ is divisible by d). Correspondingly, $S = S_1 \cup S_2 \cup \cdots \cup S_r$. We label each satisfying assignment to E' a type $\boldsymbol{t} = (t_1, t_2, ..., t_r) \in \{0, 1, ..., d\}^r$, where t_i or $(d - t_i)$ denotes the number of occurrences of $(=_2)(1, 1)$ or $(=_2)(0, 0)$ in S_i', respectively. Then $Z(G') = \sum_t \rho_t \prod_{i=1}^r (1)^{d-t_i} (1)^{t_i}$, where ρ_t is the sum of products of the functions in $(V' - S')$ under satisfying assignments with type $\boldsymbol{t}$. We define a multivariate polynomial $\mu(y_1, y_2, ..., y_n) = \sum_t \rho_t \prod_{i=1}^r (y_i)^{t_i}$ on variables $y_1, y_2, ..., y_n \in \mathbb{C}$, and $Z(G') = \mu(1, 1, ..., 1)$. We want to obtain the values of μ at $(d+1)^r$ distinct points, to recover all coefficients ρ_t by Lagrange interpolation.

We build a series of new instances G_l with $\boldsymbol{l} = (l_1, l_2, ..., l_r) \in [d+1]^r$ from G, by replacing each $(=_2)$ in S_i by a l_i-length chain $Q, Q, ..., Q$. Since $Q^{l_i} = T(\Lambda)^{l_i} T^{-1}$, G_l is the same as the instance constructed by replacing each $(=_2)$ in S_i' by Λ^{l_i}. Therefore,

$$Z(G_l) = \sum_t \rho_t \prod_{i=1}^r (\lambda_1)^{l_i(d-t_i)} (\lambda_2)^{l_i t_i} = (\prod_{i=1}^r (\lambda_1)^{dl_i}) \mu((\frac{\lambda_2}{\lambda_1})^{l_1}, (\frac{\lambda_2}{\lambda_1})^{l_2}, ..., (\frac{\lambda_2}{\lambda_1})^{l_r}).$$

Because $\frac{\lambda_2}{\lambda_1}$ is not a root of unity, we get the values of μ at $(d+1)^r$ points $((\frac{\lambda_2}{\lambda_1})^{l_1}, ..., (\frac{\lambda_2}{\lambda_1})^{l_r})$, by the oracle for $\#R_3\text{-CSP}(\mathcal{F} \cup \{Q\})$.

2. $\Lambda = \begin{pmatrix} \lambda & 1 \\ 0 & \lambda \end{pmatrix}$. The construction is the same, but we define t_i as the number of occurrences of $(=_2)(0, 1)$ in S_i'. Correspondingly, $(d - t_i)$ denotes the number of the occurrences of $(=_2)(0, 0)$ or $(=_2)(1, 1)$. $Z(G) = Z(G') = \rho_{(0,0,...,0)}$. $\Lambda^{l_i} = \begin{pmatrix} \lambda^{l_i} & l_i \lambda^{l_i - 1} \\ 0 & \lambda^{l_i} \end{pmatrix}$ so

$$Z(G_l) = \sum_t \rho_t \prod_{i=1}^r (\lambda^{l_i})^{d-t_i} (l_i \lambda^{l_i - 1})^{t_i} = (\prod_{i=1}^r \lambda^{dl_i}) \mu(\frac{l_1}{\lambda}, \frac{l_2}{\lambda}, ..., \frac{l_r}{\lambda}).$$

Analyzing as we do in the proof of Lemma 3, $\#R_3\text{-CSP}(\mathcal{F} \cup \{=_2\})$ is SERF-reducible to $\#R_3\text{-CSP}(\mathcal{F} \cup \{Q\})$.

5 Conclusion

In this article, we develop a fine-grained dichotomy of complex weighted Boolean $\#CSP$, which also holds for $\#R_3\text{-CSP}$. Besides, an essential part of this article is the pinning lemma under $\#ETH$. In the proof of pinning lemma, we can see many operators about functions we can simulate in the context of $\#CSP$ problems, for example, the addition and subtraction between functions.

This article presents the high feasibility of transforming the traditional dichotomies into fine-grained dichotomies for counting problems on general graphs. However, it is challenging if we restrict the inputs to planar graphs. The time lower bound might be $2^{o(\sqrt{N})}$. We need to build the reductions which cost $2^{o(\sqrt{N})}$ time and generate planar instances with linearly size growth.

Acknowledgements. The author is very grateful to Prof. Mingji Xia for his beneficial guidance and advise.

References

1. Brand, C., Dell, H., Roth, M.: Fine-grained dichotomies for the Tutte plane and Boolean #CSP. Algorithmica **81**(2), 541–556 (2019). https://doi.org/10.1007/s00453-018-0472-z
2. Bulatov, A., Dyer, M., Goldberg, L.A., Jalsenius, M., Richerby, D.: The complexity of weighted Boolean #CSP with mixed signs. Theor. Comput. Sci. **410**(38–40), 3949–3961 (2009)
3. Bulatov, A.A.: The complexity of the counting constraint satisfaction problem. J. ACM **60**(5), 1–41 (2013). https://doi.org/10.1145/2528400
4. Cai, J.Y., Chen, X.: Complexity Dichotomies for Counting Problems, vol. 1. Cambridge University Press, Cambridge (2017). https://doi.org/10.1017/9781107477063
5. Cai, J.Y., Chen, X.: Complexity of counting CSP with complex weights. J. ACM **64**(3), 1–39 (2017). https://doi.org/10.1145/2822891
6. Cai, J.Y., Chen, X., Lu, P.: Nonnegative weighted #CSP: an effective complexity dichotomy. SIAM J. Comput. **45**(6), 2177–2198 (2016). https://doi.org/10.1137/15M1032314
7. Cai, J.Y., Fu, Z., Girstmair, K., Kowalczyk, M.: A complexity trichotomy for k-regular asymmetric spin systems using number theory. Comput. Complex. **32**(4), 4 (2023). https://doi.org/10.1007/s00037-023-00237-w
8. Cai, J.Y., Lu, P., Xia, M.: The complexity of complex weighted Boolean #CSP. J. Comput. Syst. Sci. **80**(1), 217–236 (2014). https://doi.org/10.1016/j.jcss.2013.07.003
9. Chen, H., Curticapean, R., Dell, H.: The exponential-time complexity of counting (quantum) graph homomorphisms. In: Sau, I., Thilikos, D.M. (eds.) WG 2019. LNCS, vol. 11789, pp. 364–378. Springer, Cham (2019). https://doi.org/10.1007/978-3-030-30786-8_28
10. Creignou, N., Hermann, M.: Complexity of generalized satisfiability counting problems. Inf. Comput. **125**(1), 1–12 (1996). https://doi.org/10.1006/inco.1996.0016
11. Curticapean, R.: Block interpolation: a framework for tight exponential-time counting complexity. Inf. Comput. **261**, 265–280 (2018). https://doi.org/10.1016/j.ic.2018.02.008
12. Curticapean, R.: A full complexity dichotomy for immanant families. In: Proceedings of the 53rd Annual ACM SIGACT Symposium on Theory of Computing, pp. 1770–1783. STOC 2021, Association for Computing Machinery, New York, NY, USA (2021). https://doi.org/10.1145/3406325.3451124
13. Dell, H., Husfeldt, T., Wahlén, M.: Exponential time complexity of the permanent and the Tutte Polynomial. In: Abramsky, S., Gavoille, C., Kirchner, C., Meyer auf der Heide, F., Spirakis, P.G. (eds.) ICALP 2010. LNCS, vol. 6198, pp. 426–437. Springer, Heidelberg (2010). https://doi.org/10.1007/978-3-642-14165-2_37
14. Dyer, M., Goldberg, L.A., Jerrum, M.: The complexity of weighted Boolean #CSP. SIAM J. Comput. **38**(5), 1970–1986 (2009). https://doi.org/10.1137/070690201
15. Dyer, M.E., Richerby, D.M.: On the complexity of #CSP. In: Proceedings of the Forty-Second ACM Symposium on Theory of Computing, pp. 725–734. STOC 2010, Association for Computing Machinery, New York, NY, USA (2010). https://doi.org/10.1145/1806689.1806789

16. Impagliazzo, R., Paturi, R.: On the complexity of k-sat. J. Comput. Syst. Sci. **62**(2), 367–375 (2001). https://doi.org/10.1006/jcss.2000.1727
17. Impagliazzo, R., Paturi, R., Zane, F.: Which problems have strongly exponential complexity? J. Comput. Syst. Sci. **63**(4), 512–530 (2001). https://doi.org/10.1006/jcss.2001.1774
18. Lin, J., Wang, H.: The complexity of Boolean Holant problems with nonnegative weights. SIAM J. Comput. **47**(3), 798–828 (2018). https://doi.org/10.1137/17M113304X
19. Vadhan, S.P.: The complexity of counting in sparse, regular, and planar graphs. SIAM J. Comput. **31**(2), 398–427 (2001). https://doi.org/10.1137/S0097539797321602
20. Valiant, L.G.: The complexity of enumeration and reliability problems. SIAM J. Comput. **8**(3), 410–421 (1979). https://doi.org/10.1137/0208032
21. Valiant, L.: The complexity of computing the permanent. Theoret. Comput. Sci. **8**(2), 189–201 (1979). https://doi.org/10.1016/0304-3975(79)90044-6

Hardness and Approximation for the Star β-Hub Routing Cost Problem in Δ_β-Metric Graphs

Meng-Shiou Tsai[1], Sun-Yuan Hsieh[1], and Ling-Ju Hung[2]([✉])

[1] Department of Computer Science and Information Engineering,
National Cheng Kung University, Tainan, Taiwan
hsiehsy@mail.ncku.edu.tw
[2] Department of Creative Technologies and Product Design,
National Taipei University of Business, Taoyuan City, Taiwan
ljhung@ntub.edu.tw

Abstract. Minimizing transportation costs through the design of a hub-and-spoke network is a crucial concern in hub location problems (HLP). Within the realm of HLP, the Δ_β-STAR p-HUB ROUTING COST PROBLEM (Δ_β-SpHRP) represents an open problem stemming from the STAR p-HUB ROUTING COST PROBLEM (SpHRP) discussed in a publication by [Yeh *et al.*, *Theoretical Computer Science*, 2022]. The Δ_β-SpHRP deals with a specific vertex c, a positive integer p, and a Δ_β-metric graph denoted as G, which is an undirected, complete, and weighted graph adhering to the β-triangle inequality. The objective is to identify a spanning tree T that satisfies the following conditions: it is rooted at c, contains exactly p hubs adjacent to c, and assigns each remaining vertex to a hub while minimizing the routing cost of T. This paper expands the input instances from metric graphs to Δ_β-metric graphs. Our research demonstrates that SpHRP is NP-hard for any $\beta > \frac{1}{2}$, indicating that SpHRP remains NP-hard for various subclasses of metric graphs. For approximation algorithms, we introduce two approaches that improve upon previous results, particularly when β is close to $\frac{1}{2}$.

1 Introduction

The hub location problem (HLP) is a renowned optimization problem in the field of transportation planning. In the HLP, some nodes are selected to be hubs, and the remaining nodes (called non-hubs) are allocated to the hubs. It involves the strategic placement of hub facilities in a transportation network to optimize various objectives, such as minimizing transportation costs, maximizing service coverage, or minimizing travel distances. If each non-hub is connected to exactly one hub, it is called *single allocation*. Conversely, if a non-hub is connected to more than one hub, it is called *multi-allocation*.

The earliest research on HLPs was initiated by O'Kelly [18], who derived the first mathematical formulation for the HLP in 1987. Over the past three decades, HLP has had various applications in real life, including: airlines [1,

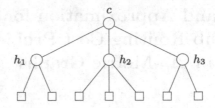

Fig. 1. A feasible solution of the Δ_β-SpHRP for $p = 3$, where c is given by input, h_i denotes i-th hub, and the rectangles are non-hubs.

21,22], delivery systems [1], and transportation systems [3]. Various approaches proposed in the literature of this well-studied problem: linear-programming [12, 16,23,26], Benders decomposition [17,19], branch-and-cut methods [10], genetic algorithm [4], and simulated annealing [20]. Interested readers can refer to the surveys [2,11,13].

Although HLP is a well-studied problem, the research results on designing approximation algorithms are still very few. Iwasa *et al.* [14] investigated the single allocation p-HUB MEDIAN problem in metric graphs and proposed a 3-approximation algorithm and a randomized 2-approximation algorithm. A variation of the p-Hub Median problem, known as the SINGLE ALLOCATION AT MOST p-HUB CENTER ROUTING problem was studied in [8]. In this variant, all origin-destination pairs have an identical unit traffic (flow) cost. It was shown in [8] the NP-hardness of the SINGLE ALLOCATION AT MOST p-HUB CENTER ROUTING problem and for any $\beta > \frac{1}{2}$, polynomial-time 2β-approximation algorithms were given, whose approximation ratio is at least $\Omega(\beta)$. Yeh *et al.* [25] studied the STAR p-HUB ROUTING COST PROBLEM in metric graphs with the assumption that $|V| \geq 2p + 1$. They proved the NP-hardness of this problem and derived a 3-approximation algorithm.

Minimizing the diameter of the specific network topology is another issue in HLPs. Chen *et al.* [7] investigated the p-HUB CENTER PROBLEM in metric graphs, where they showed the NP-hardness of approximating this problem within a ratio $1.5 - \epsilon$ for any $\epsilon > 0$ and derived a $\frac{5}{3}$-approximation algorithm. This results were extended to investigate the parameterized metric graphs instances [9]. For any $\beta > \frac{1}{2}$, they derived the corresponding approximation results. Yaman and Elloumi [24] proved the NP-hardness of the STAR p-HUB CENTER problem. Liang [15] studied this problem in metric graphs and showed the NP-hardness of approximating this problem within a ratio of $\frac{5}{4} - \epsilon$ and derived a $\frac{7}{2}$-approximation algorithm. It was shown that for any $\epsilon > 0$, to approximate the STAR p-HUB CENTER problem to a ratio $g(\beta) - \epsilon$ is NP-hard and $r(\beta)$-approximation algorithms were given in the same paper where $g(\beta)$ and $r(\beta)$ are functions of β [6].

In this paper, we study the Δ_β-STAR p-HUB ROUTING COST PROBLEM (Δ_β-SpHRP), which is a general version of SpHRP [25]. A Δ_β-metric graph $G = (V, E, w)$ is an undirected, complete, weighted graph satisfying the following conditions: (i) $w(v, v) = 0$ for any $v \in V$; (ii) $w(u, v) = w(v, u)$ for any $u, v \in V$;

(iii) for any three vertices $u, v, x \in V$, $w(u,v) \le \beta \cdot (w(u,x) + w(x,v))$, known as β-triangle inequality. (If $\beta > 1$, it is called relaxed triangle inequality. If $\beta < 1$, it is called sharpen triangle inequality.) Notice that for $\beta = \frac{1}{2}$, all edges are with the same cost which is a trivial instance. The goal of Δ_β-SpHRP is to find a spanning tree T of depth-2 that minimizes the routing cost subject to the following constraints: T is rooted at c, which is adjacent to exactly p vertices called hubs, and each remaining vertex is adjacent to a hub. Fig. 1 presents a feasible solution of this problem. In transportation systems, the transportation cost from location A to location B may be different than the cost in the opposite direction, which makes the problem even more complicated. To simplify this problem, in this paper, we assumed that the input graph is undirected. The primary motivation behind this research is to explore the possibilities and limitations of solving the Δ_β-SpHRP problem for $\beta < 1$. The goal is to determine if there exists a significant subclass of input instances that can be solved efficiently, i.e., polynomial-time approximation algorithms with a reasonable approximation ratio. On the other hand, for cases where $\beta \ge 1$, a noteworthy concern is to investigate the feasibility of developing a polynomial-time approximation algorithm with an approximation ratio which is a function of β. This investigation aims to address the question of whether it is possible to achieve efficient approximation algorithms for larger values of β while maintaining an acceptable level of accuracy.

Before we give the formal definition of the Δ_β-SpHRP, we list a property regarding to the Δ_β-metric graphs.

Lemma 1 ([5]). *Let $G = (V, E)$ be a Δ_β-metric graph for $\frac{1}{2} \le \beta < 1$. For any two edges $(u, x), (v, x)$ with a common endvertex x in G, $w(v, x) \le \frac{\beta}{1-\beta} \cdot w(u, x)$.*

In the following, we give some notations and definitions. For $u, v \in V$, we denote the distance between u, v in graph H as $d_H(u, v)$, and the length of the path between u, v in tree T as $d_T(u, v)$. The routing cost of tree T, denoted as $r(T)$, is defined as $r(T) = \sum_{u \in V} \sum_{v \in V} d_T(u, v)$. The formal definition of Δ_β-SpHRP is listed below.

Δ_β-STAR p-HUB ROUTING COST PROBLEM (Δ_β-SpHRP)
Input: A Δ_β-metric graph $G = (V, E, w)$, a specific vertex c, and a positive integer p.
Output: A depth-2 spanning tree T^* rooted at c with exactly p children such that the routing cost $r(T^*)$ is minimized.

We expand the input graph set of SpHRP from metric graphs to Δ_β-metric graphs, and improve the NP-hardness for any $\beta > \frac{1}{2}$. For approximation algorithms, we derive an approximation algorithms for solving Δ_β-SpHRP and analyze the algorithm given in [25]. The main results of this paper are listed in Table 1. Notice that for β close to $\frac{1}{2}$, our approximation ratios are very close to one. In the cases that $|V| \ge 2p + 1$, we improve the ratio to $2\beta + 1$ for $\beta < 1$, $(2\beta^3 + 1)$ for $\beta \ge 1$.

The organization of this paper is as follows. In Sect. 2, we derive an approximation algorithms for solving Δ_β-SpHRP and analyze the algorithm given in [25]

Table 1. The Approximation results of Δ_β-SpHRP in this paper.

β	Approximation ratio for general Δ_β-SpHRP	Approximation ratio for Δ_β-SpHRP with $|V| \geq 2p+1$
$(\frac{1}{2}, 1)$	$\min((\frac{\beta}{1-\beta})^2, 2\beta+2)$	$2\beta+1$
$[1, \infty)$	$2\beta^3+2$	$2\beta^3+1$

for $\beta < 1$. In Sect. 3, we analyze the algorithm given in [25] for solving Δ_β-SpHRP for $\beta \geq 1$. Finally, we give a concluding remark in Sect. 4.

We end this section with the following NP-hardness theorem. Due to limitation of space, the proof is omitted.

Theorem 1. *For any $\beta > \frac{1}{2}$, Δ_β-SpHRP is NP-hard.*

2 Approximation Algorithms for Δ_β-SpHRP for $\beta < 1$

For $\beta = 1$, Yeh *et al.* [25] gave a 3-approximation algorithm with the assumption that $|V| \geq 2p+1$. In this paper, we solve a more general problem which has no such constraint. In this section, we design an algorithm, Algorithm 1 and analyze the algorithm, Algorithm 2 given in [25] to solve Δ_β-SpHRP for $\beta < 1$.

Let $(G = (V, E, w), c, p)$ be the input graph of Δ_β-SpHRP. We use n to denote the number of the vertices, T to be the output of proposed algorithms, h_i to be the i-th hub in T, and l_i to be the i-th non-hub in T. Lemma 2 is needed for the situation that $|V| \geq 2p+1$.

Lemma 2 ([25]). *If $|V| \geq 2p+1$, $r(T^*) \geq 2(2n-2-p) \cdot \sum_{i=2}^{p} d_G(c, h_i)$.*

In the following, we show that the approximation ratio of Algorithm 1 is $(\frac{\beta}{1-\beta})^2$ for $\frac{1}{2} < \beta < 1$.

Algorithm 1. APX1

1: Choose any vertex to be h_1, and connect c to h_1 in T.
2: Choose any $p-1$ vertices to be $\{h_2, h_3, \ldots, h_p\}$.
3: Connect all vertices in non-hubs to h_1 in T.
4: Return T.

Theorem 2. *Algorithm 1 is a $(\frac{\beta}{1-\beta})^2$-approximation algorithm for $\frac{1}{2} < \beta < 1$, and the time complexity is $O(n)$.*

Proof. It is evident that Algorithm 1 returns a feasible solution of the Δ_β-SpHRP in $O(n)$ time. We now prove that $r(T) \leq (\frac{\beta}{1-\beta})^2 \cdot r(T^*)$.

Let e_1, e_2 be the cost of the shortest and longest edges in G. We construct two graphs G_1 and G_2, where all the edge costs of G_1 are e_1 and all the edge costs of G_2 are e_2. Let T_1 be the optimal solution of G_1, and T_2 be the optimal solution of G_2. The idea is that we find the lower bound of the optimal solution and the upper bound of the output of Algorithm 1 by constructing two new graphs with the shortest and longest edges in G.

It is clear that, for all vertices $u, v \in V$,

$$d_{G_1}(u, v) \leq d_G(u, v) \leq d_{G_2}(u, v). \tag{1}$$

According to (1),

$$r(T_1) \leq r(T^*) \leq r(T) \leq r(T_2). \tag{2}$$

If e_1 and e_2 have common endvertex, by Lemma 1,

$$e_2 \leq \left(\frac{\beta}{1-\beta}\right) \cdot e_1 \leq \left(\frac{\beta}{1-\beta}\right)^2 \cdot e_1 \quad (\text{since} \frac{1}{2} < \beta < 1).$$

If e_1 and e_2 have no common end vertex, there exists e_3 which has one end vertex incident to e_1 and the other end vertex incident to e_2.

According to Lemma 1, $e_2 \leq (\frac{\beta}{1-\beta}) \cdot e_3$ and $e_3 \leq (\frac{\beta}{1-\beta}) \cdot e_1$, we see that

$$e_2 \leq \left(\frac{\beta}{1-\beta}\right) \cdot e_3 \leq \left(\frac{\beta}{1-\beta}\right)^2 \cdot e_1.$$

This implies

$$r(T_2) \leq \left(\frac{\beta}{1-\beta}\right)^2 \cdot r(T_1). \tag{3}$$

Therefore, we find the approximation ratio

$$\begin{aligned}
r(T) &\leq r(T_2) \quad (\text{By (2)}) \\
&\leq \left(\frac{\beta}{1-\beta}\right)^2 \cdot r(T_1) \quad (\text{By (3)}) \\
&\leq \left(\frac{\beta}{1-\beta}\right)^2 \cdot r(T^*) \quad (\text{By (2)}).
\end{aligned}$$

This completes the proof.   □

Yeh et al. [25] gave Algorithm 2 for solving the Δ_β-SpHRP for $\beta = 1$.

In the following, we analyze the approximation ratio of Algorithm 2 for $\frac{1}{2} < \beta < 1$. Before we start to prove that the approximation ratio of Algorithm 2 is $(2\beta + 2)$ for $\frac{1}{2} < \beta < 1$, we first prove technical Lemma 3 and Lemma 4.

Lemma 3. $r(T^*) \geq \frac{1}{\beta} \cdot (n - 1) \cdot \sum_{v \in V} d_G(c_s, v)$

Algorithm 2. APX2 [25]

1: Find c_s, where $c_s = \arg\min_{u \in V \setminus \{c\}} (\sum_{v \in V} d_G(u, v))$.
2: Let vertex c_s be h_1 and connect c to h_1 in T.
3: Select $p - 1$ vertices $\{h_2, h_3, \ldots, h_p\}$ closest to c from $V \setminus \{c, h_1\}$, and connect the hubs to c in T.
4: Connect all non-hubs to h_1 in T.
5: Return T.

Algorithm 3. ONE HUB

1: $r(T^*) \leftarrow \infty$
2: $i \leftarrow 0$
3: **while** $i \neq n$ **do**
4:     Let v_i be the hub h_1^* adjacent to all non-hubs.
5:     For $v \in V \setminus \{c, h_1^*\}$, select the first $p - 1$ vertices to be hubs $h_2^*, \ldots, h_p^*$ whose $d_G(u, v) - d_G(c, v)$ are the greatest.
6:     Connect all non-hubs to h_1^*.
7:     $i \leftarrow i + 1$
8:     **if** $r(T^*) > r(T)$ **then**
9:         $T^* = T$
10:     **end if**
11: **end while**

Proof. We denote H to be the set containing all hubs, L to be the set containing all non-hubs. The routing cost of T^* is $D_{T^*}(L, V) + D_{T^*}(H, V) + D_{T^*}(c, V)$.

Assume that there exist at least two hubs adjacent to the non-hubs. If all non-hubs are adjacent to the same hub, then T^* can be computed in $O(n^2 \lg n)$-time by Algorithm 3. In step 4, we iterate all nodes to be the hub adjacent to all non-hubs, and we select $h_2^*, \ldots, h_p^*$ by sorting $d_G(u, v) - d_G(c, v)$ in step 5. Therefore, the output of Algorithm 3 is the optimal solution.

For a non-hub l_i, we calculate the routing cost $\sum_{l_i \in L} D_{T^*}(l_i, V)$ in the following.
For a hub v adjacent to l_i,

$$D_{T^*}(l_i, v) = d_G(l_i, v). \tag{4}$$

For $v = c$ or a non-hub v such that l_i, v are adjacent to the same hub,

$$D_{T^*}(l_i, v) = d_G(l_i, f^*(l_i)) + d_G(f^*(l_i), v)$$
$$\geq \frac{1}{\beta} \cdot d_G(l_i, v) \quad \text{(by } \beta\text{-triangle inequality)}. \tag{5}$$

For a hub v not adjacent to l_i,

$$D_{T^*}(l_i, v) = D_{T^*}(f^*(l_i), v) + d_G(l_i, f^*(l_i))$$

$$\geq \frac{1}{\beta} \cdot d_G(f^*(l_i), v) + d_G(l_i, f^*(l_i))$$

$$\text{(by } \beta\text{-triangle inequality)} \tag{6}$$

$$\geq \frac{1}{\beta} \cdot d_G(l_i, v) + (\frac{1}{\beta} - 1) \cdot d_G(f^*(l_i), v)$$

$$\text{(by } \beta\text{-triangle inequality)}.$$

For a non-hub v such that l_i, v are not adjacent to the same hub,

$$D_{T^*}(l_i, v) = D_{T^*}(f^*(l_i), f^*(v)) + d_G(l_i, f^*(l_i)) + d_G(v, f^*(v))$$

$$\geq \frac{1}{\beta} \cdot d_G(f^*(l_i), f^*(v)) + d_G(l_i, f^*(l_i)) + d_G(v, f^*(v))$$

$$\text{(by } \beta\text{-triangle inequality)}$$

$$\geq \frac{1}{\beta} \cdot d_G(l_i, f^*(v)) + d_G(v, f^*(v)) + (\frac{1}{\beta} - 1) \cdot d_G(f^*(l_i), f^*(v))$$

$$\text{(by } \beta\text{-triangle inequality)} \tag{7}$$

$$\geq \frac{1}{\beta} \cdot d_G(l_i, v) + (\frac{1}{\beta} - 1) \cdot [d_G(l_i, f^*(v)) + d_G(f^*(l_i), f^*(v))]$$

$$\text{(by } \beta\text{-triangle inequality)}$$

$$\geq \frac{1}{\beta} \cdot d_G(l_i, v) + \frac{1}{\beta} \cdot (\frac{1}{\beta} - 1) d_G(l_i, f^*(l_i))$$

$$\text{(by } \beta\text{-triangle inequality)}.$$

By the fact that

$$\left[1 + \frac{1}{\beta} \cdot (\frac{1}{\beta} - 1)\right] \cdot d_G(l_i, f^*(l_i)) \geq \frac{1}{\beta} \cdot d_G(l_i, f^*(l_i)), \tag{8}$$

and according to (4)–(8),

$$\sum_{l_i \in L} D_{T^*}(l_i, V) \geq \frac{1}{\beta} \cdot \sum_{l_i \in L} \sum_{v \in V} d_G(l_i, v). \tag{9}$$

For c, we simplify $D_{T^*}(c, V)$

$$D_{T^*}(c, V) > \sum_{h_i \in H} d_G(c, h_i) + \sum_{l_i \in L} d_G(l_i, f^*(l_i)). \tag{10}$$

For a hub h_i, we calculate the routing cost $\sum_{h_i \in H} D_{T^*}(h_i, V)$ in the following. For v is a hub such that $v \neq h_i$ or $f^*(v) \neq h_i$

$$D_{T^*}(h_i, v) \geq \frac{1}{\beta} \cdot d_G(h_i, v). \tag{11}$$

For $v = c$ or $f^*(v) = h_i$,

$$D_{T^*}(h_i, v) + D_{T^*}(c, V) \geq 2d_G(h_i, v)$$

$$\geq \frac{1}{\beta} \cdot d_G(h_i, v) \quad (\text{since } \beta > \frac{1}{2}). \tag{12}$$

According to (10)–(12),

$$\sum_{h_i \in H} D_{T^*}(h_i, V) + D_{T^*}(c, V) \geq \frac{1}{\beta} \cdot \sum_{h_i \in H} \sum_{v \in V} d_G(h_i, v). \tag{13}$$

We calculate the routing cost of T^* to complete the proof.

$$r(T^*) = D_{T^*}(L, V) + D_{T^*}(H, V) + D_{T^*}(c, V)$$

$$\geq \frac{1}{\beta} \cdot \left(\sum_{l_i \in L} \sum_{v \in V} d_G(l_i, v) + \sum_{h_i \in H} \sum_{v \in V} d_G(h_i, v) \right)$$

$$\geq \frac{1}{\beta} \cdot \sum_{u \in V \setminus \{c\}} \sum_{v \in V} d_G(u, v)$$

$$\geq \frac{1}{\beta} \cdot (n - 1) \cdot \sum_{v \in V} d_G(c_s, v) \quad (\text{by the selection of } c_s \text{ in Algorithm 1}).$$

This completes the proof. □

Lemma 4. $r(T^*) \geq 2(n - 1) \cdot \sum_{i=2}^{p} d_G(c, h_i)$

Proof. Note that $r(T^*)$ can be calculated by the sum of the occurrences of each edge. If there is no child adjacent to a hub h_i^* in T^*, then the occurrences of $d_G(c, h_i^*)$ is $2(n-1)$. In this proof, we only consider the occurrences of $d_G(c, h_i^*)$.

$$r(T^*) \geq 2(n - 1) \cdot \sum_{i=2}^{p} d_G(c, h_i^*) \geq 2(n - 1) \cdot \sum_{i=2}^{p} d_G(c, h_i)$$

(due to the selection of $h_2, \ldots, h_p$ in Algorithm 2)

This completes the proof. □

According to Lemma 3 and Lemma 4, we now show that the approximation ratio of Algorithm 2 is $(2\beta + 2)$ for $\frac{1}{2} \leq \beta < 1$.

Theorem 3. *Algorithm 2 is a $(2\beta + 2)$-approximation algorithm running in $O(n^2)$ time for $\frac{1}{2} < \beta < 1$.*

Proof. It is not difficult to verify that Algorithm 2 returns a feasible solution of the Δ_β-SpHRP in $O(n^2)$ time. We now prove that $r(T) \leq (2\beta + 2) \cdot r(T^*)$.

$$r(T) = 2p(n-p)d_G(h_1, c) + 2(n-1)\left[\sum_{i=1}^{n-p-1} d_G(h_1, l_i) + \sum_{i=2}^{p} d_G(c, h_i)\right]$$

$$= 2(n-1)d_G(h_1, c) + 2(n-1-p)(p-1)d_G(h_1, c)$$

$$+ 2(n-1)\left[\sum_{i=1}^{n-p-1} d_G(h_1, l_i) + \sum_{i=2}^{p} d_G(c, h_i)\right]$$

$$\leq 2(n-1)d_G(h_1, c) + 2(n-1-p)\sum_{i=2}^{p}[d_G(h_1, h_i) + d_G(h_i, c)]$$

$$+ 2(n-1)\left[\sum_{i=1}^{n-p-1} d_G(h_1, l_i) + \sum_{i=2}^{p} d_G(c, h_i)\right]$$

(by β-triangle inequality)

$$= 2(n-1)\sum_{v\in V} d_G(h_1, v) - 2p\sum_{i=2}^{p} d_G(h_1, h_i) + 2(2n-2-p)\sum_{i=2}^{p} d_G(c, h_i)$$

$$\leq 2\beta \cdot r(T^*) - 2p\sum_{i=2}^{p} d_G(h_1, h_i) + 2(2n-2-p)\sum_{i=2}^{p} d_G(c, h_i)$$

(by Lemma 3)

$$\leq 2\beta \cdot r(T^*) - 2p\sum_{i=2}^{p} d_G(h_1, h_i) + 2r(T^*) - 2p\sum_{i=2}^{p} d_G(c, h_i)$$

(by Lemma 4)

$$\leq (2\beta + 2) \cdot r(T^*)$$

This completes the proof.   □

Corollary 1. *If* $|V| \geq 2p+1$, *Algorithm 2 is a* $(2\beta+1)$-*approximation algorithm running in* $O(n^2)$ *time for* $\frac{1}{2} < \beta < 1$.

Proof. By Lemma 2, we improve the approximation ratio in Algorithm 2 for $\frac{1}{2} < \beta < 1$ to $2\beta + 1$.

$$
\begin{aligned}
r(T) =& 2p(n-p)d_G(h_1,c) + 2(n-1)\left[\sum_{i=1}^{n-p-1} d_G(h_1,l_i) + \sum_{i=2}^{p} d_G(c,h_i)\right] \\
=& 2(n-1)d_G(h_1,c) + 2(n-1-p)(p-1)d_G(h_1,c) \\
&+ 2(n-1)\left[\sum_{i=1}^{n-p-1} d_G(h_1,l_i) + \sum_{i=2}^{p} d_G(c,h_i)\right] \\
\leq& 2(n-1)d_G(h_1,c) + 2(n-1-p)\sum_{i=2}^{p}[d_G(h_1,h_i)+d_G(h_i,c)] \\
&+ 2(n-1)\left[\sum_{i=1}^{n-p-1} d_G(h_1,l_i) + \sum_{i=2}^{p} d_G(c,h_i)\right]
\end{aligned}
$$

(by the β-triangle inequality)

$$
\begin{aligned}
=& 2(n-1)\sum_{v\in V} d_G(h_1,v) - 2p\sum_{i=2}^{p} d_G(h_1,h_i) + 2(2n-2-p)\sum_{i=2}^{p} d_G(c,h_i) \\
\leq& 2\beta \cdot r(T^*) - 2p\sum_{i=2}^{p} d_G(h_1,h_i) + 2(2n-2-p)\sum_{i=2}^{p} d_G(c,h_i)
\end{aligned}
$$

(by Lemma 3)

$$
\begin{aligned}
\leq& 2\beta \cdot r(T^*) - 2p\sum_{i=2}^{p} d_G(h_1,h_i) + r(T^*) \quad \text{(by Lemma 2)} \\
\leq& (2\beta+1) \cdot r(T^*)
\end{aligned}
$$

This completes the proof. □

3 Approximation Algorithms for Δ_β-SpHRP for $\beta \geq 1$

In this section, we analyze the approximation ratio of Algorithm 2 for $\beta \geq 1$. We first prove Lemma 5 to show that the approximation ratio of Algorithm 2 is $(2\beta^3 + 2)$ for $\beta \geq 1$.

Lemma 5. $r(T^*) \geq \frac{1}{\beta^3} \cdot (n-1) \cdot \sum_{v\in V} d_G(c_s,v)$.

Proof. For two non-hubs u, v such that u, v connect to the different hubs,

$$D_{T^*}(u, v) \geq \frac{1}{\beta} \cdot d_G(u, c) + d_G(c, f^*(v)) + d_G(f^*(v), v)$$

(by β-triangle inequality)

$$\geq \frac{1}{\beta^2} \cdot d_G(u, f^*(v)) + d_G(f^*(v), v) \qquad (14)$$

(by β-triangle inequality)

$$\geq \frac{1}{\beta^3} \cdot d_G(u, v) \quad \text{(by β-triangle inequality)}.$$

According to (14), we know that for any two vertices u, v,

$$D_{T^*}(u, v) \geq \frac{1}{\beta^3} \cdot d_G(u, v). \qquad (15)$$

We now prove $r(T^*) \geq \frac{1}{\beta^3} \cdot (n-1) \cdot \sum_{v \in V} d_G(c_s, v)$.

$$r(T^*) = D_{T^*}(L, V) + D_{T^*}(H, V) + D_{T^*}(c, V)$$

$$\geq \frac{1}{\beta^3} \cdot \left(\sum_{l_i \in L} \sum_{v \in V} d_G(l_i, v) + \sum_{h_i \in H} \sum_{v \in V} d_G(h_i, v) \right) \quad \text{(according to (15))}$$

$$\geq \frac{1}{\beta^3} \cdot \sum_{u \in V \setminus \{c\}} \sum_{v \in V} d_G(u, v)$$

$$\geq \frac{1}{\beta^3} \cdot (n-1) \cdot \sum_{v \in V} d_G(c_s, v) \quad \text{(by the selection of c_s in Algorithm 2)}.$$

This completes the proof. □

According to Lemma 4 and Lemma 5, we now show that the approximation ratio of Algorithm 2 is $2\beta^3 + 2$.

Theorem 4. *Algorithm 2 is a $(2\beta^3 + 2)$-approximation algorithm running in $O(n^2)$ time for $\beta \geq 1$.*

Proof. It is not difficult to verify that Algorithm 2 returns a feasible solution of the Δ_β-SpHRP in $O(n^2)$ time. In the following, we prove $r(T) \le (2\beta^3 + 2) \cdot r(T^*)$.

$$r(T) = 2p(n-p)d_G(h_1, c) + 2(n-1) \left[\sum_{i=1}^{n-p-1} d_G(h_1, l_i) + \sum_{i=2}^{p} d_G(c, h_i) \right]$$

$$= 2(n-1)d_G(h_1, c) + 2(n-1-p)(p-1)d_G(h_1, c)$$

$$+ 2(n-1) \left[\sum_{i=1}^{n-p-1} d_G(h_1, l_i) + \sum_{i=2}^{p} d_G(c, h_i) \right]$$

$$\le 2(n-1)d_G(h_1, c) + 2(n-1-p) \sum_{i=2}^{p} [d_G(h_1, h_i) + d_G(h_i, c)]$$

$$+ 2(n-1) \left[\sum_{i=1}^{n-p-1} d_G(h_1, l_i) + \sum_{i=2}^{p} d_G(c, h_i) \right]$$

(by the β-triangle inequality)

$$= 2(n-1) \sum_{v \in V} d_G(h_1, v) - 2p \sum_{i=2}^{p} d_G(h_1, h_i) + 2(2n - 2 - p) \sum_{i=2}^{p} d_G(c, h_i)$$

$$\le 2\beta^3 \cdot r(T^*) - 2p \sum_{i=2}^{p} d_G(h_1, h_i) + 2(2n - 2 - p) \sum_{i=2}^{p} d_G(c, h_i)$$

(by Lemma 4)

$$\le 2\beta^3 \cdot r(T^*) - 2p \sum_{i=2}^{p} d_G(h_1, h_i) + 2r(T^*) - 2p \sum_{i=2}^{p} d_G(c, h_i)$$

(by Lemma 5)

$$\le (2\beta^3 + 2) \cdot r(T^*)$$

This completes the proof. $\square$

Corollary 2. *If $|V| \ge 2p+1$, Algorithm 2 is a $(2\beta^3+1)$-approximation algorithm running in $O(n^2)$ time for $\beta \ge 1$.*

Proof. By Lemma 2, we improve the approximation ratio in Algorithm 2 for $\beta > 1$ to $2\beta^3 + 1$.

$$r(T) = 2p(n-p)d_G(h_1,c) + 2(n-1)\left[\sum_{i=1}^{n-p-1} d_G(h_1,l_i) + \sum_{i=2}^{p} d_G(c,h_i)\right]$$

$$= 2(n-1)d_G(h_1,c) + 2(n-1-p)(p-1)d_G(h_1,c)$$

$$+ 2(n-1)\left[\sum_{i=1}^{n-p-1} d_G(h_1,l_i) + \sum_{i=2}^{p} d_G(c,h_i)\right]$$

$$\leq 2(n-1)d_G(h_1,c) + 2(n-1-p)\sum_{i=2}^{p}[d_G(h_1,h_i) + d_G(h_i,c)]$$

$$+ 2(n-1)\left[\sum_{i=1}^{n-p-1} d_G(h_1,l_i) + \sum_{i=2}^{p} d_G(c,h_i)\right]$$

(by the β-triangle inequality)

$$= 2(n-1)\sum_{v \in V} d_G(h_1,v) - 2p\sum_{i=2}^{p} d_G(h_1,h_i) + 2(2n-2-p)\sum_{i=2}^{p} d_G(c,h_i)$$

$$\leq 2\beta^3 \cdot r(T^*) - 2p\sum_{i=2}^{p} d_G(h_1,h_i) + 2(2n-2-p)\sum_{i=2}^{p} d_G(c,h_i)$$

(by Lemma 5)

$$\leq 2\beta^3 \cdot r(T^*) - 2p\sum_{i=2}^{p} d_G(h_1,h_i) + r(T^*) \quad \text{(by Lemma 2)}$$

$$\leq (2\beta^3 + 1) \cdot r(T^*)$$

This completes the proof.   □

4   Conclusion

In this paper, we proved that Δ_β-SpHRP is NP-hard for any $\beta > \frac{1}{2}$. For $\frac{1}{2} < \beta < 1$, we designed a $\min\{(\frac{\beta}{1-\beta})^2, 2\beta+2\}$-approximation algorithm to solve Δ_β-SpHRP. For $\beta \geq 1$, we derive a $(2\beta^3+2)$-approximation algorithm for solving the same problem. One may argue that the solution returned by our approximation algorithms always allocates all non-hubs to exactly one hub which is not very practical in real application scenarios. However, the solution, say T, returned by our algorithm is with approximation ratios guaranteed. Suppose that there exists an AI algorithm $\mathcal{A}$ which computes a solution, T', with the same input instance. Usually, we don't know how good the solution T' is. If the routing cost of T' is less than the routing cost of T, then we can definitely say that T' is an approximation solution with a certain approximation ratio guarantee. In future works, it is still open to show the inapproximability of Δ_β-SpHRP for any $\beta > \frac{1}{2}$. Moreover, in real application scenario, some infrastructures (hubs) may already exist, the goal is to expand the whole networks with some known hubs such that the routing cost is minimized. Thus, it is also very interesting to consider the extension problem of Δ_β-SpHRP which parts of hubs are specified in advance.

References

1. Alumur, S., Kara, B.Y.: A hub covering network design problem for cargo applications in Turkey. J. Oper. Res. Soc. **60**(10), 1349–1359 (2009)
2. Alumur, S., Kara, B.Y.: Network hub location problems: the state of the art. Eur. J. Oper. Res. **190**(1), 1–21 (2008)
3. Aversa, R., Botter, R.C., Haralambides, H., Yoshizaki, H.: A mixed integer programming model on the location of a hub port in the east coast of south America. Marit. Econ. Logist. **7**, 1–18 (2005)
4. Bashiri, M., Rezanezhad, M., Tavakkoli-Moghaddam, R., Hasanzadeh, H.: Mathematical modeling for a p-mobile hub location problem in a dynamic environment by a genetic algorithm. Appl. Math. Model. **54**, 151–169 (2018)
5. Böckenhauer, H.J., Hromkovič, J., Klasing, R., Seibert, S., Unger, W.: Approximation algorithms for the tsp with sharpened triangle inequality. Inf. Process. Lett. **75**, 133–138 (2000)
6. Chen, L.-H., et al.: Approximability and inapproximability of the star p-hub center problem with parameterized triangle inequality. J. Comput. Syst. Sci. **92**, 92–112 (2018)
7. Chen, L.-H., Cheng, D.-W., Hsieh, S.-Y., Hung, L.-J., Lee, C.-W., Wu, B.-Y.: Approximation algorithms for single allocation k-hub center problem. In: Proceedings of the 33rd Workshop on Combinatorial Mathematics and Computation Theory (CMCT 2016), pp. 13–18 (2016)
8. Chen, L.-H., Hsieh, S.-Y., Hung, L.-J., Klasing, R.: Approximation algorithms for the p-hub center routing problem in parameterized metric graphs. Theoret. Comput. Sci. **806**, 271–280 (2020)
9. Chen, L.-H., Hsieh, S.-Y., Hung, L.-J., Klasing, R.: On the approximability of the single allocation p-hub center problem with parameterized triangle inequality. Algorithmica **84**, 1993–2027 (2022)
10. Espejo, I., Marín, A., Muñoz-Ocaña, J.M., Rodríguez-Chía, A.M.: A new formulation and branch-and-cut method for single-allocation hub location problems. Comput. Oper. Res. **155**, 106241 (2023)
11. Farahani, R.Z., Hekmatfar, M., Arabani, A.B., Nikbakhsh, E.: Hub location problems: a review of models, classification, solution techniques, and applications. Comput. Ind. Eng. **64**(4), 1096–1109 (2013)
12. Ghaffarinasab, N.: Stochastic hub location problems with Bernoulli demands. Comput. Oper. Res. **145**, 105851 (2022)
13. Hsieh, S.-Y., Kao, S.-S.: A survey of hub location problems. J. Interconnect. Netw. **19**(01), 1940005 (2019)
14. Iwasa, M., Saito, H., Matsui, T.: Approximation algorithms for the single allocation problem in hub-and-spoke networks and related metric labeling problems. Discret. Appl. Math. **157**, 2078–2088 (2009)
15. Liang, H.: The hardness and approximation of the star p-hub center problem. Oper. Res. Lett. **41**, 138–141 (2013)
16. Lüer-Villagra, A., Eiselt, H., Marianov, V.: A single allocation p-hub median problem with general piecewise-linear costs in arcs. Comput. Ind. Eng. **128**, 477–491 (2019)
17. Mokhtar, H., Krishnamoorthy, M., Ernst, A.T.: The 2-allocation p-hub median problem and a modified benders decomposition method for solving hub location problems. Comput. Oper. Res. **104**, 375–393 (2019)

18. O'kelly, M.E.: A quadratic integer program for the location of interacting hub facilities. Eur. J. Oper. Res. **32**, 393–404 (1987)
19. Oliveira, F.A., de Sá, E.M., de Souza, S.R.: Benders decomposition applied to profit maximizing hub location problem with incomplete hub network. Comput. Oper. Res. **142**, 105715 (2022)
20. Rodríguez, V., Alvarez, M., Barcos, L.: Hub location under capacity constraints. Transp. Res. Part E: Logist. Transp. Rev. **43**(5), 495–505 (2007)
21. Sharma, A., Kohar, A., Jakhar, S.K.: Sonia: profit maximizing hub location problem in the airline industry under coopetition. Comput. Ind. Eng. **160**, 107563 (2021)
22. Soylu, B., Katip, H.: A multiobjective hub-airport location problem for an airline network design. Eur. J. Oper. Res. **277**(2), 412–425 (2019)
23. Wang, C., Liu, Y., Yang, G.: Adaptive distributionally robust hub location and routing problem with a third-party logistics strategy. Socioecon. Plann. Sci. **87**, 101563 (2023)
24. Yaman, H., Elloumi, S.: Star p-hub center problem and star p-hub median problem with bounded path lengths. Comput. Oper. Res. **39**(11), 2725–2732 (2012)
25. Yeh, H.-P., Wei, L., Chen, L.-H., Hung, L.-J., Klasing, R., Hsieh, S.-Y.: Hardness and approximation for the star p-hub routing cost problem in metric graphs. Theoret. Comput. Sci. **922**, 13–24 (2022)
26. Yin, F., Chen, Y., Song, F., Liu, Y.: A new distributionally robust p-hub median problem with uncertain carbon emissions and its tractable approximation method. Appl. Math. Model. **74**, 668–693 (2019)

Graph Algorithms

Linear Time Algorithms for NP-Hard Problems Restricted to GATEX Graphs

Marc Hellmuth[1] and Guillaume E. Scholz[2]([⊠])

[1] Department of Mathematics, Faculty of Science, Stockholm University,
10691 Stockholm, Sweden
`marc.hellmuth@math.su.se`
[2] Bioinformatics Group, Department of Computer Science and Interdisciplinary
Center for Bioinformatics, Universität Leipzig, 04107 Leipzig, Germany
`guillaume@bioinf.uni-leipzig.de`

Abstract. The class of GAlled-Tree Explainable (GATEX) graphs has just recently been discovered as a natural generalization of cographs. Cographs are precisely those graphs that can be uniquely represented by a rooted tree where the leaves of the tree correspond to the vertices of the graph. As a generalization, GATEX graphs are precisely those graphs that can be uniquely represented by a particular rooted directed acyclic graph (called galled-tree).

We consider here four prominent problems that are, in general, NP-hard: computing the size $\omega(G)$ of a maximum clique, the size $\chi(G)$ of an optimal vertex-coloring and the size $\alpha(G)$ of a maximum independent set of a given graph G as well as determining whether a graph is perfectly orderable. We show here that $\omega(G)$, $\chi(G)$, $\alpha(G)$ can be computed in linear-time for GATEX graphs G. The crucial idea for the linear-time algorithms is to avoid working on the GATEX graphs G directly, but to use the galled-trees that explain G as a guide for the algorithms to compute these invariants. In particular, we show first how to employ the galled-tree structure to compute a perfect ordering of GATEX graphs in linear-time which is then used to determine $\omega(G)$, $\chi(G)$, $\alpha(G)$.

Keywords: modular decomposition · perfect order · galled-tree · cograph · NP-hard problems · linear-time algorithms

1 Introduction

Modular decomposition is a general technique to display nested "substructures" (modules) of a given graph in form of a rooted tree (the *modular decomposition tree of G*) whose inner vertices are labeled with "0", "1", and "prime". Cographs are precisely those graphs for which the modular decomposition tree has no prime vertices. In this case, complete structural information of the underlying cograph, i.e., the knowledge of whether two vertices are linked by an edge or not, is provided by their modular decomposition tree. As a consequence, these modular decomposition trees serve as perfect guide for algorithms to efficiently solve many computationally hard problems on cographs (e.g. the graph-isomorphism problem or classical NP-hard problems as "minimum independent set", "maximum clique" or "minimum vertex coloring") [4,5]. In general, however, "prime" vertices get in the way and refute the algorithmic utility of modular decomposition trees. To circumvent this issue, we recently introduced the concept of graphs that are explained by rooted networks instead of trees [13]. In particular, we focused on so-called galled-trees that are obtained from the modular decomposition tree by replacing prime-vertices by rooted 0/1-labeled cycles. A graph $G = (X, E)$ is GAlled-Tree EXplainable (GATEx) if there is a 0/1-labeled galled-tree (N, t) such that $\{x, y\} \in E$ if and only if the label $t(\mathrm{lca}_N(x, y))$ of the unique least-common ancestor of x and y in N is "1". GATEx graphs, thus, naturally generalize the concept of cographs.

We consider here the problems of determining the size $\omega(G)$ of a maximum clique, the size $\chi(G)$ of an optimal vertex-coloring and the size $\alpha(G)$ of a maximum independent set of a given graph G. In general, determining the invariants $\omega(G)$, $\chi(G)$ and $\alpha(G)$ for arbitrary graphs G is an NP-hard task [10]. All these invariants are not only of interest from a theoretical point of view but also have many practical applications in case the underlying graph models real-world structures, e.g. social networks [18], gene/protein-interaction networks [1,22], job/time-slots assignments in scheduling problems [19] and many more. In addition, we consider the problem of determining a perfect ordering of GATEx graphs, i.e., an ordering of the vertices of G such that a greedy coloring algorithm with that ordering optimally colors every induced subgraph of G. As shown by Middendorf and Pfeiffer [21], the problem of deciding whether a graph is perfectly orderable is NP-complete. As we will argued below, the problem of finding a perfect ordering remains NP-hard even for perfectly orderable graphs.

We show here that $\omega(G)$, $\chi(G)$, $\alpha(G)$ as well as a perfect ordering can be computed in linear-time for GATEx graphs G. The crucial idea for the linear-time algorithms is to avoid working directly on the GATEx graphs G, but rather to utilize the galled-trees that explain G as a guide for the algorithms to compute these invariants. In particular, we show first how to employ the galled-tree structure to compute a perfect ordering of GATEx graphs. This result is then used to determine $\omega(G)$, $\chi(G)$, $\alpha(G)$.

2   Preliminaries

Graphs. We consider graphs $G = (V, E)$ with vertex set $V(G) := V \neq \emptyset$ and edge set $E(G) := E$. A graph G is *undirected* if E is a subset of the set of two-element subsets of V and G is *directed* if $E \subseteq V \times V \setminus \{(v, v) \mid v \in V\}$. Thus, edges $e \in E$ in an undirected graph G are of the form $e = \{x, y\}$ and in directed graphs of the form $e = (x, y)$ with $x, y \in V$ being distinct. We write $H \subseteq G$ if H is a subgraph of G and $G[W]$ for the subgraph in G that is induced by some subset $W \subseteq V$. A P_4 denotes an induced undirected path on four vertices. We often write $a - b - c - d$ for an induced P_4 with vertices a, b, c, d and edges $\{a, b\}, \{b, c\}, \{c, d\}$. An undirected graph is *connected* if, for every two vertices $u, v \in V$, there is a path connecting u

and v. A directed graph G is *connected* if its underlying undirected graph (i.e., the undirected graph obtained from G by ignoring its edge-directions) is connected. A (directed or undirected) graph G is *biconnected* if it contains no vertex whose removal disconnects G. A *biconnected component* of G is a maximal biconnected subgraph. If such a biconnected component is not a single vertex or an edge, then it is called *non-trivial*.

From here on, we will call an undirected graph simply graph.

A *clique* of a graph G is an inclusion-maximal complete subgraph of G. The size of a maximum clique of G is called the *clique number* and denoted by $\omega(G)$. A *coloring* of a graph G is a map $\sigma\colon V(G) \to S$, where S denotes a set of colors, such that $\sigma(u) \neq \sigma(v)$ for all $\{u, v\} \in E(G)$. The minimum number of colors needed for a coloring of G is called the *chromatic number* of G and denoted by $\chi(G)$. A subset $W \subseteq V(G)$ of pairwise non-adjacent vertices in G is called *independent set*. The size of a maximum independent set in G is called the *independence number* of G and denoted by $\alpha(G)$.

We consider total orders $\zeta = v_1 \ldots v_{|V|}$ of graphs $G = (V, E)$ and assume that $v_i <_\zeta v_j$ precisely if v_i is left of v_j in this sequence ζ (or equivalently, if $i < j$ in case indices are provided). We denote with $\zeta_{|H}$ the order ζ that is restricted to $V(H)$. Let X and Y be two disjoint sets. If $\zeta_1 = x_1, x_2, \ldots x_l$ and $\zeta_2 = y_1, y_2, \ldots y_m$ are two total orderings on X and Y, respectively, then we denote with $\zeta_1\zeta_2$ the total ordering on $X \cup Y$ given by concatenating ζ_1 and ζ_2, i.e., $\zeta_1\zeta_2 = x_1, x_2, \ldots x_l y_1, y_2, \ldots y_m$.

For a given total order ζ of G, a *greedy coloring algorithm* scans the vertices in the order ζ and assigns to each vertex v the smallest positive integer (color) assigned to none of the vertices $w <_\zeta v$ that are adjacent to v. A coloring of G obtained with such an algorithm is called *greedy coloring*. A total order ζ of G is *perfect* if, for all induced subgraphs H of G, a greedy coloring algorithm that scans the vertices in order $\zeta_{|H}$ uses the minimum number of colors to color H. A graph G is *perfectly orderable* if it admits a *perfect order* ζ. A total order ζ on G contains an *obstruction* (w.r.t. G) if there is an induced P_4 $a - b - c - d$ in G such that $a <_\zeta b$ and $c >_\zeta d$.

Proposition 1 ([3]). *A total order ζ on a graph G is a perfect order if and only if ζ does not contain any obstructions w.r.t. G.*

Perfectly orderable graphs are NP-complete to recognize [21]. By Proposition 1, one can test in polynomial-time whether a given order is perfect: simply check if one of the $O(|V|^4)$ induced P_4s yields an obstruction. This, in particular, implies that the problem to find a perfect ordering of a graph remains NP-hard, even if the graph is already known to be perfectly orderable.

Trees, Galled-Trees and GaTEx Graphs. Phylogenetic trees and galled-trees are particular directed acyclic graphs (DAGs). To be more precise, a *galled-tree* $N = (V, E)$ on X is such that either

(N0) $V = X = \{x\}$ and, thus, $E = \emptyset$.

or N satisfies the following four properties

(N1) There is a unique *root* ρ_N with indegree 0 and outdegree at least 2; and
(N2) $x \in X$ if and only if x has outdegree 0 and indegree 1 (x is a *leaf*); and
(N3) Every vertex $v \in V^0 := V \setminus X$ with $v \neq \rho_N$ has
   (i) indegree 1 and outdegree at least 2 (*tree-vertex*) or
   (ii) indegree 2 and outdegree at least 1 (*hybrid-vertex*).
(N4) Each biconnected component C contains at most one hybrid-vertex v for
   which the two vertices v_1, v_2 with $(v_1, v), (v_2, v) \in E$ belong to C.

We note that in [13] galled-trees have been called level-1 networks. By definition, every non-trivial biconnected component in a galled-tree N forms a (rooted) "*cycle*" C in N [2,17] that is composed of two directed paths $P^1(C)$ and $P^2(C)$ in N (called *sides* of C) with the same start-vertex ρ_C (the root of C) and end-vertex η_C (the hybrid-vertex of C) and whose internal vertices are pairwise distinct. *Trees* are galled-trees without hybrid-vertices.

Let $N = (V, E)$ be a galled-tree on X. A vertex $u \in V$ is called an *ancestor* of $v \in V$ and v a *descendant* of u, in symbols $v \preceq_N u$, if there is a directed path (possibly reduced to a single vertex) in N from u to v. We write $v \prec_N u$ if $v \preceq_N u$ and $u \neq v$. If $(u, v) \in E$, then the vertex v is a *child of* u and u is a *parent of* v. The set of children, resp., parents of a vertex w in N is denoted by $\mathrm{child}_N(w)$, resp., $\mathrm{par}_N(w)$. For a non-empty subset of leaves $A \subseteq V$ of N, we define a *least common ancestor* $\mathrm{lca}_N(A)$ *of* A to be a $\preceq_N$-minimal vertex of N that is an ancestor of every vertex in A. Note that in trees and galled-trees the $\mathrm{lca}_N(\{x, y\})$ is uniquely determined for all leaves x and y [12,17]. For simplicity we put $\mathrm{lca}_N(x, y) := \mathrm{lca}_N(\{x, y\})$.

A galled-tree N with leaf-set L is *elementary* if it contains a single rooted cycle C of length $|L| + 1$ with root $\rho_C = \rho_N$ and single hybrid-vertex $\eta_C \in V(C)$ and additional edges $\{v_i, x_i\}$ such that every vertex $v_i \in V(C) \setminus \{\rho_C\}$ is adjacent to a unique vertex $x_i \in L$. A galled-tree is *strong* if it *does not* contain cycles of the following form: (i) $P^1(C)$ or $P^2(C)$ consist of ρ_C and η_C only or (ii) both $P^1(C)$ and $P^2(C)$ contain only one vertex distinct from ρ_C and η_C.

The tuple (N, t) denotes a galled-tree $N = (V, E')$ on X that is equipped with a *(vertex-)labeling* t i.e., a map $t: V \to \{0, 1, \odot\}$ such that $t(x) = \odot$ if and only if $x \in X$ is a leaf in N. A labeling t (or equivalently (N, t)) is *quasi-discriminating* if $t(u) \neq t(v)$ for all $(u, v) \in E'$ with v not being a hybrid-vertex. The graph $\mathcal{G}(N, t) = (X, E)$ with vertex set X and edges $\{x, y\} \in E$ precisely if $t(\mathrm{lca}_N(x, y)) = 1$ is said to be *explained* by (N, t). A graph $G = (X, E)$ is *GAlled-Tree EXplainable* (*GATEx*)) if there is a labeled galled-tree (N, t) such that $G \simeq \mathcal{G}(N, t)$.

Proposition 2 ([13]). GATEX *graphs can be recognized in linear-time and a galled-tree (N, t) that explains a* GATEX *graph can be constructed in linear-time as well.*

Moreover, GATEx graphs are characterized by a finite set of forbidden subgraphs [15]. GATEx graphs that are explained by labeled trees (T, t) are precisely the cographs and, therefore, those graphs that do not contain induced P_4s [4].

Modular Decomposition (MD). A *module* M of a graph $G = (X, E)$ is a subset $M \subseteq V(G) = X$ such that for all $x, y \in M$ it holds that $N_G(x) \setminus M = N_G(y) \setminus M$, where $N_G(x)$ is the set of all vertices of X that are adjacent to x in G. A module M of G is *strong* if M does not *overlap* with any other module of G, that is, $M \cap M' \in \{M, M', \emptyset\}$ for all modules M' of G. The set of strong modules $\mathbb{M}_{\mathrm{str}}(G) \subseteq \mathbb{M}(G)$ is uniquely determined [9, 16] and forms a hierarchy which gives rise to a unique tree representation $\mathcal{T}_G$ of G, known as the *modular decomposition tree* (*MDT*) of G. Uniqueness and the hierarchical structure of $\mathbb{M}_{\mathrm{str}}(G)$ implies that there is a unique partition $\mathbb{M}_{\mathrm{max}}(G) = \{M_1, \ldots, M_k\}$ of X into inclusion-maximal strong modules $M_j \neq X$ of G [7, 8].

Similar as for galled-trees, one can equip $\mathcal{T}_G$ with a vertex-labeling t_G such that, for $M \in \mathbb{M}_{\mathrm{str}}(G) = V(\mathcal{T}_G)$, we have

$t_G(M) = \odot$ if $|M| = 1$;
$t_G(M) = 0$ if $|M| > 1$ and $G[M]$ is disconnected;
$t_G(M) = 1$ if $|M| > 1$ and $G[M]$ is connected but $\overline{G}[M]$ is disconnected;
$t_G(M) = prime$ in all other cases.

Strong modules of G are called *series, parallel* and *prime* if $t_G(M) = 1, t_G(M) = 0$ and $t_G(M) = prime$, respectively. Efficient linear-time algorithms to compute $(\mathcal{T}_G, t)$ have been proposed e.g. in [6, 20, 23]. The *quotient graph* $G/\mathbb{M}_{\mathrm{max}}(G)$ has $\mathbb{M}_{\mathrm{max}}(G)$ as its vertex set and edges $\{M_i, M_j\} \in E(G/\mathbb{M}_{\mathrm{max}}(G))$ if and only if there are $x \in M_i$ and $y \in M_j$ that are adjacent in G. As argued in [11], this quotient graph is well-defined.

From Modular Decomposition Trees to Galled-Trees. Galled-trees that explain a given GATEx graph G can be obtained from the modular decomposition trees $(\mathcal{T}_G, t_G)$ by replacing its prime vertices locally by simple rooted cycles. To this end, we first compute for prime vertices v and the corresponding prime modules $M = L(\mathcal{T}_G(v))$ the quotient $H = G[M]/\mathbb{M}_{\mathrm{max}}(G[M])$ which can be explained by a strong elementary quasi-discriminating galled-tree (N_v, t_v) (cf. [13, Thm. 6.10]). We then use (the rooted cycle in) (N_v, t_v) to replace v in $(\mathcal{T}_G, t_G)$, see Fig. 1 for an illustrative example. The latter is formalized as follows.

Definition 1 (prime-vertex replacement (pvr) networks). *Let G be a GATEx graph and $\mathcal{P}$ be the set of all prime vertices in $(\mathcal{T}_G, t_G)$. A prime-vertex replacement (pvr) networks (N, t) of G (or equivalently, of $(\mathcal{T}_G, t_G)$) is obtained by the following procedure:*

1. *For all $v \in \mathcal{P}$, let (N_v, t_v) be a strong quasi-discriminating elementary galled-tree with root v that explains $G[M]/\mathbb{M}_{\mathrm{max}}(G[M])$ with $M = L(\mathcal{T}_G(v))$.*
2. *For all $v \in \mathcal{P}$, remove all edges (v, u) with $u \in \mathrm{child}_{\mathcal{T}_G}(v)$ from $\mathcal{T}_G$ to obtain the forest (T', t_G) and add N_v to T' by identifying the root of N_v with v in T' and each leaf M' of N_v with the corresponding child $u \in \mathrm{child}_{\mathcal{T}_G}(v)$ for which $M' = L(\mathcal{T}_G(u))$.*

This results in the pvr graph N.

3. *Define the labeling* $t \colon V(N) \to \{0, 1, \odot\}$ *by putting, for all* $w \in V(N)$,

$$t(w) = \begin{cases} t_G(v) & \text{if } v \in V(\mathfrak{T}_G) \setminus \mathcal{P} \\ t_v(w) & \text{if } w \in V(N_v) \setminus X \text{ for some } v \in \mathcal{P} \end{cases}$$

The construction of a pvr network for a GATEx graph can be done in linear-time, cf. [13, Alg. 4 & Thm. 9.4]. By [13, Prop. 7.4 & 8.3], a pvr-network (N, t) of a GATEx graph G is a galled-tree that explains G. Moreover, there is a 1:1 correspondence between cycles C in N and prime modules M of G. By the latter result, we can define C_M as the unique cycle in N corresponding to prime module M.

Observation 1. *Let v be a prime vertex associated with the prime module $M_v = L(\mathfrak{T}_G(v))$ module and let $C := C_{M_v}$. Since we used strong elementary networks for the replacement of v, one easily verifies that:*

- *C has a unique root ρ_C and a unique hybrid-vertex η_C.*
- *η_C has precisely one child and precisely two parents.*
- *All vertices $v \neq \eta_C$ in C have two children and one parent.*
  In particular, all vertices $v \neq \eta_C, \rho_C$ in C have one child v_1 located in C and one child v_2 that is not located in C and these children satisfy $L(N(v_1)) \cap L(N(v_2)) = \emptyset$ and it holds that $\operatorname{lca}_N(x, y) = v$ for all $x \in L(N(v_1))$ and $y \in L(N(v_2))$.
  Both children u' and u'' of ρ_C are located in C and satisfy $L(N(u')) \cap L(N(u'')) = L(N(\eta_C))$. Moreover, $L(N(\eta_C)) \cap L(N(v_2)) = \emptyset$ for the child v_2 of $v \neq \eta_C, \rho_C$ that is not located in C.

Lemma 1. *Let $P = a - b - c - d$ be an induced P_4 in a GATEx graph G and (N, t) a pvr-network that explains G. Moreover, let M be the inclusion-minimal strong module of G that contains $V(P)$, i.e., $V(P) \subseteq M$ and there is no strong module M' of G that satisfies $V(P) \subseteq M' \subsetneq M$. Then, M is a prime module of G. Moreover, in the unique cycle C_M in N that corresponds to M, there are vertices $u_a, u_b, u_c, u_d \in V(C_M)$ that satisfy the following conditions:*

1. *For $x \in \{a, b, c, d\}$ it holds that $x \in L(N(u'_x))$ where u'_x is the unique child of u_x that is not located in C_M.*
2. *The vertices u_a, u_b, u_c, u_d are pairwise distinct.*
3. *The vertices u_a, u_b, u_c, u_d do not all belong to the same side of C_M.*
4. *One of u_a, u_b, u_c, u_d coincides with the unique hybrid η_{C_M} of C_M.*

Proof. The full proof is available in [14, Lemma 1]. □

3 Linear-Time Algorithms for Hard Problems

We provide here linear-time algorithms to compute the clique number $\omega(G)$, the chromatic number $\chi(G)$ and an optimal coloring, the independence number $\alpha(G)$ as well as a perfect ordering of a given GATEx graph G. For this purpose, we show first how to employ the structure of labeled galled-trees (N, t) to determine a perfect ordering of GATEx graphs in linear time (cf. Algorithm 1).

Algorithm 1 Perfect ordering of G

Input: A GaTEx graph $G = (V, E)$
Output: A perfect ordering ζ of the vertices of $V(G)$
1: Construct $(\mathcal{T}_G, t_G)$ and pvr-network (N, t) of G
2: Initialize $\zeta(v) := v$ for all leaves v in $\mathcal{T}_G$
3: **for all** $v \in V(\mathcal{T}_G) \setminus L(\mathcal{T}_G)$ in postorder **do**
4:     **if** $t_G(v) \in \{0, 1\}$ **then**
5:        Put $\zeta(v) := \zeta(v_1) \ldots \zeta(v_k)$ arbitrarily for the $k = |\text{child}_{\mathcal{T}_G}(v)|$ children
       $v_1, \ldots, v_k$ of v in $\mathcal{T}_G$
6:     **else**   ▷ $t_G(v) = $ prime
7:        Let C be the unique cycle in N with $L(N(\rho_C)) = L(\mathcal{T}_G(v))$
8:        For all vertices $w \in V(C) \setminus \{\rho_C\}$, put $\zeta(w) := \zeta(w')$, where w' is
       the unique child of w in N that is not a vertex of C.
9:        Put $\zeta^*(v) := \zeta(v_1) \ldots \zeta(v_k)$ arbitrarily for the $k = |V(C)| - 2$ vertices
       $v_1, \ldots, v_k$ in $V(C) \setminus \{\rho_C, \eta_C\}$
10:       **if** $t(\rho_C) = 0$ **then**
11:          $\zeta(v) := \zeta(\eta_C)\zeta^*(v)$
12:       **else**   ▷ $t(\rho_C) = 1$
13:          $\zeta(v) := \zeta^*(v)\zeta(\eta_C)$
14: **return** $\zeta(v)$

Before studying Algorithm 1 in detail, we illustrate this algorithm on the graph G as shown in Fig. 1. We first compute the modular decomposition tree $(\mathcal{T}_G, t_G)$ and a pvr-network (N, t) that explains G (Line 1). For all leaves v of $\mathcal{T}_G$ (and thus, of N), we initialize the perfect order $\zeta(v) = v$ of the induced subgraph $G[\{v\}]$ (Line 2). We then traverse $\mathcal{T}_G$ in postorder and choose in our example the order v_1, v_2, v_3, v_4, v_5 in which the vertices are visited (Line 3). Note that postorder-traversal ensures that all children of a given vertex v in $\mathcal{T}_G$ are visited before this vertex v is processed. Since v_1 is a non-prime vertex of $\mathcal{T}_G$, we put $\zeta(v_1) = \zeta(a)\zeta(b) = ab$ (Line 4). Similarly, we put $\zeta(v_2) = \zeta(v_1)\zeta(c) = abc$. The vertex v_3 is a prime vertex in $\mathcal{T}_G$. The unique cycle C in N such that $L(N(\rho_C)) = L(\mathcal{T}_G(v_3)) = \{f, g, h, i, j\}$ is the cycle with root ρ_1 (Line 7). Note that $u_h = \eta$ where η is the unique hybrid of C. In Line 8, we put $\zeta(u_x) = \zeta(x) = x$ for each $x \in \{f, g, h, i, j\}$. In Line 9 we can choose an arbitrary ordering $\zeta^*(v_3)$ and decide, in this example, for $\zeta^*(v_3) = \zeta(u_g)\zeta(u_f)\zeta(u_i)\zeta(u_j) = gfij$. Since $u_h = \eta$ and $t(\rho_1) = 1$, we put $\zeta(v_3) = \zeta^*(v_3)\zeta(\eta) = gfijh$ (Line 13). Then, the non-prime vertex v_4 is processed, and we put $\zeta(v_4) = \zeta(k)\zeta(l) = kl$ (Line 4). Finally, the prime vertex v_5 is processed. The unique cycle C in N such that $L(N(\rho_C)) = L(\mathcal{T}_G(v_5)) = V(G)$ is the cycle with root ρ_2 (Line 7). We put $\zeta(w_1) = \zeta(v_2) = abc$, $\zeta(w_2) = \zeta(v_4) = kl$, $\zeta(w_3) = \zeta(v_3) = gfijh$, $\zeta(w_4) = \zeta(e) = e$ and $\zeta(w_5) = \zeta(d) = d$ (Line 8). Again, we can choose an arbitrary ordering $\zeta^*(v_5)$ in Line 9 and decide, in this example, for $\zeta^*(v_5) = \zeta(w_1)\zeta(w_4)\zeta(w_3)\zeta(w_2) = abcegfijhkl$. Finally, since $t(\rho_1) = 0$ and $\eta = w_5$, we put $\zeta(v_5) = \zeta(\eta)\zeta^*(v_5) = dabcegfijhkl$ (Line 11). Since v_5 is the root of $\mathcal{T}_G$, the algorithm stops there, and returns the ordering $\zeta = \zeta(v_5) = dabcegfijhkl$. As we shall show in Proposition 3, the ordering returned by Algorithm 1 is a perfect ordering provided the input graph is GaTEx.

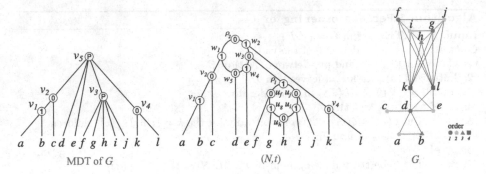

Fig. 1. Shown is a GATEX graph G (right), its modular decomposition tree $(\mathcal{T}_G, t_G)$ (left) and a pvr network (N, t) that explains G (middle). The inner vertices of $(\mathcal{T}_G, t_G)$ and (N, t) have label "1", "0" and "P" (prime). One easily observes that (N, t) is obtained from $(\mathcal{T}_G, t_G)$ by replacing the prime vertices by rooted 0/1-labeled cycles. Using Algorithm 1, we obtain a perfect order $\zeta = dabcegfijhkl$ of G (see text for details). The graph G is colored w.r.t. the greedy coloring based on the order ζ (the order of colors is shown in the figure.) Since $G[\{g, h, i, k\}] \simeq K_4$, this coloring is optimal, i.e., it uses $\chi(G) = 4$ colors. (Color figure online)

Proposition 3. *Algorithm 1 determines a perfect ordering of* GATEX *graphs.*

Proof. Let $G = (V, E)$ be a GATEX graph that serves as input for Algorithm 1. We first compute $(\mathcal{T}_G, t_G)$ and a pvr-network (N, t) of G (Line 1). In this proof, we put $L_w := L(N(w))$ for $w \in V(N)$. Let $\zeta(w)$ be the ordering computed with Algorithm 1 for the subgraph $G[L_w]$ induced by the vertices in L_w. We then initialize $\zeta(v) = v$ for all leaves v in $\mathcal{T}_G$ (Line 2). Clearly, $\zeta(v)$ is a perfect ordering of $G[\{v\}]$. We then continue to traverse the remaining vertices in $\mathcal{T}_G$ in postorder. This ensures that, whenever we reach a vertex v in $\mathcal{T}_G$, all its children have been processed and thus, that $\zeta(v)$ is well-defined in each step.

To verify that the ordering ζ returned by Algorithm 1 is a perfect order of G, we must show that ζ does not contain any obstructions w.r.t. G (cf. Proposition 1). If G does not contain any induced P_4, then any ordering is perfect. Thus, assume that G contains an induced P_4, say $P = a - b - c - d$. Put $Y = \{a, b, c, d\}$.

We first remark that Algorithm 1 builds ζ by successively concatenating sub-orderings of the form $\zeta(w)$, $w \in V(\mathcal{T}(G))$. In particular $\zeta_{|Y} = \zeta(w)_{|Y}$ holds for all $w \in V(\mathcal{T}_G)$ for which $Y \subseteq M_w$ where $M_w := L(\mathcal{T}_G(w))$. Let M be the inclusion-minimal strong module of G that contains Y. By Lemma 1, M is a prime module of G. Hence, there is the unique cycle $C := C_M$ in N corresponding to M. For all vertices $v \in V(C) \setminus \{\rho_C\}$, we denote with v' the unique child of v that is not in $V(C)$. By Lemma 1, there are four vertices $u_a, u_b, u_c, u_d \in V(C)$ that satisfy the Condition (1)–(4). Hence, for $x \in \{a, b, c, d\}$ it holds that $x \in L_{u'_x}$. Moreover, the vertices u_a, u_b, u_c, u_d are pairwise distinct, do not all belong to the same side of C_M and one of u_a, u_b, u_c, u_d coincides with the unique hybrid $\eta := \eta_C$ of C. The latter arguments, in particular, allow us to denote by P^- (resp., P^+) the side of C such that the set $V(P^-) \setminus \{\eta\}$ (resp., $V(P^+) \setminus \{\eta\}$) contains one (resp., two) of u_a, u_b, u_c and u_d. In the following, let v be the prime vertex in $\mathcal{T}_G$ with $L(\mathcal{T}_G(v)) = M$. We now distinguish between two cases: (1) $t(\rho_C) = 0$ and (2) $t(\rho_C) = 1$.

Case (1): $t(\rho_C) = 0$. Let $x \in Y$ be the vertex such that $u_x \in V(P^-) \setminus \{\eta\}$. Then, for all $y \in Y \setminus \{x\}$ with $u_y \in V(P^+) \setminus \{\eta\}$, we have $\operatorname{lca}_N(x, y) = \rho_C$ and thus, x and y are not joined by an edge in $G[Y]$. In particular, x has degree at most one in $G[Y]$. Since $G[Y] = P$, it follows that x has degree exactly one in $G[Y]$, and that the unique vertex $z \in Y$ adjacent to x in N satisfies $u_z = \eta$. Due to the "symmetry" of $G[Y] = P = a - b - c - d$, we can assume w.l.o.g. that $x = a$ and thus, $z = b$. By construction of $\zeta(v)$ in Line 11, we have $\zeta(v) = \zeta(\eta)\zeta^*(v)$. Since vertex b appears in the order $\zeta(\eta)$ and vertex a appears in the order $\zeta^*(v)$, we have in the final order ζ of G always $b <_\zeta a$. In this case, P does not yield an obstruction of ζ.

Case (2): $t(\rho_C) = 1$. Let $x \in Y$ be the vertex such that $u_x \in V(P^-) \setminus \{\eta\}$. Then for all $y \in Y \setminus \{x\}$ such that $u_y \in V(P^+) \setminus \{\eta\}$, we have $\operatorname{lca}_N(x, y) = \rho_C$ and thus, x and y are joined by an edge in $G[Y]$. In particular, x has degree at least two in $G[Y]$. Since $G[Y] = P$, it follows that x has degree exactly two in $G[Y]$, and that the unique vertex $z \in Y$ that is *not* adjacent to x in N satisfies $u_z = \eta$. Again, by "symmetry" of $G[Y] = P = a - b - c - d$, we can assume w.l.o.g. that $x = c$ and thus, $z = a$. Now consider the unique vertex b that is adjacent to a in $G[Y]$. By assumption, $b \in V(P^+) \setminus \{\eta\}$. Furthermore, by construction of $\zeta(v)$ in Line 13, we have $\zeta(v) = \zeta^*(v)\zeta(\eta)$. Since vertex a appears in the order $\zeta(\eta)$ and vertex b appears in the order $\zeta^*(v)$, we have in the final order ζ of G always $b <_\zeta a$. In this case, P does not yield an obstruction of ζ.

In summary, the ordering ζ returned by Algorithm 1 does not contain any obstructions w.r.t. G. By Proposition 1, ζ is a perfect order of G. □

Proposition 4. *Algorithm 1 can be implemented to run in $O(|V| + |E|)$ time where $G = (V, E)$ is the input GATEX graph.*

Proof. We show now that Algorithm 1 can be implemented to run in $O(|V|+|E|)$ time for a given GATEX graph $G = (V, E)$. The modular decomposition tree $(\mathcal{T}_G, t_G)$ can be computed in $O(|V|+|E|)$ time [11]. By [13, Thm. 9.4 and Alg. 4], the pvr-network (N, t) of G can be computed within the same time complexity. Thus, Line 1 takes $O(|V| + |E|)$ time. Initializing $\zeta(v) := v$ for all leaves v (and thus, the vertices of G) in Line 2 can be done in $O(|V|)$ time.

We then traverse each of the $O(|V|)$ vertices in $(\mathcal{T}_G, t_G)$ in postorder. To compute the final prefect order, we consider an auxiliary directed graph H that, initially, just consists of the vertices in V and is edge-less. Whenever we concatenate ζ' and ζ'', we simply add an edge (u, v) from the maximal element u in ζ' to the minimal element v in ζ'' and define the minimal element of this now order $\zeta''' = \zeta'\zeta''$ as the minimal element of ζ' and the maximal element of ζ''' as the maximal element of ζ''. Since we can keep track of these maximal and minimal elements (starting with $\zeta(v) := v$ for all leaves v and defining v as the maximal and minimal element of $\zeta(v)$) in each of the steps, the concatenation of two orders ζ' and ζ'' and updating the maximal and minimal element of $\zeta''' = \zeta'\zeta''$

can be done in constant time. The final graph H is then isomorphic to a single directed path that traverses each vertex in V once. If $t_G(v) \in \{0,1\}$, then we pick an arbitrary ordering of the children of v and define $\zeta(v) = \zeta(v_1)\ldots\zeta(v_k)$ by concatenating the orderings of its k children $v_1,\ldots,v_k$ (Line 4–5). By the latter arguments, this task can be done in $O(|\,\mathrm{child}_{\mathcal{T}_G}(v)|)$ time for each non-prime vertex v. Otherwise, if $t_G(v) = $ prime, we consider the unique cycle C in N that satisfies $L(N(\rho_C)) = L(\mathcal{T}_G(v))$ in Line 7. We note that we can keep track of C and its correspondence to v when constructing the pvr-network (N, t) based on $(\mathcal{T}_G, t_G)$ and thus have constant-time access to these cycles C in N. The assignment $\zeta(w) = \zeta(w')$ for all $w \in V(C) \setminus \{\rho_C\}$ can be done in $O(|V(C)|)$ time (Line 8). By the latter arguments, construction of $\zeta^*(v)$ in Line 9 can be done in $O(|V(C)|)$ time. Note that $O(|V(C)|) = O(|\,\mathrm{child}_{\mathcal{T}_G}(v)|)$, since the elementary galled-tree N_v that is used to replace v and the edges to its children in $\mathcal{T}_G$, contains C and has $2\,\mathrm{child}_{\mathcal{T}_G}(v) + 1$ edges and vertices. The tasks in Line 10–13 can be done in constant time. Hence, the time-complexity of the Lines 6 to 13 is in $O(|\,\mathrm{child}_{\mathcal{T}_G}(v)|)$ for each prime vertex v.

To obtain the overall time complexity of the *for* loop starting in Line 3, observe that the degrees of vertices in $\mathcal{T}_G$ sum up to $2|E(\mathcal{T}_G)| = 2(|V(\mathcal{T}_G)| - 1)$. By the latter arguments and by iterating over each vertex $v \in V(\mathcal{T}_G) \setminus L(\mathcal{T}_G)$, we obtain $\sum_{v \in V(\mathcal{T}_G) \setminus L(\mathcal{T}_G)} O(|\,\mathrm{child}_{\mathcal{T}_G}(v)|) = O(|V(\mathcal{T}_G)|) = O(|V|)$.

Hence, the overall time-complexity of Algorithm 1 is dominated by the time-complexity to compute $(\mathcal{T}_G, t_G)$ and (N, t) in Line 1 and is, therefore, in $O(|V| + |E|)$. □

As an immediate consequence of Proposition 3 and 4, we obtain

Theorem 1. *Every* GATEx *graph is perfectly orderable and this ordering can be determined in linear-time.*

For a given graph $G = (V, E)$, a greedy coloring algorithm can be implemented to run in $O(|V| + |E|)$ time, see e.g. [24, Sec. 6.4]. This together with Theorem 1 implies

Theorem 2. *The chromatic number $\chi(G)$ and an optimal coloring of a* GATEx *graph G can be determined in linear-time.*

A graph G is *perfect*, if the chromatic number of every induced subgraph equals the size of the largest clique of that subgraph. Every perfectly orderable graph is a perfect graph [3]. Hence, GATEx graphs are perfect and we have $\omega(G) = \chi(G)$. Thus, we obtain

Theorem 3. *The clique number $\omega(G)$ of a* GATEx *graph G can be determined in linear-time.*

We consider now the problem of determining the independence number $\alpha(G)$ of GATEx graphs G, i.e., a maximum subset $W \subseteq V(G)$ of pairwise non-adjacent vertices. Suppose that a GATEx graph G is explained by the network (N, t) and let $\bar{t}: V(N) \to \{0, 1, \odot\}$ where $\bar{t}(v) = \odot$ for all leaves v of N and

$\bar{t}(v) = 1$ if and only if $t(v) = 0$. Since $L(N) = V(G)$ and by [2, Prop. 1], we have $O(|V(N)|) = O(|V(G)|)$ and thus, this labeling can be computed in $O(|V(G)|)$ time. It is easy to verify that $(N, \bar{t})$ explains the complement $\overline{G}$ of G. The latter arguments imply that the complement of every GATEx graph is a GATEx graph as well. Since maximum cliques in $\overline{G}$ are precisely the maximum independent sets in G, we obtain

Theorem 4. *The independence number $\alpha(G)$ of a* GATEx *graph can be computed in linear-time.*

4 Outlook and Summary

GATEx graphs form a natural generalization of cographs and are precisely those graphs that can be explained by a 0/1-labeled galled-trees. We have shown here that several NP-hard problems become linear-time solvable when restricted to GATEx graphs. In particular, we showed how to determine a perfect order, the chromatic number and an optimal coloring, the clique number and the independence number of GATEx graphs in linear-time. We assume that the underlying idea of letting algorithms be guided by the galled-trees that explains a GATEx graph G instead of working directly on G has the potential to show that many other computationally difficult problems as e.g. the graph isomorphism problem or other NP-hard problems become tractable on GATEx graphs. This will, however, be part of future work.

References

1. Ben-Dor, A., Shamir, R., Yakhini, Z.: Clustering gene expression patterns. J. Comput. Biol. **6**(3–4), 281–297 (1999). https://doi.org/10.1089/106652799318274. pMID: 10582567
2. Cardona, G., Rosselló, F., Valiente, G.: Comparison of tree-child phylogenetic networks. IEEE/ACM Trans. Comput. Biol. Bioinf. **6**, 552–569 (2007)
3. Chvátal, V.: Perfectly ordered graphs. In: Berge, C., Chvátal, V. (eds.) Topics on Perfect Graphs. North-Holland Mathematics Studies, vol. 88, pp. 63–65. North-Holland (1984). https://doi.org/10.1016/S0304-0208(08)72923-2. https://www.sciencedirect.com/science/article/pii/S0304020808729232
4. Corneil, D.G., Lerchs, H., Stewart Burlingham, L.K.: Complement reducible graphs. Discr. Appl. Math. **3**, 163–174 (1981)
5. Corneil, D.G., Perl, Y., Stewart, L.K.: A linear recognition algorithm for cographs. SIAM J. Comput. **14**(4), 926–934 (1985). https://doi.org/10.1137/0214065
6. Dahlhaus, E., Gustedt, J., McConnell, R.M.: Efficient and practical algorithms for sequential modular decomposition. J. Algorithms **41**(2), 360–387 (2001)
7. Ehrenfeucht, A., Rozenberg, G.: Theory of 2-structures, part I: clans, basic subclasses, and morphisms. Theor. Comp. Sci. **70**, 277–303 (1990)
8. Ehrenfeucht, A., Rozenberg, G.: Theory of 2-structures, part II: representation through labeled tree families. Theor. Comp. Sci. **70**, 305–342 (1990)
9. Ehrenfeucht, A., Gabow, H.N., Mcconnell, R.M., Sullivan, S.J.: An $O(n^2)$ divide-and-conquer algorithm for the prime tree decomposition of two-structures and modular decomposition of graphs. J. Algorithms **16**(2), 283–294 (1994)

10. Garey, M.R., Johnson, D.S.: Computers and Intractability, vol. 174. Freeman, San Francisco (1979)
11. Habib, M., Paul, C.: A survey of the algorithmic aspects of modular decomposition. Comput. Sci. Rev. **4**(1), 41–59 (2010)
12. Hellmuth, M., Schaller, D., Stadler, P.F.: Clustering systems of phylogenetic networks. Theory Biosci. (2023). https://doi.org/10.1007/s12064-023-00398-w
13. Hellmuth, M., Scholz, G.E.: From modular decomposition trees to level-1 networks: pseudo-cographs, polar-cats and prime polar-cats. Discret. Appl. Math. **321**, 179–219 (2022). https://doi.org/10.1016/j.dam.2022.06.042
14. Hellmuth, M., Scholz, G.E.: Linear time algorithms for NP-hard problems restricted to GaTEx graphs (2023). arXiv:2306.04367
15. Hellmuth, M., Scholz, G.E.: Resolving prime modules: the structure of pseudo-cographs and galled-tree explainable graphs (2023). arXiv:2211.16854
16. Hellmuth, M., Stadler, P.F., Wieseke, N.: The mathematics of xenology: di-cographs, symbolic ultrametrics, 2-structures and tree-representable systems of binary relations. J. Math. Biol. **75**(1), 199–237 (2017). https://doi.org/10.1007/s00285-016-1084-3
17. Huber, K.T., Scholz, G.E.: Beyond representing orthology relations with trees. Algorithmica **80**(1), 73–103 (2018)
18. Luce, R.D., Perry, A.D.: A method of matrix analysis of group structure. Psychometrika **14**(2), 95–116 (1949). https://doi.org/10.1007/BF02289146
19. Marx, D.: Graph colouring problems and their applications in scheduling. Periodica Polytechnica Electr. Eng. **48**, 11–16 (2004)
20. McConnell, R.M., Spinrad, J.P.: Modular decomposition and transitive orientation. Discret. Math. **201**(1–3), 189–241 (1999)
21. Middendorf, M., Pfeiffer, F.: On the complexity of recognizing perfectly orderable graphs. Discret. Math. **80**(3), 327–333 (1990). https://doi.org/10.1016/0012-365X(90)90251-C. https://www.sciencedirect.com/science/article/pii/0012365X9090251C
22. Spirin, V., Mirny, L.A.: Protein complexes and functional modules in molecular networks. Proc. Natl. Acad. Sci. **100**(21), 12123–12128 (2003). https://doi.org/10.1073/pnas.2032324100
23. Tedder, M., Corneil, D., Habib, M., Paul, C.: Simpler linear-time modular decomposition via recursive factorizing permutations. In: Aceto, L., Damgård, I., Goldberg, L.A., Halldórsson, M.M., Ingólfsdóttir, A., Walukiewicz, I. (eds.) ICALP 2008. LNCS, vol. 5125, pp. 634–645. Springer, Heidelberg (2008). https://doi.org/10.1007/978-3-540-70575-8_52
24. Turau, V., Weyer, C.: Algorithmische Graphentheorie. De Gruyter, Berlin, München, Boston (2015). https://doi.org/10.1515/9783110417326

Polynomial Turing Compressions for Some Graph Problems Parameterized by Modular-Width

Weidong Luo[1,2(✉)] [iD]

[1] Université de Sherbrooke, Sherbrooke, Canada
weidong.luo@yahoo.com
[2] Humboldt-Universität zu Berlin, Berlin, Germany

Abstract. A polynomial Turing compression (PTC) for a parameterized problem L is a polynomial-time Turing machine that has access to an oracle for a problem L' such that a polynomial in the input parameter bounds each query. Meanwhile, a polynomial (many-one) compression (PC) can be regarded as a restricted variant of PTC where the machine can query the oracle exactly once and must output the same answer as the oracle. Bodlaender et al. (ICALP 2008) and Fortnow and Santhanam (STOC 2008) initiated an impressive hardness theory for PC under the assumption coNP $\not\subseteq$ NP/poly. Since PTC is a generalization of PC, we define $\mathcal{C}$ as the set of all problems that have PTCs but have no PCs under the assumption coNP $\not\subseteq$ NP/poly. Based on the hardness theory for PC, Fernau et al. (STACS 2009) found the first problem LEAF OUT-TREE(k) in $\mathcal{C}$. However, very little is known about $\mathcal{C}$, as only a dozen problems were shown to belong to the complexity class in the last ten years. Several problems are open, for example, whether CNF-SAT(n) and k-path are in $\mathcal{C}$, and novel ideas are required to better understand the fundamental differences between PTCs and PCs.

In this paper, we enrich our knowledge about $\mathcal{C}$ by showing that several problems parameterized by modular-width (mw) belong to $\mathcal{C}$. More specifically, exploiting the properties of the well-studied structural graph parameter mw, we demonstrate 17 problems parameterized by mw are in $\mathcal{C}$, such as CHROMATIC NUMBER(mw) and HAMILTONIAN CYCLE(mw). In addition, we develop a general recipe to prove the existence of PTCs for a large class of problems, including our 17 problems.

Keywords: Turing compression · modular-width · Turing kernel · structural graph parameter · fixed parameter tractable

1 Introduction

Preprocessing, such as compression (kernelization) and Turing compression (kernelization), is a core research topic in parameterized complexity [11,13,17]. Let

Supported by CSC 201708430114 fellowship.
The full version of this paper can be found on https://arxiv.org/abs/2201.04678.

$Q \subseteq \Sigma^* \times \mathbb{N}$ be a parameterized problem, and $f : \mathbb{N} \to \mathbb{N}$ be a computable function. A *compression* for Q is a polynomial-time algorithm that, given an instance (x, k) of Q, returns an instance l of a problem L with length at most $f(k)$, such that $(x, k) \in Q$ if and only if $l \in L$. We say Q admits a *polynomial compression* (PC) if f is a polynomial function. If L equals Q, then the compression is called a *kernelization*. Turing compression is a generalization of compression. A *Turing compression* for Q of size $f(k)$ is a polynomial-time algorithm A with access to an oracle for a problem L such that, for any input (x, k), A can decide whether $(x, k) \in Q$ sending queries of length at most $f(k)$ to the oracle. We say Q has a *polynomial Turing compression* (PTC) if f is a polynomial function. If L equals Q, then the Turing compression is called a *Turing kernelization*. Q is fixed-parameter tractable (FPT) if there is an $f(k) \cdot |x|^{O(1)}$ algorithm deciding whether $(x, k) \in Q$. Note that a PTC for Q is not sufficient for an FPT algorithm for Q since L can be undecidable.

The upper bounds and lower bounds for PCs have been studied extensively and a large number of results were achieved [17]. In particular, Bodlaender et al. [5,18] initiated an impressive hardness theory to refute the existence of PCs for a large class of problems under the assumption coNP $\not\subseteq$ NP/poly. Since Turing compression generalizes compression [35], it is possible that some natural problems without a PC admit a PTC. Guo [4] was the first to introduce the concept of Turing compression by asking whether some problems, such as the important and still open problem about k-path [17], have PTCs but have no PCs unless coNP $\subseteq$ NP/poly. More than ten years ago, Fernau et al. [14] found the first problem of this kind, by showing that LEAF OUT-TREE(k) has a PTC but has no PCs unless coNP $\subseteq$ NP/poly. However, by now, only about a dozen problems of this kind are known [1,6,12,24–26,32,34]. In addition, despite a few results on the non-existence of PTCs [22,29] and PTCs of restricted types [8,15], negative results on PTCs are much sparser and generally harder to obtain than positive results. In fact, developing a framework for refuting the existence of PTCs (under widely believed assumptions) is a significant open problem in parameterized complexity, and is referred to as "a big research challenge" in the textbook [17]. In order to tackle this ambitious challenge, more knowledge on PTCs is required.

In this work, we focus on the PTC versus PC question for problems parameterized by modular-width (mw). The modular-width is a well-studied structural parameter first proposed in [9] and introduced into parameterized complexity in [19]. Let $G = (V, E)$ be a graph. A *module* of G is a subset of vertices $M \subseteq V$ such that, for every $v \in V \setminus M$, either $M \cap N(v) = \emptyset$ or $M \subseteq N(v)$. The empty set, V, and every singleton $\{v\}$ for $v \in V$ are the *trivial modules*. G is called a *prime* graph if all modules of G are trivial modules. The modular-width of G, denoted by $mw(G)$ or mw, is the number of vertices of the largest prime induced subgraph of G. In addition, mw can also be defined as the number of children of the largest prime node of the *modular decomposition tree*, whose definition can be found in the preliminaries. Moreover, the dynamic programming technique can be used to design algorithms for some problems over modular decomposition trees in a bottom-up fashion. The solution of each node is obtained by combining the partial solutions of its children, where the small number of children for each node leads to an efficient algorithm. The usage of this technique can be dated back to the 1980s [31], where some efficient parallel algorithms for problems such as CLIQUE, MAX-CUT, and CHROMATIC NUMBER are provided. In fact, from the perspective of parameterized complexity, which appears after that, the algorithms in [31] for CLIQUE(mw) and CHROMATIC NUMBER(mw) are FPT, and the algorithm in [31] for MAX-CUT(mw) is XP[1]. Recently, the technique is also used in designing FPT algorithms for graph problems in structural parameters [2,9,27]. Observe that the process of combining the partial solutions in the dynamic programming over a modular decomposition tree can be replaced by a query to an oracle with length at most a function of the largest degree of the tree. Thus, we can use Turing compression to solve a problem by simulating the dynamic programming process over the modular decomposition tree for the problem. Consequently, this technique can also help us to obtain PTCs for graph problems parameterized by mw, all of which coincidentally have no PCs unless coNP $\subseteq$ NP/poly.

Our Results. Exploiting the well-studied technique of dynamic programming algorithm over modular decomposition trees, we provide PTCs for 17 fundamental graph problems parameterized by mw, which have no PCs unless coNP $\subseteq$ NP/poly (some of the PC lower bounds are provided in [28]). Thus, we largely enrich the class of problems that admit PTC but do not admit PC. In addition, by capturing the characteristics of constructing PTCs using the technique of dynamic programming algorithm over modular decomposition tree, we develop a recipe to facilitate the development of PTCs for a large class of problems, including all the 17 problems. In particular, our study gives rise to the following result.

Theorem 1. *The following problems parameterized by mw have PTCs but have no PCs unless coNP $\subseteq$ NP/poly:* INDEPENDENT SET, CLIQUE, VERTEX COVER, CHROMATIC NUMBER, DOMINATING SET, HAMILTONIAN CYCLE, HAMILTONIAN PATH, FEEDBACK VERTEX SET, ODD CYCLE TRANSVERSAL, CONNECTED VERTEX COVER, INDUCED MATCHING, NONBLOCKER, MAXIMUM INDUCED FOREST, PARTITIONING INTO PATHS, LONGEST INDUCED PATH, INDEPENDENT TRIANGLE PACKING, *and* INDEPENDENT CYCLE PACKING, *where the results of the PC lower bounds for the first 11 problems are demonstrated in [28].*

2 Preliminaries

We denote $\Sigma = \{0,1\}$ and $[n] = \{1,\ldots,n\}$. Let $\overline{G}$ denote the complement of a graph G. Unless otherwise specified, $V(G)$ and $E(G)$ indicate the vertex and edge sets of G, respectively. For $v \in V(G)$, $N(v)$ consists of all neighbors of v. We denote $N[v] = N(v) \cup \{v\}$. For $M \subseteq V(G)$, $G[M]$ denotes the subgraph induced in G by M, and $N(M)$ consists of all vertices that are not in M but are adjacent to some vertex of M. We denote $N[M] = N(M) \cup M$. The cardinality of a set S is denoted by $|S|$. Symbols $\mathcal{G}$ and $\mathbb{N}$ denote the sets of undirected graphs and the natural numbers, respectively. For two disjoint vertex sets M and M' of a graph,

---

[1] polynomial-time algorithm for any fixed parameter.

the edges between M and M' refer to all edges uv such that $u \in M$ and $v \in M'$. In addition, we say M and M' are adjacent if all possible edges between M and M' exist, and M and M' are non-adjacent if there are no edges between M and M'. K_n and C_n denote a complete graph and a cycle with n vertices, respectively. For $X \subseteq V(G)$, we write $G - X$ for the subgraph induced in G by $V(G) \setminus X$. In addition, we also use $G - G'$ to represent $G - V(G')$ for a subgraph G' of G. The intersection and union of two graphs $G = (V, E)$ and $G' = (V', E')$ are denoted as $G \cap G' = (V \cap V', E \cap E')$ and $G \cup G' = (V \cup V', E \cup E')$, respectively. The O^*-notation suppresses factors that are polynomial in the input size.

Let $G = (V, E)$ be a graph. Recall that a module of G is a subset of vertices $M \subseteq V$ such that, for every $v \in V \setminus M$, either $M \cap N(v) = \emptyset$ or $M \subseteq N(v)$. A module M is a *strong module* if, for any module M', only one of the following holds: (1) $M \subseteq M'$ (2) $M' \subseteq M$ (3) $M \cap M' = \emptyset$. A module M is *maximal* if $M \subsetneq V$ and no module M' satisfies $M \subsetneq M' \subsetneq V$. Assume $P \subseteq 2^V$ is a vertex partition of V. P is a *(maximal) modular partition* if all $M \in P$ are (maximal strong) modules of G. For a modular partition P, the *quotient graph* $G_{/P} = (V_P, E_P)$ is defined as follows. The set of vertices V_P contains one vertex v_M for each module $M \in P$, so that $V_P = \{v_M : M \in P\}$. An edge $v_M v_{M'}$ is contained in E_P if and only if $M, M' \in P$ are adjacent. All strong modules M of G can be represented by an inclusion tree $MD(G)$, where each M corresponds to a vertex v_M of $MD(G)$, and, for any two strong modules M, M' of G, $v_{M'}$ is a descendant of v_M in $MD(G)$ if and only if $M' \subsetneq M$. This unique tree $MD(G)$ is called the *modular decomposition tree* of G. The internal vertices are divided into three types: a vertex v_M is *parallel* if $G[M]$ is disconnected, *series* if $\overline{G[M]}$ is disconnected, *prime* if both $G[M]$ and $\overline{G[M]}$ are connected. The *modular-width* of G can also be defined as the minimum number k such that the number of children of any prime vertex in $MD(G)$ is at most k. In addition, for a module M of G, $G[M]$ is called a *factor*. The modular decomposition tree of G can be obtained in time $O(m+n)$ [33]. Refer to [20] for more information about modular decomposition trees. The null graph and the empty module are disregarded in the proofs of this paper (the results are trivial for these cases).

Next, we give the definitions of the problems of Theorem 1, all of which are NP-hard. Given a graph G, HAMILTONIAN CYCLE (HAMILTONIAN PATH) asks whether G has a cycle (path) that visits each vertex of G exactly once. Let (G, k) be the input of the following problems, where $G = (V, E)$ is a graph and k is an integer. CHROMATIC NUMBER asks whether V can be colored by at most k colors such that no two adjacent vertices share the same color. CLIQUE asks whether V has a subset of size at least k such that any two vertices in it are adjacent. VERTEX COVER asks whether V has a subset, called vertex cover, of size at most k such that every edge of G has at least one endpoint in it. CONNECTED VERTEX COVER asks whether G has a vertex cover X such that $|X| \leq k$ and $G[X]$ is connected. DOMINATING SET asks whether V has a subset of size at most k such that every vertex not in it is adjacent to at least one vertex of it. FEEDBACK VERTEX SET asks whether V has a subset X of size at most k such that $G - X$ is a forest. INDEPENDENT CYCLE (TRIANGLE) PACKING asks whether G contains an induced subgraph consisting of at least k pairwise vertex-disjoint cycles (triangles). INDEPENDENT SET asks whether V has a subset of size at least k such that any two vertices in it are not adjacent. INDUCED MATCHING asks whether V has a subset X of size at least $2k$ such that $G[X]$ is a matching with at least k edges. LONGEST INDUCED PATH asks whether G contains the path on k vertices as an induced subgraph. MAX LEAF SPANNING TREE asks whether G has a spanning tree with at least k leaves. NONBLOCKER asks whether V has a subset of size at least k such that every vertex in it is adjacent to a vertex outside of it. ODD CYCLE TRANSVERSAL asks whether V has a subset X of size at most k such that $G - X$ is a bipartite graph. PARTITIONING INTO PATHS asks whether G contains k vertex disjoint paths whose union includes every vertex of G. The function (optimization) versions of all these problems are defined in their natural ways, for example, for the function (optimization) version of CLIQUE, the input is G and the output is the number of vertices of the largest clique of G.

3 Recipe for Polynomial Turing Compression in Parameter Modular-Width

Suppose we are given graphs $G = (V, E)$ and $H = (V_H, E_H)$, as well as a module M of G. Assume G_S is a supergraph of G, which is obtained as follows: (1) add G and H into G_S, (2) add uv to G_S for all $v \in N(M), u \in V_H$. Let G' be the subgraph induced in G_S by $(V \setminus M) \cup V_H$. We say G' is obtained from G by replacing $G[M]$ with H, and the process of obtaining G' from G is called a *modular replacement*. Clearly, V_H is a module of G'. Recall that symbols $\mathcal{G}$ and $\mathbb{N}$ denote the sets of undirected graphs and the natural numbers, respectively.

Lemma 2. *Let each F_i be a function from $\mathcal{G}$ to $\mathbb{N}$ for $i \in [r]$. For any graphs G and H, as well as any module M of G, suppose $F_i(G[M]) = F_i(H)$ for all i implies $F_i(G) = F_i(G')$ for all i, where G' is obtained from G by replacing $G[M]$ with H. Then, for any graph G and any modular partition P of $V(G)$, the quotient graph $G_{/P}$ together with $F_1(G[M]), \ldots, F_r(G[M])$ for all modules M in P completely determine $F_1(G), \ldots, F_r(G)$.*

Proof. Let tuple $T(G) = (F_1(G), \ldots, F_r(G))$. For a graph G and a modular partition P of $V(G)$, we say X is a *values-attached quotient graph* generated from P if X is the quotient graph $G_{/P} = (V_P, E_P)$ with each vertex $v_M \in V_P$ attached the tuple $T(G[M])$. Suppose $\mathcal{X}$ consists of all possible X generated from any P of $V(G)$, where $G \in \mathcal{G}$. Assume the binary relation R over $\mathcal{X}$ and $\mathcal{G}$ consists of all pairs (X, G) such that X is generated from some modular partition of $V(G)$. For each $i \in [r]$, let binary relation f_i be the composition relation of R and F_i over $\mathcal{X}$ and $\mathbb{N}$, which means that $f_i = R; F_i = \{(X, n) \mid \text{there exists } G \in \mathcal{G} \text{ such that } (X, G) \in R \text{ and } (G, n) \in F_i\}$.

Consider every f_i. According to the definition of R, for any $X \in \mathcal{X}$, there is at least a $G \in \mathcal{G}$ such that $(X, G) \in R$. Moreover, as F_i is a function, for any $G \in \mathcal{G}$, there is an $n \in \mathbb{N}$ such that $(G, n) \in F_i$. Therefore, for any $X \in \mathcal{X}$, there is at least an $n \in \mathbb{N}$ such that $(X, n) \in f_i$, so f_i is left-total. For an $X \in \mathcal{X}$, assume $\mathcal{G}'$ consists of all G such that $(X, G) \in R$. We claim that $F_i(G_1) = F_i(G_2)$ for any $G_1, G_2 \in \mathcal{G}'$. Since $(X, G_1), (X, G_2) \in R$, there exist modular partitions P_1 of G_1 and P_2 of G_2 such that X is not only generated from P_1 but also generated from P_2. Thus, the quotient graphs G_{1/P_1} and G_{2/P_2} are isomorphic, and there exists an edge-preserving bijection g from $V(G_{1/P_1})$ to $V(G_{2/P_2})$ such that $T(G[M_1]) = T(G[M_2])$ for every v_{M_1} of $V(G_{1/P_1})$ and $v_{M_2} = g(v_{M_1})$ of $V(G_{2/P_2})$. Here, we say $G[M_2]$ of G_2 corresponds to $G[M_1]$ of G_1 if $v_{M_2} = g(v_{M_1})$. Now, consider every module M_1 of P_1. We replace each $G[M_1]$ of G_1 with its corresponding factor $G[M_2]$ of G_2 one by one. According to the prerequisite of this lemma,[2] since $T(G[M_1]) = T(G[M_2])$, the value of F_i for the new obtained graph after every modular replacement does not change. After the final modular replacement, G_2 is obtained, so we have $F_i(G_1) = F_i(G_2)$. Hence, $F_i(G)$ is a fixed number for any $G \in \mathcal{G}'$. This means that f_i is right-unique (note that a relation is called right-unique if each element on the right side of the relation is mapped to a unique element on the left side). Consequently, f_i is a function from $\mathcal{X}$ to $\mathbb{N}$, moreover, $F_i(G)$ equals $f_i(X)$, where $(X, G) \in R$.

Note that, for each i, we say an algorithm solves $F_i(G)$ if it outputs $F_i(G)$ with the input G. The decision version of $F_i(G)$ is as follows: given a graph G and an integer k, decide whether $F_i(G) \leq k$ (or $F_i(G) \geq k$).

Lemma 3. *Let each F_i be a function from $\mathcal{G}$ to $\mathbb{N}$ for $i \in [r]$, where r is a constant. Assume the following statements hold.*

1. *For any graphs G and H, as well as any module M of G, $F_i(H) = F_i(G[M])$ for all i implies $F_i(G) \leq F_i(G')$ (or $F_i(G) \geq F_i(G')$) for all i,[3] where G' is obtained from G by replacing $G[M]$ with H.*
2. *For each i, there is a $2^{mw^{O(1)}}|G|^{O(1)}$ time algorithm to solve $F_i(G)$.*
3. *$F_i(G) \leq |G|^{O(1)}$ for each i.*

Then each $F_i(G)$ can be solved by a polynomial-time algorithm together with an oracle for a problem Q that can decide in one step whether a string of length $mw^{O(1)}$ is in Q. Moreover, the decision version of each $F_i(G)$ has a PTC parameterized by mw.

Proof. For all i, assume $F_i(G) \leq |G|^c$ and $F_i(G)$ can be solved in $2^{mw^c}|G|^c$ for some constant c. Suppose w.l.o.g. that $mw \geq \log^{\frac{1}{c}}|G|$ henceforth (otherwise, each $F_i(G)$ can be solved in polynomial time based on statement 2 and the consequence of this lemma holds). Consider statement 1. Since $V(H)$ is a module of G', we can obtain $F_i(G) = F_i(G')$ by exchanging the position of G and G' as follows. For the graphs G' and $G[M]$, as well as the module $V(H)$ of G', that $F_i(G[M]) = F_i(H)$ for all i also implies that $F_i(G) \geq F_i(G')$ (or $F_i(G) \leq F_i(G')$) for all i, where G is obtained from G' by replacing H with $G[M]$. Thus, functions $F_1, \ldots, F_r$ fulfill the conclusion of Lemma 2. Let the

---

[2] The prerequisite of this lemma is the second sentence of Lemma 2.

[3] Here, we only require every F_i is monotone under module-substitution.

definitions of values-attached quotient graph X, set $\mathcal{X}$, binary relation R and functions $f_1, \ldots, f_r$ be the same as that in the proof of Lemma 2. Assume Q consists of all the strings $(X, f_1(X), \ldots, f_r(X))$ for $X \in \mathcal{X}$.

Recall that every vertex v_M of the modular decomposition tree $MD(G)$ of a graph G corresponds to the strong module M of G, where $MD(G)$ can be constructed in linear time. Here, we call $G[M]$ the corresponding graph of v_M. Roughly speaking, to obtain $F_1(G), \ldots, F_r(G)$, we compute in a bottom-up fashion the values of all F_i for the graphs that correspond to the vertices of $MD(G)$. First, the corresponding graph of each leaf of $MD(G)$ is the singleton graph K_1, where $F_i(K_1)$ can be solved in $O(1)$ time according to statement 2. Secondly, consider an internal vertex v_M of $MD(G)$. Let P be the maximal modular partition of M. Then every child $v_{M'}$ of v_M corresponds to the module M' of P and the quotient graph of v_M is $G[M]_{/P}$. Suppose v_M is prime. Assume $X \in \mathcal{X}$ is the values-attached quotient graph generated from P. Since $F_1(G[M']), \ldots, F_r(G[M'])$ for all M' of P and $G[M]_{/P}$ are given, X can be obtained immediately. Let $k_1, \ldots, k_r$ be non-negative integers at most $|G|^c$. Exhaustively generate string $(X, k_1, \ldots, k_r)$ and query the oracle whether it is in Q. Based on Lemma 2, we have $f_i(X) = F_i(G[M]) \leq |G|^c$ for all i. Hence, we can obtain $F_1(G[M]), \ldots, F_r(G[M])$ by finding the string $(X, k_1, \ldots, k_r) = (X, f_1(X), \ldots, f_r(X))$ after querying the oracle at most $(|G|^c+1)^r = |G|^{O(1)}$ times. Moreover, the length of any $(X, k_1, \ldots, k_r)$ is at most $O(mw^2 \log |G|) = mw^{O(1)}$. Assume v_M is parallel. Suppose P contains t modules, where $2 \leq t \leq |V(G)|$. Then, the quotient graph $G[M]_{/P}$ is $\overline{K}_t$. Consider any two modules M_1' and M_2' of P. Let M_{12}' be $M_1' \cup M_2'$ and $G[M_{12}']$ be the subgraph induced by M_{12}' in $G[M]$. Then, $P' = \{M_1', M_2'\}$ is a modular partition of M_{12}'. Since the values-attached quotient graph generated from P' are given, we can obtain $F_1(G[M_{12}']), \ldots, F_r(G[M_{12}'])$ using the same method as that of the prime vertex. Now, consider the new modular partition $P = \{M_{12}'\} \cup (P \setminus \{M_1', M_2'\})$ of M. It contains only $t - 1$ modules and the new quotient graph $G[M]_{/P}$ is $\overline{K}_{t-1}$. Moreover, $F_1(G[M']), \ldots, F_r(G[M'])$ are known for every module M' in P. Clearly, this process decreases the vertex number of the quotient graph by one. We can repeat $t - 1$ times the same process on every newly generated quotient graph. Finally, $F_1(G[M]), \ldots, F_r(G[M])$ can be obtained. Assume v_M is a series vertex. $F_1(G[M]), \ldots, F_r(G[M])$ can be obtained using the same strategy as that of the parallel vertex. Therefore, for any internal vertex v_M of $MD(G)$, we obtain $F_1(G[M]), \ldots, F_r(G[M])$ in $|G|^{O(1)}$ time with each query length at most $mw^{O(1)}$. In addition, the vertex number of $MD(G)$ is $O(|G|)$. As a result, $F_1(G), \ldots, F_r(G)$ can be solved by a polynomial-time algorithm with an oracle for Q that can decide whether a string of length $mw^{O(1)}$ is in Q. Furthermore, according to the definition of PTC, the decision version of each $F_i(G)$ has a PTC parameterized by mw.

4   Polynomial Turing Compressions for Problems

Obviously, the function versions of all problems in Theorem 1 fulfill statement 3 of Lemma 3. CLIQUE, FEEDBACK VERTEX SET, LONGEST INDUCED

PATH, INDUCED MATCHING, INDEPENDENT TRIANGLE PACKING, INDEPENDENT CYCLE PACKING can be solved in $O^*(1.74^{mw})$ time [16]. CHROMATIC NUMBER can be solved in $O^*(2^{mw})$ time [19]. HAMILTONIAN CYCLE and PARTITIONING INTO PATHS can be solved in $2^{O(mw^2 \log mw)}n^{O(1)}$ time [19]. In addition, clique-width (cw) [10] is a generalized parameter of mw such that $cw \leq mw + 2$ in a graph (the cw and mw of a cograph are two and zero, respectively). CONNECTED VERTEX COVER [3], DOMINATING SET [7], and ODD CYCLE TRANSVERSAL [21,23] can be solved in $2^{O(cw)}n^{O(1)}$ time, thus also in $2^{O(mw)}n^{O(1)}$ time. Therefore, the function versions of all the above-mentioned problems fulfill statement 2 of Lemma 3.

Clearly, a PTC of each problem in Theorem 1 can be obtained if we can obtain a PTC of the problem with connected input graphs. So we assume w.l.o.g. the input graph of every problem is connected. In this section, unless otherwise specified, assume $G = (V, E)$ and $H = (V_H, E_H)$ are connected graphs, $M \neq \emptyset$ is a module of G, $M' = N(M)$ in G, and graph $G' = (V', E')$ is obtained from G by replacing $G[M]$ with H. Now, we only need to prove the function versions of the above-mentioned problems, all of which will be discussed in this section, fulfill statement 1 of Lemma 3 to provide PTCs for the problems parameterized by mw. More specifically, we will prove that $F_i(H) = F_i(G[M])$ for all i implies that $F_i(G) \leq F_i(G')$ (or $F_i(G) \geq F_i(G')$) for all i, where functions F_i are the function versions of the problems discussed in some lemma of this section. Obviously, the statement is true for any function F_i if $M = V$. So assume $M \neq V$ henceforth. In addition, $M' \neq \emptyset$ since G is connected and $M \notin \{\emptyset, V\}$. In this section, assume function $F_v(I)$ denotes the vertex number of I for any $I \in \mathcal{G}$. Let min-DS, min-CVC, min-VC, min-FVS, min-OCT, max-IM, max-ITP, and max-ICP be the abbreviations of the minimum dominating set, minimum connected vertex cover, minimum vertex cover, minimum feedback vertex set, minimum odd cycle transversal, maximum induced matching, maximum independent triangle packing, and maximum independent cycle packing, respectively.

Lemma 4. CHROMATIC NUMBER(mw) *has a PTC.*

Proof. Let function $F(I)$ denote the chromatic number for any $I \in \mathcal{G}$. Suppose $F(G[M]) = F(H)$. Then, there is a coloring $c : V \to C$ for G, where $C = [F(G)]$. Let $C_M = \{c(v) \mid v \in M\}$ and $C_{M'} = \{c(v) \mid v \in M'\}$. Since $M' = N(M)$, $C_M \cap C_{M'} = \emptyset$. Consider G'. Since $F(H) = F(G[M]) \leq |C_M|$, there is a coloring $c_H : V_H \to C_M$ for H. Suppose $c' : V' \to C$ is a function such that $c'(v) = c(v)$ for all $v \in V' \setminus V_H$ and $c'(v) = c_H(v)$ for all $v \in V_H$. Since $N(V_H) = M'$ and $C_M \cap C_{M'} = \emptyset$, c' is a coloring for G'. Thus, $F(G') \leq |C| = F(G)$.

Lemma 5. *Let D be a min-DS of G. Then $M \cap D$ is either $\emptyset$, $\{v\}$, or a min-DS of $G[M]$.*

Proof. Assume, for contradiction, $|D \cap M| \geq 2$ and $D \cap M$ is not a min-DS of $G[M]$. If $D \cap M$ is a dominating set of $G[M]$ that is not minimum, then let X be a smaller one, and $(D \setminus M) \cup X$ is a smaller dominating set of G, a contradiction. Hence, $D \cap M$ is not a dominating set of $G[M]$, so there must be an $x \in D \cap M'$.

Then $M \subseteq N(x)$ and every vertex in M has the same neighborhood outside of M, so D is still a dominating set of G by removing from D all but one vertex of $M \cap D$, contradicting the minimality of D.

Lemma 6. DOMINATING SET(mw) has a PTC.

Proof. Let function $F(I)$ denote the size of the min-DS for any $I \in \mathcal{G}$. Suppose $F(G[M]) = F(H)$ and D is a min-DS of G. Assume $D \cap M = \emptyset$. Then there exists an $x \in D \cap M'$ such that $V_H \subseteq N(x)$ in G'. Thus, D is a dominating set of G' and $F(G') \leq F(G)$. Assume $D \cap M = \{u\}$. Suppose $F(G[M]) = F(H) \geq 2$. Then $\{u\}$ is not a dominating set of $G[M]$, so there is a $v \in D \cap M'$ such that $V_H \subseteq N(v)$ in G'. Clearly, $\{w\} \cup (D \setminus \{u\})$ is a dominating set of G' for any $w \in V_H$, so $F(G') \leq F(G)$. Suppose $F(G[M]) = F(H) = 1$. We may assume $\{v\}$ is a dominating set of H. Clearly, $\{v\} \cup (D \setminus \{u\})$ is a dominating set of G', so $F(G') \leq F(G)$. Now, according to Lemma 5, we only need to consider that $D \cap M$ is a min-DS of $G[M]$. Suppose D_H is a min-DS of H. Clearly, $D_H \cup (D \setminus M)$ is a dominating set of G', so $F(G') \leq F(G)$.

Lemma 7. *Assume $I_{G[M]}$ and I_H with the same size are independent sets of $G[M]$ and H, respectively. Suppose $S \subseteq V \setminus M$. Then, the subgraph in G induced by $S \cup I_{G[M]}$ and the subgraph in G' induced by $S \cup I_H$ are isomorphic.*

Proof. It is trivial if S or $I_{G[M]}$ is empty. Assume $I_{G[M]}$ and S are not empty. Since $|I_H| = |I_{G[M]}|$, we may assume $S = \{v_1, \ldots, v_r\}$, $I_{G[M]} = \{u_1, \ldots, u_s\}$, and $I_H = \{w_1, \ldots, w_s\}$. Suppose f is a bijection from $S \cup I_{G[M]}$ to $S \cup I_H$ such that $f = \{[v_1, v_1], \ldots, [v_r, v_r], [u_1, w_1], \ldots, [u_s, w_s]\}$. Clearly, f is an edge-preserving bijection.

Lemma 8. *G has an edge if and only if $G[V \setminus S]$ has an edge, where S is a min-FVS or a min-OCT of G.*

Proof. Let S be a min-FVS of G. For the forward direction, suppose G has an edge. $G[V \setminus S]$ has an edge if $S = \emptyset$. Assume $S \neq \emptyset$. After deleting any $|S| - 1$ vertices of S from G, there exists a cycle in G, otherwise, G has a feedback vertex set of size $|S| - 1$. Hence, $G[V \setminus S]$ has at least one edge since the edges of a cycle cannot be entirely removed by deleting one vertex. The reverse direction is trivial. The proof goes the same way if S is a min-OCT of G

Lemma 9. *Assume C, F, O, R are a min-VC, a min-FVS, a min-OCT, and a min-CVC of G, respectively. Let v be a vertex of M. The following statements hold. (1) $M \cap C$ is either M or a min-VC of $G[M]$. (2) $M \cap F$ is either M, $M \setminus \{v\}$, a min-VC of $G[M]$, or a min-FVS of $G[M]$. (3) $M \cap O$ is either M, a min-VC of $G[M]$, or a min-OCT of $G[M]$. (4) $M \cap R$ is either M, a min-VC of $G[M]$, or $\{v\}$.*

Proof. Recall that $M' \neq \emptyset$. (1) If $M \cap C \neq M$, then $M' \subseteq C$ and $M \cap C$ is a min-VC of $G[M]$. (2) Assume $G[M \setminus F]$ has an edge. Then $M' \subseteq F$ and $M \cap F$ is a min-FVS of $G[M]$. Assume $G[M \setminus F]$ has no edges but has at least two vertices. Then $M' \setminus F$ contains at most one vertex, otherwise, there exists a C_4 in $G[V \setminus F]$. Hence, $F \cap M$ is a min-VC of $G[M]$. Assume $M \setminus F$ equals $\{v\}$ or $\emptyset$. Then $M \cap F$ is $M \setminus \{v\}$ or M. (3) Clearly, $M \cap O = M$ if $M \setminus O = \emptyset$. Assume $G[M \setminus O]$ contains an edge. Then $M' \subseteq O$, so $M \cap O$ is a min-OCT of $G[M]$. Assume $G[M \setminus O]$ contains a vertex but no edges. Then $G[M' \setminus O]$ contains no edges, so $M \cap O$ is a min-VC of $G[M]$. (4) Clearly, $M \subseteq R$ if $M' \not\subseteq R$. Assume $M' \subseteq R$ henceforth. Then, $M \cap R$ is a min-VC of $G[M]$ if $G[R \setminus M]$ is connected or $G[M]$ contains an edge, otherwise, $M \cap R$ is a vertex of M to ensure the connectivity of $G[R]$.

Lemma 10. VERTEX COVER(mw), CONNECTED VERTEX COVER(mw), FEEDBACK VERTEX SET(mw), and ODD CYCLE TRANSVERSAL(mw) have PTCs.

Proof. Suppose functions $F_{oct}(I)$, $F_{fvs}(I)$, $F_{cvc}(I)$, and $F_{vc}(I)$ represent the sizes of min-OCT, min-FVS, min-CVC, and min-VC of any $I \in \mathcal{G}$, respectively. Suppose $F_v(G[M]) = F_v(H)$, $F_{vc}(G[M]) = F_{vc}(H)$, $F_{cvc}(G[M]) = F_{cvc}(H)$, $F_{fvs}(G[M]) = F_{fvs}(H)$, and $F_{oct}(G[M]) = F_{oct}(H)$. Let C, R, F, and O represent a min-VC, a min-CVC, a min-FVS, and a min-OCT of G, respectively. Let C_H, F_H, and O_H represent a min-VC, a min-FVS, and a min-OCT of H, respectively. Obviously, $F_v(G') \leq F_v(G)$.

We claim $F_{vc}(G') \leq F_{vc}(G)$. Based on Lemma 9, $C \cap M$ is either M or a min-VC of $G[M]$. Let S denote $V \setminus (C \cup M)$. Assume $C \cap M = M$. Clearly, $(C \setminus M) \cup V_H$ is a vertex cover (VC) of G'. Assume $C \cap M$ is a min-VC of $G[M]$. Then, $M \setminus C$ and $V_H \setminus C_H$ are independent sets of $G[M]$ and H, respectively. Moreover, $|M \setminus C| = |V_H \setminus C_H|$ since $F_v(G[M]) = F_v(H)$ and $F_{vc}(G[M]) = F_{vc}(H)$. According to Lemma 7, the subgraph induced by $S \cup (M \setminus C) = V \setminus C$ in G and the subgraph induced by $S \cup (V_H \setminus C_H)$ in G' are isomorphic, so $S \cup (V_H \setminus C_H)$ is an independent set of G'. Hence, $(C \setminus M) \cup C_H$ is a VC of G'.

We claim $F_{cvc}(G') \leq F_{cvc}(G)$. According to Lemma 9, $M \cap R$ is either M, a min-VC of $G[M]$, or $\{v\}$. Suppose $M \cap R = M$. Clearly, $(R \setminus M) \cup V_H$ is a connected vertex cover (CVC) of G'. Suppose $M \cap R$ is a min-VC of $G[M]$. Let $S = V \setminus (R \cup M)$. According to Lemma 7, $G'[S \cup (V_H \setminus C_H)]$ and $G[S \cup (M \setminus R)] = G[V \setminus R]$ are isomorphic. Hence, $S \cup (V_H \setminus C_H)$ is an independent set of G', and $(R \setminus M) \cup C_H$ is a VC of G'. Clearly, $G'[(R \setminus M) \cup C_H]$ is connected since $G[R]$ is connected. Therefore, $(R \setminus M) \cup C_H$ is a CVC of G'. Suppose $M \cap R = \{v\}$. Assume $u \in M \setminus R$ (the case $M = \{v\}$ has been discussed). Since G is connected, $M' \subseteq N(u)$. Thus, $M' \subseteq R$. Since $F_{vc}(G[M]) = F_{vc}(H)$, $F_v(G[M]) = F_v(H)$, and $\{v\}$ is a VC of $G[M]$, there exists $w \in V_H$ that covers all edges of H. Therefore, $(R \setminus \{v\}) \cup \{w\}$ is a VC of G'. In addition, according to Lemma 7, $G[R]$, which is connected, and the subgraph induced by $(R \setminus \{v\}) \cup \{w\}$ in G' are isomorphic. Thus, $(R \setminus \{v\}) \cup \{w\}$ is a CVC of G'.

We claim $F_{fvs}(G') \leq F_{fvs}(G)$. Based on Lemma 9, $F \cap M$ is either M, $M \setminus \{v\}$, a min-VC of $G[M]$, or a min-FVS of $G[M]$. Let $S = V \setminus (F \cup M)$. Clearly, $(F \setminus M) \cup V_H$ is an FVS of G' if $F \cap M = M$. Suppose $F \cap M = M \setminus \{v\}$. Let $u \in V_H$. Based on Lemma 7, $G[S \cup \{v\}]$ and $G'[S \cup \{u\}]$ are isomorphic.

Hence, $G'[S \cup \{u\}]$ is a forest, and $(F \setminus M) \cup (V_H \setminus \{u\})$ is an FVS of G'. Suppose $F \cap M$ is a min-VC of $G[M]$. Based on Lemma 7, $G[S \cup (M \setminus F)] = G[V \setminus F]$ and $G'[S \cup (V_H \setminus C_H)]$ are isomorphic. So $G'[S \cup (V_H \setminus C_H)]$ has no cycles, and $(F \setminus M) \cup C_H$ is an FVS of G'. Suppose $F \cap M$ is a min-FVS of $G[M]$. Assume $G[M \setminus F]$ has no edges. Based on Lemma 8, $G[M]$ has no edges. Thus, H has no edges, moreover, G and G' are isomorphic according to Lemma 7. Assume $G[M \setminus F]$ has an edge. Then $M' \subseteq F$. Hence, $(F \setminus M) \cup F_H$ is an FVS of G'.

We claim $F_{oct}(G') \leq F_{oct}(G)$. Based on Lemma 9, $O \cap M$ is either M, a min-VC of $G[M]$, or a min-OCT of $G[M]$. Let $S = V \setminus (O \cup M)$. Clearly, $(O \setminus M) \cup V_H$ is an OCT of G' if $O \cap M = M$. Assume $O \cap M$ is a min-VC of $G[M]$. According to Lemma 7, $G[S \cup (M \setminus O)] = G[V \setminus O]$ and $G'[S \cup (V_H \setminus C_H)]$ are isomorphic. Therefore, $G'[S \cup (V_H \setminus C_H)]$ has no odd cycles, and $(O \setminus M) \cup C_H$ is an OCT of G'. Assume $O \cap M$ is a min-OCT of $G[M]$. Suppose $G[M \setminus O]$ has no edges. Based on Lemma 8, $G[M]$ has no edges. Thus, H has no edges, moreover, G and G' are isomorphic according to Lemma 7. Suppose $G[M \setminus O]$ has an edge, then $M' \subseteq O$. Hence, $(O \setminus M) \cup O_H$ is an OCT of G'.

Due to space constraints, we omit the proofs of the following three lemmas, which can be found in the full version of this paper [30].

Lemma 11. PARTITIONING INTO PATHS(mw) *and* HAMILTONIAN CYCLE(mw) *have PTCs.*

Lemma 12. LONGEST INDUCED PATH(mw) *has a PTC.*

Lemma 13. INDUCED MATCHING(mw), INDEPENDENT TRIANGLE PACKING (mw), *and* INDEPENDENT CYCLE PACKING(mw) *have PTCs.*

Corollary 14 holds according to Lemma 6, 10, 11 and the following reasons. (1) An independent set, a nonblocker, and the vertex set of a maximum induced forest of G are complements of a vertex cover, a dominating set, and a feedback vertex set of G, respectively. (2) mw does not change under graph complementation, and an independent set in G is a clique in the complement graph of G. (3) The mw of the output graph equals that of the input graph using the routine reduction from HAMILTONIAN PATH to HAMILTONIAN CYCLE.

Corollary 14. INDEPENDENT SET(mw), CLIQUE(mw), MAXIMUM INDUCED FOREST(mw), NONBLOCKER(mw), *and* HAMILTONIAN PATH(mw) *have PTCs.*

5 Polynomial Compression Lower Bounds for Problems

Polynomial compression (PC) lower bounds for the first 11 problems of Theorem 1 are provided in [28] that include DOMINATING SET(mw), FEEDBACK VERTEX SET(mw), and HAMILTONIAN PATH(mw), so PC lower bounds for NONBLOCKER(mw), MAXIMUM INDUCED FOREST(mw), and PARTITIONING INTO PATHS(mw) are obtained immediately. Next, we use the cross-composition

[6] to prove LONGEST INDUCED PATH(mw), INDEPENDENT TRIANGLE PACK-
ING(mw), and INDEPENDENT CYCLE PACKING(mw) have no PCs unless NP $\subseteq$
coNP/poly.

Due to space constraints, we only provide sketches of Lemma 15 as well as
16, and omit the proofs of Lemma 17 and 18. The proofs of them can be found
in the full version of this paper [30].

We define a new problem INDEPENDENT TRIANGLE PACKING REFINEMENT
(ITPR) as follows: the input is a graph G and an independent triangle packing
(ITP) of G with k triangles, decide whether G has an ITP with $k + 1$ triangles?

Lemma 15. *ITPR is NP-hard under Karp reductions.*

Proof (sketch). We provide a Karp reduction from INDEPENDENT TRIANGLE
PACKING to ITPR. Given an instance (G, k), where $G = (V, E)$ and $V = \{v_1, \ldots, v_n\}$. Assume w.l.o.g. that $2 \leq k \leq \frac{n}{3}$. Construct $G' = (V', E')$ as follows.
First, add G, vertices $x_1, \ldots, x_{n-k+1}$, and n triangles $u_1\text{-}w_1\text{-}w_1', \ldots, u_n\text{-}w_n\text{-}w_n'$
into G'. Then, connect u_i with all vertices of V for each $i \in [n]$. Finally, connect
x_i with w_i, w_i' for each $i \in [n-k]$ and connect x_{n-k+1} with $w_{n-k+1}, \ldots, w_n$,
$w_{n-k+1}', \ldots, w_n'$. Clearly, $T = \{u_1\text{-}w_1\text{-}w_1', \ldots, u_n\text{-}w_n\text{-}w_n'\}$ is an ITP of G'. Thus,
(G', T) is an instance of ITPR with n triangles. Then, we can prove that (G, k)
is a yes instance of INDEPENDENT TRIANGLE PACKING if and only if (G', T) is
a yes instance of ITPR.

Lemma 16. INDEPENDENT TRIANGLE PACKING(mw) *has no PCs unless NP*
$\subseteq$ *coNP/poly.*

Proof (sketch). We can provide an or-cross-composition from ITPR to it, where
the output instance is the disjoint union of the input instances.

Lemma 17. INDEPENDENT CYCLE PACKING(mw) *has no PCs unless NP $\subseteq$*
coNP/poly.

Lemma 18. LONGEST INDUCED PATH(mw) *has no PCs unless NP $\subseteq$*
coNP/poly.

6   Conclusions

We conclude this paper by proposing some open questions. Does k-PATH have a
PTC parameterized by mw? In addition, Fomin et al. [16] gives a meta-theorem
that proves a family of problems is FPT parameterized by mw. Can we also give
a meta-theorem to prove the problems in that family have PTCs?

Acknowledgements. I thank Manuel Lafond for his careful reading and constructive comments to improve this manuscript, as well as his valuable help in many other
aspects. I thank the anonymous referees for their valuable comments on the improvement of this manuscript.

References

1. Ambalath, A.M., et al.: On the kernelization complexity of colorful motifs. In: Raman, V., Saurabh, S. (eds.) IPEC 2010. LNCS, vol. 6478, pp. 14–25. Springer, Heidelberg (2010). https://doi.org/10.1007/978-3-642-17493-3_4
2. Belmonte, R., Hanaka, T., Lampis, M., Ono, H., Otachi, Y.: Independent set reconfiguration parameterized by modular-width. Algorithmica **82**(9), 2586–2605 (2020)
3. Bergougnoux, B., Kanté, M.M.: Fast exact algorithms for some connectivity problems parameterized by clique-width. Theor. Comput. Sci. **782**, 30–53 (2019)
4. Bodlaender, H.L., et al.: Open problems in parameterized and exact computation. In: IWPEC 2008 (2008)
5. Bodlaender, H.L., Downey, R.G., Fellows, M.R., Hermelin, D.: On problems without polynomial kernels (extended abstract). In: Aceto, L., Damgård, I., Goldberg, L.A., Halldórsson, M.M., Ingólfsdóttir, A., Walukiewicz, I. (eds.) ICALP 2008. LNCS, vol. 5125, pp. 563–574. Springer, Heidelberg (2008). https://doi.org/10.1007/978-3-540-70575-8_46
6. Bodlaender, H.L., Jansen, B.M.P., Kratsch, S.: Kernelization lower bounds by cross-composition. SIAM J. Discret. Math. **28**(1), 277–305 (2014)
7. Bodlaender, H.L., van Leeuwen, E.J., van Rooij, J.M.M., Vatshelle, M.: Faster algorithms on branch and clique decompositions. In: Hliněný, P., Kučera, A. (eds.) MFCS 2010. LNCS, vol. 6281, pp. 174–185. Springer, Heidelberg (2010). https://doi.org/10.1007/978-3-642-15155-2_17
8. Burjons, E., Rossmanith, P.: Lower bounds for conjunctive and disjunctive turing kernels. In: IPEC 2021. Schloss Dagstuhl-Leibniz-Zentrum für Informatik (2021)
9. Courcelle, B., Makowsky, J.A., Rotics, U.: Linear time solvable optimization problems on graphs of bounded clique-width. Theory Comput. Syst. **33**(2), 125–150 (2000)
10. Courcelle, B., Olariu, S.: Upper bounds to the clique width of graphs. Discret. Appl. Math. **101**(1–3), 77–114 (2000)
11. Cygan, M., et al.: Parameterized Algorithms. Springer, Cham (2015). https://doi.org/10.1007/978-3-319-21275-3
12. Donkers, H., Jansen, B.M.P.: A Turing kernelization dichotomy for structural parameterizations of f-minor-free deletion. J. Comput. Syst. Sci. **119**, 164–182 (2021)
13. Downey, R.G., Fellows, M.R.: Fundamentals of Parameterized Complexity. TCS, Springer, London (2013). https://doi.org/10.1007/978-1-4471-5559-1
14. Fernau, H., Fomin, F.V., Lokshtanov, D., Raible, D., Saurabh, S., Villanger, Y.: Kernel(s) for problems with no kernel: on out-trees with many leaves. In: STACS 2009, February 26–28, Freiburg, Germany, Proceedings. LIPIcs, vol. 3, pp. 421–432 (2009)
15. Fluschnik, T., Heeger, K., Hermelin, D.: Polynomial Turing kernels for clique with an optimal number of queries. CoRR abs/2110.03279 (2021)
16. Fomin, F.V., Liedloff, M., Montealegre, P., Todinca, I.: Algorithms parameterized by vertex cover and modular width, through potential maximal cliques. Algorithmica **80**(4), 1146–1169 (2017)
17. Fomin, F.V., Lokshtanov, D., Saurabh, S., Zehavi, M.: Kernelization: Theory of Parameterized Preprocessing. Cambridge University Press, Cambridge (2013)
18. Fortnow, L., Santhanam, R.: Infeasibility of instance compression and succinct PCPS for NP. In: STOC 2008, Victoria, British Columbia, Canada, May 17–20, pp. 133–142. ACM (2008)

19. Gajarský, J., Lampis, M., Ordyniak, S.: Parameterized algorithms for modular-width. In: Gutin, G., Szeider, S. (eds.) IPEC 2013. LNCS, vol. 8246, pp. 163–176. Springer, Cham (2013). https://doi.org/10.1007/978-3-319-03898-8_15
20. Habib, M., Paul, C.: A survey of the algorithmic aspects of modular decomposition. Comput. Sci. Rev. **4**(1), 41–59 (2010)
21. Hegerfeld, F., Kratsch, S.: Towards exact structural thresholds for parameterized complexity. In: IPEC 2022, September 7–9, 2022, Potsdam, Germany. LIPIcs, vol. 249, pp. 17:1–17:20. Schloss Dagstuhl - Leibniz-Zentrum für Informatik (2022)
22. Hermelin, D., Kratsch, S., Soltys, K., Wahlström, M., Wu, X.: A completeness theory for polynomial (Turing) kernelization. Algorithmica **71**(3), 702–730 (2015)
23. Jacob, H., Bellitto, T., Defrain, O., Pilipczuk, M.: Close relatives (of feedback vertex set), revisited. In: IPEC 2021, September 8–10, Lisbon, Portugal. LIPIcs, vol. 214, pp. 21:1–21:15. Schloss Dagstuhl - Leibniz-Zentrum für Informatik (2021)
24. Jansen, B.M.P.: Turing kernelization for finding long paths and cycles in restricted graph classes. J. Comput. Syst. Sci. **85**, 18–37 (2017)
25. Jansen, B.M.P., Marx, D.: Characterizing the easy-to-find subgraphs from the viewpoint of polynomial-time algorithms, kernels, and Turing kernels. In: SODA 2015, San Diego, CA, USA, January 4–6, 2015. pp. 616–629 (2015)
26. Jansen, B.M.P., Pilipczuk, M., Wrochna, M.: Turing kernelization for finding long paths in graph classes excluding a topological minor. Algorithmica **81**(10), 3936–3967 (2019)
27. Kratsch, S., Nelles, F.: Efficient and adaptive parameterized algorithms on modular decompositions. In: ESA 2018, August 20–22, 2018, Helsinki, Finland. LIPIcs, vol. 112, pp. 55:1–55:15. Schloss Dagstuhl - Leibniz-Zentrum für Informatik (2018)
28. Lafond, M., Luo, W.: Preprocessing complexity for some graph problems parameterized by structural parameters. CoRR abs/2306.12655 (2023)
29. Luo, W.: On some FPT problems without polynomial Turing compressions. Theor. Comput. Sci. **905**, 87–98 (2022)
30. Luo, W.: Polynomial Turing compressions for some graph problems parameterized by modular-width. CoRR abs/2201.04678 (2022)
31. Novick, M.B.: Fast parallel algorithms for the modular decomposition. Cornell University, Technical report (1989)
32. Schäfer, A., Komusiewicz, C., Moser, H., Niedermeier, R.: Parameterized computational complexity of finding small-diameter subgraphs. Optim. Lett. **6**(5), 883–891 (2012)
33. Tedder, M., Corneil, D., Habib, M., Paul, C.: Simpler linear-time modular decomposition via recursive factorizing permutations. In: Aceto, L., Damgård, I., Goldberg, L.A., Halldórsson, M.M., Ingólfsdóttir, A., Walukiewicz, I. (eds.) ICALP 2008. LNCS, vol. 5125, pp. 634–645. Springer, Heidelberg (2008). https://doi.org/10.1007/978-3-540-70575-8_52
34. Thomassé, S., Trotignon, N., Vuskovic, K.: A polynomial Turing-kernel for weighted independent set in bull-free graphs. Algorithmica **77**(3), 619–641 (2017)
35. Witteveen, J., Bottesch, R., Torenvliet, L.: A hierarchy of polynomial kernels. In: Catania, B., Královič, R., Nawrocki, J., Pighizzini, G. (eds.) SOFSEM 2019. LNCS, vol. 11376, pp. 504–518. Springer, Cham (2019). https://doi.org/10.1007/978-3-030-10801-4_39

Shortest Longest-Path Graph Orientations

Yuichi Asahiro[1], Jesper Jansson[2(✉)], Avraham A. Melkman[3], Eiji Miyano[4], Hirotaka Ono[5], Quan Xue[6], and Shay Zakov[7]

[1] Kyushu Sangyo University, Fukuoka, Japan
asahiro@is.kyusan-u.ac.jp
[2] Kyoto University, Kyoto, Japan
jj@i.kyoto-u.ac.jp
[3] Ben-Gurion University of the Negev, Be'er Sheva, Israel
melkmana@gmail.com
[4] Kyushu Institute of Technology, Iizuka, Japan
miyano@ai.kyutech.ac.jp
[5] Nagoya University, Nagoya, Japan
ono@nagoya-u.jp
[6] The University of Hong Kong, Hong Kong, China
quan.xue@connect.polyu.hk
[7] Ruppin Academic Center, Kfar Monash, Israel
Zakov.Shay@ruppin365.net

Abstract. We consider a graph orientation problem that can be viewed as a generalization of Minimum Graph Coloring. Our problem takes as input an undirected graph $G = (V, E)$ in which every edge $\{u, v\} \in E$ has two (potentially different and not necessarily positive) weights representing the lengths of its two possible directions (u, v) and (v, u), and asks for an orientation, i.e., an assignment of a direction to each edge of G, such that the length of a longest simple directed path in the resulting directed graph is minimized. A longest path in a graph is not always a maximal path when some edges have negative lengths, so the problem has two variants depending on whether *all* simple directed paths or maximal simple directed paths *only* are taken into account in the definition. We prove that the problems are NP-hard to approximate even if restricted to subcubic planar graphs, and develop fast polynomial-time algorithms for both problem variants for three classes of graphs: path graphs, cycle graphs, and star graphs.

Keywords: Algorithm · Computational complexity · Graph orientation · Graph coloring · Path graph · Cycle graph · Star graph

Y. Asahiro—Funded by KAKENHI grant number JP22K11915.
E. Miyano—Funded by KAKENHI grant number JP21K11755.

W. Wu and G. Tong (Eds.): COCOON 2023, LNCS 14422, pp. 141–154, 2024.
https://doi.org/10.1007/978-3-031-49190-0_10

1   Introduction

1.1   Background

An *orientation* of an undirected graph is an assignment of a direction to each of its edges. Based on this natural concept, many kinds of algorithmic problems with applications to telecommunications, scheduling, data structures for supporting fast adjacency queries in sparse graphs, bioinformatics, etc. can be defined (see, e.g., [2,3,5,7,8,18,20]). Certain graph orientation problems have turned out to be equivalent to well-known classic graph algorithmic problems, and modifying the definitions of these graph orientation problems may then yield new generalizations of their classic counterparts. To illustrate, recall the Minimum Vertex Cover problem and the Maximum Independent Set problem, which take as input an undirected graph $G = (V, E)$ and ask for a smallest possible subset $V' \subseteq V$ such that every edge in E is incident to at least one vertex in V', and a largest possible subset $V' \subseteq V$ such that no two vertices in V' are adjacent in G, respectively. As shown in [1], to orient an undirected graph while minimizing the number of vertices with outdegree at least 1 is in fact the Minimum Vertex Cover problem; by replacing the number "1" by a parameter W, one obtains a relaxed variant of Minimum Vertex Cover in which every vertex of the input graph is allowed to cover up to $W - 1$ of its incident edges without having to be placed in the output vertex cover. Similarly, maximizing the number of vertices that get outdegree at most W is the Maximum Independent Set problem when $W = 0$ (see [1]).

In this paper, we study a graph orientation problem that can be viewed as a generalization of another classic problem, namely Minimum Graph Coloring. Our starting point is the following problem, which we call Unweighted Shortest Longest-Path Orientation (USLPO): *Given an undirected, unweighted graph G, find an orientation of G that minimizes the length of a longest simple directed path.*

For any undirected graph G, let $H(G)$ and $\chi(G)$ denote the length of a longest simple directed path in an optimal solution to USLPO for G and the chromatic number of G, respectively. In the 1960s and 1970s, several researchers independently proved that $H(G)+1 = \chi(G)$ [10,13,19,21] and that this equality still holds when only acyclic orientations are allowed [7]. Note that since Minimum Graph Coloring is NP-hard [15], this immediately implies that USLPO is NP-hard. Moreover, known inapproximability results apply directly as well; e.g., Theorem 1.2 of [22] shows that Minimum Graph Coloring, and hence USLPO, cannot be approximated within a ratio of $n^{1-\varepsilon}$ for any constant $\varepsilon > 0$ in polynomial time, where n is the number of vertices in the input graph, unless P = NP. Also, $H(G)+1 = \chi(G)$ implies that even if USLPO is restricted to 4-regular planar graphs, it cannot be approximated within a ratio of $(3/2-\varepsilon)$ for any constant $\varepsilon > 0$ in polynomial time, unless P = NP, because it is an NP-complete problem to determine if a 4-regular planar graph G satisfies $\chi(G) \leq 3$ or $\chi(G) = 4$ [6].

1.2   New Results

We generalize USLPO (thus making the problem even harder) to *bi-weighted* edges, which means that every edge $\{u, v\}$ in G has two (potentially different and not necessarily positive) weights representing the lengths of its two possible directions (u, v) and (v, u). Our goal is then to determine whether the generalized problem, from here on simply referred to as Shortest Longest-Path Orientation (SLPO), becomes efficiently solvable in some special cases. Observe that in a directed graph, if some edge lengths are negative then a longest path is not necessarily a maximal path. For this reason, we consider two variants of the problem called $SLPO_s$ and $SLPO_m$, in which the longest path is taken, respectively, among *all* simple directed paths and among maximal simple directed paths *only*.

An undirected graph G is *subcubic* if every vertex in G has degree at most three. We first prove that $SLPO_s$ and $SLPO_m$ are NP-hard to approximate even if restricted to subcubic planar graphs. This result is important in view of the close connection between USLPO and Minimum Graph Coloring described above and the fact that the latter is polynomial-time solvable for subcubic graphs [11].

Motivated by the hardness of the general case, we then focus on special cases. Throughout the paper, n denotes the number of vertices in the input graph. As a first step, note that if G is a tree then one can root G in an arbitrarily selected vertex and apply dynamic programming over rooted subtrees: For every node v of G (in bottom-up order) and every possible triple (W_u, W_d, W_ℓ) of weights, check if there exists an orientation of the subtree rooted at v whose longest paths to, from, and not passing through v have lengths W_u, W_d, and W_ℓ, respectively. Since G has $\Theta(n^2)$ many paths, one can precompute the set of all possible path weights and use this set to prune the dynamic programming table, resulting in a polynomial-time algorithm. However, the degree of the polynomial will be large because the table can have $\Omega(n \cdot n^2 \cdot n^2 \cdot n^2) = \Omega(n^7)$ entries and computing any entry involving a node v may take $\Omega(n^6 \cdot \deg(v))$ time, where $\deg(v)$ is the number of children of v, if done by directly looking at the entries for (W_u, W_d, W_ℓ)-triples for each child of v. The rest of the paper is devoted to developing much faster algorithms for solving $SLPO_s$ and $SLPO_m$ on path graphs, cycle graphs, and star graphs. See the following table for a summary of our algorithms' time complexities.

Graph class	$SLPO_s$	$SLPO_m$	Section	Theorems
Tree	(Polynomial time)	(Polynomial time)	1.2	–
Path graph	$O(n)$	$O(n \log n)$	4	4 and 5
Cycle graph	$O(n)$	$O(n^2 \log n)$	5	6 and 7
Star graph	$O(n \log n)$	$O(n \log n)$	6	9 and 8

The paper is organized as follows. Section 2 defines $SLPO_s$ and $SLPO_m$ formally and lists some useful properties. The NP-hardness result is presented in Sect. 3. We describe our fast algorithms for path graphs in Sect. 4, for cycle graphs in Sect. 5, and for star graphs in Sect. 6. Due to space constraints, several correctness proofs have been omitted from the conference version of this paper. They will be available in the journal version.

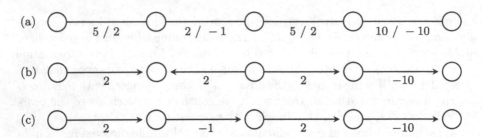

Fig. 1. An example. Let G be the undirected graph in (a). For every edge $\{u, v\}$, its two weights $w(u, v)$ and $w(v, u)$ are specified as $w(v, u) \, / \, w(u, v)$. The orientation in (b) satisfies $h_s(\tilde{G}) = h_m(\tilde{G}) = 2$, while the one in (c) satisfies $h_s(\tilde{G}) = 3$, $h_m(\tilde{G}) = -7$. They are optimal under h_s and h_m, respectively, so $H_s(G) = 2$ and $H_m(G) = -7$.

2   Preliminaries

An *orientation* of an undirected graph $G = (V, E)$ is a directed graph $\tilde{G}$ obtained by replacing each undirected edge $\{u, v\} \in E$ by either the directed edge (u, v) or the directed edge (v, u). Denote by $\mathcal{O}(G)$ the set of all orientations of G.

An *edge-bi-weighted graph* is an undirected graph $G = (V, E)$ in which every edge $\{u, v\} \in E$ has a pair of possibly nonpositive weights $w(u, v)$ and $w(v, u)$ associated with the two directions (u, v) and (v, u), respectively. The *length* of a directed edge (u, v) in an orientation of an edge-bi-weighted graph is $w(u, v)$, and the *length* of a path $\vec{P} = \langle v_1, \ldots, v_q \rangle$ is:

$$W(\vec{P}) = \sum_{k=1}^{q-1} w(v_k, v_{k+1}). \tag{1}$$

A path in a directed graph is said to be *maximal* if it is not contained in any path with more edges. As pointed out in Sect. 1.2, if some edge lengths are negative in a directed graph then a longest path is not necessarily a maximal path. We therefore employ two ways of measuring the cost of orienting a graph.

Definition 1. *Define the following two cost measures for an oriented graph $\tilde{G}$:*

$$h_s(\tilde{G}) = \max\{W(\vec{P}) \mid \vec{P} \text{ is a simple directed path in } \tilde{G}\}, \tag{2}$$

$$h_m(\tilde{G}) = \max\{W(\vec{P}) \mid \vec{P} \text{ is a maximal simple directed path in } \tilde{G}\}. \tag{3}$$

The corresponding cost functions for orienting an undirected graph are:

$$H_s(G) = \min\{h_s(\tilde{G}) \mid \tilde{G} \in \mathcal{O}(G)\}, \tag{4}$$

$$H_m(G) = \min\{h_m(\tilde{G}) \mid \tilde{G} \in \mathcal{O}(G)\}. \tag{5}$$

Figure 1 illustrates the difference between $H_s(G)$ and $H_m(G)$.

The two problem variants that we consider in this paper are the following:

The Shortest Longest-Path Orientation Problem, variants $SLPO_s$ & $SLPO_m$:

Input: An undirected, edge-bi-weighted graph G.

Output: An orientation $\widetilde{G}$ of G such that $h_s(\widetilde{G}) = H_s(G)$ (for $SLPO_s$) or $h_m(\widetilde{G}) = H_m(G)$ (for $SLPO_m$).

In other words, $SLPO_s$ and $SLPO_m$ ask for orientations of G such that the length of a longest simple path in the resulting directed graph is minimized, using the two alternative definitions of a "longest path".

We now state some easily verified properties of the cost measures. Note that the empty path participates in determining the value of $h_s(\widetilde{G})$, so $h_s(\widetilde{G}) \geq 0$ for any $\widetilde{G}$ and $H_s(G) \geq 0$. Another useful property of H_s is its monotonicity:

Lemma 1. *If G' is a subgraph of an edge-bi-weighted graph G then $H_s(G') \leq H_s(G)$.*

H_m is not monotone in general, but if all weights are nonnegative then H_m is monotone as well. In fact, if all weights are nonnegative then the definitions of H_m and H_s coincide:

Lemma 2. *If all weights of an edge-bi-weighted graph G are nonnegative then $h_s(\widetilde{G}) = h_m(\widetilde{G})$ for any orientation $\widetilde{G}$ of G. In particular, $H_s(G) = H_m(G)$, and $\widetilde{G}$ is an optimal orientation of G with respect to h_s if and only if it is optimal with respect to h_m.*

3 NP-Hardness for Subcubic Planar Graphs

The following theorem is proved by a polynomial-time reduction from PLANAR 3-SAT restricted to instances where each variable occurs in at most four clauses, which is known to be NP-hard [14]. For the details of the reduction, the reader is referred to the journal version of this paper.

Theorem 1. *$SLPO_s$ is NP-hard even if restricted to subcubic planar graphs where all edge weights belong to $\{0, 1, 2\}$.*

By Lemma 2, we immediately obtain:

Corollary 1. *$SLPO_m$ is NP-hard even if restricted to subcubic planar graphs where all edge weights belong to $\{0, 1, 2\}$.*

The reduction in the proof of Theorem 1 constructs a subcubic planar graph G such that if one could determine whether $H_s(G) = 2$ or $H_s(G) \geq 3$ then one would know if the instance of PLANAR 3-SAT is satisfiable or not. This yields:

Theorem 2. *If, for any constant $\varepsilon > 0$, there exists a polynomial-time $(3/2 - \varepsilon)$-approximation algorithm for $SLPO_s$ or $SLPO_m$ (even if restricted to subcubic planar graphs) then $P = NP$.*

4   Algorithms for Path Graphs

A *path graph* is an undirected, connected graph in which two vertices have degree 1 and all other vertices have degree 2. In Sect. 4.1 below, we first give a generic dynamic programming algorithm for finding the cost of an optimal orientation of an edge-bi-weighted path graph L. A straightforward implementation of the algorithm runs in $O(n^2)$ time. This high-level version makes no use of the details of the cost function, be it H_m or H_s. Then, Sects. 4.2 and 4.3 present faster implementations that take into account the specifics of H_s and H_m, thereby improving the time complexity to $O(n)$ and $O(n \log n)$, respectively.

4.1   A Generic Algorithm for Path Graphs

Given a path graph L with n vertices, assume without loss of generality that its vertices are numbered from 1 to n. For simplicity, we present an algorithm that computes the cost of an optimal orientation only; a corresponding optimal orientation can be found by adding a traceback step at no increase in the asymptotic time complexity. The algorithm is named BestOrientPath$_x(L)$ and its pseudocode is listed in Algorithm 1.

Algorithm 1: BestOrientPath$_x(L)$

Input: an edge-bi-weighted path graph L with n vertices
Output: an optimal orientation of L under H_x
1 **if** $n = 1$ **then**
2 $\quad \rfloor$ **return** 0;
3 $H_x^{\rightarrow}(1) = H_x^{\leftarrow}(1) = -\infty$;
4 **for** $j = 2$ *to* n **do**
5 $\quad \big\lvert \quad H_x^{\rightarrow}(j) = \min_{1 \le i < j} \max\{H_x^{\leftarrow}(i), h_x(\vec{L}_{i,j})\}$;
6 $\quad \big\lfloor \quad H_x^{\leftarrow}(j) = \min_{1 \le i < j} \max\{H_x^{\rightarrow}(i), h_x(\overleftarrow{L}_{i,j})\}$;
7 **return** $\min\{H_x^{\rightarrow}(n), H_x^{\leftarrow}(n)\}$;

We use the following notation.

- $L_{i,j}$ is the subgraph of L induced by the vertices $i, \ldots, j$.
- $\vec{L}_{i,j}$ and $\overleftarrow{L}_{i,j}$ are the oriented versions of $L_{i,j}$ in which all edges are directed towards larger numbered vertices (i.e., of the form $(i, i+1)$) and towards smaller numbered vertices (i.e., of the form $(i+1, i)$), respectively.
- The subscript x satisfies $x \in \{s, m\}$.
- $H_x^{\rightarrow}(i)$ is the value of an optimal orientation of $L_{1,i}$ under H_x assuming that edge $\{i-1, i\}$ is directed towards i. Analogously, $H_x^{\leftarrow}(i)$ is the value of an optimal orientation of $L_{1,i}$ under H_x assuming that edge $\{i-1, i\}$ is directed towards $i-1$. In particular, the cost of an optimal orientation of L under H_x is given by $\min\{H_x^{\rightarrow}(n), H_x^{\leftarrow}(n)\}$.

To compute $H_x^{\rightarrow}(n)$ and $H_x^{\leftarrow}(n)$, we compute $H_x^{\rightarrow}(j)$ and $H_x^{\leftarrow}(j)$ using dynamic programming from $j = 2$ up to $j = n$. The idea is to locate, in an optimal orientation of $L_{1,j}$, the largest vertex i at which there is a change in direction given that the last edge $\{j - 1, j\}$ has a specified direction.

Theorem 3. *BestOrientPath$_x$ finds the cost of an optimal orientation of L.*

4.2 Running Time Under Cost Function H_s

Here, we show that Algorithm BestOrientPath$_s$ can be made to run in $O(n)$ time.

The first issue is computing $h_s(\vec{L}_{i,j})$ and $h_s(\overleftarrow{L}_{i,j})$ for given $1 \leq i < j \leq n$. The weight of an edge $\{i, i+1\}$ when oriented as $(i, i+1)$ is denoted by $w(i, i+1)$ and when oriented as $(i + 1, i)$ by $w(i + 1, i)$. Equations (1) and (2) give:

$$h_s(\vec{L}_{i,j}) = \max \left\{ \sum_{t=i'}^{j'-1} w(t, t+1) \mid i \leq i' \leq j' \leq j \right\}. \qquad (6)$$

Computing $h_s(\vec{L}_{i,j})$ for a pair (i, j) is therefore an instance of the Range Maximum-sum Segment Online Query problem, RMSOQ for short [4]:

Problem 1 (Range Maximum-sum Segment Online Query).
Input to be preprocessed: A nonempty sequence $a_1, \ldots, a_n$ of real numbers.
Online query: respond to a query of the form $RMSOQ(i, j)$ by returning a pair of indices (i', j') that maximizes $\sum_{t=i'}^{j'} a_t$ over all $i \leq i' \leq j' \leq j$.

Chen and Chao [4] presented a method for answering each such query in constant time after A has been preprocessed in $O(n)$ time. This gives:

Lemma 3. *Suppose $w(i, i + 1), 1 \leq i < n$ and $w(i + 1, i), 1 \leq i < n$ have been preprocessed in linear time for the RMSOQ problem. After $H_s^{\leftarrow}(i)$ and $H_s^{\rightarrow}(i)$ have been computed for $1 \leq i < j$, each value of the form $\max\{H_s^{\leftarrow}(i), h_s(\vec{L}_{i,j})\}$ and $\max\{H_s^{\rightarrow}(i), h_s(\overleftarrow{L}_{i,j})\}$ appearing in steps 5 and 6 in iteration j can be evaluated in constant time.*

Next, we address the second issue: what is the time needed to find all minimum values in steps 5 and 6 in Algorithm BestOrientPath$_s(L)$? To answer the question, first define two $(n \times n)$-matrices $M^{\rightarrow}(i, j)$ and $M^{\leftarrow}(i, j)$ by:

- $M^{\rightarrow}[1, j] = h_s(\vec{L}_{1,j})$ and $M^{\leftarrow}[1, j] = h_s(\overleftarrow{L}_{1,j})$ for $2 \leq j \leq n$
- $M^{\rightarrow}[i, j] = M^{\leftarrow}[i, j] = \infty$ for $1 \leq j \leq i \leq n$
- $M^{\rightarrow}[i, j] = \max\{H_s^{\leftarrow}(i), h_s(\vec{L}_{i,j})\}$ and $M^{\leftarrow}[i, j] = \max\{H_s^{\rightarrow}(i), h_s(\overleftarrow{L}_{i,j})\}$ for $2 \leq i < j \leq n$

In these terms, the algorithm computes the minimum value in column j of the matrices $M^{\rightarrow}$ and $M^{\leftarrow}$ for $1 \leq j \leq n$. Two features of this computation deserve particular attention. The first is that before computing the minimum value in column j of the matrix $M^{\rightarrow}$, or $M^{\leftarrow}$, the minimum values of all previous columns $j' < j$, $H_s^{\leftarrow}[j']$, respectively $H_s^{\rightarrow}[j']$, must have been computed already. The second feature we want to point out is that it follows from Eq. (6) that both matrices $M^{\rightarrow}$ and $M^{\leftarrow}$ are of the form $M[i,j] = \max\{f(i), g(i,j)\}$, with g non-increasing in i and non-decreasing in j. This leads to the next proposition, where a matrix M is called *totally r-monotone* if it has the following property: If $M[i_1, j_1] \geq M[i_2, j_1]$ then $M[i_1, j_2] \geq M[i_2, j_2]$ for all $i_1 < i_2$ and $j_1 < j_2$.

Proposition 1. *If $g(i,j)$ is non-increasing in i and non-decreasing in j and $M(i,j) = \max\{f(i), g(i,j)\}$ then M is totally r-monotone.*

Proof. Suppose that $M[i_1, j_1] \geq M[i_2, j_1]$ holds. Then $\max\{f(i_1), g(i_1, j_1)\} \geq \max\{f(i_2), g(i_2, j_1)\}$. In particular, we have $f(i_2) \leq \max\{f(i_1), g(i_1, j_1)\} \leq \max\{f(i_1), g(i_1, j_2)\}$ since g is non-decreasing in j. Moreover, $g(i_1, j_2) \geq g(i_2, j_2)$. Hence $\max\{f(i_1), g(i_1, j_2)\} \geq \max\{f(i_2), g(i_2, j_2)\}$. □

To find all minimum values in steps 5 and 6 of the algorithm, we apply a solution to the following problem.

Problem 2 (Online Column Minima of a Totally r-Monotone Matrix). For $1 \leq j \leq n$, compute $H(j) = \min\{M(i,j) \mid 1 \leq i \leq n\}$, where M is totally r-monotone and the values of $H(j'), j' < j$ have to be computed before $M(i,j)$ can be evaluated.

Several linear-time online algorithms for solving Problem 2 exist [9,16,17].

Theorem 4. *BestOrientPath$_x$ with cost function H_s runs in $O(n)$ time.*

4.3   Running Time Under Cost Function H_m

We now make BestOrientPath$_m$ run in $O(n \log n)$ time. Combining (1) and (3):

$$h_m(\vec{L}_{i,j}) = W_j - W_i = \sum_{t=i}^{j-1} w(t, t+1), \tag{7}$$

where $W_1 = 0$ and $W_j = \sum_{t=1}^{j-1} w(t, t+1)$ for $j \geq 2$. A similar equality holds for $h_m(\overleftarrow{L}_{i,j})$. Consequently, finding the minimum values in steps 5 and 6 takes on a form different from the one obtained for H_s; more precisely, Eq. (7) shows that for $2 \leq j \leq n$, $H_m^{\rightarrow}(j) = \min_{1 \leq i < j} \max\{H_m^{\leftarrow}(i), W_j - W_i\}$. We split the points of the interval $1 \leq i < j$ into those that satisfy $H_m^{\leftarrow}(i) + W_i \geq W_j$ and the remainder. Since $H_m^{\leftarrow}(1) = -\infty$, the point $i = 1$ belongs to the latter. Also, $H_m^{\rightarrow}(2) = W_2$, and for $j \geq 3$, $H_m^{\rightarrow}(j) = \min\{M_1(j), M_2(j)\}$ with:

$$M_1(j) = \min_{2 \leq i < j}\{H_m^{\leftarrow}(i) \mid H_m^{\leftarrow}(i) + W_i \geq W_j\}, \tag{8}$$

$$M_2(j) = \min_{1 \leq i < j}\{W_j - W_i \mid H_m^{\leftarrow}(i) + W_i < W_j\}. \tag{9}$$

Next, we consider how to efficiently compute the minima in Eq. (8). Equation (9) can be handled similarly. We rephrase the problem as:

Problem 3 (Minima of Sequence Prefixes under Key-Bounds).

Given: A sequence KV of n pairs of the form $(key, value)$, and a sequence of lower bounds W_j, $1 \le j \le n$,

Compute: $min_j(W_j) = \min\{value \mid (key, value) \in Pre_j \text{ and } key \ge W_j\}$, for $1 \le j \le n$, where Pre_j is the prefix of length j of KV.

Problem 3 can be solved by red-black trees [12]; see the journal version for details. This gives:

Theorem 5. *BestOrientPath$_x$ with cost function H_m runs in $O(n \log n)$ time.*

5   Algorithms for Cycle Graphs

A *cycle graph* is an undirected, connected graph in which all vertices have degree 2. Given a cycle graph C on n vertices, we fix a numbering of its vertices by assigning the number 0 to an arbitrarily selected vertex and then assigning the numbers 1 through $n-1$ to the remaining vertices in the order that they are visited by a depth-first traversal starting at 0. Denote the weight of a directed edge $(i, i+1)$ by $w(i, i+1)$. When a vertex j with $j \ge n$ is referred to, it should be understood as referring to vertex $j \bmod n$. For example, $w(n-1, n) = w(n-1, 0)$.

5.1   An Algorithm for Cost Function H_s

Our algorithm for cycle graphs under H_s, BestOrientCycle$_s$, employs the following strategy. It first creates a special path graph L from the input cycle graph C and then applies BestOrientPath$_s$ from Sect. 4.2 to L to find an optimal orientation $\tilde{L}^*$ of L. An optimal orientation of C will always be one of five possible orientations that depends on the structure of $\tilde{L}^*$, so the algorithm simply checks which one of five conditions holds and outputs the corresponding orientation of C.

The pseudocode is given in Algorithm 2. It uses the following notation.

Definition 2. *Let C be a cycle graph, L the path graph constructed in step 1 of BestOrientCycle$_s(C)$, and $\tilde{L}^*$ an optimal orientation of L. The five orientations of C denoted by $\tilde{C}^{*1way}$, $\tilde{C}^{2+xx}$, $\tilde{C}^i$, $\tilde{C}^{rflip_i}$, and $\tilde{C}^{lflip_i}$ are defined as:*

- *$\tilde{C}^{*1way}$ is a one-way orientation of C whose cost, $OneWayCost_s$, is the least.*

- *For a cycle graph of odd length, $\tilde{C}^{2+xx}$ is the orientation of C obtained as follows: among all possible directed paths of length two find one, $\overrightarrow{L_2}$, whose weight is minimal; starting with this path, direct each successive edge of the cycle graph in the direction opposite to that of its predecessor.*

- $\widetilde{C}^i$ is the orientation of C obtained by copying from $\widetilde{L}^*$ the directions of the edges $\{k, k+1\}$, $i \le k < i+n$.

- $\widetilde{C}^{rflip_i}$ is created when $0 \le i \le 2n$ and $\widetilde{L}^*$ has the directed edges $(i, i+1), (i+1, i+2), (i+3, i+2)$. Its first directed edge is $(i+1, i)$, and the directions of the following edges $\{k, k+1\}$, $i+1 \le k < i+n$, are copied from $\widetilde{L}^*$.

- $\widetilde{C}^{lflip_i}$ is created when $n \le i \le 3n$ and $\widetilde{L}^*$ has the directed edges $(i, i-1), (i-1, i-2), (i-3, i-2)$. Its first directed edge is $(i-1, i)$, and the directions of the following edges $\{k, k-1\}$, $i-n < k \le i-1$, are copied from $\widetilde{L}^*$.

The algorithm's time complexity is linear because running BestOrientPath$_s$ in step 2 takes $O(n)$ time according to Theorem 4 and each of the other steps is easy to implement in $O(n)$ time.

Theorem 6. *BestOrientCycle$_s$ returns an orientation for a cycle graph that is optimal under the cost function H_s in $O(n)$ time.*

5.2   An Algorithm for Cost Function H_m

A simple method that works for cycle graphs under the cost function H_m is shown in Algorithm 3 (BestOrientCycle$_m$). It tries all ways of breaking the input cycle graph into a path graph by cutting one edge, applies BestOrientPath$_m$ to each such obtained path graph, and chooses one with the least cost. (This approach

---

Algorithm 2: BestOrientCycle$_s(C)$

Input: an edge-bi-weighted cycle graph C
Output: an optimal orientation of C under H_s

1 create a path graph L of length $3n$ by unrolling the cycle graph three times starting from vertex 0, numbering its vertices 0 to $3n$, and assigning each edge $\{i, i+1\}$ of L the same weights as edge $\{i, i+1\}$ of C;

2 let $\widetilde{L}^*$ be the oriented graph returned by BestOrientPath$_s(L)$;

3 **if** $h_s(\widetilde{L}^*) \ge OneWayCost_s$ **then** set $\widetilde{C}$ to $\widetilde{C}^{*1way}$;

4 **else if** *n is odd and every two consecutive edges in $\widetilde{L}^*$ have opposite directions* **then** set $\widetilde{C}$ to $\widetilde{C}^{2+xx}$;

5 **else if** *two edges $\{i, i+1\}$ and $\{i+n-1, i+n\}$ have opposite directions in $\widetilde{L}^*$* **then** set $\widetilde{C}$ to $\widetilde{C}^i$;

6 **else if** *there is an i, $0 \le i \le 2n$, such that the directed edges $(i, i+1), (i+1, i+2), (i+3, 1+2)$ appear in $\widetilde{L}^*$* **then** set $\widetilde{C}$ to $\widetilde{C}^{rflip_i}$;

7 **else** find i, $n \le i \le 3n$, such that $\widetilde{L}^*$ contains the directed edges $(i, i-1), (i-1, i-2), (i-3, i-2)$, and set $\widetilde{C}$ to $\widetilde{C}^{lflip_i}$;

8 **return** $\widetilde{C}$;

---

would also work for the cost function H_s, but the resulting time complexity would be worse than the one in Theorem 6.)

In the pseudocode, $\widetilde{C}^{*1way}$ denotes the one-way orientation of C whose cost, $OneWayCost_m$, is the least. Also, $L_{i+1,i-1}$ for $0 \leq i \leq n$ is the subgraph of C induced by vertices $i+1, i+2, \ldots, i-1$, with sums taken mod n.

Theorem 7. *Algorithm BestOrientCycle$_m(C)$ returns an optimal orientation of C in $O(n^2 \log n)$ time.*

Proof. Let $\widetilde{C}^*$ be an optimal orientation of C. If $\widetilde{C}^*$ is a directed cycle then an optimal solution will be found in step 1. Otherwise, $\widetilde{C}^*$ has at least one vertex i such that both edges $\{i-1, i\}$ and $\{i, i+1\}$ are oriented away from i. Consider the path graph L^i constructed in iteration i. Denote by $\widetilde{L}^i$ the orientation of L^i induced by breaking the cycle at vertex i, and note that $h_m(\widetilde{L}^i) = h_m(\widetilde{C}^*)$. Let $\widehat{L}^{i^*}$ be an optimal orientation of L^i. $\widehat{L}^{i^*}$ induces an orientation $\widetilde{C}$ of C by identifying i' with i''. Then $h_m(\widetilde{C}) = h_m(\widehat{L}^{i^*}) \leq h_m(\widetilde{L}^i) = h_m(\widetilde{C}^*)$. Since $h_m(\widetilde{C}) \geq h_m(\widetilde{C}^*)$, it follows that $h_m(\widehat{L}^{i^*}) = h_m(\widetilde{C}^*)$.

Because one call to BestOrientPath$_m$ takes $O(n \log n)$ time by Theorem 5, the algorithm runs in $O(n^2 \log n)$ time. □

Algorithm 3: BestOrientCycle$_m(C)$

Input: an edge-bi-weighted cycle graph C
Output: an optimal orientation of C under H_m

1 set $\widetilde{C}$ to $\widetilde{C}^{*1way}$ and set $BestCost$ to $OneWayCost_m$;
2 **for** $i = 0$ *to* $n-1$ **do**
3   construct a path graph L^i by adding two vertices i' and i'' and two edges $\{i+1, i'\}$ and $\{i-1, i''\}$ to $L_{i+1,i-1}$ with edge weights $w_{L^i}(i+1, i') = \infty$, $w_{L^i}(i', i+1) = w(i, i+1)$, $w_{L^i}(i-1, i'') = \infty$, and $w_{L^i}(i'', i-1) = w(i, i-1)$;
4   let $\widehat{L}^i$ be the graph returned by BestOrientPath$_m(L^i)$;
5   **if** $h_m(\widehat{L}^i) < BestCost$ **then**
6   set $\widetilde{C}$ to the orientation of C induced by $\widehat{L}^i$ and $BestCost$ to $h_m(\widehat{L}^i)$;

7 **return** $\widetilde{C}$;

---

6 Algorithms for Star Graphs

A *star graph* is a tree with exactly one internal node and at least two leaves. In this section, the internal node of a given star graph is denoted by c.

6.1 An Algorithm for Cost Function H_m

Algorithm BestOrientStar$_m$ for SLPO$_m$ on star graphs is given in Algorithm 4. It initially orients each edge so that it points in its lighter direction and then refines this solution to obtain an optimal orientation. To do so, it either flips edges that were initially pointing inwards only or edges that were initially pointing outwards only. The correctness of this approach is guaranteed by:

Lemma 4. *There is no optimal orientation of a star graph in which both the largest inward edge weight is less than $w(u_1, c)$ and the largest outward edge weight is less than $w(c, v_1)$.*

Proof. The cost of the initial orientation is $h_m(\widetilde{S}) = w(u_1, c) + w(c, v_1)$. For the purpose of obtaining a contradiction, suppose that there is an optimal orientation $\widetilde{S}^*$ with largest inward weight $w(x, c) < w(u_1, c)$ and largest outward weight $w(c, y) < w(c, v_1)$. Then $h_m(\widetilde{S}^*) = w(x, c) + w(c, y)$. The initial orientation implies that $w(c, u_1) > w(u_1, c)$, and $w(v_1, c) \geq w(c, v_1)$. Since $\{u_1, c\}$ is an outward edge in $\widetilde{S}^*$ and $\{v_1, c\}$ an inward edge, $h_m(\widetilde{S}^*) \geq w(v_1, c) + w(c, u_1) > w(u_1, c) + w(c, v_1) = h_m(\widetilde{S})$, which is impossible. ☐

Theorem 8. *BestOrientStar$_m$ solves SLPO$_m$ on star graphs in $O(n \log n)$ time.*

6.2 An Algorithm for Cost Function H_s

Based on the observation in the next lemma, we can use BestOrientStar$_m$ from Sect. 6.1 as a subroutine to obtain a solution for SLPO$_s$ on star graphs, as shown in Algorithm 5 (BestOrientStar$_s$).

Algorithm 4: BestOrientStar$_m(S)$

    Input: an edge-bi-weighted star graph S
    Output: an optimal orientation of S under H_m
1 orient each edge $\{u, c\}$ inwards to c if $w(u, c) < w(c, u)$, and outwards from c
    otherwise;
2 denote by $\widetilde{S}$ the resulting directed graph, by *BestCost* its cost,
3 by $E_{in} = \{(u_1, c), \ldots, (u_\ell, c)\}$ the list of its inward edges,
4 and by $E_{out} = \{(c, v_1), \ldots, (c, v_r)\}$ the list of its outward edges;
5 reorder each of E_{in} and E_{out} so that the weights of its edges are in
    non-increasing order;
6 set $\widetilde{S}'$ to $\widetilde{S}$;
7 **for** $k = 1$ *to* ℓ **do**
8   ⌊ flip the direction of edge (u_k, c) in $\widetilde{S}'$ and update *BestCost* if necessary;
9 set $\widetilde{S}'$ to $\widetilde{S}$;
10 **for** $k = 1$ *to* r **do**
11   ⌊ flip the direction of edge (c, v_k) in $\widetilde{S}'$ and update *BestCost* if necessary;
12 **return** an orientation whose cost is *BestCost*;

Lemma 5. *Suppose S has a vertex u with an edge to or from c with non-positive weight. Let S' be the star graph obtained by removing u from S. Then an optimal orientation of S is obtained by first optimally orienting S', and then adding to it the vertex u with its edge $\{u, c\}$ oriented in a direction with non-positive weight.*

Proof. The number of edges on a directed path in an oriented star graph is no more than two. An edge with non-positive weight is therefore either the first or the last edge on any directed path, and does not contribute to its cost.   □

Theorem 9. *$BestOrientStar_s$ solves $SLPO_s$ on star graphs in $O(n \log n)$ time.*

Algorithm 5: $\text{BestOrientStar}_s(S)$

Input: an edge-bi-weighted star graph S
Output: an optimal orientation of S under H_s
1   remove from S every edge with a direction of non-positive weight, and denote the resulting star graph S';
2   set $\widetilde{S}'$ to $\text{BestOrientStar}_m(S')$;
3   direct all edges removed in step 1 in a direction of non-positive weight and add them to $\widetilde{S}'$, and denote the resulting directed star graph $\widetilde{S}$;
4   **return** $\widetilde{S}$;

References

1. Asahiro, Y., Jansson, J., Miyano, E., Ono, H.: Graph orientations optimizing the number of light or heavy vertices. J. Graph Algorithms Appl. **19**(1), 441–465 (2015)
2. Asahiro, Y., Jansson, J., Miyano, E., Ono, H., Zenmyo, K.: Approximation algorithms for the graph orientation minimizing the maximum weighted outdegree. J. Comb. Optim. **22**(1), 78–96 (2011)
3. Borradaile, G., Iglesias, J., Migler, T., Ochoa, A., Wilfong, G., Zhang, L.: Egalitarian graph orientations. J. Graph Algorithms Appl. **21**(4), 687–708 (2017)
4. Chen, K.Y., Chao, K.M.: On the range maximum-sum segment query problem. Discret. Appl. Math. **155**(16), 2043–2052 (2007)
5. Chrobak, M., Eppstein, D.: Planar orientations with low out-degree and compaction of adjacency matrices. Theoret. Comput. Sci. **86**(2), 243–266 (1991)
6. Dailey, D.P.: Uniqueness of colorability and colorability of planar 4-regular graphs are NP-complete. Discret. Math. **30**(3), 289–293 (1980)
7. Deming, R.W.: Acyclic orientations of a graph and chromatic and independence numbers. J. Comb. Theory B **26**(1), 101–110 (1979)
8. Elberfeld, M., et al.: On the approximability of reachability-preserving network orientations. Internet Math. **7**(4), 209–232 (2011)
9. Galil, Z., Park, K.: A linear-time algorithm for concave one-dimensional dynamic programming. Inf. Process. Lett. **33**(6), 309–311 (1990)
10. Gallai, T.: On directed graphs and circuits. In: Theory of Graphs (Proceedings of the Colloquium held at Tihany 1966), pp. 115–118. Akadémiai Kiadó (1968)

11. Garey, M., Johnson, D.: Computers and Intractability - A Guide to the Theory of NP-Completeness. W. H. Freeman and Company, New York (1979)
12. Guibas, L.J., Sedgewick, R.: A dichromatic framework for balanced trees. In: Proceedings of the Nineteenth Annual Symposium on Foundations of Computer Science (FOCS 1978), pp. 8–21. IEEE Computer Society (1978)
13. Hasse, M.: Zur algebraischen Begründung der Graphentheorie. I. Mathematische Nachrichten **28**(5–6), 275–290 (1965)
14. Jansen, K., Müller, H.: The minimum broadcast time problem for several processor networks. Theoret. Comput. Sci. **147**(1–2), 69–85 (1995)
15. Karp, R.M.: Reducibility among combinatorial problems. In: Proceedings of Complexity of Computer Computations, pp. 85–103. The IBM Research Symposia Series, Plenum Press (1972)
16. Klawe, M.M.: A simple linear time algorithm for concave one-dimensional dynamic programming. Technical Report 89–16, Department of Computer Science, University of British Columbia (1989)
17. Larmore, L.L., Schieber, B.: On-line dynamic programming with applications to the prediction of RNA secondary structure. J. Algorithms **12**(3), 490–515 (1991)
18. Medvedovsky, A., Bafna, V., Zwick, U., Sharan, R.: An algorithm for orienting graphs based on cause-effect pairs and its applications to orienting protein networks. In: Crandall, K.A., Lagergren, J. (eds.) WABI 2008. LNCS, vol. 5251, pp. 222–232. Springer, Heidelberg (2008). https://doi.org/10.1007/978-3-540-87361-7_19
19. Roy, B.: Nombre chromatique et plus longs chemins d'un graphe. Revue française d'informatique et de recherche opérationnelle **1**(5), 129–132 (1967)
20. Venkateswaran, V.: Minimizing maximum indegree. Discret. Appl. Math. **143**(1–3), 374–378 (2004)
21. Vitaver, L.M.: Determination of minimal coloring of vertices of a graph by means of Boolean powers of the incidence matrix (in Russian). In: Proceedings of the USSR Academy of Sciences, vol. 147, pp. 758–759. Nauka (1967)
22. Zuckerman, D.: Linear degree extractors and the inapproximability of Max Clique and Chromatic Number. Theory Comput. **3**(1), 103–128 (2007)

Sink Location Problems in Dynamic Flow Grid Networks

Yuya Higashikawa, Ayano Nishii[(✉)], Junichi Teruyama, and Yuki Tokuni

Graduate School of Information Science, University of Hyogo, Kobe, Japan
{higashikawa,junichi.teruyama}@gsis.u-hyogo.ac.jp,
{ad23f047,af23v006}@guh.u-hyogo.ac.jp

Abstract. A *dynamic flow network* consists of a directed graph, where nodes called *sources* represent locations of evacuees, and nodes called *sinks* represent locations of evacuation facilities. Each source and each sink are given *supply* representing the number of evacuees and *demand* representing the maximum number of acceptable evacuees, respectively. Each edge is given *capacity* and *transit time*. Here, the capacity of an edge bounds the rate at which evacuees can enter the edge per unit time, and the transit time represents the time which evacuees take to travel across the edge. The *evacuation completion time* is the minimum time at which each evacuees can arrive at one of the evacuation facilities. Given a dynamic flow network without sinks, once sinks are located on some nodes or edges, the evacuation completion time for this sink location is determined. We then consider the problem of locating sinks to minimize the evacuation completion time, called the *sink location problem*. The problems have been given polynomial-time algorithms only for limited networks such as paths [1,2,10], cycles [1], and trees [4,9,18], but no polynomial-time algorithms are known for more complex network classes. In this paper, we prove that the 1-sink location problem can be solved in polynomial-time when an input network is a grid with uniform edge capacity and transit time.

Keywords: facility location problem · dynamic flow · quickest transshipment problem · evacuation problem · polynomial-time algorithm

1 Introduction

In recent years, natural disasters such as earthquakes, tsunamis and hurricanes have been occurring more frequently, and in Japan, the revision of the Basic

Supported by JSPS KAKENHI Grant Numbers 19H04068, 23H03349.
The full version of this paper [12] is available at the following link:
https://doi.org/10.48550/arXiv.2308.12651.

Act on Disaster Management[1] has accelerated the development of evacuation facilities, i.e., locating tsunami evacuation towers and setting tsunami evacuation buildings, which had been delayed. On the other hand, when considering evacuation planning for urgent large-scale disasters such as tsunamis, floods, and nuclear power plant accidents, one of the most important issues to be considered is the delay in evacuation time due to traffic congestion. In fact, it is known that many people were killed in the Great East Japan Earthquake due to delayed evacuation caused by traffic congestion [17].

To attack this issue, the *dynamic flow network model* proposed by Ford and Fulkerson [6] can be applied. Since the dynamic flow network model can handle the movement of people or objects over time, it is possible to develop an evacuation plan that quantitatively takes traffic congestion into account. A *dynamic flow network* consists of a directed graph, where nodes called *sources* represent locations of evacuees, and nodes called *sinks* represent locations of evacuation facilities. Each source and each sink are given *supply* representing the number of evacuees and *demand* representing the maximum number of acceptable evacuees, respectively. Each edge is given *capacity* and *transit time*. The capacity of an edge bounds the rate at which evacuees can enter the edge per unit time, and the transit time represents the time which evacuees take to travel across the edge.

One of the most fundamental problems in dynamic flow networks is the *quickest transshipment problem*. The objective of this problem is to compute the minimum time by which each evacuees can arrive at one of sinks, i.e., the *evacuation completion time*, and to find the optimal flow of evacuees that achieves the evacuation completion time. For the quickest transshipment problem, several strongly polynomial-time algorithms have been developed so far [13,16,19].

Given a dynamic flow network without sinks, once sinks are located on some nodes or edges, the evacuation completion time for this sink location is determined. We then consider the problem of locating sinks to minimize the evacuation completion time, called the *sink location problem*. The problems have been given polynomial-time algorithms only for limited networks such as paths [1,2,10], cycles [1], and trees [4,9,18], but no polynomial-time algorithms are known for more complex network classes. In this paper, we address the sink location problem on grid networks. A grid network can model an actual road network better than the networks studied so far, in that it consists of a number of cycles. We present the first polynomial-time algorithm for the 1-sink location problem in dynamic flow grid networks with uniform edge capacity and transit time.

We describe the basic ideas of this study. In our model, a sink can be located on a node or an edge. When a sink is located at a node, the evacuation completion time for that sink can be computed in polynomial-time using the algorithms [13, 16,19] for the quickest transshipment problem. Therefore, if one can compute in polynomial-time the optimal sink location for each edge, i.e., the sink location that minimizes the evacuation completion time over the points on the edge, the

---

[1] https://www.bousai.go.jp/taisaku/kihonhou/index.html (in Japanese).

optimal sink location for the entire network can be obtained in polynomial-time. In the following, we deal with the 1-sink location problem on a particular edge of the input network.

This paper is organized as follows. In Sect. 2, we introduce notations, models and known related results used throughout the paper. In Sect. 3, we will give a polynomial-time algorithm for the 1-sink location problem on a particular edge. Combining this algorithm and previous results, we will give a polynomial-time algorithm for the 1-sink location problem on a grid network. We conclude this paper in Sect. 4 with some discussions.

2 Preliminaries

2.1 Models

Let $\mathbb{R}$ and $\mathbb{R}_+$ denote the sets of real values and non-negative real values. A dynamic flow network $\mathcal{N}$ is given as a 6-tuple $\mathcal{N} = (G = (V, E), S^+, S^-, w, c, \tau)$, where $G = (V, E)$ is a directed graph with node set V and edge set E, $S^+ \subseteq V$ and $S^- \subseteq V$ are sets of sources and sinks, respectively, function $w\colon S^+ \cup S^- \to \mathbb{R}$ represents supply of a source or demand of a sink, function $c\colon E \to \mathbb{R}_+$ is capacity of an edge, and function $\tau\colon E \to \mathbb{R}_+$ is transit time of an edge. As for function w, for a source $s \in S^+$, $w(s)(\geq 0)$ represents supply of source s, and for a sink $s \in S^-$, $w(s)(< 0)$ represents demand of sink s. We define $w(X) := \sum_{s \in X} w(s)$ for a subset $X \subseteq S^+ \cup S^-$.

In this paper, we consider the problem of locating a sink in a dynamic flow grid network. Here, let us describe an input network $\mathcal{N}$ in our problem. First, a graph $G = (V, E)$ is a grid, where V consists of $n = N \times N$ nodes and E consists of bidirected edges in between each adjacent nodes of V (see Fig. 1). Furthermore, we assume that all edge capacities and transit times are uniform, thus c and τ are constant functions. In the sink location problem, input network $\mathcal{N}$ has no sinks and all nodes are treated as sources. Therefore, $\mathcal{N}$ is represented as $(G = (V, E), S^+ = V, S^- = \emptyset, w, c, \tau)$. In the following, we abuse c and τ as non-negative real constants.

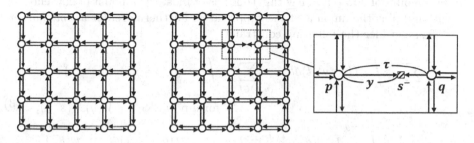

Fig. 1. The input network $\mathcal{N}$ **Fig. 2.** $\mathcal{N}(y)$

Let us consider the problem of locating a sink s^- in between two adjacent nodes $p, q \in V$ at a distance of y $(0 < y < \tau)$ from node p. For a given dynamic flow grid network $\mathcal{N}$ without sinks, we apply the following operations to $\mathcal{N}$:

1. Remove directed edges (p, q) and (q, p).
2. Add a sink s^- with demand $w(s^-) = -\sum_{v \in V} w(v)$.
3. Add directed edges (p, s^-) with capacity c and transit time y, and (q, s^-) with capacity c and transit time $\tau - y$.

The dynamic flow network obtained from the above operations is denoted as $\mathcal{N}(y)$ (see Fig. 2). Hereafter, we denote the sets of nodes and edges in $\mathcal{N}(y)$ as $V(y)$ and $E(y)$, respectively. In the following, we consider embedding network $\mathcal{N}(y)$ in a Cartesian coordinate system so that p and q are mapped at $(0, 0)$ and $(\tau, 0)$, respectively, and the other nodes are mapped at points $(i\tau, j\tau)$ with some integers i, j. For simplicity, a node mapped at $(i\tau, j\tau)$ is called node (i, j). We refer to the directions from $(0, 0)$ to $(0, 1)$, from $(0, 1)$ to $(0, 0)$, from $(0, 0)$ to $(1, 0)$, and from $(1, 0)$ to $(0, 0)$ as *upward*, *downward*, *rightward*, and *leftward*, respectively.

2.2   Evacuation Completion Time

A *dynamic flow* f is defined as a function $f \colon E(y) \times \mathbb{R}_+ \to \mathbb{R}_+$, where $f(e, \theta)$ represents the flow rate entering edge $e \in E(y)$ at time $\theta \in \mathbb{R}_+$. In this paper, we deal with dynamic flows in continuous-time model [5, 11]. Let us consider the following constraints for a dynamic flow f:

$$0 \le f(e, \theta) \le c \quad \text{for any } e \in E(y), \text{ and for any } \theta \in \mathbb{R}_+, \tag{1}$$

$$\sum_{(v,u) \in E(y)} \int_0^\theta f((v, u), t) dt - \sum_{(u,v) \in E(y)} \int_0^{\theta - \tau} f((u, v), t) dt \le w(v) \tag{2}$$

$$\text{for any } v \in S^+ (= V(y) \setminus \{s^-\}), \text{ and for any } \theta \in \mathbb{R}_+.$$

The constraints (1) and (2) are called the capacity constraint and the conserve constraint. The conserve constraint (2) means that for any time θ and any source v, the amount of flow out of v within time θ is at most the amount of flow entering v within θ plus the amount of the supply at v. Furthermore, for a *time horizon* $T \in \mathbb{R}_+$, consider the following constraint:

$$\sum_{(v,u) \in E(y)} \int_0^T f((v, u), t) dt - \sum_{(u,v) \in E(y)} \int_0^{T - \tau} f((u, v), t) dt = w(v)$$

$$\text{for any } v \in S^+ (= V(y) \setminus \{s^-\}), \quad (3)$$

$$\int_0^{T-y} f((p, s^-), t) dt + \int_0^{T-(\tau-y)} f((q, s^-), t) dt = -w(s^-).$$

The constraint (3) implies that for each node v, the net amount of flow accumulated at v within time T equals its supply or demand. For a time horizon T, if a dynamic flow f on $\mathcal{N}(y)$ satisfies the above constraints (1), (2), and (3), then f is said to be *feasible* for T. Moreover, a time horizon T for which a dynamic flow is feasible is said to be feasible. The evacuation completion time denotes the minimum time for which a feasible dynamic flow exists. Letting $\Theta^*(y)$ denote the evacuation completion time in $\mathcal{N}(y)$, the 1-sink location problem between p and q is formulated as follows:

$$\text{minimize} \quad \Theta^*(y)$$
$$\text{subject to} \quad 0 < y < \tau,$$

which we call SINK-LOCATION-ON-EDGE (for short, SLE). Next, we describe known properties of the evacuation completion time. Given a dynamic flow network $\mathcal{N}(y)$, for a subset of the sources $X \subseteq S^+$ and a time horizon $\theta \in \mathbb{R}_+$, let $o_\theta(X, y)$ be the maximum amount of dynamic flow that can reach the sink s^- from X within time horizon θ (more formal definition in our model will be given in (7)). The following theorem by Hoppe and Tardos [13] says a property of a feasible time horizon.

Theorem 1 ([13]). *Given a dynamic flow network $\mathcal{N}(y)$ for a real value y with $0 < y < \tau$, and a time horizon $\theta \in \mathbb{R}_+$, there exists a feasible dynamic flow for θ on $\mathcal{N}(y)$ if and only if*

$$\min\{o_\theta(X, y) - w(X) \mid X \subseteq S^+\} \geq 0 \qquad (4)$$

holds.

Therefore, the evacuation completion time $\Theta^*(y)$ in $\mathcal{N}(y)$ is the minimum θ that satisfies (4). Here, we define

$$\Theta(X, y) := \min\{\theta \mid o_\theta(X, y) - w(X) \geq 0\} \qquad (5)$$

and then, because $o_\theta(X, y) - w(X)$ is monotonically non-decreasing in $\theta \in \mathbb{R}_+$ for any $X \subseteq S^+$ and any y, we have

$$\Theta^*(y) = \max\{\Theta(X, y) \mid X \subseteq S^+\}. \qquad (6)$$

2.3 Residual Networks

The dynamic flow network $\mathcal{N}(y)$ can be treated as a static flow network $\overline{\mathcal{N}}(y)$ by considering the transit time as the cost of each edge. We say that a static flow $\bar{f}: E(y) \rightarrow \mathbb{R}_+$ is feasible with respect to $X \subseteq S^+$ if $\bar{f}$ satisfies the following conditions:

$$0 \leq \bar{f}(e) \leq c \quad \text{for any } e \in E(y),$$

$$\sum_{(v,u) \in E(y)} \bar{f}(v, u) - \sum_{(u,v) \in E(y)} \bar{f}(u, v) = 0 \quad \text{for any } v \in V(y) \setminus (X \cup \{s^-\}),$$

$$\sum_{v \in X} \sum_{(v,u) \in E(y)} \bar{f}(v, u) - \sum_{u \in \{p,q\}} \bar{f}(u, s^-) = 0.$$

When a feasible static flow $\bar{f}\colon E(y) \to \mathbb{R}_+$ is given to $\overline{\mathcal{N}}(y)$, the *residual network* with respect to $\bar{f}$, denoted by $\overline{\mathcal{N}}(y)_{\bar{f}}$, is constructed as follows. For each edge e with $\bar{f}(e) > 0$, a reverse edge $\overleftarrow{e}$ is added, and the edge set obtained by adding such reverse edges to $E(y)$ is denoted by $E(y)_{\bar{f}}$. The capacity of residual edges $c_{\bar{f}}\colon E(y)_{\bar{f}} \to \mathbb{R}_+$ is defined so that $c_{\bar{f}}(e) = c - \bar{f}(e)$ for original edges $e \in E(y)$, and $c_{\bar{f}}(\overleftarrow{e}) = \bar{f}(e)$ for reverse edges $\overleftarrow{e} \in E(y)_{\bar{f}} \setminus E(y)$. The cost of residual edges $\tau_{\bar{f}}\colon E(y)_{\bar{f}} \to \mathbb{R}$ is defined as follows: For edges (p, s^-), (s^-, p), (q, s^-), and (s^-, q), $\tau_{\bar{f}}(p, s^-) = y$, $\tau_{\bar{f}}(s^-, p) = -y$, $\tau_{\bar{f}}(q, s^-) = \tau - y$, and $\tau_{\bar{f}}(s^-, q) = -\tau + y$ respectively. For other original edges $e \in E(y)$, $\tau_{\bar{f}}(e) = \tau$, and for reverse edges $\overleftarrow{e} \in E(y)_{\bar{f}} \setminus E(y)$, $\tau_{\bar{f}}(\overleftarrow{e}) = -\tau$.

2.4   Envelope of Two-Dimensional Line Segments

Let $\mathcal{F}$ be a family of one-variable linear functions $f_1, \ldots, f_n$ defined on closed intervals $[a_1, b_1], \ldots, [a_n, b_n]$, respectively. We simply refer to such $\mathcal{F}$ as a set of two-dimensional line segments. Moreover, for each function $f \in \mathcal{F}$ defined on a closed interval $[a, b]$, we define $\underline{f}(x)$ as follows:

$$\underline{f}(x) := \begin{cases} f(x) \ (x \in [a, b]), \\ -\infty \ (x \notin [a, b]). \end{cases}$$

The *upper envelope* $U_{\mathcal{F}}$ of the two-dimensional line segments $\mathcal{F}$ is defined as

$$U_{\mathcal{F}}(x) := \max_{1 \le i \le n} \underline{f_i}(x)$$

that is defined on $\bigcup_{1 \le i \le n}[a_i, b_i]$. A point $(x', U_{\mathcal{F}}(x'))$ in the upper envelope $U_{\mathcal{F}}$ is called a *break point* if a function constituting $U_{\mathcal{F}}$ switches at x', that is, if

$$\lim_{x \to x'+0} \mathrm{argmax}_{1 \le i \le n} \underline{f_i}(x) \ne \lim_{x \to x'-0} \mathrm{argmax}_{1 \le i \le n} \underline{f_i}(x).$$

It is known that the following theorem holds for the upper envelope $U_{\mathcal{F}}$ of the set of two-dimensional line segments $\mathcal{F}$.

Theorem 2 ([7,8]). *Given a set $\mathcal{F}$ containing n two-dimensional line segments, the upper envelope $U_{\mathcal{F}}$ has at most $O(n\alpha(n))$ break points, and they can be computed in $O(n \log n)$ time. Here, the function α denotes the inverse function of the Ackermann function.*

In the upper envelope $U_{\mathcal{F}}$ of a two-dimensional line segments $\mathcal{F}$, a line segment connects each pair of adjacent break points. Then, this implies that some of break points gives the minimum value of $U_{\mathcal{F}}$, and we have the following corollary.

Corollary 1. *Given a two-dimensional line segments $\mathcal{F}$ containing n line segments, the break point that minimizes the upper envelope $U_{\mathcal{F}}$ can be computed in $O(n \log n)$ time.*

3 Sink Location on an Edge

In this section, we propose a polynomial-time algorithm for problem SLE. According to (6), if we can compute the function $\Theta(X, y)$ in y, for all $X \subseteq S^+$, then the objective function $\Theta^*(y)$ can be represented as the upper envelope of $\Theta(X, y)$ over $0 < y < \tau$. However, this brute-force method does not provide a polynomial-time algorithm since there are exponentially many subsets X. On the contrary, we give a family of $O(\sqrt{n})$ source sets that must contain X maximizing $\Theta(X, y)$ for any y with $0 < y < \tau$, and based on this property, provide a polynomial-time algorithm. In Sect. 3.1, we show properties of $\Theta(X, y)$, in Sect. 3.2, we show properties of X that maximizes $\Theta(X, y)$, and in Sect. 3.3, a polynomial-time algorithm is provided.

3.1 Properties of $\Theta(X, y)$

First, we introduce the notation used in the following. For any two sources $u, v \in S^+$, we define $d(u, v)$ as the length of the shortest path from u to v in the static flow network $\overline{N}(y)$. Note that the shortest path from u to v does not use edges (p, s^-) and (q, s^-) in $E(y)$. For any source set $X \subseteq S^+$ and source $v \in S^+$, we define $d(X, v)$ as the length of the shortest path from X to v in $\overline{N}(y)$, that is,

$$d(X, v) := \min\{d(x, v) \mid x \in X\}.$$

Here, we discuss the properties of $o_\theta(X, y)$ which represents the maximum amount of dynamic flow that can reach the sink s^- from X within time horizon θ. It is known that $o_\theta(X, y)$ is characterized by a minimum-cost flow from X to S^- in the static flow network $\overline{N}(y)$ [6]. Specifically, given $X \subseteq S^+ \cup S^-$, $o_\theta(X, y)$ as a function in θ can be computed by applying the successively shortest path algorithm [3,14,15] for the minimum-cost flow problem in the following manner. Initially, set $\bar{f}$ as a zero static flow, i.e., $\bar{f} : E(y) \to 0$. At each step i (≥ 1), execute the following two procedures.

1. Find the shortest (i.e., minimum-cost) path $P_i(X, y)$ from X to s^- in the current residual network $\overline{N}(y)_{\bar{f}}$. If there is no such path, then break the iteration.
2. Update $\bar{f}$ by adding a static flow of amount c along path $P_i(X, y)$.

Note that the number of paths obtained when the above iteration halts is exactly two. Letting $|P_1(X, y)|$ and $|P_2(X, y)|$ be the lengths of paths $P_1(X, y)$ and $P_2(X, y)$, respectively, $o_\theta(X, y)$ is represented by the following equation.

$$o_\theta(X, y) = \begin{cases} 0 & 0 \leq \theta < |P_1(X, y)|, \\ c(\theta - |P_1(X, y)|) & |P_1(X, y)| \leq \theta < |P_2(X, y)|, \\ c(\theta - |P_1(X, y)|) + c(\theta - |P_2(X, y)|) & |P_2(X, y)| \leq \theta. \end{cases}$$

$$(7)$$

Since (7) implies that $o_\theta(X, y)$ is a piecewise linear convex function in θ, it can be transformed into the following:

$$o_\theta(X, y) = \max\{0,\ c(\theta - |P_1(X, y)|), c(\theta - |P_1(X, y)|) + c(\theta - |P_2(X, y)|)\}. \quad (8)$$

Note that $P_1(X, y)$ clearly does not contain any reverse edge since $P_1(X, y)$ is a path in $\overline{\mathcal{N}}(y)$. In fact, not only that, there exists $P_2(X, y)$ not containing any reverse edge, which follows Lemma 1 below. Thus, it is not necessary to consider the residual network when calculating $o_\theta(X, y)$. Here, we define $d_1(X, y)$ as the minimum of the shortest path lengths that pass through p or q, respectively, from X to s^- on $\overline{\mathcal{N}}(y)$. We define $d_2(X, y)$ as the maximum of these shortest path lengths. We thus have

$$
\begin{aligned}
d_1(X, y) &:= \min\{d(X, p) + y, d(X, q) + \tau - y\}, \text{ and} \\
d_2(X, y) &:= \max\{d(X, p) + y, d(X, q) + \tau - y\}.
\end{aligned}
\tag{9}
$$

As for paths in $\overline{\mathcal{N}}(y)$ corresponding to $d_1(X, y)$ and $d_2(X, y)$, we present the following lemma. See the full version of this paper [12] for the proof.

Lemma 1. *For any source set $X \subseteq S^+$, there exists on $\overline{\mathcal{N}}(y)$ a pair of edge-disjoint paths from X to y whose lengths are $d_1(X, y)$ and $d_2(X, y)$, respectively.*

As mentioned above, by Lemma 1, we have $|P_1(X, y)| = d_1(X, y)$ and $|P_2(X, y)| = d_2(X, y)$. Substituting these into (8), $o_\theta(X, y)$ is represented by

$$
o_\theta(X, y) = \max\left\{0, c\left(\theta - d_1(X, y)\right), c\left(\theta - d_1(X, y)\right) + c\left(\theta - d_2(X, y)\right)\right\}.
\tag{10}
$$

By the definition of $d_1(X, y)$, we obtain

$$
\begin{aligned}
c\left(\theta - d_1(X, y)\right) &= c\left(\theta - \min\{d(X, p) + y, d(X, q) + \tau - y\}\right) \\
&= \max\left\{c(\theta - d(X, p) - y), c(\theta - d(X, q) - \tau + y)\right\}.
\end{aligned}
\tag{11}
$$

Since

$$
d_1(X, y) + d_2(X, y) = d(X, p) + d(X, q) + \tau
$$

holds by (9), we have

$$
c\left(\theta - d_1(X, y)\right) + c\left(\theta - d_2(X, y)\right) = c\left(2\theta - d(X, p) - d(X, q) - \tau\right).
\tag{12}
$$

Substituting (11) and (12) into (10), we obtain

$$
\begin{aligned}
o_\theta(X, y) = \max\big\{&0, c(\theta - d(X, p) - y), c(\theta - d(X, q) - \tau + y), \\
&c(2\theta - d(X, p) - d(X, q) - \tau)\big\}.
\end{aligned}
\tag{13}
$$

Here, we describe the properties of $\Theta(X, y)$. According to its definition (5), $\Theta(X, y)$ is the value of θ that satisfies $o_\theta(X, y) = w(X)$. Regarding $\Theta(X, y)$ as a function in y, we have the following theorem (see Fig. 3 for the shape of $\Theta(X, y)$ for $0 < y < \tau$).

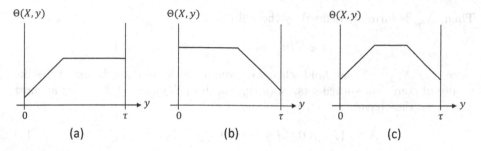

Fig. 3. The shape of $\Theta(X, y)$ for $0 < y < \tau$ in the cases where (a) $d(X, p) < d(X, q)$, (b) $d(X, p) > d(X, q)$, and (c) $d(X, p) = d(X, q)$, respectively.

Theorem 3. *For any source set $X \subseteq S^+$, $\Theta(X, y)$ is a piecewise linear function in y with $0 < y < \tau$ of at most three line segments, represented by*

$$\Theta(X, y) = \min\left\{ y + \frac{w(X)}{c} + d(X, p), -y + \tau + \frac{w(X)}{c} + d(X, q), \right.$$
$$\left. \frac{w(X)}{2c} + \frac{d(X, p) + d(X, q) + \tau}{2} \right\}.$$

Proof. Let us denote the linear functions in θ that compose $o_\theta(X, y)$ as

$$g_1(\theta) = c\left(\theta - d(X, p) - y\right),$$
$$g_2(\theta) = c\left(\theta - d(X, q) - \tau + y\right),$$
$$g_3(\theta) = c\left(2\theta - d(X, p) - d(X, q) - \tau\right).$$

By Eqs. (5) and (13), $\Theta(X, y)$ is the minimum value of θ among values that make each of $g_1(\theta)$, $g_2(\theta)$, and $g_3(\theta)$ equal to $w(X)$. By solving $g_1(\theta) = w(X)$ for θ, we obtain $\theta = y + \frac{w(X)}{c} + d(X, p)$. Similarly, $g_2(\theta) = w(X)$ when $\theta = -y + \tau + \frac{w(X)}{c} + d(X, q)$, and $g_3(\theta) = w(X)$ when $\theta = \frac{w(X)}{2c} + \frac{d(X,p)+d(X,q)+\tau}{2}$. Therefore, the minimum value among these three values is $\Theta(X, y)$. Additionally, with respect to y, each of the above θ is linear function. Hence, $\Theta(X, y)$ is a piecewise linear function in y consisting of at most three segments. □

3.2 Dominant Source Sets

For two subsets of S^+, X and X', X is said to *dominate* X' if $\Theta(X, y) \geq \Theta(X', y)$ holds for all y. Let us consider the sufficient condition for a source set X to dominate another source set X'. If $d(X, p) = d(X', p)$ and $d(X, q) = d(X', q)$, then functions $o_\theta(X, y)$ and $o_\theta(X', y)$ coincide according to (13). In this case, if $w(X) \geq w(X')$, then $\Theta(X, y) \geq \Theta(X', y)$ by definition (5). Among source sets X such that $d(X, p)$ and $d(X, q)$ remain the same respectively, one with the maximum $w(X)$ is called a *dominant source set*. For two integers i and j, let $X_{i,j}$ be the dominant source set such that $d(X_{i,j}, p) = i\tau$ and $d(X_{i,j}, q) = j\tau$.

Then, $X_{i,j}$ is formally defined by the following:

$$X_{i,j} := \{x \in V(y) \mid d(x,p) \geq i\tau, \; d(x,q) \geq j\tau\}.$$

Note that $X_{i,j} = \emptyset$ may hold when i or/and j are large enough. Let $\mathcal{X}$ be the family of dominant source sets. Note that both $d(X,p)$ and $d(X,q)$ are at most $2\tau\sqrt{n}$, we thus have

$$\mathcal{X} := \{X_{i,j} \mid 0 \leq i \leq 2\sqrt{n}, 0 \leq j \leq 2\sqrt{n}\} \setminus \{\emptyset\}, \tag{14}$$

and also

$$\Theta^*(y) = \max\{\Theta(X_{i,j}, y) \mid X_{i,j} \in \mathcal{X}\}. \tag{15}$$

Although the number of source sets in $\mathcal{X}$ seems to be $O(n)$ by definition (14), we can provide a more precise upper bound on the number of $X_{i,j}$. The following lemma implies that when considering $X_{i,j}$ for a fixed i, only the constant number of j are enough to be considered. See the full version of this paper [12] for the proof.

Lemma 2. *For any source set $X \subseteq S^+$, it holds in $\overline{\mathcal{N}}(y)$*

$$d(X,p) - d(X,q) = i\tau,$$

where i is one of integers $-3, -2, -1, 0, 1, 2, 3$.

Recall that $d(X,p)$ can take at most $O(\sqrt{n})$ values, and for a fixed value of $d(X,p)$, the number of values which $d(X,q)$ can take is at most seven by Lemma 2. Therefore, we obtain the following theorem.

Theorem 4. *The number of dominant source sets in $\mathcal{X}$ is $O(\sqrt{n})$.*

3.3 Algorithms

Our algorithm consists of three steps. First, we compute $w(X_{i,j})$ for all $X_{i,j} \in \mathcal{X}$. Next, we calculate the segments that compose $\Theta(X_{i,j}, y)$ for all $X_{i,j} \in \mathcal{X}$. Finally, we find the value of y that minimizes the upper envelope of the function family $\{\Theta(X_{i,j}, y) \mid X_{i,j} \in \mathcal{X}\}$.

Let us first describe a method to compute $w(X_{i,j})$ for all $X_{i,j} \in \mathcal{X}$. Since the number of nodes is n, we can compute $w(X_{i,j})$ in $O(n)$ time for any i, j. Moreover, we have $|\mathcal{X}| = O(\sqrt{n})$ by Theorem 4, so all $w(X_{i,j})$ can be computed in $O(n\sqrt{n})$ time. However, $w(X_{i,j})$ can be calculated more efficiently by using dynamic programming. See the following lemma.

Lemma 3. *Values $w(X_{i,j})$ for all $X_{i,j} \in \mathcal{X}$ can be computed in $O(n)$ time.*

Proof. Basically we calculate $w(X_{i,j})$ by subtracting the sum of supplies of $O(\sqrt{n})$ nodes from $w(X_{i-1,j})$ or $w(X_{i,j-1})$ as shown in Fig. 4, which can be done in $O(\sqrt{n})$ time. We then show the order of calculation of $w(X_{i,j})$ as follows. First of all, we calculate $w(X_{0,0})$ in $O(n)$ time, and successively updating from

$w(X_{0,0})$, we calculate $w(X_{1,0})$ and $w(X_{2,0})$. For $i = 0, 1, 2$, successively updating from $w(X_{i,0})$, we calculate $w(X_{i,1}), \ldots, w(X_{i,i+3})$. For $i \geq 3$, we update $w(X_{i-1,i-3})$ to $w(X_{i,i-3})$, and successively updating from $w(X_{i,i-3})$, calculate $w(X_{i,i-2}), \ldots, w(X_{i,i+3})$. Note that for each i, we consider only at most seven subsets $X_{i,j}$ according to Lemma 2.

By Theorem 4, the total number of subsets $X_{i,j}$ considered is $O(\sqrt{n})$, therefore, all values of $w(X_{i,j})$ can be computed in $O(n) + O(\sqrt{n}) \times O(\sqrt{n}) = O(n)$ time. □

In the second step, we calculate the segments that compose $\Theta(X_{i,j}, y)$ for all $X_{i,j} \in \mathcal{X}$. Recall that $d(X_{i,j}, p) = i\tau$ and $d(X_{i,j}, q) = j\tau$. By Theorem 3, $\Theta(X_{i,j}, y)$ is piecewise linear function represented by

$$\Theta(X_{i,j}, y) = \min\left\{ y + \frac{w(X_{i,j})}{c} + i\tau, -y + \tau + \frac{w(X_{i,j})}{c} + j\tau, \right.$$
$$\left. \frac{w(X_{i,j})}{2c} + \frac{i+j+1}{2}\tau \right\}. \tag{16}$$

Using $w(X_{i,j})$ obtained at the first step, $\Theta(X_{i,j}, y)$ can be calculated in $O(1)$ time. According to Eq. (15), $\Theta^*(y)$ is the upper envelope of the function family $\{\Theta(X_{i,j}, y) \mid X_{i,j} \in \mathcal{X}\}$. By Eq. (16), $\Theta(X_{i,j}, y)$ consists of at most three line segments, hence by Theorem 4, $\Theta^*(y)$ is determined by the upper envelope of $O(\sqrt{n})$ line segments, which can be computed in $O(\sqrt{n} \log n)$ time by Corollary 1. Thus, the value of y that minimizes $\Theta^*(y)$ can also be found in $O(\sqrt{n} \log n)$ time in the third step.

Summarizing the above discussions, the first step requires $O(n)$ time, the second step requires $O(\sqrt{n})$ time, and the third step requires $O(\sqrt{n} \log n)$ time. Therefore, we obtain the following theorem.

Theorem 5. *Given a dynamic flow grid network of $N \times N$ nodes (without any sink) with uniform edge capacity and transit time, the 1-sink location problem on a particular edge can be solved in $O(n)$ time.*

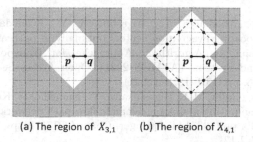

(a) The region of $X_{3,1}$ (b) The region of $X_{4,1}$

Fig. 4. The regions of $X_{3,1}$ and $X_{4,1}$ consist of nodes within the gray area respectively. The nodes located on the dashed line in (b) represent the difference between $X_{3,1}$ and $X_{4,1}$.

Solving problem SLE for each edge, the 1-sink location problem for all edges can be solved in $O(n^2)$ time. In general, the evacuation completion time can be computed in $\tilde{O}(m^2 k^4 + m^2 nk)$ time using the algorithm developed by Schlöter et al. [19]. Here, m is the number of edges and k is the total number of sources and sinks, and the $\tilde{O}$ omits the logarithmic factor from Big O notation. Adapting this algorithm to our problem, the evacuation completion time for each node can be computed in $\tilde{O}(n^6)$ time since $m = O(n)$ and $k = O(n)$. If we focus on the sink location on only nodes, by applying this algorithm to each node, the optimal 1-sink location can be computed in $\tilde{O}(n^7)$ time. Therefore, we obtain the following corollary.

Corollary 2. *Given a dynamic flow grid network of $N \times N$ nodes (without any sink) with uniform edge capacity and transit time, the 1-sink location problem can be solved in $\tilde{O}(n^7)$ time.*

4 Conclusion

The sink location problem is a kind of evacuation problems that seeks to determine the location of sinks that minimize the time it takes for each evacuee to arrive at one of the evacuation facilities. In this paper, we proposed a polynomial-time algorithm for the 1-sink location problem on a dynamic flow grid network with uniform edge capacity and transit time. This is the first polynomial-time algorithm for networks that contain a number of cycles. We remark that our provided approach can be extended to the 1-sink location problem on $M \times N$ grid networks, which gives a polynomial-time algorithm even when $M \neq N$.

It would be interesting to develop polynomial-time algorithms for the sink location problems on more complex networks, especially, a grid networks with multiple number of capacities and transit times. In addition, from the viewpoint of real world applications, developing polynomial-time algorithms for grid networks with holes or the case of multiple sinks is a future challenge.

References

1. Benkoczi, R., Bhattacharya, B., Higashikawa, Y., Kameda, T., Katoh, N., Teruyama, J.: Locating evacuation centers optimally in path and cycle networks. In: 21st Symposium on Algorithmic Approaches for Transportation Modelling, Optimization, and Systems (ATMOS 2021). Open Access Series in Informatics (OASIcs), vol. 96, pp. 13:1–13:19 (2021)
2. Bhattacharya, B., Golin, M.J., Higashikawa, Y., Kameda, T., Katoh, N.: Improved algorithms for computing k-sink on dynamic flow path networks. In: WADS 2017. LNCS, vol. 10389, pp. 133–144. Springer, Cham (2017). https://doi.org/10.1007/978-3-319-62127-2_12
3. Busacker, R.G., Gowen, P.J.: A procedure for determining a family of minimum-cost network flow patterns. Technical report, RESEARCH ANALYSIS CORP MCLEAN VA (1960)

4. Chen, D., Golin, M.: Sink evacuation on trees with dynamic confluent flows. In: 27th International Symposium on Algorithms and Computation (ISAAC 2016). Schloss Dagstuhl-Leibniz-Zentrum fuer Informatik (2016)
5. Fleischer, L., Tardos, É.: Efficient continuous-time dynamic network flow algorithms. Oper. Res. Lett. **23**(3–5), 71–80 (1998)
6. Ford, L.R., Jr., Fulkerson, D.R.: Constructing maximal dynamic flows from static flows. Oper. Res. **6**(3), 419–433 (1958)
7. Hart, S., Sharir, M.: Nonlinearity of Davenport-Schinzel sequences and of generalized path compression schemes. Combinatorica **6**(2), 151–177 (1986)
8. Hershberger, J.: Finding the upper envelope of n line segments in $O(n \log n)$ time. Inf. Process. Lett. **33**(4), 169–174 (1989)
9. Higashikawa, Y., Golin, M.J., Katoh, N.: Minimax regret sink location problem in dynamic tree networks with uniform capacity. In: Pal, S.P., Sadakane, K. (eds.) WALCOM 2014. LNCS, vol. 8344, pp. 125–137. Springer, Cham (2014). https://doi.org/10.1007/978-3-319-04657-0_14
10. Higashikawa, Y., Golin, M.J., Katoh, N.: Multiple sink location problems in dynamic path networks. Theoret. Comput. Sci. **607**, 2–15 (2015)
11. Higashikawa, Y., Katoh, N.: A survey on facility location problems in dynamic flow networks. Rev. Socionetw. Strateg. **13**(2), 163–208 (2019)
12. Higashikawa, Y., Nishii, A., Teruyama, J., Tokuni, Y.: Sink location problems in dynamic flow grid networks. arXiv e-prints arXiv:2308.12651 (2023). https://doi.org/10.48550/arXiv.2308.12651
13. Hoppe, B., Tardos, É.: The quickest transshipment problem. Math. Oper. Res. **25**(1), 36–62 (2000)
14. Iri, M.: A new method of solving transportation-network problems. J. Oper. Res. Soc. Jpn. **3**(1), 2 (1960)
15. Jewell, W.S.: Optimal flow through networks. In: Operations Research, vol. 6, pp. 633–633 (1958)
16. Kamiyama, N.: Discrete newton methods for the evacuation problem. Theoret. Comput. Sci. **795**, 510–519 (2019)
17. Kinjo, K., Matsumoto, M.: An analysis of the impact of individuals and society to evacuation based on the Great East Japan Earthquake survey (in Japanese). Jpn. Soc. Urbanol. Annu. Rep. **45**, 104–112 (2011)
18. Mamada, S., Uno, T., Makino, K., Fujishige, S.: An $O(n \log^2 n)$ algorithm for the optimal sink location problem in dynamic tree networks. Discret. Appl. Math. **154**(16), 2387–2401 (2006)
19. Schlöter, M., Skutella, M., Van Tran, K.: A faster algorithm for quickest transshipments via an extended discrete Newton method. In: Proceedings of the 2022 Annual ACM-SIAM Symposium on Discrete Algorithms (SODA), pp. 90–102. SIAM (2022)

List 3-Coloring on Comb-Convex and Caterpillar-Convex Bipartite Graphs

Banu Baklan Şen[1] , Öznur Yaşar Diner[1] , and Thomas Erlebach[2]($\boxtimes$)

[1] Computer Engineering Department, Kadir Has University, Istanbul, Turkey
oznur.yasar@khas.edu.tr
[2] Department of Computer Science, Durham University, Durham, UK
thomas.erlebach@durham.ac.uk

Abstract. Given a graph $G = (V, E)$ and a list of available colors $L(v)$ for each vertex $v \in V$, where $L(v) \subseteq \{1, 2, \ldots, k\}$, LIST k-COLORING refers to the problem of assigning colors to the vertices of G so that each vertex receives a color from its own list and no two neighboring vertices receive the same color. The decision version of the problem LIST 3-COLORING is NP-complete even for bipartite graphs, and its complexity on comb-convex bipartite graphs has been an open problem. We give a polynomial-time algorithm to solve LIST 3-COLORING for caterpillar-convex bipartite graphs, a superclass of comb-convex bipartite graphs. We also give a polynomial-time recognition algorithm for the class of caterpillar-convex bipartite graphs.

1 Introduction

Graph coloring is the problem of assigning colors to the vertices of a given graph in such a way that no two adjacent vertices have the same color. *List coloring* [16,27] is a generalization of graph coloring in which each vertex must receive a color from its own list of allowed colors. In this paper, we study the list coloring problem with a fixed number of colors in subclasses of bipartite graphs. We give a polynomial-time algorithm for the list 3-coloring problem for caterpillar-convex bipartite graphs, a superclass of comb-convex bipartite graphs. We also give a polynomial-time recognition algorithm for the class of caterpillar-convex bipartite graphs. Our results resolve the open question regarding the complexity of list 3-coloring for comb-convex bipartite graphs stated in [5,6].

We consider finite simple undirected graphs $G = (V, E)$ with vertex set V and edge set E. By $N_G(v)$ (or by $N(v)$ if the graph is clear from the context) we denote the neighborhood of v in G, i.e., the set of vertices that are adjacent to v. A *k-coloring* of G is a labeling that assigns colors to the vertices of G from the set $[k] = \{1, 2, \ldots, k\}$. A coloring is *proper* if no two adjacent vertices have the same color. A *list assignment* of a graph $G = (V, E)$ is a mapping $\mathcal{L}$ that assigns each vertex $v \in V$ a list $\mathcal{L}(v) \subseteq \{1, 2, \ldots\}$ of admissible colors for v. When $\mathcal{L}(v) \subseteq [k] = \{1, 2, \ldots k\}$ for every $v \in V$ we say that $\mathcal{L}$ is a k-list assignment of G. The total number of available colors is bounded by k in a k-list assignment.

W. Wu and G. Tong (Eds.): COCOON 2023, LNCS 14422, pp. 168–181, 2024.
https://doi.org/10.1007/978-3-031-49190-0_12

On the other hand, when the only restriction is that $|\mathcal{L}(v)| \leq k$ for every $v \in V$, then we say that $\mathcal{L}$ is a list k-assignment of G. *List coloring* is the problem of deciding, for a given graph $G = (V, E)$ and list assignment $\mathcal{L}$, whether G has a proper coloring where each vertex v receives a color from its list $\mathcal{L}(v)$. If $\mathcal{L}$ is a k-list assignment for a fixed value of k, the problem becomes the list k-coloring problem:

LIST k-COLORING (LI k-COL)
Instance: A graph $G = (V, E)$ and a k-list assignment $\mathcal{L}$.
Question: Does G have a proper coloring where each vertex v receives a color from its list $\mathcal{L}(v)$?

If $\mathcal{L}$ is a list k-assignment instead of a k-list assignment, the problem is called k-LIST COLORING.

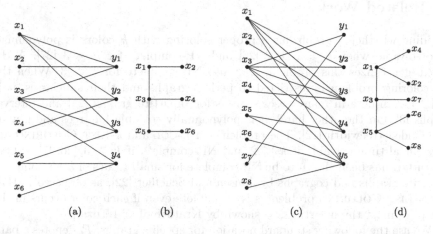

Fig. 1. (a) A comb-convex bipartite graph G_1, (b) a comb representation of G_1, (c) a caterpillar-convex bipartite graph G_2, and (d) a caterpillar representation for G_2.

The classes of bipartite graphs of interest to us are defined via a convexity condition for the neighborhoods of the vertices on one side of the graph with respect to a tree defined on the vertices of the other side. The following types of trees are relevant here: A *star* is a tree of diameter 2. A *comb* is a tree that consists of a chordless path P, called the *backbone*, with a single leaf neighbor attached to each backbone vertex [10]. A *caterpillar* is a tree that consists of a chordless path P, called the *backbone*, with an arbitrary number (possibly zero) of leaf vertices attached to each vertex on P. Note that if a caterpillar has exactly one leaf vertex attached to each vertex on P, then that caterpillar is a comb.

A bipartite graph $G = (X \cup Y, E)$ is called a *star-convex* (or *comb-convex*, or *caterpillar-convex*) bipartite graph if a star (or comb, or caterpillar) $T = (X, F)$ can be defined on X such that for each vertex $y \in Y$, its neighborhood $N_G(y)$ induces a subtree of T. The star (or comb, or caterpillar) $T = (X, F)$ is then called a *star representation* (or *comb representation*, or *caterpillar representation*) of G.

Figure 1 shows an example of a comb-convex bipartite graph and its comb representation, and a caterpillar-convex bipartite graph and its caterpillar representation. Both the comb (in part (b)) and the caterpillar (in part (d)) have the path $P = x_1 x_3 x_5$ as backbone.

The remainder of this paper is organized as follows. In Sect. 2, we discuss related work. In Sect. 3, we give a polynomial-time algorithm for LI 3-COL for caterpillar-convex bipartite graphs (and thus also for comb-convex bipartite graphs). In Sect. 4, we give a polynomial-time recognition algorithm for caterpillar-convex bipartite graphs. In Sect. 5, we give concluding remarks. Proofs omitted due to space restrictions can be found in the full version [1].

2   Related Work

Deciding whether a graph has a proper coloring with k colors is polynomial-time solvable when $k = 1$ or 2 [26] and NP-complete for $k \geq 3$ [27]. As LI k-COL generalizes this problem, it is also NP-complete for $k \geq 3$. When the list coloring problem is restricted to perfect graphs and their subclasses, it is still NP-complete in many cases such as for bipartite graphs [25] and interval graphs [3]. On the other hand, it is polynomially solvable for trees and graphs of bounded treewidth [22]. The problems LI k-COL and k-LIST COLORING are polynomial-time solvable if $k \leq 2$ and NP-complete if $k \geq 3$ [26,27]. k-LIST COLORING has been shown to be NP-complete for small values of k for complete bipartite graphs and cographs by Jansen and Scheffler [22], as observed in [17]. The 3-LIST COLORING problem is NP-complete even if each color occurs in the lists of at most three vertices, as shown by Kratochvil and Tuza [24].

We use the following standard notation for specific graphs: P_t denotes a path with t vertices; K_t denotes a clique with t vertices; $K_{\ell,r}$ denotes a complete bipartite subgraph with parts of sizes ℓ and r; $K_{1,s}^1$ denotes the 1-subdivision of $K_{1,s}$ (i.e., every edge $e = \{u,v\}$ of $K_{1,s}$ is replaced by two edges $\{u, w_e\}$ and $\{w_e, v\}$, where w_e is a new vertex); and $sP_1 + P_5$ is the disjoint union of s isolated vertices and a P_5. LI k-COL is known to be NP-complete even for $k = 3$ within the class of 3-regular planar bipartite graphs [23]. On the positive side, for fixed $k \geq 3$, LI k-COL is polynomially solvable for P_5-free graphs [19]. LI 3-COL is polynomial for P_6-free graphs [8] and for P_7-free graphs [4]. LI 3-COL is polynomial-time solvable for $(K_{1,s}^1, P_t)$-free graphs for every $s \geq 1$ and $t \geq 1$ [11]. LI k-COL is polynomial-time solvable for $(sP_1 + P_5)$-free graphs, which was proven for $s = 0$ by Hoàng et al. [19] and for every $s \geq 1$ by Couturier et al. [12].

An overview of complexity results for LI k-COL in some subclasses of bipartite graphs is shown in Fig. 2. The computational complexity of LI 3-COL for chordal bipartite graphs has been stated as an open problem in 2015 [21] and has been of interest since then [13]. In [13] a partial answer is given to this question by showing that LI 3-COL is polynomial in the class of biconvex bipartite graphs and convex bipartite graphs. LI 3-COL is solvable in polynomial time when it is

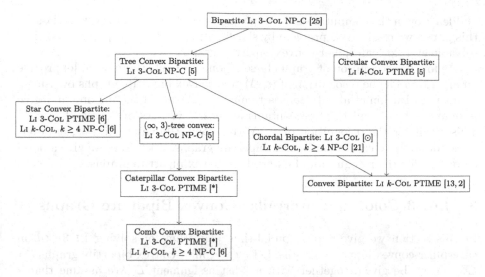

Fig. 2. Computational complexity results for LI k-COL on subclasses of bipartite graphs. [*] refers to this paper, [⊙] refers to an open problem, NP-C denotes NP-complete problem, PTIME denotes polynomial-time solvable problem.

restricted to graphs with all connected induced subgraphs having a multichain ordering [15]. This result can be applied to permutation graphs and interval graphs. In [13], it is shown that connected biconvex bipartite graphs have a multichain ordering, implying a polynomial time algorithm for LI 3-COL on this graph class. They also provide a dynamic programming algorithm to solve LI 3-COL in the class of convex bipartite graphs and show how to modify the algorithm to solve the more general LI H-COL problem on convex bipartite graphs. The computational complexity of LI 3-COL for P_8-free bipartite graphs is open [3]. Even the restricted case of LI 3-COL for P_8-free chordal bipartite graphs is open. Golovach et al. [17] survey results for LI k-COL on H-free graphs in terms of the structure of H.

So-called *width parameters* play a crucial role in algorithmic complexity. For various combinatorial problems, it is possible to find a polynomial-time solution by exploiting bounded width parameters such as mim-width, sim-width and clique-width. Given a graph class, it is known that when mim-width is bounded, then LI k-COL is polynomial-time solvable [7]. Brettell et al. [7] proved that for every $r \geq 1, s \geq 1$ and $t \geq 1$, the mim-width is bounded and quickly computable for $(K_r, K_{1,s}^1, P_t)$-free graphs. This result further implies that for every $k \geq 1, s \geq 1$ and $t \geq 1$, LI k-COL is polynomial-time solvable for $(K_{1,s}^1, P_t)$-free graphs. Most recently, Bonomo-Braberman et al. [5] showed that mim-width is unbounded for star-convex and comb-convex bipartite graphs. On the other hand, LI 3-COL is polynomial-time solvable for star-convex bipartite graphs whereas LI k-COL is NP-complete for $k \geq 4$ [6]. Furthermore, Bonomo-Braberman et al. [6] show that for comb-convex bipartite graphs, LI k-COL remains NP-complete for $k \geq 4$

and leave open the computational complexity of LI 3-COL for this graph class. In this paper, we resolve this problem by showing that LI 3-COL is polynomial-time solvable even for caterpillar-convex bipartite graphs.

As for the recognition of graph classes, Bonomo-Braberman et al. [6] provide an algorithm for the recognition of (t, Δ)-tree convex bipartite graphs by using a result from Buchin et al. [9]. Here, a tree is a (t, Δ)-tree if the maximum degree is bounded by Δ and the tree contains at most t vertices of degree at least 3. This result for the recognition of (t, Δ)-tree convex bipartite graphs, however, does not apply to caterpillar-convex bipartite graphs. Therefore, we give a novel algorithm for the recognition of caterpillar-convex bipartite graphs.

3   List 3-Coloring Caterpillar-Convex Bipartite Graphs

In this section we give a polynomial-time algorithm for solving LI 3-COL in caterpillar-convex bipartite graphs. Let a caterpillar-convex bipartite graph $G = (X \cup Y, E)$ be given, together with a 3-list assignment $\mathcal{L}$. We assume that a caterpillar $T = (X, F)$ is also given, where $N_G(y)$ induces a subtree of T for each $y \in Y$. If the caterpillar is not provided as part of the input, we can compute one in polynomial time using the recognition algorithm that we present in Sect. 4.

Let T consist of a backbone B with vertices $b_1, b_2, \ldots, b_n$ (in that order) and a set of leaves $L(b_i)$, possibly empty, attached to each $b_i \in B$. We use L to denote the set of all leaves, i.e., $L = \bigcup_{i=1}^n L(b_i)$. Furthermore, for any $1 \leq i \leq j \leq n$, we let $B_{i,j} = \{b_i, b_{i+1}, \ldots, b_j\}$ and $L_{i,j} = \bigcup_{k=i}^j L(b_k)$.

The idea of the algorithm is to define suitable subproblems that can be solved in polynomial time, and to obtain the overall coloring as a combination of solutions to subproblems. Roughly speaking, the subproblems consider stretches of the backbone in which all backbone vertices are assumed to be assigned the same color in a proper list 3-coloring. More precisely, a subproblem $SP(i, j, c_1, c_2, c_3)$ is specified via two values i, j with $1 \leq i \leq j \leq n$ and three colors c_1, c_2, c_3 with $c_1 \neq c_2$ and $c_2 \neq c_3$ where $c_i \in [3]$, for $i = 1, 2, 3$. Hence, there are $O(n^2)$ subproblems.

The subproblem $S = SP(i, j, c_1, c_2, c_3)$ is concerned with the subgraph G_S of G induced by $B_{i-1,j+1} \cup L_{i,j} \cup \{y \in Y \mid N(y) \cap (B_{i,j} \cup L_{i,j}) \neq \emptyset\}$. It assumes that color c_1 is assigned to b_{i-1}, color c_2 to the backbone vertices from b_i to b_j, and color c_3 to b_{j+1}. See Fig. 3 for an illustration of $SP(i, j, 2, 1, 2)$. Solving the subproblem S means determining whether this coloring of $B_{i-1,j+1}$ can be extended to a proper list 3-coloring of G_S. The result of the subproblem is False if this is not possible, or True (and a suitable proper list 3-coloring of G_S) otherwise. If $c_1 \notin \mathcal{L}(b_{i-1})$, or $c_3 \notin \mathcal{L}(b_{j+1})$, or $c_2 \notin \mathcal{L}(b_k)$ for some $i \leq k \leq j$, then the result of the subproblem is trivially False.

We will show that this subproblem can be solved in polynomial time as it can be reduced to the 2-list coloring problem, which is known to be solvable in linear time [14,18]. Furthermore, solutions to consecutive 'compatible' subproblems can be combined, and a proper list 3-coloring of G exists if and only if there

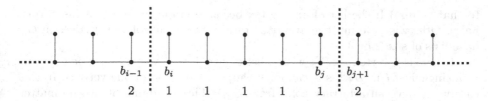

Fig. 3. Illustration of subproblem $SP(i, j, c_1, c_2, c_3)$ for the case $SP(i, j, 2, 1, 2)$.

is a collection of subproblems whose solutions can be combined into a list 3-coloring of G. For example, the colorings of two subproblems $SP(i, j, c_1, c_2, c_3)$ and $SP(j + 1, k, c_2, c_3, c_4)$ can be combined because they agree on the colors of backbone vertices that are in both subproblems, they do not share any leaf vertices, and the vertices $y \in Y$ that have neighbors in both $B_{i,j} \cup L_{i,j}$ and $B_{j+1,k} \cup L_{j+1,k}$ must be adjacent to b_j and b_{j+1}, which are colored with colors c_2 and c_3 (where $c_2 \neq c_3$) in the colorings of both subproblems, and hence must have received the same color (the only color in $\{1, 2, 3\} \setminus \{c_2, c_3\}$) in both colorings. To check whether there is a collection of compatible subproblems whose solutions can be combined into a list 3-coloring of G, we will show that it suffices to search for a directed path between two vertices in an auxiliary directed acyclic graph (DAG) on the subproblems whose result is True.

For a subproblem $S = SP(i, j, c_1, c_2, c_3)$, if $i = 1$, there is no vertex b_{i-1}, and we write $*$ for c_1; similarly, if $j = n$, there is no vertex b_{j+1}, and we write $*$ for c_3. The graph G_S considered when solving such a subproblem does not contain b_{i-1} or b_{j+1}, respectively, but is otherwise defined analogously. If $i = 1$ and $j = n$, then G_S contains neither b_{i-1} nor b_{j+1}.

Lemma 1. *There is a linear-time algorithm for solving any subproblem of the form $SP(i, j, c_1, c_2, c_3)$.*

Proof. Consider the subproblem $S = SP(i, j, c_1, c_2, c_3)$. Let G_S be the subgraph of G defined by S, and let $X_S \subseteq X$, $Y_S \subseteq Y$ be such that the vertex set of G_S is $X_S \cup Y_S$. First, we check whether $c_1 \in \mathcal{L}(b_{i-1})$ (only if $i > 1$), $c_3 \in \mathcal{L}(b_{j+1})$ (only if $j < n$), and $c_2 \in \mathcal{L}(b_k)$ for all $i \leq k \leq j$. If one of these checks fails, we return False. Otherwise, we assign color c_1 to b_{i-1}, color c_2 to all vertices in $B_{i,j}$, and color c_3 to b_{j+1}.

For every vertex $y \in Y_S$, we check if $N(y)$ contains any vertices of $B_{i-1,j+1}$ and, if so, remove the colors of those vertices from $\mathcal{L}(y)$ (if they were contained in $\mathcal{L}(y)$). If the list of any vertex $y \in Y_S$ becomes empty in this process, we return False.

Let B_S denote the backbone vertices in X_S and L_S the leaf vertices in X_S (with respect to the caterpillar T). If there is a vertex in L_S or Y_S with a list of size 1, assign the color in that list to that vertex and remove that color from the lists of its neighbors (if it is contained in their lists). Repeat this operation until there is no uncolored vertex with a list of size 1. (If an uncolored vertex with a list of size 1 is created later on in the algorithm, the same operation is applied to that vertex.) If the list of any vertex becomes empty in this process, return False. Otherwise, we must arrive at a state where all uncolored vertices in G_S have lists of size 2 or 3.

If there is a vertex $y \in Y_S$ with a list of size 3, that vertex must be adjacent to a single leaf ℓ in L_S (as it cannot be adjacent to a backbone vertex). In this case we remove an arbitrary color from $\mathcal{L}(y)$: This is admissible as, no matter what color ℓ receives in a coloring, vertex y can always be colored with one of the two colors that have remained in its list.

If there is a vertex $\ell \in L_S$ with a list of length 3, assign color c_2 to ℓ (and remove color c_2 from the lists of vertices in $N(\ell)$). This color assignment does not affect the existence of a proper list 3-coloring for the following reasons (where we let b_k denote the backbone vertex with $\ell \in L(b_k)$):

- If a vertex $y \in N(\ell)$ is adjacent to more than one vertex, it must be adjacent to b_k, which has been colored with c_2, and hence it cannot receive color c_2 in any case.
- If a vertex $y \in N(\ell)$ is adjacent only to ℓ and no other vertex, then y can still be colored after ℓ is assigned color c_2, because we cannot have $\mathcal{L}(y) = \{c_2\}$; this is because, if y had the list $\mathcal{L}(y) = \{c_2\}$, it would have been colored c_2 and the color c_2 would have been removed from $\mathcal{L}(\ell)$.

If at any step of this process, an uncolored vertex with an empty list is created, return False. Otherwise, we arrive at an instance I of LI 3-COL where all uncolored vertices have lists of size 2. Such an instance can be solved in linear time [14,18] (via reduction to a 2SAT problem). If I admits a proper list 3-coloring, that coloring gives a proper list 3-coloring of G_S, and we return True and that coloring. Otherwise, we return False.

Correctness of the algorithm follows from its description, and the algorithm can be implemented to run in linear time using standard techniques.   □

Call a subproblem $S = SP(i, j, c_1, c_2, c_3)$ *valid* if its answer is True (and a proper list 3-coloring of G_S has been produced), and *invalid* otherwise. To check whether the colorings obtained from valid subproblems can be combined into a list 3-coloring of G, we make use of an auxiliary DAG H constructed as follows. The existence of a proper list 3-coloring of G can then be determined by checking whether H contains a directed path from s to t.

Definition 1. *The auxiliary DAG $H = (V_H, A)$ has vertices s, t, and a vertex for each valid subproblem $SP(i, j, c_1, c_2, c_3)$. Its arc set A contains the following arcs: An arc $(s, SP(1, i, *, c_2, c_3))$ for each $i < n$ and $c_2, c_3 \in [3]$ such that $SP(1, i, *, c_2, c_3)$ is valid; an arc $(SP(i, n, c_1, c_2, *), t)$ for each $i > 1$ and $c_1, c_2 \in [3]$ such that $SP(i, n, c_1, c_2, *)$ is valid; arcs $(s, SP(1, n, *, c_2, *))$ and $(SP(1, n, *, c_2, *), t)$ if $SP(1, n, *, c_2, *)$ is valid; an arc $(SP(i, j, c_1, c_2, c_3), SP(j + 1, k, c_2, c_3, c_4))$ for each $i \leq j \leq k - 1$ and each $c_1, c_2, c_3, c_4 \in [3]$ (or $c_1 = *$ or $c_4 = *$ if $i = 1$ or $k = n$, respectively) such that $SP(i, j, c_1, c_2, c_3)$ and $SP(j + 1, k, c_2, c_3, c_4)$ are both valid.*

Algorithm 1 . List-3-Coloring Algorithm for Caterpillar-Convex Bipartite Graphs

---

Require: A caterpillar-convex bipartite graph $G = (X \cup Y, E)$ (with caterpillar $T = (X, F)$) and a list assignment $\mathcal{L}$.
Ensure: A proper coloring that obeys $\mathcal{L}$, or False if no such coloring exists.
1: ▷ Compute solutions to all subproblems
2: **for** $i = 1$ to n **do**
3:     **for** $j = i$ to n **do**
4:         **for** $c_i \in [3], i = 1, 2, 3$ with $c_1 \neq c_2$ and $c_2 \neq c_3$ **do**
5:             ▷ let $c_1 = *$ if $i = 1$ and $c_3 = *$ if $j = n$
6:             Solve $SP(i, j, c_1, c_2, c_3)$ (Lemma 1)
7:         **end for**
8:     **end for**
9: **end for**
10: ▷ Check if solutions of subproblems can be combined into a list 3-coloring of G
11: Build a DAG H whose vertices are s, t, and a vertex $SP(i, j, c_1, c_2, c_3)$ for each subproblem with answer True (Definition 1)
12: **if** H contains a directed path P from s to t **then**
13:     Return the coloring obtained as union of the colorings of the subproblems on P
14: **else**
15:     Return False
16: **end if**

Theorem 1. Lɪ 3-ᴄᴏʟ *can be solved in polynomial time for caterpillar-convex bipartite graphs.*

The resulting algorithm is shown in Algorithm 1. As comb-convex bipartite graphs are a subclass of caterpillar-convex bipartite graphs, we obtain:

Corollary 1. Lɪ 3-ᴄᴏʟ *can be solved in polynomial time for comb-convex bipartite graphs.*

Combining Corollary 1 with Theorem 4 in [6] and the polynomial-time solvability of Lɪ k-ᴄᴏʟ for $k \leq 2$ [16,27] yields a complexity dichotomy: Lɪ k-ᴄᴏʟ is polynomial-time solvable on comb-convex bipartite graphs when $k \leq 3$; otherwise, it is NP-complete.

4 Recognition of Caterpillar-Convex Bipartite Graphs

We give a polynomial-time recognition algorithm for caterpillar-convex bipartite graphs. We are given a bipartite graph $G = (X \cup Y, E)$ and want to decide whether it is caterpillar-convex and, if so, construct a caterpillar representation $T = (X, F)$. First, we assume that a specific partition of the vertex set into independent sets X and Y is given as part of the input, and we want to decide whether there is a caterpillar representation $T = (X, F)$ with respect to that given bipartition (i.e., the vertex set of the caterpillar is the independent set X that was specified in the input). At the end of this section, we will discuss how to handle the case that the bipartite graph is given without a specific bipartition of the vertex set and we want to decide whether the vertex set can be partitioned into independent sets X and Y in such a way that there is a caterpillar representation with respect to that bipartition.

The main idea of the algorithm for recognizing caterpillar-convex bipartite graphs is to construct an auxiliary DAG D on vertex set X in such a way that the sinks in D can be used as the backbone vertices of T. To make this work, it turns out that we first need to remove some vertices from G that have no effect on whether G is caterpillar-convex. First, we show that we can remove isolated vertices from X and vertices of degree 0 or 1 from Y.

Lemma 2. *Let $x \in X$ be a vertex with degree 0, and let G' be the graph obtained from G by removing x. Then G' is caterpillar-convex if and only if G is caterpillar-convex. Furthermore, a caterpillar representation of G can be constructed from a caterpillar representation of G' by adding x in a suitable location.*

Lemma 3. *Let $y \in Y$ be a vertex with degree 0 or 1, and let G' be the graph obtained from G by removing y. Then G' is caterpillar-convex if and only if G is caterpillar-convex. Any caterpillar representation of G' is also a caterpillar representation of G.*

We call a pair of vertices x_i and x_j twins if $N(x_i) = N(x_j)$. The twin relation on X partitions X into equivalence classes, such that $x_1, x_2 \in X$ are twins if and only if they are in the same class. We say that two twins x, x' are *special twins* if $\{x, x'\}$ is an equivalence class of the twin relation on X and if there is $y \in Y$ with $N_G(y) = \{x, x'\}$. Now, we show that removing a twin from X (with some additional modification in the case of special twins) has no effect on whether the graph is caterpillar-convex or not.

Lemma 4. *Let $x, x' \in X$ be twins of non-zero degree, and let $G' = (X' \cup Y', E')$ be the graph obtained from G by deleting x. If x, x' are special twins in G, then modify G' by adding a new vertex $\bar{x}$ to X', a new vertex $\bar{y}$ to Y', and the edges $\{x', \bar{y}\}$ and $\{\bar{x}, \bar{y}\}$ to E'. Then G is caterpillar-convex if and only if G' is caterpillar-convex. Furthermore, a caterpillar representation of G can be constructed from a caterpillar representation of G' by adding x in a suitable location (and removing $\bar{x}$ if it has been added to G').*

We remark that the special treatment of special twins in Lemma 4 seems necessary because there is a graph $G = (X \cup Y, E)$ with special twins that does not have a caterpillar representation $T = (X, F)$, while simply removing one of the two special twins (without adding the extra vertices $\bar{x}$ and $\bar{y}$) would produce a graph $G' = (X' \cup Y', E')$ that has a caterpillar representation $T' = (X', F')$. An example of such a graph is the graph with $X = \{a, b, c, f, g, x, x'\}$ where the neighborhoods of the vertices in Y are $\{a, f\}, \{a, b\}, \{b, x, x'\}, \{x, x'\}, \{b, c\}, \{c, g\}$. Here, the vertices x and x' are special twins, and the graph obtained after removing x has the caterpillar representation with backbone path abc and leaf f attached to a, leaf x' attached to b, and leaf g attached to c.

Let $G_1 = (X_1 \cup Y_1, E_1)$ be the graph obtained from $G = (X \cup Y, E)$ by repeatedly removing vertices of degree 0 from X, vertices of degree 0 or 1 from Y, and twins from X (with the extra modification detailed in Lemma 4 in case of special twins) as long as such vertices exist. Lemmas 2–4 imply:

Corollary 2. G_1 *is caterpillar-convex if and only if G is caterpillar-convex.*

We now define a directed graph $D = (X_1, A)$ based on G_1: For every pair of distinct vertices $x, x' \in X_1$, we let D contain the arc (x, x') if and only if $N_{G_1}(x) \subseteq N_{G_1}(x')$, i.e., we add the arc (x, x') if and only if every vertex in y that is adjacent to x in G_1 is also adjacent to x' in G_1. Note that D is transitive: If it contains two arcs (x, x') and (x', x''), it must also contain (x, x'').

Lemma 5. D *is a directed acyclic graph.*

Proof. Assume there is a cycle on vertices $x_i, x_{i+1}, \ldots, x_j$ in D. Then, $N(x_i) \subseteq N(x_{i+1}) \subseteq \cdots \subseteq N(x_j) \subseteq N(x_i)$. Thus $N(x_i) = N(x_{i+1}) = \cdots = N(x_j)$, and so $x_i, x_{i+1}, \ldots, x_j$ are twins, a contradiction because there are no twins in X_1. Thus D is acyclic.   □

The following lemma can be proved along the following lines: If there is a caterpillar representation in which two backbone vertices are connected by an arc in D, then there must be two adjacent backbone vertices u and v with an arc (u, v) in D. The caterpillar can then be modified by attaching u and the vertices in $L(u)$ as leaves to v, giving another valid caterpillar representation with fewer arcs between backbone vertices.

Lemma 6. *If $G_1 = (X_1 \cup Y_1, E_1)$ is caterpillar-convex, there is a caterpillar representation $T_1 = (X_1, F)$ in which no two backbone vertices are connected by an arc in D.*

Lemma 7. *If $G_1 = (X_1 \cup Y_1, E_1)$ is caterpillar-convex, there is a caterpillar representation $T_1 = (X_1, F)$ such that the set of backbone vertices is exactly the set of sinks in D.*

Proof. By Lemma 6, there exists a caterpillar representation T_1 of G_1 in which no two backbone vertices are connected by an arc in D. Furthermore, every leaf attached to a backbone vertex (in T_1) has an arc (in D) to that backbone vertex (because every $y \in Y$ has degree at least 2). A backbone vertex cannot have an arc (in D) to a leaf attached to it (in T_1), as D is acyclic (Lemma 5). Finally, a backbone vertex b cannot have an arc (in D) to a leaf vertex ℓ attached to a different backbone vertex b' because that would imply that b has an arc to b' (since every vertex in y that is adjacent to b is also adjacent to ℓ and hence, as $N(y)$ induces a tree in T_1, also to b'). Therefore, the backbone vertices of T_1 are exactly the sinks (vertices without outgoing edges) of D.   □

Theorem 2. *Algorithm 2 decides in polynomial time whether a given bipartite graph $G = (X \cup Y, E)$ is caterpillar-convex and, if so, outputs a caterpillar representation $T = (X, F)$.*

Algorithm 2. Recognition Algorithm for Caterpillar-Convex Bipartite Graphs

Require: A bipartite graph $G = (X \cup Y, E)$
Ensure: Either return a caterpillar representation $T = (X, F)$ of G, or decide that G is not caterpillar-convex and return 'fail'

1: Obtain $G_1 = (X_1 \cup Y_1, E_1)$ from G by removing vertices of degree 0 from X, vertices of degree 0 or 1 from Y, and twins from X (with the extra modification stated in Lemma 4 in case of special twins), as long as any such vertex exists (Lemmas 2–4)
2: Create a directed graph $D = (X_1, A)$ that contains the arc (x, x') if and only if $N_{G_1}(x) \subseteq N_{G_1}(x')$
3: $B =$ the set of sinks in D, $L =$ all other vertices in D
4: Use an algorithm for consecutive ones [20] to order B. If it fails, return 'fail'.
5: Form caterpillar T_1 by taking the ordered backbone B and attaching each vertex in L as leaf to an arbitrary vertex in B to which it has an arc in D
6: Obtain T from T_1 by adding the vertices that were deleted from X in Step 1 (and removing vertices that have been added when special twins were processed) (Lemmas 2 and 4).

Proof. Let $G = (X \cup Y, E)$ be a bipartite graph and $n = |X \cup Y|$. First, the algorithm removes vertices of degree 0 from X, vertices of degree 0 or 1 from Y, and twins from X (with the extra modification of Lemma 4 in case of special twins) as long as such vertices exist. The resulting graph is $G_1 = (X_1 \cup Y_1, E_1)$. By Corollary 2, G_1 is caterpillar-convex if and only if G is caterpillar-convex.

Next, the algorithm constructs the directed graph $D = (X_1, A)$ from $G_1 = (X_1 \cup Y_1, E_1)$ by adding an arc (x_i, x_j) for $x_i, x_j \in X_1$ if $N_{G_1}(x_i) \subseteq N_{G_1}(x_j)$. Once D has been constructed, the set B of sinks and the set L of remaining vertices are determined. Then, we create a set system $\mathcal{S}$ containing for every $y \in Y$ the set $N(y) \cap B$ and apply an algorithm for checking the consecutive ones property [20] to check if B can be ordered in such a way that every set in $\mathcal{S}$ consists of consecutive vertices. If so, the resulting order is used to determine the order in which B forms the backbone path. Otherwise, G_1 (and hence G) cannot be caterpillar-convex (cf. Lemma 7), and the algorithm returns 'fail'.

Next, the algorithm attaches each vertex $\ell \in L$ as a leaf to an arbitrary vertex $b \in B$ to which it has an arc in D. Every $\ell \in L$ must indeed have at least one arc to a vertex in B: As D is acyclic, every vertex ℓ that is not a sink must have a directed path leading to some sink b, and as D is transitive, the arc (ℓ, b) must exist. Attaching ℓ to b yields a valid caterpillar representation for the following reason: As every neighbor y of ℓ is also adjacent to b, and as $N(y) \cap B$ is a contiguous segment of B, it is clear that $N(y)$ induces a tree in the resulting caterpillar T_1. Hence, T_1 is a caterpillar representation of G_1.

Finally, the vertices that have been deleted in the first step are added back (and vertices that have been added when special twins were processed are removed) in order to extend the caterpillar T_1 to a caterpillar representation T of G (Lemmas 2 and 4). By Corollary 2, T is a caterpillar representation of G.

It can be shown that the algorithm can be implemented to run in $O(n^3)$ time (see the full version [1] for details). □

Finally, we discuss the case that the bipartition of the vertex set V of the input graph $G = (V, E)$ into independent sets X and Y is not provided as part of the input. First, if $G = (V, E)$ is a connected bipartite graph, note that there is a unique partition of V into two independent sets Q and R. We can then run the recognition algorithm twice, once with $X = Q$ and $Y = R$ and once with $X = R$ and $Y = Q$. G is caterpillar-convex if and only if at least one of the two runs of the algorithm produces a caterpillar representation. If $G = (V, E)$ is not connected, let $H_1, \ldots, H_r$ for some $r > 1$ be its connected components. As just discussed, we can check in polynomial time whether each connected component H_j, $1 \le j \le r$, is a caterpillar-convex bipartite graph. If all r connected components are caterpillar-convex, the whole graph G is caterpillar-convex, and a caterpillar representation can be obtained by concatenating the backbones of the caterpillar representations of the connected components in arbitrary order. If at least one of the connected components, say, the component H_j, is not caterpillar-convex, then G is not caterpillar-convex either. This can be seen as follows: Assume for a contradiction that G is caterpillar-convex while H_j is not caterpillar-convex. Then let $T = (X, F)$ be a caterpillar representation of G. Observe that the subgraph of T induced by $V(H_j) \cap X$, where $V(H_j)$ denotes the vertex set of H_j, must be connected. Therefore, that subgraph of T provides a caterpillar representation of H_j, a contradiction to our assumption. Thus we get:

Corollary 3. *There is a polynomial-time algorithm that decides whether a given bipartite graph $G = (V, E)$ is caterpillar-convex, i.e., whether it admits a bipartition of V into independent sets X and Y such that there is a caterpillar representation $T = (X, F)$.*

5   Conclusion

Determining the computational complexity of LI k-COL for $k \ge 3$ when restricted to comb-convex bipartite graphs was stated as an open problem by Bonomo-Braberman et al. [5]. Subsequently, the same authors proved that the problem is NP-complete for $k \ge 4$ [6], but the complexity for $k = 3$ was still left open. In this paper, we resolve this question by showing that LI 3-COL is solvable in polynomial time even for caterpillar-convex bipartite graphs, a superclass of comb-convex bipartite graphs.

Recall that if mim-width is bounded for a graph class $\mathcal{G}$, then LI k-COL is polynomially solvable when it is restricted to $\mathcal{G}$. Polynomial-time solvability of LI k-COL on circular convex graphs is shown by demonstrating that mim-width is bounded for this graph class [5]. On the other hand, there are graph classes for which LI 3-COL is tractable but mim-width is unbounded, such as star-convex bipartite graphs [6]. By combining our result with Theorem 3 in [5], we conclude that caterpillar-convex bipartite graphs and comb-convex bipartite graphs also belong to this type of graph classes. On a much larger graph class, chordal bipartite graphs, the computational complexity of LI 3-COL is still open [21].

Finally, as for future work, it would be interesting to see whether one can modify and extend Algorithm 2 to recognize comb-convex bipartite graphs.

References

1. Baklan Şen, B., Diner, Ö.Y., Erlebach, T.: List 3-coloring on comb-convex and caterpillar-convex bipartite graphs. CoRR **abs/2305.10108** (2023). https://doi.org/10.48550/arXiv.2305.10108
2. Belmonte, R., Vatshelle, M.: Graph classes with structured neighborhoods and algorithmic applications. Theor. Comput. Sci. **511**, 54–65 (2013). https://doi.org/10.1016/j.tcs.2013.01.011
3. Biró, M., Hujter, M., Tuza, Z.: Precoloring extension. I. Interval graphs. Discret. Math. **100**(1–3), 267–279 (1992). https://doi.org/10.1016/0012-365X(92)90646-W
4. Bonomo, F., Chudnovsky, M., Maceli, P., Schaudt, O., Stein, M., Zhong, M.: Three-coloring and list three-coloring of graphs without induced paths on seven vertices. Comb. **38**(4), 779–801 (2018). https://doi.org/10.1007/s00493-017-3553-8
5. Bonomo-Braberman, F., Brettell, N., Munaro, A., Paulusma, D.: Solving problems on generalized convex graphs via Mim-width. In: Lubiw, A., Salavatipour, M. (eds.) WADS 2021. LNCS, vol. 12808, pp. 200–214. Springer, Cham (2021). https://doi.org/10.1007/978-3-030-83508-8_15
6. Bonomo-Braberman, F., Brettell, N., Munaro, A., Paulusma, D.: Solving problems on generalized convex graphs via mim-width. CoRR **abs/2008.09004**, September 2022. https://arxiv.org/abs/2008.09004
7. Brettell, N., Horsfield, J., Munaro, A., Paulusma, D.: List k-colouring P_t-free graphs: a mim-width perspective. Inf. Process. Lett. **173**, 106168 (2022). https://doi.org/10.1016/j.ipl.2021.106168
8. Broersma, H., Fomin, F.V., Golovach, P.A., Paulusma, D.: Three complexity results on coloring P_k-free graphs. Eur. J. Comb. **34**(3), 609–619 (2013). https://doi.org/10.1016/j.ejc.2011.12.008
9. Buchin, K., van Kreveld, M.J., Meijer, H., Speckmann, B., Verbeek, K.: On planar supports for hypergraphs. J. Graph Algorithms Appl. **15**(4), 533–549 (2011). https://doi.org/10.7155/jgaa.00237
10. Chen, H., Lei, Z., Liu, T., Tang, Z., Wang, C., Xu, K.: Complexity of domination, hamiltonicity and treewidth for tree convex bipartite graphs. J. Comb. Optim. **32**(1), 95–110 (2016). https://doi.org/10.1007/s10878-015-9917-3
11. Chudnovsky, M., Spirkl, S., Zhong, M.: List 3-coloring P_t-free graphs with no induced 1-subdivision of $K_{1,s}$. Discret. Math. **343**(11), 112086 (2020). https://doi.org/10.1016/j.disc.2020.112086
12. Couturier, J.F., Golovach, P.A., Kratsch, D., Paulusma, D.: List coloring in the absence of a linear forest. Algorithmica **71**(1), 21–35 (2015). https://doi.org/10.1007/s00453-013-9777-0
13. Díaz, J., Diner, Ö., Serna, M., Serra, O.: On list k-coloring convex bipartite graphs. Graphs Comb. Optim. Theory Appl. **5**, 15–26 (2021)
14. Edwards, K.: The complexity of colouring problems on dense graphs. Theor. Comput. Sci. **43**, 337–343 (1986). https://doi.org/10.1016/0304-3975(86)90184-2
15. Enright, J.A., Stewart, L., Tardos, G.: On list coloring and list homomorphism of permutation and interval graphs. SIAM J. Discret. Math. **28**(4), 1675–1685 (2014). https://doi.org/10.1137/13090465X
16. Erdős, P., Rubin, A.L., Taylor, H.: Choosability in graphs. In: Proceedings of the West Coast Conference on Combinatorics, Graph Theory and Computing. Congressus Numerantium, vol. 26, pp. 125–157 (1979)
17. Golovach, P.A., Paulusma, D.: List coloring in the absence of two subgraphs. Discret. Appl. Math. **166**, 123–130 (2014). https://doi.org/10.1016/j.dam.2013.10.010

18. Gravier, S., Kobler, D., Kubiak, W.: Complexity of list coloring problems with a fixed total number of colors. Discret. Appl. Math. **117**(1–3), 65–79 (2002). https://doi.org/10.1016/S0166-218X(01)00179-2

19. Hoàng, C.T., Kaminski, M., Lozin, V.V., Sawada, J., Shu, X.: Deciding k-colorability of P_5-free graphs in polynomial time. Algorithmica **57**(1), 74–81 (2010). https://doi.org/10.1007/s00453-008-9197-8

20. Hsu, W.L.: A simple test for the consecutive ones property. J. Algorithms **43**(1), 1–16 (2002). https://doi.org/10.1006/jagm.2001.1205

21. Huang, S., Johnson, M., Paulusma, D.: Narrowing the complexity gap for coloring (C_s, P_t)-free graphs. Comput. J. **58**(11), 3074–3088 (2015)

22. Jansen, K., Scheffler, P.: Generalized coloring for tree-like graphs. Discret. Appl. Math. **75**(2), 135–155 (1997). https://doi.org/10.1016/S0166-218X(96)00085-6

23. Kratochvil, J.: Precoloring extension with fixed color bound. Acta Math. Univ. Comen. **62**, 139–153 (1993)

24. Kratochvíl, J., Tuza, Z.: Algorithmic complexity of list colorings. Discret. Appl. Math. **50**(3), 297–302 (1994). https://doi.org/10.1016/0166-218X(94)90150-3

25. Kubale, M.: Some results concerning the complexity of restricted colorings of graphs. Discret. Appl. Math. **36**(1), 35–46 (1992). https://doi.org/10.1016/0166-218X(92)90202-L

26. Lovász, L.: Coverings and coloring of hypergraphs. In: Proceedings of the 4th Southeastern Confeernce on Combinatorics, Graph Theory, and Computing, Utilitas Math, pp. 3–12 (1973)

27. Vizing, V.: Coloring the vertices of a graph in prescribed colors. Diskret. Analiz **101**(29), 3–10 (1976)

Parameterized Algorithms for Cluster Vertex Deletion on Degree-4 Graphs and General Graphs

Kangyi Tian[1]($\boxtimes$), Mingyu Xiao[1], and Boting Yang[2]

[1] School of Computer Science and Engineering,
University of Electronic Science and Technology of China, Chengdu, China
tkysss@outlook.com
[2] University of Regina, Regina, Canada

Abstract. In the CLUSTER VERTEX DELETION problem, we are given a graph G and an integer k, and the goal is to determine whether we can delete at most k vertices from G to make the remaining graph a cluster, i.e., a graph with each connected component being a complete graph. In this paper, we show that CLUSTER VERTEX DELETION can be solved in $O^*(1.7549^k)$ time, improving the previous result of $O^*(1.811^k)$. To obtain this result, one crucial step is to show CLUSTER VERTEX DELETION on graphs of maximum degree at most 4 can be solved in $O^*(1.7485^k)$ time. After this, we know that the graph will always have a vertex of degree at least 5. Then by adopting the previous method of automated generation of searching trees, we can get the result on general graphs.

1 Introduction

In clustering problems, there is a set of elements and a value of similarity between each pair of elements in it. The goal is to partition these elements into distinct groups based on their similarities. This problem has wide-ranging applications in various fields, such as machine learning [1] and computational biology [2]. One effective approach to model this problem is by constructing a similarity graph, where the elements form the vertices and edges exist only if the similarity between two vertices exceeds a specified threshold. Therefore, the clustering problem transforms into partitioning the similarity graph into parts where each part induces a complete graph. Occasionally, the given similarity data may deviate slightly from the actual clustering model due to noise or other factors, resulting in improper no-instance. For such cases, modifications are permitted within the clustering problem, leading to cluster graph modification problems. Depending on the type of modifications allowed, different problems can be defined. If we can only delete edges to create a cluster, i.e., a graph where each connected component forms a complete graph, the problem is referred to as Cluster Edge Deletion [7,8,10,11,15]. If both deleting and adding edges are allowed to form a cluster, the problem is known as Cluster Editing [3,4,7,10,11]. When solely the deletion of vertices is allowed, the problem becomes Cluster Vertex Deletion (CVD) [5,6,10,12,14,16]. These problems have been extensively examined

W. Wu and G. Tong (Eds.): COCOON 2023, LNCS 14422, pp. 182–194, 2024.
https://doi.org/10.1007/978-3-031-49190-0_13

in the literature. In this paper, we focus on studying parameterized and exact algorithms for CVD.

Table 1. Parameterized algorithms for CLUSTER VERTEX DELETION

References	Results	Main Methods
Cai, 1996 [6]	$O^*(3^k)$	forbidden graphs
Gramm et al., 2004 [10]	$O^*(2.26^k)$	automated generation of searching trees
Wahlström, 2007 [16]	$O^*(2.076^k)$	3-hitting
Hüffer et al., 2010 [12]	$O^*(2^k)$	iterative compression
Boral et al., 2016 [5]	$O^*(1.911^k)$	auxiliary graph H^v
Tsur, 2021 [14]	$O^*(1.811^k)$	careful analysis by considering twins
This paper	$O^*(1.7549^k)$	dealing with low-degree graphs

In the CLUSTER VERTEX DELETION (CVD) problem, we are given a graph $G = (V, E)$ and an integer k. The goal is to determine whether there is a set of vertices S of size at most k such that after deleting S from G the remaining graph is a cluster. One important property is that a graph is a cluster if and only if there is no induced 3-path (i.e., a path with three vertices such that there is no edge between the first and the last vertices) in the graph. In other words, induced 3-path is the forbidden graph of a cluster. By using this property and applying the general branching algorithm of Cai [6], we can get an $O^*(3^k)$-time branching algorithm for this problem. Later, Gramm et al. [10] improved this result to $O^*(2.26^k)$ by using automated generation techniques for search trees. Since hitting all induced 3-paths is a special case of the 3-hitting problem, the $O^*(2.076^k)$-time algorithm for the 3-hitting problem [16] directly implies the same complexity for CVD. In 2010, Hüffer et al. [12] improved the runtime to $O^*(2^k)$ using the iterative compression method. However, the iterative compression method naturally provides a lower bound of 2^k in complexity, limiting further improvement. The next breakthrough came in 2016 when Boral et al. [5] proposed a novel idea of selecting a vertex v and considering vertex sets that hit all induced 3-paths containing v. They constructed an auxiliary graph and used a Python script, similar to the one in [10], to automate the generation of search trees, resulting in an improved $O^*(1.911^k)$-time algorithm. Tsur [14] further refined the algorithm, achieving the current best runtime of $O^*(1.811^k)$.

In this paper, we further improve the running time bound to $O^*(1.7549^k)$ for CVD. To get this result, we first show that the problem can be solved in $O^*(1.7485^k)$ when the input graph has maximum degree at most 4. After that, we adopt the technique of the auxiliary graph H^v and a method to automatically generate branching vectors used in [5,14]. Since the degree of the vertex considered is at least 5, we have more vertices in the local structure and the previous bottlenecks can be avoided. The history of parameterized algorithms for CVD was listed in Table 1. We also remark that by using the general approach to obtain exact algorithms via fast parameterized algorithms in [9], we can also solve CVD in $O(1.4302^n)$ time, improving all previous results.

One of the most important contributions in this paper is to obtain the improved result for the problem in degree-4 graphs. For vertices of degree ≥ 5, we just modify previous algorithms to get our desired result. Due to the space limitation, we mainly introduce our algorithm for degree-4 graphs in the main body and the part for high-degree vertices and general graphs can be found in the full version of this paper.

2   Notations

Let $G = (V, E)$ denote an undirected graph. For a vertex $v \in V$, we use $N_G(v)$ to represent the set of neighbors of v in graph G, i.e., vertices adjacent to v in G. Define $deg_G(v) = |N_G(v)|$ as the degree of the vertex v. For a subset V' of vertices, let $N_G(V') = (\cup_{v \in V'} N_G(v)) \setminus V'$ and $N_G[V'] = N_G(V') \cup V'$. We also use $N_G^2(v) = N_G(N_G(v)) \setminus \{v\}$ to represent the vertices with distance exactly 2 to vertex v. In the above notations, we may omit the subscript G if the graph G is clear from the context. A singleton $\{v\}$ may be simply written as v. For a graph G', we use $V(G')$ and $E(G')$ to denote the set of vertices and the set of edges in G', respectively. For $V' \subseteq V$ and $E' \subseteq E$, a graph $G' = (V', E')$ is called an induced subgraph of G if any edge (x, y) in E is in E' iff $x \in V'$ and $y \in V'$. Let $G[V']$ denote the induced subgraph of G with vertices V' and $G - V'$ denote $G[V \setminus V']$.

A vertex set S of a graph G is called a *vertex cover* if for any edge in G there is at least one endpoint in S. A *3-path* is a path of 3 vertices. A 3-path (u, v, w) is called an *induced 3-path* if there is no edge between u and w. A complete graph is also called a *clique*. A graph is called a *cluster* if each connected component is a clique. The CLUSTER VERTEX DELETION problem (CVD) is to check whether we can delete at most k vertices from a graph to make the remaining graph a cluster. An *instance* of CVD is denoted by (G, k). A subset S of vertices is called a *CVD-set* if $G - S$ is a cluster. A CVD-set is also called a *solution set* to CVD and a minimum CVD-set is called an *optimal solution set* to CVD. The following forbidden-graph property has been frequently used in algorithms for CVD [5,6,10,12,14,16].

Lemma 1. *A graph is a cluster iff the graph contains no induced 3-path.*

3   Properties for Algorithm of Degree-4 Graph

3.1   The CVD-Dominating Family

Definition 1. *Let V' be a subset of vertices of the graph G. Define $\Delta_G(V') = N_G(V \setminus V') \cap V'$, i.e., the set of vertices in V' adjacent to some vertices not in V'. Let T be a CVD-set of $G[V']$. Define $\Delta_G(V', T)$ as the set of vertices of cliques in $G[V' - T]$ containing some vertices in $\Delta_G(V')$. See Fig. 1 for an illustration of these concepts.*

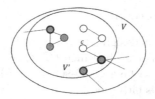

Fig. 1. An example for Definition 1.

In Fig. 1, vertices in $\Delta_G(V')$ are bold. A CVD-set T of $G[V']$ is $\{c\}$. After removing T from V', there are 3 cliques and the vertices in $\Delta_G(V', T)$ are all marked as gray vertices.

Definition 2. *Let V' be a subset of vertices of the graph G. For two CVD-sets X and Y of induced graph $G[V']$, we say that Y dominates X under $G[V']$ if*

$$|\Delta_G(V', Y) \setminus \Delta_G(V', X)| \leq |X| - |Y|.$$

Lemma 2. *Let $V' \subseteq V$ be a vertex subset of G, and X and Y be two CVD-sets of $G[V']$ such that Y dominates X under $G[V']$. If there is a CVD-set S of G containing X then there is a CVD-set S' of G containing Y such that $|S'| \leq |S|$.*

Proof. Let S be a CVD-set of G containing X. Let $Y' = Y \cup (\Delta_G(V', Y) \setminus \Delta_G(V', X))$. We show that $S' = (S \setminus X) \cup Y'$ is a CVD-set of G containing Y such that $|S'| \leq |S|$.

Since $Y \subseteq Y' \subseteq S'$, we know that S' contains Y. Next, we prove $G - S'$ is a cluster. Assume to the contrary that $G - S'$ is not a cluster. By Lemma 1, we know there is an induced 3-path P in graph $G - S'$.

Since Y is a CVD-set of $G[V']$ and $Y \subseteq S'$, we know that $G[V' \setminus S']$ is a cluster. Then the induced 3-path P must contain at least one vertex in $V \setminus V'$. Thus, all the three vertices of P must be in $(V \setminus (V' \cup S')) \cup \Delta_G(V', Y')$, which is a superset of $(V \setminus (V' \cup S')) \cup \Delta_G(V', X)$, by the definition of Y'. This implies that the induced 3-path P also exists in $G - S$, a contradiction. So $G - S'$ is a cluster. Last, we consider the size of S'. We can see that $|S'| = |(S \setminus X) \cup Y'| = |(S \setminus X) \cup Y \cup (\Delta_G(V', Y) \setminus \Delta_G(V', X))| \leq |S| - |X| + |Y| + |\Delta_G(V', Y) \setminus \Delta_G(V', X)| \leq |S|$. So S' is a satisfied set. □

Definition 3. *A CVD-dominating family $\mathcal{F}_G(V')$ is a set of CVD-sets of $G[V']$ such that for every CVD-set X of $G[V']$, there is a CVD-set in $\mathcal{F}_G(V')$ dominating X under $G[V']$.*

Based on Definition 3, to search for a solution, instead of listing all CVD-sets of V', we can only consider a CVD-dominating family.

Definition 4. *A vertex subset $V' \subseteq V$ is called an important set if there is a minimum CVD-set C of induced subgraph $G[V']$ such that $\Delta_G(V') \subseteq C$. The set C is called an important cut.*

Corollary 1 (of Lemma 2). *Let $V' \subseteq V$ be an important set with an important cut C. There is a minimum CVD-set of G containing C.*

3.2 Core Branching Processing

We introduce a technique to design branching rules. The idea is to relax the concept of important sets. For an instance (G, k), we consider a set of vertices C that induces a clique. Let S be a CVD-set of (G, k). Since $G - S$ is a cluster, we know that $C \cap (V - S)$ is either empty or forms a clique. When designing a branching rule, we can only consider all possible cliques in $G - S$ intersecting with C. Furthermore, we can also reduce some dominated cases. This processing to generate branching rules is called core branching processing and the selected set C is called the *core clique*.

Definition 5. *Given an reduced instance (G, k), the core branching processing will produce a branching rule by the following steps:*
1. Choose an induced clique of G as the core clique C;
2. Let $\mathcal{T}$ be the set of cliques intersected with C and $\mathcal{C} = \{C\} \cup \{N(T) | T \in \mathcal{T}\}$;
3. If there exist T_1 and $T_2 \in \mathcal{T}$ such that $|N(T_1)| \geq |N(T_2)|$ and $N[T_1] \subseteq N[T_2]$, remove $N(T_1)$ from $\mathcal{C}$;
4. If there exist $T_1 \in \mathcal{T}$ such that $|N(T_1)| \geq |C|$ and $N[T_1] \subseteq C$ (resp., $|C| \geq |N(T_1)|$ and $C \subseteq N[T_1]$), remove $N(T_1)$ (resp., C) from $\mathcal{C}$.

The correctness of Step 3 of the core branching processing is based on the fact that if there is a CVD-set of G containing $N(T_1)$, then there must be a CVD-set containing $N(T_2)$. Here we use the concept of dominating. Let $V' = N[T_2]$. We know that $\Delta_G(V') \subseteq N(T_2)$. Assume that S is a CVD-set of G containing $N(T_1)$. Let $X = S \cap V'$. Then S is also a CVD-set of V'. Since $N[T_1] \subseteq N[T_2]$, we know that $|X| \geq |N(T_1)| \geq |N(T_2)|$. By Definition 2 and Lemma 2, we know that $N(T_2)$ dominates X and there must be a CVD-set containing $N(T_2)$.

4 Reduction Rules

We first introduce some reduction rules that will be applied in our algorithms to reduce the instance in polynomial time.

R-Rule 1. *If $k < 0$, return 'no' to indicate that the instance is a no-instance.*

R-Rule 2. *If the graph is an empty graph, return 'yes' to indicate that the instance is a yes-instance.*

R-Rule 3. *If there is a connected component of G that is a clique, delete it from the graph.*

R-Rule 4. *If there is a connected component C of G and a vertex $u \in C$ such that $G[C] - u$ is a cluster, delete u from the graph and decrease k by 1.*

R-Rule 5. *If there is an important set $V' \subseteq V$ with an important cut C such that $|C| \leq 10$, then return the instance $(G - C, k - |C|)$.*

In R-Rule 5, we require that $|C| \leq 10$. So the important sets V' and important cuts C can be found in polynomial time if they exist. We mention a special case of the important sets that will be reduced by this rule. Let u be a degree-2 vertex with one degree-1 neighbor v and another neighbor w. Then $\{u, v, w\}$ is an important set satisfying the condition in R-Rule 5. We may delete w directly from the graph.

5 The Algorithm for Degree-4 Graphs

First of all, we present our algorithm for the case that the input graph is a graph with maximum degree at most 4. In this section, we mainly use the concept of the CVD-dominating family and the method of core branching processing. The algorithm contains the following seven main steps. When introduce one step, we will assume all previous steps can not be applied.

1. Iteratively apply the above five reduction rules to get a reduced instance.
2. If there is a degree-1 vertex v, execute the branching operations in Sect. 5.1.
3. If there is a 4-clique, execute the branching operations in Sect. 5.2.
4. If there is a degree-4 vertex v, execute the branching operations in Sect. 5.3.
5. If there are two triangles sharing one edge, execute the branching operations in Sect. 5.4.
6. If there is a triangle, execute the branching operations in Sect. 5.5.
7. Reduce the problem to the 3-path vertex cover problem.

5.1 Degree-1 Vertices

Assume there is a degree-1 vertex u. We let v denote the unique neighbor of u. We know that v is a vertex of degree at least three otherwise if v is of degree 2 then R-Rule 5 would be applied on the important set $N[v]$ and if v is of degree 1 then R-Rule 3 would be applied. We pick $\{u\}$ as the core clique and apply the core branching processing. The set of intersecting cliques is $\mathcal{T} = \{\emptyset, \{u\}, \{u, v\}\}$ and thus we will get three initial branches: (1) deleting $\{u\}$ and decreasing k by 1; (2) deleting $\{v\}$ and decreasing k by 1; (3) deleting $N(u, v)$ and decreasing k by $|N(u, v)| \geq 2$. However, the first branch is dominated by the second branch and then the first branch will be removed according to the last step of the core branching processing. Thus, we get a branching vector at least $(1, 2)$.

5.2 4-Cliques

Assume there is a 4-clique $\{v, u, w, t\}$. We show all possible configurations of the clique in Fig. 2, where dark vertices mean all neighbors of them are already drawn on the figure and a dotted edge incident on a vertex means there may be some further edges incident on this vertex. Note that a vertex has degree at most 4. Each vertex in $\{v, u, w, t\}$ can be adjacent to at most one vertex not in this set. Thus $|N(v, u, w, t)| \leq 4$. Let $N_c = |N(v, u, w, t)|$. The cases in Fig. 2 are listed

188   K. Tian et al.

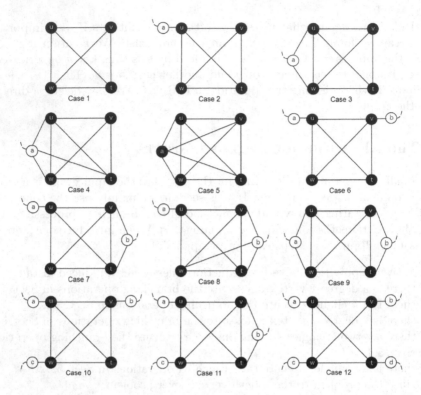

Fig. 2. The cases with 4-cliques

according to the value of N_c. In Case 1, $N_c = 0$; In Cases 2–5, $N_c = 1$; In Cases 6–9, $N_c = 2$; In Cases 10–11, $N_c = 3$; In Case 12, $N_c = 4$. For Cases 1 and 5, the graph is a component of clique and it would be reduced by R-Rule 3. For Cases 2–4, $\{v, u, w, t, a\}$ is an important set with the important cut $\{a\}$ and it would be reduced by R-Rule 5. For Cases 6, 7 and 9, $\{v, u, w, t, a, b\}$ is an important set with the important cut $\{a, b\}$ and it would be reduced by R-Rule 5. For Case 8, we consider the subgraph G' induced by $\{v, u, w, t, a, b\}$. All CVD-sets in G' of size at least 2 are dominated by $\{a, b\}$ and there is only one CVD-set $\{u\}$ of size less than 2 in G'. Thus, we branch into two branches: deleting $\{a, b\}$ and decreasing k by 2; deleting $\{u\}$ and decreasing k by 1. We get a branching vector $(2, 1)$. For Case 10, $\{v, u, w, t, a, b, c\}$ is an important set with the important cut $\{a, b, c\}$ and it would be reduced by R-Rule 5. For Case 11, we consider the subgraph G' induced by $\{v, u, w, t, a, b, c\}$. All CVD-sets in G' of size at least 3 are dominated by $\{a, b, c\}$ and there is only one CVD-set $\{u, w\}$ of size less than 3 in G'. Thus, we branch into two branches: deleting $\{a, b, c\}$ and decreasing k by 3; deleting $\{u, w\}$ and decreasing k by 2. We get a branching vector $(3, 2)$. For Case 12, we first consider the subgraph G' induced by $\{v, u, w, t, a, b, c, d\}$. All CVD-sets in G' of size at least 4 are dominated by $\{a, b, c, d\}$. Next, we assume that S is a CVD-set such that $S' = S \cap \{v, u, w, t, a, b, c, d\}$ contains at most 3 vertices. Then, there are only four possibilities: $S' = \{u, v, w\}, \{u, v, t\}, \{v, t, w\}$ or $\{t, w, u\}$. We consider $S' = \{u, v, w\}$. For this case, both t and d are left in a component of a clique of size 2 and thus $N(t, d)$ is in S. After Sect. 4.1, there is no degree-1 vertex, and then d is adjacent to a vertex $d' \neq t$. Furthermore, $d' \notin \{u, v, w\}$ since the graph is a degree-4 graph. For this case, $\{u, v, w, d'\} \subseteq N(t, d)$ and at least $|N(t, d)| \geq 4$ vertices can be deleted. We can decrease k by at least 4. By the same reason, for the cases $S' = \{u, v, t\}, \{v, t, w\}$ and $\{t, w, u\}$, we can also decrease k by at least 4. Thus, we can branch into five branches and in each branch the parameter k decreases by at least 4. We get a branching vector $(4, 4, 4, 4, 4)$.

Among all the cases, the worst branching vector is $(2, 1)$, the branching factor of which is 1.6182.

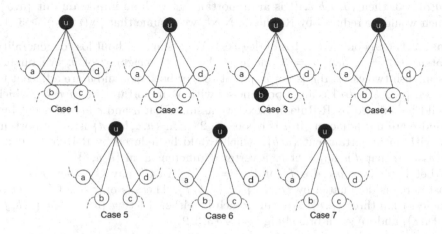

Fig. 3. The cases with 4-degree vertices

5.3 Degree-4 Vertices

Assume there is a degree-4 vertex u with four neighbors $\{a, b, c, d\}$. Let E_l denote the set of edges with both endpoints in $\{a, b, c, d\}$ and l be the size of E_l. Note that the edges in E_l will not form a triangle, since there is no cliques of size 4 after Sect. 4.2. We consider different cases according to the value of l. All the cases are shown in Fig. 3.

Case 1. $l = 4$. Since there is no triangle, the four edges in E_l will form a cycle of size 4. Without loss of generality, we assume that $E_l = \{(a, b), (b, c), (c, d), (d, a)\}$. Let X be the set with the least neighbors among $\{u, a, b\}, \{u, b, c\}, \{u, c, d\}$ and $\{u, d, a\}$. If $|N(X)| \leq 3$, then $N[X]$ is an important set with the important cut being $N(X)$, because any CVD-set of the graph induced by $N[X]$ will contain at least $|N(X)|$ vertices. For this case, it will be reduced by R-Rule 5. Next, we assume that $|N(X)| \geq 4$. Let a', b', c' and d' be the four neighbor of a, b, c and d, respectively, where $a' \neq b'$, $b' \neq c'$, $c' \neq d'$ and $d' \neq a'$.

We choose $\{u\}$ as the core clique and apply the core branching processing. The intersecting cliques are $\emptyset$, $\{u\}$, $\{u,a\}$, $\{u,b\}$, $\{u,c\}$, $\{u,d\}$, $\{u,a,b\}$, $\{u,b,c\}$, $\{u,c,d\}$, and $\{u,d,a\}$. We get initial branches of deleting $\{u\}$, $N(u)$, $N(u,a)$, $N(u,b)$, $N(u,c)$, $N(u,d)$, $N(u,a,b)$, $N(u,b,c)$, $N(u,c,d)$, and $N(u,d,a)$. The branches of deleting $N(u)$, $N(u,a)$ and $N(u,b)$ are dominated by the branch of deleting $N(u,a,b)$. The branches of deleting $N(u,c)$ and $N(u,d)$ are dominated by the branch of deleting $N(u,c,d)$. We can branch by deleting $\{u\}$, $N(u,a,b)$, $N(u,b,c)$, $N(u,c,d)$ or $N(u,d,a)$, where $|N(u,a,b)|, |N(u,b,c)|, |N(u,c,d)|$ and $|N(u,d,a)| = 4$ since $|N(X)| \geq 4$ and the maximum degree of the graph is 4. The branching vector is (1,4,4,4,4) with a branching factor 1.7485.

Case 2. $l = 3$. There are three edges in E_l and they form a path of length 3. Without loss of generality, we assume that $E_l = \{(a,b),(b,c),(c,d)\}$. If $deg(b) = deg(c) = 3$, then $\{u,a,b,c,d\}$ is an important set with an important cut $\{a,d\}$, which would be reduced by R-Rule 5. Next, we assume that $|N(u,b,c)| \geq 3$.

Case 2.1. Only one of $\{b,c\}$ is of degree 4. We assume without loss of generality that $deg(b) = 4$, $deg(c) = 3$ and e is the fourth neighbor of b. We further distinguish two cases. If a is not adjacent to e or both of a and d are adjacent to e, then $\{u,a,b,c,d,e\}$ is an important set with an important cut $\{a,d,e\}$, which would be reduced by R-Rule 5. Next, we assume that a and e are adjacent but d and e are not adjacent. If d is of degree 2, then $\{u,a,b,c,d\}$ is an important set with an important cut $\{a,b\}$, which would be reduced by R-Rule 5. Thus, we assume that d is adjacent to a vertex f different from $\{e,u,c\}$.

Let $V' = \{u,a,b,c,d,e,f\}$ and $G' = G[V']$. In G', any CVD-set of size at least three is dominated by $\{a,d,e\}$ or $\{a,b,f\}$. There is only one CVD-set of size less than three $\{u,c\}$. We can branch by either deleting $\{a,d,e\}$ or $\{a,b,f\}$ or $\{u,c\}$, and we get a branching vector $(3,3,2)$.

Case 2.2. Both of $\{b,c\}$ are degree-4 vertices. Let e be the fourth neighbor of b and f be the fourth neighbor of d. We know that $e,f \notin \{u,a,b,c,d\}$. First we show that $e \neq f$. If $e = f$, then e is not adjacent to either a or d. Otherwise, this case will be solved in Case 1. Thus, there is an important set $V' = \{u,a,b,c,d,e\}$ with an important cut $\{a,e,d\}$, which should be reduced by R-Rule 5. Next, we assume that $e \neq f$.

If a is not adjacent to a vertex different from u,b and e, then there is an important set $\{u,a,b,c,d,e\}$ with an important cut $\{c,d,e\}$. If d is not adjacent to a vertex different from u,c and f, then there is an important set $\{u,a,b,c,d,f\}$ with an important cut $\{a,b,f\}$. Thus, we can assume that $N(u,a,b), N(u,c,d) \geq 4$.

We choose $\{u\}$ as the core clique and apply the core branching processing. The intersecting cliques are $\emptyset$, $\{u\}$, $\{u,a\}$, $\{u,b\}$, $\{u,c\}$, $\{u,d\}$, $\{u,a,b\}$, $\{u,b,c\}$, $\{u,c,d\}$, and $\{u,d,a\}$. The branches of deleting $N(u)$, $N(u,b)$ and $N(u,c)$ are dominated by the branch of deleting $N(u,b,c)$. The branch of deleting $N(u,a)$ is dominated by the branch of deleting $N(u,a,b)$. The branch of deleting $N(u,d)$ is dominated by the branch of deleting $N(u,c,d)$. We can branch by deleting $\{u\}$,

$N(u, a, b)$, $N(u, b, c)$ or $N(u, c, d)$, where $|N(u, a, b)|, |N(u, b, c)|, |N(u, c, d)| \geq 4$. The branching vector is $(1, 4, 4, 4)$ with a branching factor 1.6851.

Case 3. $l = 3$ and the three edges in E_l have a common endpoint. Without loss of generality, we assume that $E_l = \{(a, b), (b, c), (b, d)\}$. Let $X = \{u, a, b\}, \{u, b, c\}$, or $\{u, b, d\}$. If $|N(X)| \leq 2$, then $N[X]$ is an important set with the important cut being $N(X)$. It will be reduced by R-Rule 5. Next, we assume that $|N(X)| \geq 3$. Let a', c' and d' be the three neighbors of a, c and d, respectively.

We choose $\{u\}$ as the core clique and apply the core branching processing. The intersecting cliques are $\emptyset$, $\{u\}$, $\{u, a\}$, $\{u, b\}$, $\{u, c\}$, $\{u, d\}$, $\{u, a, b\}$, $\{u, b, c\}$, and $\{u, b, d\}$. We get initial branches of deleting $\{u\}$, $N(u)$, $N(u, a)$, $N(u, b)$, $N(u, c)$, $N(u, d)$, $N(u, a, b)$, $N(u, b, c)$ and $N(u, b, d)$. The branches of deleting $N(u)$ and $N(u, a)$ are dominated by the branch of deleting $N(u, a, b)$. The branches of deleting $N(u, c)$ and $N(u, d)$ are dominated by the branches of deleting $N(u, b, c)$ and $N(u, b, d)$, respectively. The branch of deleteing $N(u, b)$ will be dominated by the branching deleting X if $|N(X)| \leq 3$. Notice that u and b are twins of each other and by Lemma 9 in [15], we can delete b at the same time if we delete u. We can branch by deleting $\{u, b\}$, $N(u, b)$, $N(u, a, b)$, $N(u, b, c)$ or $N(u, b, d)$, where $|N(u, a, b)|, |N(u, b, c)|$ and $|N(u, b, d)| = 4$ or deleting $\{u, b\}$, $N(u, a, b)$, $N(u, b, c)$ or $N(u, b, d)$, where $|N(u, a, b)|, |N(u, b, c)|$ or $|N(u, b, d)| = 3$. The branching vector is $(2,3,4,4,4)$ or $(2,3,3,3)$ with a branching factor 1.6472 or 1.6717.

Case 4. $l = 2$ and the two edges in E_l have a common endpoint. Without loss of generality, we assume that $E_l = \{(a, b), (b, c)\}$. After Sect. 5.1, there is no degree-1 vertex. We assume that d is adjacent to a vertex $e \neq u$.

Case 4.1. $deg(b) = 3$. For this case, neither a or c can be a degree-2 vertex, otherwise there would be an important set $\{u, a, b, c, d\}$ with an important cut $\{c, d\}$ or $\{a, d\}$. Thus, we have $|N(a, b, u)| \geq 3$ and $|N(c, d, u)| \geq 3$. We choose $\{b\}$ as the core clique and apply the core branching processing. The intersecting cliques are $\emptyset$, $\{b\}$, $\{b, a\}$, $\{b, u\}$, $\{b, c\}$, $\{b, a, u\}$ and $\{b, c, u\}$. The branch of deleting $N(b)$ and deleting $N(b, a)$ is dominated by the branch of deleting $N(b, a, u)$. The branch of deleting $N(b, c)$ is dominated by the branch of deleting $N(b, c, u)$. After ignoring the branching of deleting $N(b)$, $N(b, a)$ and $N(b, c)$, we may be able to get a branching vector $(1, 3, 3, 3)$. However, it is not good enough. We further analyze several cases. If $deg(a) = 3$ (resp., $deg(c) = 3$), then the branching of deleting $N(b, u)$ is dominated by the branch of deleting $N(b, u, a)$ (resp., $N(b, u, c)$). For this case, we can branch with $(1, 3, 3)$ at least, the branching factor of which is 1.6957. Otherwise, $deg(a) \geq 4$ and $deg(c) \geq 4$. The branching vector is at least $(1, 3, 4, 4)$ with a branching factor 1.7254.

Case 4.2. $deg(b) = 4$. Let f be the fourth neighbor of b. We choose $\{u\}$ as the core clique and apply the core branching processing. The intersecting cliques are $\emptyset$, $\{u\}$, $\{u, a\}$, $\{u, b\}$, $\{u, c\}$, $\{u, d\}$, $\{u, a, b\}$, and $\{u, b, c\}$. The branch of deleting $N(u)$ is dominated by the branch of deleting $N(u, b)$. The branch of deleting $N(u, a)$ is dominated by the branch of deleting $N(u, a, b)$. The branch of deleting $N(u, c)$ is dominated by the branch of deleting $N(u, b, c)$. We only need to consider five branches of deleting $\{u\}$, $N(u,b)$, $N(u,d)$, $N(u,a,b)$, and $N(u,b,c)$. We further consider several subcases. If $deg(a) = deg(c) = 4$, we have $|N(a,u,b)| = |N(b,u,c)| = 5$ because a and b (resp., b and c) have no common neighbor other than u after Case 1. Then, we will get a branching vector $(1,4,4,5,5)$ with a branching factor 1.6770. Next, we assume that either a or c is a vertex of degree at most 3. Then the branch of deleting $N(u,b)$ is dominated by the branch of deleting $N(u,a,b)$ or $N(u,c,d)$. There are four branches left: deleting $\{u\}$, $N(u,d)$, $N(u,a,b)$, and $N(u,b,c)$. If $deg(a) = 3$ or $deg(c) = 3$, we can get a branching vector $(1,4,4,3)$ with a branching factor 1.7254. The only left case is that $deg(a) = deg(c) = 2$. For this case, the branch of deleting $N(u,b,c)$ is dominated by the branch of deleting $N(u,a,b)$. we get a branching vector $(1,4,3)$ with a branching factor 1.6181.

Please see the details of analysis for Cases 5,6 and 7 in the full version of this paper.

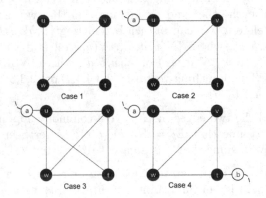

Fig. 4. two triangles sharing an edge

5.4   Two Triangles Sharing One Edge

In this step, there is no vertex of degree ≥ 4. Assume there are two triangles $\{u,w,v\}$ and $\{t,w,v\}$ sharing one edge (w,v). All possible configurations are shown in Fig. 4. Since the maximum degree is at most 3 now, we know that only u and t are possibly adjacent to a vertex out of $\{u,v,w,t\}$. Case 1 is a component of 4 vertices. We can simply put u to the solution. For Case 2, let $V' = \{u,v,w,t,a\}$ and $G' = G[V']$. We can see that $\{u\}$ is a CVD-set dominates all other CVD-sets in G'. Thus, we can also simply put u to the solution. For Case 3, let $V' = \{u,v,w,t,a\}$ and $G' = G[V']$. We can see that $\{a,u\}$ is a CVD-set dominates all other CVD-sets in G'. Thus, we can also simply put $\{a,u\}$ to the solution. For Case 4, let $V' = \{u,v,w,t,a,b\}$ and $G' = G[V']$. Any CVD-set in G' is dominated by either $\{a,t\}$ or $\{u,b\}$. We can branch with a branching vector $(2,2)$.

5.5 Triangles

Assume that there is still a triangle $\{u, v, w\}$ in the graph. If $|N(u, v, w)| \leq 2$, then $N[u, v, w]$ is an important set with an important cut $N(u, v, w)$, which should be reduced by R-Rule 5. Thus, it holds that $|N(u, v, w)| = 3$. Note that after Sect. 5.4, no two triangles share an edge and there are exact three edges between $\{u, v, w\}$ and $\{u', v', w'\}$. Let $\{u', v', w'\} = N(u, v, w)$, where there are three edges (u, u'), (v, v'), (w, w') between $\{u, v, w\}$ and $\{u', v', w'\}$. We choose $\{u, v, w\}$ as the core clique and apply the core branching processing. The intersecting cliques will be $\emptyset$, $\{u\}$, $\{v\}$, $\{w\}$, $\{u, v\}$, $\{u, w\}$, $\{v, w\}$,$\{u, u'\}$, $\{v, v'\}$, $\{w, w'\}$ and $\{u, v, w\}$. Any branch will be dominated by one of the branches of deleting $N(u, u'), N(v, v'), N(w, w')$ and $\{u', v', w'\}$. Since there is no degree-1 vertex, we can get a branching vector at least $(3, 3, 3, 3)$ with the branching factor 1.5875.

5.6 The Remaining Case

Now the graph has the maximum degree at most 3 and there is no triangle in it. CVD is equal to delete at most k vertices from the graph to make the each connected component containing at most 2 vertices. The latter problem is the 3-path vertex cover problem. We directly use the $O^*(1.713^k)$-time algorithm for the 3-path vertex cover problem in [13] to solve our problem.

We have considered all the cases and the worst branching factor is 1.7485.

Theorem 1. CLUSTER VERTEX DELETION *in graphs with maximum degree at most 4 can be solved in* $O^*(1.7485^k)$ *time and polynomial space.*

With this result, we can improve the running time bound for CLUSTER VERTEX DELETION in general graphs to $O^*(1.7549^k)$. The details can be found in the full version of this paper.

6 Conclusion

In this paper, we proposed an improved parameterized algorithm for the classic CLUSTER VERTEX DELETION. Except using previous techniques such as the auxiliary graph, twins, and automated generation of searching trees, we introduce two new concepts: CVD-dominating family and core branching processing. A significant contribution of our work is to tackle the problem in low-degree graphs as the first step. Subsequently, we focus solely on vertices with high degrees, which potentially contain more valuable information. This allows us to design specific branching rules to achieve better branching factors. It is worth noting that even for graphs with maximum degree 4, our algorithm still requires some new techniques and deep analyses to get the desired improvement.

Acknowledgments. The author is grateful to all the anonymous reviewers for fruitful and insightful comments to improve the presentation of the paper. The work is supported by the National Natural Science Foundation of China, under the grants 62372095 and 61972070.

References

1. Bansal, N., Blum, A., Chawla, S.: Correlation clustering. Mach. Learn. **56**(1–3), 89–113 (2004)
2. Ben-Dor, A., Shamir, R., Yakhini, Z.: Clustering gene expression patterns. J. Comput. Biol. **6**(3/4), 281–297 (1999)
3. Böcker, S.: A golden ratio parameterized algorithm for cluster editing. J. Disc. Algor. **16**, 79–89 (2012)
4. Böcker, S., Briesemeister, S., Bui, Q.B.A., Truß, A.: Going weighted: parameterized algorithms for cluster editing. Theor. Comput. Sci. **410**(52), 5467–5480 (2009)
5. Boral, A., Cygan, M., Kociumaka, T., Pilipczuk, M.: A fast branching algorithm for cluster vertex deletion. Theory Comput. Syst. **58**(2), 357–376 (2016)
6. Cai, L.: Fixed-parameter tractability of graph modification problems for hereditary properties. Inf. Process. Lett. **58**(4), 171–176 (1996)
7. Cygan, M., Fomin, F.V., Kowalik, Ł, Lokshtanov, D., Marx, D., Pilipczuk, M., Pilipczuk, M., Saurabh, S.: Parameterized Algorithms. Springer, Cham (2015). https://doi.org/10.1007/978-3-319-21275-3
8. Damaschke, P.: Bounded-degree techniques accelerate some parameterized graph algorithms. In: Chen, J., Fomin, F.V. (eds.) IWPEC 2009. LNCS, vol. 5917, pp. 98–109. Springer, Heidelberg (2009). https://doi.org/10.1007/978-3-642-11269-0_8
9. Fomin, F.V., Gaspers, S., Lokshtanov, D., Saurabh, S.: Exact algorithms via monotone local search. In: Wichs, D., Mansour, Y. (eds.) Proceedings of the 48th Annual ACM SIGACT Symposium on Theory of Computing, STOC 2016, Cambridge, MA, USA, 18–21 June 2016, pp. 764–775. ACM (2016)
10. Gramm, J., Guo, J., Hüffner, F., Niedermeier, R.: Automated generation of search tree algorithms for hard graph modification problems. Algorithmica **39**(4), 321–347 (2004)
11. Gramm, J., Guo, J., Hüffner, F., Niedermeier, R.: Graph-modeled data clustering: exact algorithms for clique generation. Theory Comput. Syst. **38**(4), 373–392 (2005)
12. Hüffner, F., Komusiewicz, C., Moser, H., Niedermeier, R.: Fixed-parameter algorithms for cluster vertex deletion. Theory Comput. Syst. **47**(1), 196–217 (2010)
13. Tsur, D.: Parameterized algorithm for 3-path vertex cover. Theor. Comput. Sci. **783**, 1–8 (2019)
14. Tsur, D.: Faster parameterized algorithm for cluster vertex deletion. Theory Comput. Syst. **65**(2), 323–343 (2021)
15. Tsur, D.: Cluster deletion revisited. Inf. Process. Lett. **173**, 106171 (2022)
16. Wahlström, M.: Algorithms, measures and upper bounds for satisfiability and related problems. PhD thesis, Linköping University, Sweden (2007)

Sum-of-Local-Effects Data Structures for Separable Graphs

Xing Lyu[1], Travis Gagie[2](✉) [ID], Meng He[2] [ID], Yakov Nekrich[3] [ID],
and Norbert Zeh[2] [ID]

[1] Halifax West High School, Halifax, Canada
[2] Dalhousie University, Halifax, Canada
{travis.gagie,meng.he,norbert.zeh}@dal.ca
[3] Michigan Technological University, Houghton, USA
yakov@mtu.edu

Abstract. It is not difficult to think of applications that can be mod-
elled as graph problems in which placing some facility or commodity at a
vertex has some positive or negative effect on the values of all the vertices
out to some distance, and we want to be able to calculate quickly the
cumulative effect on any vertex's value at any time or the list of the most
beneficial or most detrimential effects on a vertex. In this paper we show
how, given an edge-weighted graph with constant-size separators, we can
support the following operations in time polylogarithmic in the number
of vertices and the number of facilities placed on the vertices, where
distances between vertices are measured with respect to edge weights:

ADD (v, f, w, d) places a facility of weight w and with effect radius d
onto vertex v.

REMOVE (v, f) removes a facility f previously placed on v using ADD
from v.

SUM (v) or SUM(v, d) returns the total weight of all facilities affecting
v or, with a distance parameter d, the total weight of all facilities
whose effect region intersects the "circle" with radius d around v.

TOP(v, k) or TOP(v, k, d) returns the k facilities of greatest weight
that affect v or, with a distance parameter d, whose effect region
intersects the "circle" with radius d around v.

The weights of the facilities and the operation that SUM uses to "sum"
them must form a semigroup. For TOP queries, the weights must be
drawn from a total order.

Keywords: Graph data structures · Treewidth · Branchwidth · Graph
decompositions · Tree decompositions · Sum of local effects

1 Introduction

Even people who have never heard of Baron Samuel of Wych Cross may have
heard a saying often attributed to him, that there are three things that matter in

This work is supported by NSERC.

W. Wu and G. Tong (Eds.): COCOON 2023, LNCS 14422, pp. 195–206, 2024.
https://doi.org/10.1007/978-3-031-49190-0_14

real estate: location, location, location. This means that the value of a property may increase or decrease depending on whether it is close to a bus stop, a good school, a supermarket or a landfill, for example. Of course, "close" may not mean the same thing for a bus stop as it does for a landfill, and the positive effect of the former may not offset the negative effect of the latter. In fact, "close" may not refer to Euclidean distance, since walking to a bus stop five minutes down the street is preferable to walking to one five minutes away through a landfill. To model applications in which there are such additive local effects with a non-Euclidean definition of locality, we propose in this paper a data structure for a graph G that supports the following operations:

ADD (v, f, w, d) places a facility of weight w onto vertex v. The *effect region* of f is a circle with radius d around v.

REMOVE (v, f) removes a facility f previously placed on v using ADD from v.

SUM (v) or SUM(v, d) returns the total weight of all facilities affecting v or, with a distance parameter d, the total weight of all facilities whose effect region intersects the "circle" with radius d around v.

TOP(v, k) or TOP(v, k, d) returns the k facilities of greatest weight that affect v or, with a distance parameter d, whose effect region intersects the "circle" with radius d around v.

We assume that every edge $e \in G$ has a non-negative length $\ell(e)$ and that distances between vertices are measured as the minimum total length of all edges on any path between these two vertices. A circle with radius d around some vertex v includes all vertices and (parts of) edges at distance d from v. More precisely, if f is a facility with effect radius d' placed on some vertex u, then we consider f's effect region to intersect a circle with radius d around some other vertex v if and only if $\text{dist}(u, v) \leq d + d'$.

The weights of the facilities and the operation that SUM uses to "sum" them must form a semigroup. For TOP queries, the weights must be drawn from a total order. Note that SUM(v) and TOP(v, k) can be viewed as "range stabbing queries on graphs", whereas SUM(v, d) and TOP(v, k, d) with $d > 0$ are "range intersection queries on graphs," where the ranges are the effect regions of the facilities and a query is either an individual vertex or a region of some radius d around some vertex.

We call such a data structure a sum-of-local-effects (SOLE) data structure. In Sect. 2, we show that when G is a tree on n vertices, then there is a SOLE data structure for it supporting ADD, REMOVE, and SUM operations in $O(\lg n \lg m)$ time, and TOP queries in $O(k \lg n \lg m)$ time, where m is the total number of facilities currently placed on the vertices of G. In Sect. 3, we generalize this result to t-separable graphs, for any constant t, which includes series-parallel graphs $(t \leq 2)$, graphs of constant treewidth w $(t \leq w + 1)$, and graphs of constant branchwidth b $(t \leq b)$. We show that when G is t-separable, there exists a SOLE data structure for it supporting ADD, REMOVE, and SUM operations in $O(\lg n \lg^t m)$ time, and TOP queries in $O(k \lg n \lg^t m)$ time. The costs of ADD and REMOVE operations are amortized in this case.

Our results can be extended to *directed* graphs G and our data structure can be made to support vertex and edge deletions in G. We will investigate this generalization in the full version of this paper.

2   A SOLE Data Structure for Trees

In this section, we prove that

Theorem 1. *If G is a tree on n vertices, then there is a SOLE data structure for it supporting* ADD, REMOVE, *and* SUM *operations in $O(\lg n \lg m)$ time, and* TOP(v, k, d) *operations in $O(k \lg n \lg m)$ time, where m is the number of facilities currently on the vertices of G. The size of this data structure is $O(n + m \lg n)$.*

To obtain a SOLE data structure for arbitrary trees, we can transform any tree G into a tree G' whose nodes have degree at most 3 by replacing every high-degree vertex u in G with a degree-3 subtree G'_u whose edges all have length 0 (see Figs. 1b,c). We choose an arbitrary vertex in G'_u as the representative of u in G'. This ensures that the distances between vertices in G and between their representatives in G' are the same. Thus, we can support operations on G by building a SOLE data structure on G' instead. Therefore, for the rest of this section, we assume that all vertices of G have degree at most 3.

2.1   Designing SOLE Data Structures for Trees

We choose an arbitrary vertex ρ of G and label every vertex v in G with its distance dist(ρ, v) from ρ. A *centroid edge* of G is an edge (u, v) whose removal splits G into two subtrees G_u and G_v with at most $2n/3$ vertices each. Such an edge exists because all vertices of G have degree at most 3. A *centroid decomposition* of G is a binary tree T defined inductively as follows (see Fig. 1c): If G has a single vertex v, then T has v as its only node. Otherwise, let (u, v) be an arbitrary centroid edge of G. Then the root of T is (u, v), and the two children of (u, v) are the roots of centroid decompositions of G_u and G_v. For each edge e of T, let T_e be the subtree of T below e, and let V_e be the set of vertices of G corresponding to the leaves of T_e. The height of T is at most $\log_{3/2} n = O(\lg n)$.

Our SOLE data structure for G consists of a centroid decomposition T of G where each edge e of T has an associated data structure W_e storing facilities in $V_{e'}$ that may affect the vertices in V_e, where e' is the other edge in T descending from the same node as e. Each leaf v of T (corresponding to the vertex v of G) also has an associated data structure W_v storing the facilities placed on v itself. Each facility f with weight w in W_e or W_v has an associated *radius* r and is stored as the triple (r, f, w) in W_e.

We represent each data structure W_x, where x can be an edge or a leaf of T, as two search trees R_x and F_x. R_x is a priority search tree [4] on the triples (r, f, w) in W_x, using the radii r as x-coordinates and the weights w as y-coordinates. Each node v of R_x is augmented with the total weight of all triples in the subtree below v. F_x is a standard search tree over the triples (r, f, w)

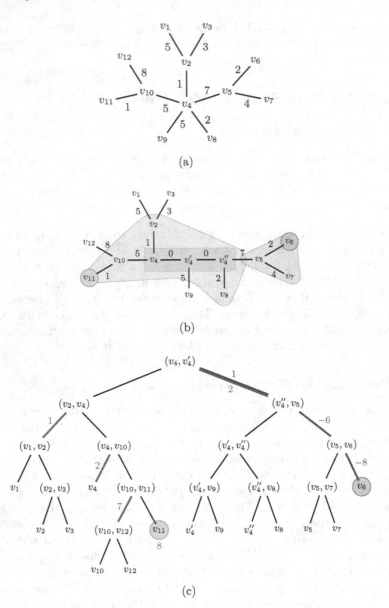

Fig. 1. A tree (a), its degree-3 version (b), and the centroid decomposition (c) of the tree in (b). The shaded subtree in (b) is the degree-3 tree replacing the high-degree vertex v_4 in (a). A facility f with effect radius 8 placed on v_{11} has the pink effect region in (b). This region overlaps the blue query region with radius 8 around v_6. In the centroid decomposition (c), f is stored in the edge data structures of the fat red edges and the node data structure of v_{11}, with the radii shown in red. A query with radius 8 around v_6 queries the node data structure of v_6 an the edge data structures of the fat blue edges, with the query radii shown in blue. In particular, f is reported as part of the query on the data structure W_e associated with the highlighted child edge of (v_4, v_4'). (Color figure online)

in W_x, using the identifiers f of facilities as keys. The two copies of (r, f, w) in R_x and F_x are linked using cross pointers. Thus, W_x supports the following operations in $O(\lg m)$ time: insertion of a new triple (r, f, w), deletion of a triple associated with facility f, and reporting of the total weight of all triples (r, f, w) with $q \leq r$, for some query radius q. It also supports, in $O(k \lg m)$ time, reporting the k triples with maximum weight among all triples (r, f, w) with $q \leq r$.

To bound the size of the data structure, note that T has size $O(n)$, each data structure W_x, with x a leaf or edge of T, has size linear in the number of triples it stores, and each facility placed on some vertex v is stored in W_v and in the data structure W_e associated with each edge e on the path $P_v = \langle x_1, \ldots, x_h, v \rangle$ from the root of T to v (which is a leaf of T). Since this path has length $O(\lg n)$, each facility is stored $O(\lg n)$ times. Thus, the SOLE data structure for trees has size $O(n + m \lg n)$.

2.2 Supporting Queries over Trees

We now use our data structures to support operations.

An $\text{ADD}(v, f, w, d)$ operation traverses the path P_v. We insert the triple (d, f, w) into W_v. Each node x_i in P_v represents an edge (x, y) of G and has two child edges e_x and e_y such that $x \in V_{e_x}$ and $y \in V_{e_y}$. Assume w.l.o.g. that $v \in V_{e_x}$, and let $d' = |\text{dist}(\rho, y) - \text{dist}(\rho, v)|$. We insert the triple $(d - d', f, w)$ into the data structure W_{e_y} associated with e_y. This is illustrated in Fig. 1c. This takes $O(\lg m)$ time per vertex in P_v, $O(\lg n \lg m)$ time in total.

A $\text{REMOVE}(v, f)$ operation traverses the path P_v and deletes the triple associated with f from the data structure W_e of every child edge e of every node x_i in P_v, and from W_v. By a similar analysis as for $\text{ADD}(v, f, w, d)$ operations, this takes $O(\lg n \lg m)$ time.

A $\text{SUM}(v, d)$ query traverses the path P_v. For each edge e on this path with top endpoint x_i, x_i represents an edge $(x, y) \in G$ such that w.l.o.g. $e = e_y$. We query W_e to report the total weight of all triples (r, f, w) in W_e with $r \geq |\text{dist}(\rho, y) - \text{dist}(\rho, v)| - d$. We also query W_v to report the total weight of all facilities placed on v itself. This is illustrated in Fig. 1c. We sum these weights retrieved from W_v and from all the edge data structures W_e along P_v and report the resulting total. Since we answer a $\text{SUM}(v, d)$ query by querying $O(\lg n)$ data structures W_e and W_v, the cost is $O(\lg n \lg m)$. A $\text{SUM}(v)$ query is the same as a $\text{SUM}(v, 0)$ query.

A $\text{TOP}(v, k, d)$ query traverses P_v. For each edge e on this path with top endpoint x_i, x_i represents an edge $(x, y) \in G$ such that w.l.o.g. $e = e_y$. We query W_e to retrieve the k triples with maximum weight among all triples (r, f, w) in W_e such that $r \geq |\text{dist}(\rho, y) - \text{dist}(\rho, v)| - d$. This takes $O(k \lg n \lg m)$ time for all edges on P_v. We also retrieve the k facilities with maximum weight from W_v, which takes $O(k \lg m)$ time. The k maximum-weight facilities affecting vertices at distance at most d from v are among the $O(k \lg n)$ facilities retrieved by these queries and can be found in $O(k \lg n)$ time using linear-time selection [2]. Thus, a $\text{TOP}(v, k, d)$ query takes $O(k \lg n \lg m)$ time. A $\text{TOP}(v, k)$ query is the same as a $\text{TOP}(v, k, 0)$ query.

To prove the correctness of $\text{SUM}(v, d)$ and $\text{TOP}(v, k, d)$ queries, note that both queries query the same data structures W_e, with the same query regions. A SUM query reports the total weight of all facilities in these query regions. A TOP query reports the k maximum weight queries in these regions. Both queries are correct if we can argue that if either query were to *report* all facilities in these query regions instead of summing their weights or picking the k facilities with maximum weight, then any facility placed on some vertex u is reported if and only if its effect radius d' satisfies $d + d' \geq \text{dist}(u, v)$, and any such facility is reported exactly once.

So consider a facility f with effect radius d' placed on some vertex $u \in G$, and let v be another vertex $v \in G$. If $v = u$, then f must be reported because it affects v no matter the query radius d. The facility f does not belong to any data structure W_e on the path $P_u = P_v$. Thus, if f is to be reported, it must be reported by the query W_v. Since placing f on u adds f to $W_u = W_v$, a $\text{SUM}(v, d)$ query reports the total weight of all facilities in W_v, and a $\text{TOP}(v, k, d)$ query reports the k facilities with maximum weight in W_v, the corresponding reporting query would report all facilities in W_v, including f.

If $v \neq u$, then let x_i be the highest vertex on the path from u to v in T. This vertex represents an edge (x, y) such that w.l.o.g. $u \in V_{e_x}$ and $v \in V_{e_y}$. In this case, f is not stored in W_v, and e_y is the only edge on the path P_v that is a pendant edge of P_u. Thus, W_{e_y} is the only data structure considered by a $\text{SUM}(v, d)$ or $\text{TOP}(v, k, d)$ query that stores f. In W_{e_y}, f is stored with radius $r = d' - |\text{dist}(\rho, y) - \text{dist}(\rho, u)| = d' - \text{dist}(u, y)$. The path from u to v in G passes through y, so $\text{dist}(u, v) = \text{dist}(u, y) + \text{dist}(y, v)$. Therefore, $\text{dist}(u, v) \leq d + d'$ if and only if $q = |\text{dist}(\rho, v) - \text{dist}(\rho, y)| - d = \text{dist}(v, y) - d \leq d' - \text{dist}(u, y) = r$. The reporting version of a $\text{SUM}(v, d)$ or $\text{TOP}(v, k, d)$ query reports all triples (r, f, w) in W_{e_y} with $r \geq q$. Thus, f is reported if and only if $d + d' \geq \text{dist}(u, v)$.

This finishes the proof of Theorem 1.

3   A SOLE Data Structures for Separable Graphs

We call a graph G *t-separable*, for some constant t, if it has a *t-separator decomposition* C of the following structure (see Figs. 2a,b):

- C is an unrooted tree with $O(n)$ nodes, all of which have degree at most 3.
- Each edge e of C has an associated subset $S_e \subseteq V$ of vertices of G of size $|S_e| \leq t$. We call S_e the *(edge) bag* associated with e.
- Every vertex of G belongs to at least one bag of C.
- Let C_1 and C_2 be the subtrees of C obtained by removing any edge e from C, and let V_i, $i \in \{1, 2\}$, be the union of the bags of all edges in C_i. Then any path from a vertex in V_1 to a vertex in V_2 includes at least one vertex in S_e. In other words, S_e separates the vertices in V_1 from the vertices in V_2.

This definition of a t-separator decomposition is similar to both a tree decomposition [5] and a branch decomposition [5], but the properties of a t-separator decomposition are weaker than those of both a tree-decomposition and a branch decomposition. In particular, a branch decomposition of width b and degree 3 is easily seen to be a b-separator decomposition, and a nice tree decomposition C [3] of width w gives rise to a $(w+1)$-separator decomposition by defining the (edge) bag associated with each edge $(v,w) \in C$ to be the union of the (node) bags associated with v and w. However, a t-separator decomposition does not require a bijection between the edges of G and the leaves of C, as required by a branch decomposition. A tree decomposition requires that for every edge (v,w) of G, there exists a (node) bag that contains both v and w, a condition not imposed by a t-separator decomposition. Thus, every graph of branchwidth b has a t-separator decomposition with $t \leq b$, every graph of treewidth w has a t-separator decomposition with $t \leq w+1$, but there may exist graphs for which these inequalities are strict.

In this section, we prove that

Theorem 2. *If G is a t-separable graph on n vertices, for some constant t, then there is a SOLE data structure for it supporting* ADD, REMOVE, *and* SUM *operations in $O(\lg n \lg^t m)$ time, and* TOP(v,k,d) *operations in $O(k \lg n \lg^t m)$ time. The size of this data structure is $O(tn \lg n + m \lg^{t-1} m \lg n)$. The costs of* ADD *and* REMOVE *operations are amortized.*

3.1   Designing SOLE Data Structures for Separable Graphs

To design our SOLE structure, let C be a t-separator decomposition for G. Since C is an unrooted tree whose nodes have degree at most 3, we can once again construct its centroid decomposition T (see Fig. 2c). Each leaf of T corresponds to a node of C, and each internal node of T corresponds to an edge e of C. Since C has size $O(n)$, the height of T is $O(\lg n)$. Since a vertex v of G may be contained in more than one bag of C, we choose an arbitrary bag S_e of C that contains v and refer to e as e_v, as indicated by the bold vertex labels in Fig. 2b.

We obtain a SOLE data structure for G by augmenting each internal node e of T with two data structures W_e and D_e and augmenting each edge a of T with a data structure W_a. We refer to W_e as a *node data structure* and to W_a as an *edge data structure*. Each data structure W_x, with x a node or an edge of T, stores a number of facilities f as tuples $(r_1, \ldots, r_{t'}, f, w)$. If $x = e$ is a node of T or $x = a$ is an edge of T with top endpoint e, then $t' = |S_e| \leq t$. Again, W_x consists of two trees R_x and F_x over the set of tuples stored in W_x. F_x stores these tuples as a binary search tree with the facility f as the key for each tuple $(r_1, \ldots, r_{t'}, f, w)$ in W_x. R_x is a "t'-dimensional range sum priority search tree" over the points defined by the coordinates $(r_1, \ldots, r_{t'})$. This is a t'-dimensional range tree [1,6] augmented to support t'-dimensional range sum queries in $O(\lg^{t'} m)$ time and t'-dimensional range top-k queries in $O(k \lg^{t'} m)$ time.

The structure D_e associated with each internal node e of T stores the distance from every vertex v such that e_v is a descendant of e in T to all vertices in S_e.

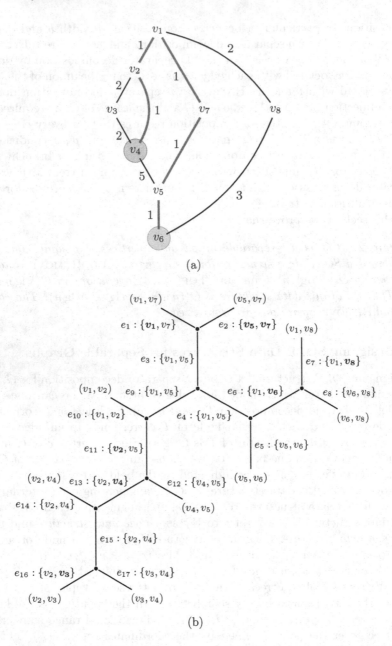

Fig. 2. A series-parallel graph G (a) and a branch decomposition C of G of width 2(b), which is also a 2-separator decomposition of G. Each edge in (a) is labelled with its length. Each edge in (b) is labelled with its name and its corresponding bag S_e. A facility with effect radius 5 placed on vertex v_6 affects vertex v_4, since the path shown in red in (a) has length 5. If we assume that $e_{v_6} = e_6$ and $e_{v_4} = e_{13}$, as indicated by the bold vertices in (b), then f is stored in the node data structure of e_6 and in the edge data structures of the red edges shown in Fig. 3. (Color figure online)

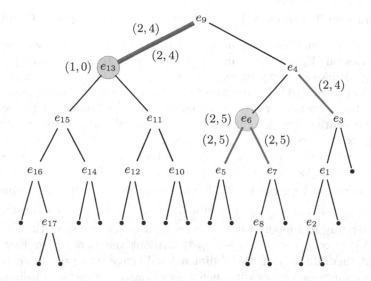

Fig. 3. The centroid decomposition T of the branch decomposition C in Fig. 2b. A facility with effect radius 5 placed on vertex v_6 is stored in the node data structure of e_6 and in the edge data structures of the red edges, with the pairs of radii shown in red. A Sum(v_4) or Top(v_4, k) query queries the vertex data structure of e_{13} and the edge data structure of the blue edge, with the pairs of radii shown in blue. The facility f is added to the query result because $2 \leq 2$ and $4 \leq 4$. One of these two conditions would have sufficed. (Color figure online)

A t'-dimensional range sum priority search tree supports insertions and deletions in $O\left(\lg^{t'} m\right)$ amortized time. Thus, each data structure W_x supports insertion of a new tuple $(r_1, \ldots, r_{t'}, f, w)$ and the deletion of the tuple associated with a facility f in $O\left(\lg^{t'} m\right) = O(\lg^t m)$ amortized time.

We now bound the size of the SOLE data structure. Once again, the tree T has size $O(n)$ and height $O(\lg n)$. For every vertex $v \in G$, the distance data structure D_e of every ancestor node e of e_v in T stores the distances from v to all $t' \leq t$ vertices in S_e. Thus, each vertex v contributes at most t to the size of each of $O(\lg n)$ distance data structures. The total size of the distance data structures is thus $O(tn \lg n)$. Each facility f placed on some vertex u is stored in the node data structures of the $O(\lg n)$ nodes along the path P_u and in one or two edge data structures of child edges of these nodes. Thus, every facility is stored in $O(\lg n)$ data structures W_x. Each such data structure is a t'-dimensional range sum priority search tree, where $t' \leq t$. Thus, if it stores s facilities, it has size $O(s \lg^{t-1} s)$. The total size of all node and edge data structures is thus $O(m \lg^{t-1} m \lg n)$.

3.2   Sum and Top-k over W_x with an Uncommon Query Range

To support SUM and TOP, we need to support range sum queries and range top-k queries on W_x, but with an uncommon query range. A query point $q = (q_1, \ldots, q_{t'})$ defines a query range $\mathbb{R}^{t'} \setminus ((-\infty, q_1) \times (-\infty, q_2) \times \cdots \times (-\infty, q_{t'}))$, i.e., the complement of t'-sided range query. We need to support range sum queries and range top-k queries for any such complement of a t'-sided range query.

An easy solution is to decompose it into t' "normal" range queries, with query ranges $[q_1, \infty) \times (-\infty, \infty) \times \cdots \times (-\infty, \infty)$, $(-\infty, q_1) \times [q_2, \infty) \times (-\infty, \infty) \times \cdots \times (-\infty, \infty)$, $(-\infty, q_1) \times (-\infty, q_2) \times [q_3, \infty) \times (-\infty, \infty) \times \cdots \times (-\infty, \infty)$, $\cdots$, $(-\infty, q_1) \times (-\infty, q_2) \times \cdots \times (-\infty, q_{t'-1}) \times [q_{t'}, \infty)$. This allows us to support range sum and range top-k queries in $O(t' \lg^{t'} m)$ and $O(t'k \lg^{t'} m)$ time, respectively, which is $O(\lg^{t'} m)$ and $O(k \lg^{t'} m)$ time because $t' \leq t$ and t is a constant.

A better way to support range queries with query ranges of the form $\mathbb{R}^{t'} \setminus ((-\infty, q_1) \times (-\infty, q_2) \times \cdots \times (-\infty, q_{t'}))$ without the factor t' overhead is to implement them directly on the t'-dimensional range sum priority search tree. To answer a range sum query with such a query range, we answer a 1-dimensional range sum query with query range $[q_1, \infty)$ on the level-1 tree of R_x. This query traverses the path corresponding to q_1 in the level-1 tree of R_x. For the root of each subtree to the left of this path, we answer a $(t'-1)$-dimensional range sum query with query range $\mathbb{R}^{t'-1} \setminus ((-\infty, q_2) \times (-\infty, q_3) \times \cdots \times (-\infty, q_{t'}))$ on the $(t'-1)$-dimensional range sum priority search tree associated with this root. The final result is the sum of the totals produced by these queries, including the 1-dimensional range sum query on the level-1 tree of R_x. Thus, a range sum query with the complement of a t'-sided range query as the query range has the same cost as a "normal" orthogonal range sum query, $O(\lg^{t'} m)$.

Similarly, to support a t'-dimensional range top-k query with query range $Q = \mathbb{R}^{t'} \setminus ((-\infty, q_1) \times (-\infty, q_2) \times \cdots \times (-\infty, q_{t'}))$, we answer a 1-dimensional range top-k query on the level-1 tree of R_x, with query range $[q_1, \infty)$. This query traverses the path corresponding to q_1 in the level-1 tree of R_x. For the root of each subtree to the left of this path, we answer a $(t'-1)$-dimensional range top-k query with query range $\mathbb{R}^{t'-1} \setminus ((-\infty, q_2) \times (-\infty, q_3) \times \cdots \times (-\infty, q_{t'}))$ on the $(t'-1)$-dimensional range sum priority search tree associated with this root. The top k tuples in the query range Q are easily seen to be among the $O(k \log n)$ elements reported by these $(t'-1)$-dimensional range top-k queries and by the 1-dimensional range top-k query on the level-1 tree. The top k tuples can now be found in $O(k \lg n)$ time using linear-time selection [2]. Thus, a range top-k query with the complement of a t'-sided range query as the query range takes $O(k \lg^{t'} m)$ time, just as a "normal" orthogonal range top-k query does.

3.3   Supporting Queries over Separable Graphs

We are ready to discuss how to support SUM and TOP queries for t-separable graphs; the support for ADD and REMOVE will be described in the full version of this paper.

A SUM(v, d) query traverses the path $P_v = \langle e_1, \ldots, e_h = e_v \rangle$. For each node e_i in P_v, let $S_{e_i} = \{v_1, \ldots, v_{t'}\}$, and let $Q = \mathbb{R}^{t'} \setminus ((-\infty, q_1) \times (-\infty, q_2) \times \cdots \times (-\infty, q_{t'}))$, where $q_j = \text{dist}(v, v_i) - d$, for all $1 \leq j \leq t'$. If $1 \leq i < h$, then we answer a range sum query with query range Q on the edge data structure W_a, where $a = (e_i, e_{i+1})$ is the child edge of e_i that belongs to P_v. If $i = h$, then we answer a range sum query with query range Q on W_{e_i}. This is illustrated in Fig. 2c. The result of the SUM(v, d) query is the sum of the results reported by all these range sum queries. Since a range sum query on each data structure W_x can be answered in $O\left(\lg^{t'} m\right) = O(\lg^t m)$ time, the cost of a SUM(v, d) query is thus $O(\lg n \lg^t m)$. A SUM(v) query is the same as a SUM$(v, 0)$ query.

A TOP(v, k, d) query traverses the path $P_v = \langle e_1, \ldots, e_h = e_v \rangle$. For each node e_i in P_v, let $S_{e_i} = \{v_1, \ldots, v_{t'}\}$, and let $Q = \mathbb{R}^{t'} \setminus ((-\infty, q_1) \times (-\infty, q_2) \times \cdots \times (-\infty, q_{t'}))$, where $q_j = \text{dist}(v, v_i) - d$, for all $1 \leq j \leq t'$. If $1 \leq i < h$, then we ask a range top-k query with query range Q on the edge data structure W_a, where $a = (e_i, e_{i+1})$ is the child edge of e_i that belongs to P_v. If $i = h$, then we answer a range top-k query with query range Q on W_{e_i}. The result of the TOP(v, k, d) query is the list of the k maximum-weight facilities among the $O(k \lg n)$ facilities reported by all these range top-k queries. These k facilities can be found in $O(k \lg n)$ time using linear-time selection [2]. Each query on a data structure W_x takes $O(k \lg^t m)$ time. Thus, the total cost of a TOP(v, k, d) query is $O(k \lg n \lg^t m)$. A TOP(v, k) query is the same as a TOP$(v, k, 0)$ query.

To establish the correctness of SUM(v, d) and TOP(v, k, d) queries, observe that, similar to Sect. 2, both queries query the same node and edge data structures, with the same query ranges. Thus, it suffices to prove that if either query reported all facilities in these query ranges, it would report any facility with effect radius d' placed on some vertex u if and only if $d + d' \geq \text{dist}(u, v)$, and each such facility is reported exactly once.

So let f be a facility with effect radius d' placed on some vertex $u \in G$, and let v by any other vertex $v \in G$. Let e be the lowest common ancestor (LCA) of e_u and e_v in T, and let $S_e = \{v_1, \ldots, v_{t'}\}$ We distinguish two cases:

If e_v is a proper descendant of e, then f is not stored in W_{e_v} and the only edge data structure in P_v that stores f is the data structure W_a corresponding to the child edge a of e on the path from e to e_v. Thus, f is reported at most once by the reporting version of a SUM(v, d) or TOP(v, k, d) query. This query queries W_a with query region $Q = \mathbb{R}^{t'} \setminus ((-\infty, q_1) \times \cdots (-\infty, q_{t'}))$, where $q_j = \text{dist}(v_j) - d$ for all $1 \leq j \leq t'$. Since e is the LCA of e_u and e_v in T, any path from u to v in G must include at least one vertex in S_e. Assume w.l.o.g. that v_1 is one such vertex. Then $\text{dist}(u, v) = \text{dist}(v, v_1) + \text{dist}(u, v_1)$ and $\text{dist}(u, v) \leq \text{dist}(v, v_j) + \text{dist}(u, v_j)$ for all $1 < j \leq t'$. The facility f is stored in W_a as the tuple $(r_1, \ldots, r_{t'}, f, w)$ with $r_j = d' - \text{dist}(u, v_j)$ for all $1 \leq j \leq t'$. Thus, $(r_1, \ldots, r_{t'}) \in Q$ if and only if there exists an index $1 \leq j \leq t'$ such that $d' - \text{dist}(u, v_j) \geq \text{dist}(v, v_j) - d$, that is, $d + d' \geq \text{dist}(u, v_j) + \text{dist}(v, v_j)$. Since $\text{dist}(u, v_1) + \text{dist}(v, v_1) = \text{dist}(u, v)$ and $\text{dist}(u, v_j) + \text{dist}(v, v_j) \geq \text{dist}(u, v)$ for all $1 \leq j \leq t'$, this is true if and only if $d + d' \geq \text{dist}(u, v)$. Thus, f is reported if and only if $d + d' \geq \text{dist}(u, v)$.

If e_v is not a proper descendant of e, then $e_v = e$ and $P_v \subseteq P_u$. Thus, f is not stored in any edge data structure along P_v, but it is stored in $W_e = W_{e_v}$, as the tuple $(r_1, \ldots, r_{t'}, f, w)$ with $r_j = d' - \text{dist}(u, v_j)$ for all $1 \leq j \leq t'$. Thus, f is reported at most once by the reporting version of a $\text{SUM}(v, d)$ or $\text{TOP}(v, k, d)$ query. This query queries W_{e_v} with query region $Q = \mathbb{R}^{t'} \setminus ((-\infty, q_1) \times \cdots (-\infty, q_{t'}))$, where $q_j = \text{dist}(v_j) - d$ for all $1 \leq j \leq t'$. Since $v \in S_{e_v}$, we can assume w.l.o.g. that $v = v_1$. Then $\text{dist}(u, v) = \text{dist}(v, v_1) + \text{dist}(u, v_1)$ and $\text{dist}(u, v) \leq \text{dist}(v, v_j) + \text{dist}(u, v_j)$ for all $1 < j \leq t'$. The same analysis as in the previous case now shows that f is reported by the query on W_{e_v} if and only if $d + d' \geq \text{dist}(u, v)$. This finishes the proof of Theorem 2.

References

1. Bentley, J.L.: Decomposable searching problems. Inf. Process. Lett. **8**(5), 244–251 (1979)
2. Blum, M., Floyd, R.W., Pratt, V., Rivest, R.L., Tarjan, R.E.: Time bounds for selection. J. Comput. Syst. Sci. **7**(4), 448–461 (1973)
3. Bodlaender, H.L., Kloks, T.: Efficient and constructive algorithms for the pathwidth and treewidth of graphs. J. Algorithms **21**(2), 358–402 (1996)
4. McCreight, E.M.: Priority search trees. SIAM J. Comput. **14**(2), 257–276 (1985)
5. Robertson, N., Seymour, P.D.: Graph minors. X. Obstructions to tree-decomposition. J. Comb. Theory B **52**(2), 153–190 (1991)
6. Willard, D.E., Lueker, G.S.: Adding range restriction capability to dynamic data structures. J. ACM **32**(3), 597–617 (1985)

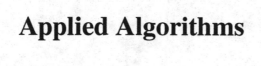

Applied Algorithms

Applied Algorithms

Variants of Euclidean k-Center Clusterings

Shin-ichi Nakano[✉]

Gunma University, Kiryu 376-8515, Japan
nakano@gunma-u.ac.jp

Abstract. Fix two constant integers k and r. Given a set P of n points on a plane, *the Euclidean k-center r-gather clustering problem* is the problem to compute a set $\{c_1, c_2, \cdots, c_k\}$ of k points on the plane and a partition $P_1 \cup P_2 \cup \cdots \cup P_k$ of P such that each P_i contains r or more points in P, and the maximum radius of $P_1, P_2, \cdots, P_k$ is minimized, where the radius of P_i (with center at c_i) is the maximum distance from $p \in P_i$ to c_i. If r is 0 then the problem is the ordinary Euclidean k-center clustering problem. This is a geometric version of the k-anonymity which is an important model for privacy preserving.

In this paper we design a polynomial-time algorithm to solve the Euclidean k-center r-gather clustering problem when k and r are constant integers. We also design polynomial-time algorithms to solve some related problems.

Keywords: k-center problem · r-gathering problem · Algorithms

1 Introduction

Given a set P of n points on a plane and a constant integer k, *the Euclidean k-center clustering problem* (also called minmax radius clustering problem, minmax Euclidean facility location problem, etc.) is the problem to compute a set C of k points such that the maximum Euclidean distance from a point in P to the nearest point in C is minimized. Intuitively our task is to locate a set of k identical disks covering P with the minimum radius. (Many similar and related problems are known. For instance, k-mean problem finds C minimizing the sum of the squared distances from a point in P to the nearest point in C, k-median problem finds C minimizing the sum of the distances from a point in P to the nearest point in C, or we need to find C either in P, a given candidate set, or anywhere in the plane. Also the distance is either arbitrary, metric, Euclidean, L_1, etc.) See surveys in Sect. 7.1 Euclidean p-Center of [1] and Sect. 8.5.1 MINMAX K-CLUSTERING of [6].

The Euclidean k-center clustering problem is NP-complete if k is a part of the input [19]. An $n^{O(\sqrt{k})}$ time exact (non-polynomial time) algorithm is known [15]. An approximation algorithm with approximation ratio at most 2 is known

© The Author(s), under exclusive license to Springer Nature Switzerland AG 2024
W. Wu and G. Tong (Eds.): COCOON 2023, LNCS 14422, pp. 209–219, 2024.
https://doi.org/10.1007/978-3-031-49190-0_15

[14], and it is NP-hard to design an approximation algorithm with approximation ratio less than 1.822 [12].

If k is a constant then one can solve the problem in $O(n^{2k+2})$ time [1], since basically every three points determine a possible disk D and every two other points determine possible two disks with the radius same to D, the number of k disks corresponding to possible solutions is at most $n^3 \cdot (2(n-3)^2)^{k-1}$, and we can check whether each set of k disks covers P or not in $O(n)$ time.

If P is a set of points on a line (1D case), one can solve (more general weighted version of) the problem in $O(n \log n)$ time [8]. If P is a set of points on a plane and we need to find the centers on a given line (1.5D case), one can solve the problem in $O(n \log^2 n)$ time [7] and $O(n \log n)$ time [16,21].

A solution $C = \{c_1, c_2, \cdots, c_k\}$ of the Euclidean k-center clustering problem generates a natural partition $P_1 \cup P_2 \cup \cdots \cup P_k$ of P in which P_i consists of points in P having the nearest point c_i in C. In the partition the size of each P_i does not matter, however in some application the size of each P_i is crucial. Thus some variants of the problem are known.

Problems with Bounded Cluster Size

Fix two constant integers k and r. Given a set P of n points on a plane, *the Euclidean k-center r-gather clustering problem* is the problem to compute a set $C = \{c_1, c_2, \cdots, c_k\}$ of k points on the plane and a partition $P_1 \cup P_2 \cup \cdots \cup P_k$ of P such that

(1a) each P_i contains r or more points in P, and
(2) the maximum radius of $P_1, P_2, \cdots, P_k$ is minimized, where the radius of P_i with center at c_i is the maximum distance from $p \in P_i$ to c_i.

For instance, this is a problem to assign n people to k shelters so that each shelter serves at least r people, where r is the minimum number of people to open a new shelter.

If r is 0 then the problem is the ordinary Euclidean k-center clustering problem.

Note that the partition may not be so natural because of (1a).

This is a geometric version of the k-anonymity, which is an important model for privacy preserving [2]. In the model to protect privacy we perturbate each data so that there are at least $k-1$ other data with the same value. Then it becomes difficult to finds the owner of each perturbated data even if one can access other open data containing the pair of the original data and its owners. Without the perturbation one may find the owner of the data, since the data may be unique before the perturbation. One can regard the perturbation as a partition of the original data into clusters (where each cluster has k or more data) and each original data is perturbated to the common data at the center to hide its owner. The data remains meaningful if the amount of perturbation is small.

By replacing the condition (1a) with the following (1b) we can define one more problem, called *the Euclidean k-center capacitated clustering problem*.
(1b) each P_i contains at most r points in P.

If r is n then the problem is the ordinary Euclidean k-center clustering problem.

This is a model for facility location in which each facility can serve at most r customers.

We can define more general problem, as follows. Given a set P of points on a plane and three constant integers k, ℓ and u, *the Euclidean k-center (ℓ, u)-clustering problem* is the problem to compute a set $C = \{c_1, c_2, \cdots, c_k\}$ of k points on the plane and a partition $P_1 \cup P_2 \cup \cdots \cup P_k$ of P such that
(1c) each P_i contains at least ℓ and at most u points in P, and
(2) the maximum radius of $P_1, P_2, \cdots, P_k$ is minimized.

If $\ell = 0$ and $u = n$ then the problem is the ordinary Euclidean k-center clustering problem.

Known Results and Our Results
Euclidean k-Center r-Gather Clustering Problem
The followings are known for the Euclidean k-center r-gather clustering problem.

If P is a set of points on a plane, the number of k is not restricted, and the distance is metric, then the problem is NP-hard [2]. If P is a set of points on a line and the number k is not restricted, then an $O(rn)$ time algorithm to solve the problem is known [20].

For some application we can regard that k and r are fixed integers, say open 4 shelters so that each shelter serves 50 or more people.

In this paper when k and r are both fixed integers we design (1) a polynomial-time algorithm to solve the Euclidean k-center r-gather clustering problem, (2) a polynomial-time algorithm to solve the problem in 1.5D, explained as follows.

Given a set P of points and a horizontal line L (1D) on a plane (2D) and two constant integers k and r, we want to compute a set $\{c_1, c_2, \cdots c_k\}$ of k points on L and a partition of P into k clusters $P_1 \cup P_2 \cup \cdots \cup P_k$ so that (1) each cluster P_i contains r or more points in P, and (2) the maximum Euclidean distance from $p \in P_i$ to c_i is minimized. We call the problem the Euclidean k-center r-gather clustering problem in 1.5D. Intuitively we need to locate shelters on the main road in a town.

Euclidean k-Center Capacitated Clustering Problem
The followings are known for the Euclidean k-center capacitated clustering problem.

If we need to find C among P (a graph version), several approximation algorithms are known [5,10,17]. An inapproximability result for more general problem is known [10,18].

For a similar problem which computes a partition of P into a minimum number of clusters such that each cluster contains at most r points in P and each cluster is covered by a unit disk, a PTAS is known [13].

Even if P is on a line (1D case), if each center has a possibly distinct capacity, say r_i, the problem is NP-hard, however if each center has identical capacity, say r, then one can solve the problem in $O(rkn)$ time [4].

In this paper, when k and r are both fixed integers, we design (1) a polynomial-time algorithm to solve the Euclidean k-center capacitated clustering problem, and (2) a polynomial-time algorithm to solve the problem in 1.5D.

Euclidean k-Center (ℓ, u)-Clustering Problem

For the Euclidean k-center (ℓ, u)-clustering problem, a constant factor approximation algorithms is known even for more general (a non-uniform upper bound) constraints [11].

In this paper, when k, ℓ and u are fixed integers, we design (1) a polynomial-time algorithm to solve the Euclidean k-center (ℓ, u)-clustering problem, (2) a polynomial-time algorithm to solve the problem in 1.5D. We also design (3) an $O(kun)$ time algorithm to solve the problem if P is a set of points on a line.

The rest of the paper is organized as follows. In Sect. 2 we give some observations. In Sect. 3 we give algorithms to solve the Euclidean k-center r-gather clustering problem. In Sect. 4 we give algorithms to solve the Euclidean k-center capacitated clustering problem. In Sect. 5 we give algorithms to solve the Euclidean k-center (ℓ, u)-clustering problem. Finally Sect. 6 is a conclusion.

2 Preliminaries

In this section we give some observations for the k-center r-gather clustering problem.

Let $_xC_y$ be the number of ways to choose y elements from a set of x distinct objects, and $_xC_y = x!/((x-y)!y!) \leq x^y$.

A solution of the k-center r-gather clustering problem consists of a set $\{c_1, c_2, \cdots, c_k\}$ of k points on a plane and a partition $P_1 \cup P_2 \cup \cdots \cup P_k$ of P.

For each solution we can construct k disks $D_1, D_2, \cdots, D_k$ with the same radius rad such that P_i is covered by disk D_i for each i.

We have the following two observations if P is a set of points on a plane (2D case).

Observation 1: For each solution one can assume that the corresponding k disks contain a disk having two or more points in P on the boundary. Otherwise there is a solution with less rad, a contradiction. Either the disk has three or more points in P on the boundary, or the disk has two points in P on the boundary with the distance equal to the diameter.

Thus the number of possible maximum radius is at most $_nC_3 +_n C_2 \leq 2n^3$.

Observation 2: For each solution one can assume that each disk in the corresponding k disks has at least two points in P on the boundary. (Otherwise we can move each disk so that two points in P appear on the boundary and the set of points in P covered by the disk remains the same.)

We have the following two observations if P is a set of points on a plane and C is on a given horizontal line (1.5D case).

Observation 3: For each solution one can assume that the corresponding k disks contain a disk having one or more points in P on the boundary. Otherwise there is a solution with less rad, a contradiction. Either the disk has two or more points in P on the boundary, or has one point in P on the boundary with exactly above the center of the disk.

Thus the number of possible maximum radius is at most $_nC_2 + n \le 2n^2$.

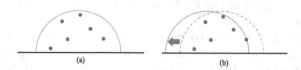

Fig. 1. An example of the move operation of a disk to the left. (a) Before the move and (b) after the move.

Observation 4: For each solution one can assume that each disk in the corresponding k disks has a point in P on the boundary. (Otherwise we can move each disk so that a point in P appears on the boundary and the set of points in P covered by the disk remains the same. See an example in Fig. 1.)

3   Algorithms for the Euclidean k-Center r-Gather Clustering Problem

In this section we design a polynomial-time algorithms to solve the Euclidean k-center r-gather clustering problem. The algorithm solves a set of the maximum flow problems with lower bounds, defined below. Then we design a similar polynomial-time algorithms to solve the problem in 1.5D.

Given a network $N = (V, A)$ with two vertices $s, t \in V$, and a capacity $u(i,j) \ge 0$ and a lower bound $\ell(i,j) \ge 0$ associated with each arc $(i,j) \in A$, a flow consisting of $f(i,j)$ for each arc $(i,j) \in A$ is *feasible* if $\sum_{(i,x) \in A, x \in V} f(i,x) = \sum_{(x,i) \in A, x \in V} f(x,i)$ for each $i \in V/\{s,t\}$ and $\ell(i,j) \le f(i,j) \le u(i,j)$ for each $(i,j) \in A$. *The maximum flow problem with lower bounds* is the problem to compute a feasible flow with the maximum $\sum_{(s,x) \in A, x \in V} f(s,x)$. One can solve the problem in polynomial time, by a reduction to the ordinary maximum flow problem [3], where the ordinary maximum flow problem is the maximum flow problem with $\ell(i,j) = 0$ for each $(i,j) \in A$. The reduction needs $O(|V| + |A|)$ time. So one can solve the maximum flow problem with lower bounds in polynomial time using an algorithm to solve the ordinary maximum flow problem, say in $O(|V|^3)$ time (Theorem 26.30 in [9]).

Now we explain a reduction from the Euclidean k-center r-gather clustering problems to a set of the maximum flow problems with lower bound.

Every two or three points in P possibly define the disk D having the points on the boundary (assuming that the three points are not on a line). Fix D. The number of possible radii is at most $2n^3$ by Observation 1. Each other two points in P possibly define the two disks having the two points on the boundary and with the radius same to D. Thus by Observation 1 and 2 the number of the sets of k disks possibly corresponding to a solution of the Euclidean k-center r-gather clustering problem is at most $2n^3(2n^2)^{k-1}$.

For each set of such possible k disks $\{D_1, D_2, \cdots, D_k\}$ with the same radius and with centers $C = \{c_1, c_2, \cdots, c_k\}$, we check if the set of k disks actually correspond to a solution of the Euclidean k-center r-gather clustering problem or not, as follows.

First we check if the k disks cover P or not in $O(n)$ time. If the k disks cover P then we construct the following instance of the maximum flow problem with lower bounds (See Fig. 2) in $O(kn)$ time, and solve it in $O(n^3)$ time.

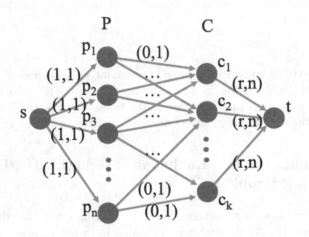

Fig. 2. The network derived from an Euclidean k-center r-gathering clustering problem.

Let $N = (V, A)$, where
$V = \{s\} \cup P \cup C \cup \{t\}$, and
$A = \{(s,p)|p \in P\} \cup \{(p,c_1)|D_1 \text{ contains } p \in P\} \cup$
$\{(p,c_2)|D_2 \text{ contains } p \in P\} \cup \cdots \cup \{(p,c_k)|D_k \text{ contains } p \in P\} \cup \{(c,t)|c \in C\}$.
$\ell(s,p) = 1$ and $u(s,p) = 1$ for each $(s,p) \in A$ with $p \in P$,
$\ell(p,c) = 0$ and $u(p,c) = 1$ for each $(p,c) \in A$ with $p \in P$ and $c \in C$,
$\ell(c,t) = r$ and $u(c,t) = n$ for each $(c,t) \in A$ with $c \in C$.

One can solve each maximum flow problem with lower bounds above in $O(n^3)$ time (Theorem 26.30 in [9]), since $|V| = 1 + n + k + 1$ and k is a constant. The number of the maximum flow problems with lower bounds is at most $2n^3 \cdot (2n^2)^{k-1}$, so the total time to solve the problems is $O(2n^3 \cdot (2n^2)^{k-1} \cdot n^3) = O(n^{3+2(k-1)+3}) = O(n^{2k+4})$.

If the maximum flow problem with lower bounds has a solution f then it generates a partition of $P = P_1 \cup P_2 \cup \cdots \cup P_k$ where $P_i = \{p|f(p,c_i) = 1\}$, and this partition and the set $\{c_1, c_2, \cdots, c_k\}$ corresponds to a possible solution of the k-center r-gather clustering problem. Note that each P_i has r or more points in P since $\ell(c_i, t) = r$ for each i.

Thus we have the following theorem.

Theorem 1. *One can solve the k-center r-gather clustering problem on a plane in $O(n^{2k+4})$ time.*

By binary search in at most $\log 2n^3$ stages to compute the minimum of the maximum radius of the k disks corresponding to a solution of the k-center r-gather clustering problem, one can improve the running time to $O(n^{2k+1} \log n)$.

Next we consider for 1.5D case. We need to find centers on the horizontal line L.

Every one or two points in P define the disk D having the points on the boundary and having the center at a point on L. The number of such D is at most $2n^2$ by Observation 3. Each other point in P define at most two disks having the point on the boundary (those are the disk having the point on the left half of the boundary and the disk having the point on the right half of the boundary) with the radius same to D, and has the center at a point on L. Thus the number of the set of k disks possibly corresponding to a solution of the Euclidean k-center r-gather clustering problem in 1.5D is at most $2n^2(2n)^{k-1}$. One can solve each maximum flow problem corresponding to each set of k disks in $O(n^3)$ time.

We have the following theorem.

Theorem 2. *One can solve the k-center r-gather clustering problem in 1.5D in $O(n^{k+4})$ time.*

4   Algorithms for the Euclidean k-Center Capacitated Clustering Problem

By slightly modifying the algorithm in Sect. 3 we can design a polynomial-time algorithms to solve the Euclidean k-center capacitated clustering problem and the Euclidean k-center capacitated clustering problem in 1.5D. The only difference is to set $\ell(c,t) = 0$ and $u(c,t) = r$ for each $(c,t) \in A$ with $c \in C$. (See Fig. 3.)

We have the following two theorem.

Theorem 3. *One can solve the k-center capacitated clustering problem on a plane in $O(n^{2k+4})$ time.*

Theorem 4. *One can solve the k-center capacitated clustering problem in 1.5D in $O(n^{k+4})$ time.*

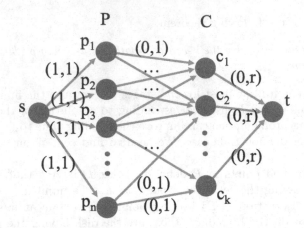

Fig. 3. The network derived from an Euclidean k-center capacitated clustering problem.

5 Algorithms for the Euclidean k-Center (ℓ, u)-Clustering Problem

By slightly modifying the algorithm in Sect. 3 we can design a polynomial-time algorithms to solve the Euclidean k-center (ℓ, u)-clustering problem and the Euclidean k-center (ℓ, u)-clustering problem in 1.5D. The only difference is to set $\ell(c, t) = \ell$ and $u(c, t) = u$ for each $(c, t) \in A$ with $c \in C$.

We have the following two theorems.

Theorem 5. *One can solve the k-center (ℓ, u)-clustering problem in $O(n^{2k+4})$ time.*

Theorem 6. *One can solve the k-center (ℓ, u)-clustering problem in 1.5D in $O(n^{k+4})$ time.*

We can design an $O(\ell u n)$ time algorithm to solve the k-center (ℓ, u)-clustering problem in 1D, that is the k-center (ℓ, u)-clustering problem when P is on a line. We assume that P is given with sorted order in x-coordinates. Our algorithm is similar to the algorithm in [20] which solve the Euclidean r-gather clustering problem on a line.

One can observe that there exists a solution in which the points in each cluster P_i are consecutive in x-coordinate. We can assume that each cluster in a solution consists of consecutive points $\{p_i, p_{i+1}, \cdots, p_j\}$ for some i and j.

Now we define the directed (acyclic) graph $D(V, E)$ and the weight of each edge, as follows.

$$V = \{v_0, v_1, v_2, \cdots, v_n\}$$
$$E = \{(v_i, v_j) | i + \ell \le j \le i + u\}$$

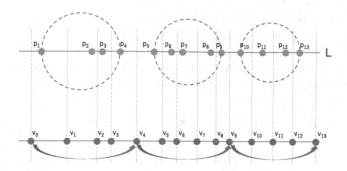

Fig. 4. The path corresponding to a solution of an Euclidean k-center ($\ell = 4, u = 5$)-clustering problem in 1D.

Intuitively each v_i corresponds to the midpoint of p_i and p_{i+1}, and each edge is directed from left to right. Note that $|V| = |P| + 1$ holds because of v_0. Also note that the number $|E|$ of edges is at most un. The weight of an edge (v_i, v_j) is the half of the distance between p_{i+1} and p_j, and denoted by $w(v_i, v_j)$.

We define the cost of a directed path from v_0 to v_n as the weight of the edge having the maximum weight in the directed path. *The min-max path* from v_0 to v_n is the directed path from v_0 to v_n with the minimum cost.

Now there is a solution of the Euclidean k-center (ℓ, u)-clustering problem with the maximum radius *rad* iff $D(V, E)$ has a directed path from p_0 to p_n with k edges and cost *rad*. See Fig. 4.

We can construct the $D(V, E)$ in $O(un)$ time. Since $D(V, E)$ is directed acyclic we can compute the min-max path from v_0 to v_n with k edges in $O(kun)$ time by a simple dynamic programming algorithm. (Let $w_{i,k'}$ be the cost of the min-max path from v_0 to v_i with k' edges. For each v_i and k', we can compute $w_{i,k'}$ by checking each incoming edge (v_x, v_i) to v_i and the cost $w_{x,k'-1}$ of the min-max path from v_0 to v_x with $k' - 1$ edges.) The total running time is $O(kun)$.

Since k and u are constants we have the following theorem.

Theorem 7. *One can solve the k-center (ℓ, u)-clustering problem in 1D in* $O(kun) = O(n)$ *time.*

6 Conclusion

In this paper we have designed a polynomial-time algorithm to solve the Euclidean k-center r-gather clustering problem, where k and r are constant integers. The running time of the algorithm is $O(n^{2k+4})$. Additionally we design a similar algorithm to solve the problems in 1.5D. The running time of the algorithm is $O(n^{k+4})$.

Also by slightly modifying the algorithm in Sect. 3 we have designed similar polynomial-time algorithms to solve the Euclidean k-center capacitated clustering problem and the Euclidean k-center (ℓ, u)-clustering problem.

Also we have designed a simple $O(kun) = O(n)$ time algorithm to solve the Euclidean k-center (ℓ, u)-clustering problem if P is on a line.

Given a radius rad can we solve the decision version of the problems in 1.5D efficiently? Note that each cluster may not consist of consecutive set of points in P with respect to their x-coordinates. See Fig. 5.

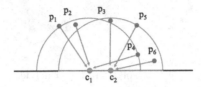

Fig. 5. An example in which a center may serve non-consecutive points with respect to their x-coordinates, where $r = 3$ and c_1 serves $\{p_1, p_2, p_4\}$ and c_2 serves $\{p_3, p_5, p_6\}$.

References

1. Agarwal, P., Sharir, M.: Efficient algorithms for geometric optimization. ACM Comput. Surv. **30**, 412–458 (1998)
2. Aggarwal, G., et al.: Achieving anonymity via clustering. Trans. Algorithms **6**(3), 1–19 (2010)
3. Ahuja, R.K., Magnanti, T.L., Orlin, J.B.: Network Flows - Theory, Algorithms and Applications. Prentice Hall, Upper Saddle River (1993)
4. Aziz, H., Chan, H., Lee, B., Li, B., Walsh, T.: Facility location problem with capacity constraints: algorithmic and mechanism design perspectives. In Proceedings of AAAI 2020, pp. 1806–1813. AAAI Press (2020)
5. Bar-Ilan, J., Kortsarz, G., Peleg, D.: How to allocate network centers. J. Algorithms **15**(3), 385–415 (1993)
6. Bern, M., Eppstein, D.: Approximation algorithms for geometric problems. Approximation algorithms for NP-hard problems, pp. 296–345 (1996)
7. Brass, P., Knauer, C., Na, H., Shin, C., Vigneron, A.: The aligned k-center problem. Int. J. Comput. Geom. Appl. **21**(2), 157–178 (2011)
8. Chen, D., Li, J., Wang, H.: Efficient algorithms for the one-dimensional k-center problem. Theor. Comput. Sci. **592**, 135–142 (2015)
9. Cormen, T.H., Leiserson, C.E., Rivest, R.L., Stein, C.: Introduction to Algorithms, 3rd Edition. MIT Press, Cambridge (2009)
10. Cygan, M., Hajiaghayi, M.T., Khuller, S.: LP rounding for k-centers with non-uniform hard capacities. In: Proceedings of FOCS 2012, pp. 273–282. IEEE Computer Society (2012)
11. Ding, H., Hu, L., Huang, L., Li, J.: Capacitated center problems with two-sided bounds and outliers. In: WADS 2017. LNCS, vol. 10389, pp. 325–336. Springer, Cham (2017). https://doi.org/10.1007/978-3-319-62127-2_28
12. Feder , T., Greene, D.H.: Optimal algorithms for approximate clustering. In: Proceedings of STOC 1988, pp. 434–444. ACM (1988)

13. Ghasemi, T., Razzazi, M.: A PTAS for the cardinality constrained covering with unit balls. Theor. Comput. Sci. **527**, 50–60 (2014)
14. Gonzalez, T.F.: Clustering to minimize the maximum intercluster distance. Theor. Comput. Sci. **38**, 293–306 (1985)
15. Hwang, R.Z., Richard C. T. Lee, and R. C. Chang. The slab dividing approach to solve the Euclidean p-center problem. Algorithmica **9**(1), 1–22 (1993). https://doi.org/10.1007/BF01185335
16. Karmakar, A., Das, S., Nandy, S.C., Bhattacharya, B.K.: Some variations on constrained minimum enclosing circle problem. J. Comb. Optim. **25**, 176–190 (2013)
17. Khuller, S., Sussmann, Y.J.: The capacitated K-center problem. SIAM J. Discret. Math. **13**(3), 403–418 (2000)
18. Kumar, A.: Capacitated k-center problem with vertex weights. In: Proceedings of FSTTCS 2016, vol. 65 of LIPIcs, pp. 8:1–8:14. Schloss Dagstuhl - Leibniz-Zentrum für Informatik (2016)
19. Megiddo, N., Supowit, K.: On the complexity of some common geometric location problems. SIAM J. Comput. **13**, 182–196 (1984)
20. Nakano, S.: A simple algorithm for r-gatherings on the line. J. Graph Algorithms Appl. **23**(5), 837–845 (2019)
21. Wang, H., Zhang, J.: Line-constrained k-median, k-means, and k-center problems in the plane. Int. J. Comput. Geom. Appl. **26**, 185–210 (2016)

Red-Black Spanners for Mixed-Charging Vehicular Networks

Sergey Bereg[1] , Yuya Higashikawa[2], Naoki Katoh[2], Junichi Teruyama[2],
Yuki Tokuni[2], and Binhai Zhu[3(✉)]

[1] Department of Computer Science, University of Texas at Dallas,
Richardson, TX 75080, USA
besp@utdallas.edu
[2] Graduate School of Information Science, University of Hyogo, Kobe, Japan
{higashikawa,naoki.katoh,junichi.teruyama,ad21o040}@gsis.u-hyogo.ac.jp
[3] Gianforte School of Computing, Montana State University,
Bozeman, MT 59717, USA
bhz@montana.edu

Abstract. Motivated by the recent trend of increasing number of e-cars
and hybrid cars, we investigate the problem of building a red-black spanner for a mixed-charging vehicular network. In such a network, we have
two kinds of gas/charging stations: electric (black) and the traditional gas
(red) stations. Our requirement is that one cannot connect two gas stations directly in the spanner (i.e., no red-red edge), and our goal is to build
a linear-size spanner with a bounded stretch factor under this requirement.
(In 2-d, it can be shown that a spanner with an optimal stretch factor could
have a quadratic size and if one is restricted to build the spanner purely
from a given road network then it is impossible to obtain a bounded stretch
factor.) Our main results are summarized as follows.

1. In 1-d, we show a linear-size red-black spanner satisfying the 'no
   red-red edge' requirement which achieves the optimal stretch factor.
2. In 2-d and under the L_2 metric, we show a linear-size red-black
   spanner satisfying the 'no red-red edge' requirement which achieves
   a stretch factor of 1.998.
3. In 2-d and under the L_1 metric, we show a linear-size red-black
   spanner satisfying the 'no red-red edge' requirement which achieves
   a stretch factor of 3.613.

Keywords: Geometric spanners · Delaunay triangulations ·
Approximation algorithms

1 Introduction

As early as in 1990, in Daniel Sperling's book [14], it was pointed out that the
traditional gas will not last forever and new technologies must be developed to
drive our vehicles. Since then, new technologies have resulted in electric vehicles

This work is partially supported by JSPS KAKENHI Grant Number 19H04068.

(e-vehicles), most notably, electric cars (e-cars) and they are becoming more and more popular nowadays, with Tesla as an eminent example. This is mostly related to the common belief that e-vehicles are more environmental friendly and are more sustainable. On the other hand, it is known that, up to this point, e-vehicles typically have relatively limited mileage (especially in cold regions), hence need to be recharged more frequently or swapped for new batteries in those regions — consequently, the charging stations should be designed and placed to capture e-car flows.

In the past decade, a lot of research has been done in this aspect; for example, Zhu and Pei studied how to locate battery swapping stations for e-bicycles [18], Kazemi et al. and Sadeghi-Barzani et al. studied how to minimize the cost of designing charging infrastructure to capture more e-car flows [9,13], Wang et al. studied how to site and size fast charging stations with budget and service capability constraints in a highway network [15], and Liang et al. studied how to site charging stations based on a Voronoi partition method [11]. For more and related references, we refer the readers to [11,15].

In this paper, we focus on hybrid cars, i.e., those that can be powered by both battery and gas. Typically, the battery power of such a car is far less compared with the regular e-cars. For example, the Chevrolet Volt can only be driven for a range of about 50 mi using a fully charged battery alone, while its gas tank can easily sustain a distance of over 250 mi. This raises an interesting question: how do we plan trips where we could charge a hybrid car more frequently by electricity when needed?

In this paper, we make the first step in answering this question. Instead of constructing new charging stations, we are more concerned with path planning using existing gas (red) and fast-charging (black) stations to form a restricted road/highway network T. Given any two sites p and q, we would like to construct a path between them in T such that every edge on the path must have at least one endpoint of black color (i.e., no red-red edge is allowed); moreover, the length of this path is no more than t times the shortest distance between p and q in a complete graph among all gas/charging stations — of course, with red-red edges removed. (Note that in a more general setting, we could require that any path between p and q in T must have at least C fraction of nodes in black. Then, the above 'no red-red edge' requirement satisfies the minimum $C = 1/3$, matching the lower bound when both p and q are red and they are connected through a black node r.)

Naturally, this problem is related to the traditional t-spanner research (t is called the *stretch factor*), with all the sites uncolored. (Clearly, in the traditional spanners if $t = 1$ then we would need the complete graph over all the sites. But usually we allow t to be greater than one, though ideally bounded from above by some constant.) To distinguish with the traditional spanners, we term our spanners as *red-black spanners*. We next briefly review the literature that is the most relevant to our research, i.e., plane t-spanners.

Chew pioneered the research on planar t-spanners by showing that for a set of planar points P there is a planar 2-spanner [5]. Subsequently, Dobkin, Friedman and Supowit showed that the Delaunay triangulation of P (under the L_2 distance) gives a 5.08-spanner [6]. This stretch factor was subsequently improved to 2.42 by Keil and Gutwin [10]. Finally, the stretch factor of the Delaunay triangulation of P was shown to be bounded by 1.998 by Xia [16]. On the other hand, the best lower bound for the stretch factor of the Delaunay triangulation of P is 1.5932 [17]. (Throughout the paper, we use $\lambda_2(P)$, or simply λ_2 when the point set P is clearly given, to represent the stretch factor of the Delaunay triangulation of P.)

For the Delaunay triangulation of P under the L_1 distance, Bonichon et al. showed a tight stretch factor of $\sqrt{4 + 2\sqrt{2}} \approx 2.613$ [2]. (Throughout the paper, we use $\lambda_1(P)$, or simply λ_1, to represent the stretch factor of the Delaunay triangulation of P under L_1 distance.) For more variants of planar t-spanners, readers are referred to the survey paper by Bose and Smid [3], or the book by Narasimhan and Smid [12]. For example, Biniaz et al. [1] studied spanners of low degree and Dumitrescu and Ghosh [7] studied lower bounds on the dilation of plane spanners.

Coming back to this paper and as discussed earlier, motivated by the above mixed-charging networks for hybrid cars, we study the red-black spanner problem for a set of points. Note that if we aim at optimal stretch factors we might need a spanner with a quadratic size. In Fig. 1, we show such an example. In this example, the building block is a square of 7 points, 3 black and 4 red; moreover, the middle column are all black points. It can be easily verified that between any two red points in column-1 and column-3 we must connect them through the black point in column-2 (which is exactly the midpoint). This immediately gives a quadratic lower bound if we aim at obtaining the optimal stretch factors. Hence, in the following we focus on linear size red-black t-spanners; of course, we just aim at constant t's.

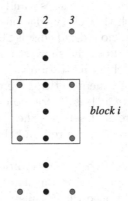

Fig. 1. An example for the optimal red-black spanner with quadratic size. (Color figure online)

We summarize our results as follows.

1. In 1-d, we show a linear size spanner satisfying the 'no red-red edge' requirement which achieves the optimal stretch factor.

2. In 2-d and under the L_2 metric, we show a linear size red-black spanner satisfying the 'no red-red edge' requirement which achieves a stretch factor of 1.998.
3. In 2-d and under the L_1 metric, we show a linear size red-black spanner satisfying the 'no red-red edge' requirement which achieves a stretch factor of 3.613.

This paper is organized as follows. In Sect. 2, we give necessary definitions. In Sect. 3 we solve the 1-d case. In Sect. 4 we handle the 2-d case under the L_2 distance and we also cover the lower bounds. We cover the L_1 case in Sect. 5, but most details are omitted due to space constraint. We close the paper in Sect. 6 with quite a few open problems.

2   Preliminaries

We make some necessary definitions. Given two points $u = (x_u, y_u)$ and $v = (x_v, y_v)$, the Euclidean (or L_2) distance between u and v is $d_2(u,v) = d(u,v) = |uv| = \sqrt{(x_u - x_v)^2 + (y_u - y_v)^2}$ and the Manhattan (or L_1) distance between them is denoted as $d_1(u,v) = |uv|_1 = |x_u - x_v| + |y_u - y_v|$. While the L_p metric could be general, we only focus on $p = 1, 2$ in this paper.

In this paper, we study spanners that can be applied to the motion planning or routing of electric/hybrid vehicles. We use two types of vertices: red (resp. black) vertices for points with gas (resp. electric) charging stations. A desired property of a spanner is that every path in the spanner enables convenient electric charging. This can be further expressed as a simple condition: every edge of the spanner has at least one endpoint of black color. We call a spanner satisfying this property a *red-black spanner* or *RB-spanner* for short. Note that any path with k vertices in an *RB-spanner* has at least $\lfloor k/2 \rfloor$ black vertices.

Let $P = R \cup B$ be a set of N points in $\mathbb{R}^d$ where R and B are sets of red and black points. We define a *complete RB-spanner* $K_{R,B}$ as $(R \cup B, E)$ where $(u,v) \in E$ iff u or v is black. Let $\delta_p(u,v)$ denote the length of the shortest path between u and v in $K_{R,B}$ under L_p metric. A general problem asks for an RB-spanner S in L_p (satisfying some condition, e.g., size and the 'no red-red edge' condition) which minimizes the *stretch factor* (sometimes also called *spanning ratio*)

$$\alpha_p(S) = \max_{u,v \in R \cup B} \frac{d_{p,S}(u,v)}{\delta_p(u,v)} \tag{1}$$

where $d_{p,S}(u,v)$ is the length of the shortest path between u and v in S under L_p. Again, when S is clear from the context, we could simply use α_p to denote the stretch factor of S. Similarly, when the L_p metric is clear from the context we would use $d_S(u,v)$ to denote $d_{p,S}(u,v)$.

Regarding some constraint on S, there is certainly a trade-off between the size of an RB-spanner and the stretch factor, reflected in Fig. 1. As another example, if an RB-spanner is a spanning tree of $R \cup B$ in the plane then the stretch factor is unbounded. This can be seen in an example where the points of $R \cup B$ are the vertices of a regular N-gon, say $v_1, v_2, \ldots, v_N$, and a vertex v_i is red/black if i is even/odd, respectively. Then (i) every two adjacent vertices along the boundary of the polygon are not both red and (ii) the same proof for the dilation of a spanning tree on the vertices of a regular polygon can be applied [8].

Also, notice that in the $\alpha_p(S)$ definition the denominator is not $d_2(u, v)$. This is different from the traditional geometric spanners, mainly because there is no red-red edge in $K_{R,B}$. In Fig. 2(I), such an example shows that had we used $d_2(u, v)$ as the denominator in the definition of $\alpha_p(S)$, the stretch factor would have been unbounded. (We could set the distance between the two red points $|pq|$ arbitrarily small, while the distance between two black points u and v is a large constant; hence to connect p, q we need a distance of at least $2|pu|$ in any spanner S, which is a large constant as well. Then, $2|pu|/|pq| = +\infty$.)

A naive though practical question is: using this new definition, can we simply use an existing road network on the given point sites to build a good spanner? Unfortunately, the answer is no. In Fig. 2(II), we augment the set P of 4 points into 6, by adding two black points r, s such that $\triangle rpq$ and $\triangle spq$ form equilateral triangles. If the existing network is a polygon $Q = \langle u, r, p, v, s, q \rangle$, then any spanner S would preserve the property that the distance of a path between p, q in S is at least $2|pu|$. However, even in $K_{R,B}$ the minimum distance between p, q is at most $2|pq|$. Hence, no matter what algorithm is used, for this example, with the existing network Q given, the stretch factor of any spanner S on these 6 points is unbounded (i.e., at least $(2|pu|)/(2|pq|) = +\infty$).

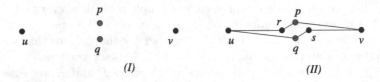

Fig. 2. (I) A set P of 4 points with $d(p, q) = \varepsilon$; and (II) An existing network which is also a polygon, with $\triangle rpq$ and $\triangle spq$ being equilateral.

Summarizing the above discussion, we focus on computing a linear-size RB-spanner in the following sections.

3   Red-Black Spanner in 1-D

Given a set R of n red points and a set B of $m(\geq 1)$ black points on the real line, the problem asks for a spanner T minimizing $\max\{d_T(x, y)/\delta_2(x, y), \forall x, y \in P\}$, s.t. there is no red-red edge in T, where $d_T(x, y)$ denotes the distance between x and y in the spanner T and $\delta_2(x, y)$ denotes the shortest path distance between x and y in $K_{R,B}$. The following algorithm constructs an optimal spanner T with at most $2n + m$ edges as follows.

1. Sort all the points from left to right. Organize them into maximal red (resp. black) blocks, say $B_1, R_1, B_2, R_2, \cdots, B_z, R_z$ (one or both of B_1 and R_z could be empty).
2. Connect all black points in B_i from left to right as a chain.
3. Connect every point in R_j to the closest black neighbors to the left and to the right (if exists). That is, we connect every point in R_j to the rightmost black point in B_j and to the leftmost one in B_{j+1} if exists.
4. Return T as the spanner.

Note that T is not a tree. We have the following lemma.

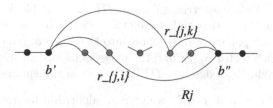

Fig. 3. Illustration for the proof of Lemma 1.

Lemma 1. *Let $R_j = \langle r_{j,1}, r_{j,2}, \cdots, r_{j,\ell} \rangle$ be a maximal block of red points. Then, $d_T(r_{j,i}, r_{j,k}) = \delta_2(r_{j,i}, r_{j,k})$ holds for $i < k$.*

Proof. We refer to Fig. 3. By the algorithm, $r_{j,i}$ is connected to the closest black point b' to its left and is also connected to the closest black point b'' to the right. Similar connections are done for $r_{j,k}$. Then the shortest path between $r_{j,i}$ and $r_{j,k}$ must be either $\langle r_{j,i}, b', r_{j,k} \rangle$ or $\langle r_{j,i}, b'', r_{j,k} \rangle$. It can be easily seen that any path connecting $r_{j,i}$ and $r_{j,k}$ and either containing both b' and b'', or containing a point to the left of b' or to the right of b'' is longer than the shorter of the length of $\langle r_{j,i}, b', r_{j,k} \rangle$ or $\langle r_{j,i}, b'', r_{j,k} \rangle$. □

Between two black points, or a red point and a black point, it is easily seen that their distance on T is exactly their Euclidean distance. Since we connect all black points as chains, the number of black-black edges is at most m. Because every red point is connected at most two black points, the number of red-black edges is at most $2n$. Therefore, we have the following theorem.

Theorem 1. *Given n red points R and $m(\geq 1)$ black points B on a line, a red-black spanner T of at most $2n+m$ edges can be constructed in $O((n+m)\log(n+m))$ time; moreover, T achieves the optimal stretch factor (i.e., one).*

4   Red-Black Spanner Under the Euclidean Distance in 2-D

4.1   Constructing the Red-Black Spanner

In this section, we investigate the red-black spanner construction for a set R of n red points and a set B of $m(\geq 1)$ black points under the Euclidean distance in 2-d. Recall that in such a spanner T there is no red-red edge, T must have a linear size and we want to minimize

$$\alpha_2(T) = \max_{x,y \in R \cup B} \frac{d_T(x,y)}{\delta_2(x,y)}.$$

As we just discussed in the previous section, a tree spanner might not be able to to achieve a bounded stretch factor in the plane. Hence we try to construct T by augmenting a Delaunay triangulation, whose stretch factor is $\lambda_2 = 1.998$ [16]. We show the algorithm for constructing a spanner T.

1. Construct the Delaunay triangulation $DT_2(B)$ of black points B. Let $\delta_{2,B}(u, v)$ be the shortest path between two black points u and v on $DT_2(B)$.
2. For each red point p, partition the plane into 13 sectors $S_p(i), i \in \{1, 2, \ldots, 13\}$ around p. In each sector $S_p(i)$, find the closest black point under L_2, if exists, and add this red-black edge into $DT_2(B)$ to form the spanner T.

To analyze the stretch factor of T, we give an algorithm for routing two points on T as follows.

1. To find a route from a red point p to a black point u, in each sector $S_p(i)$ find the closest black point v_i, then measure $|pv_i| + \delta_{2,B}(u, v_i)$. Return the minimum $\min_{i \in \{1,\ldots,13\}}\{|pv_i| + \delta_{2,B}(u, v_i)\}$.
2. To find a route from a red point p to a red point q, in each sector $S_p(i)$ find the closest black point v_i, and in each sector $S_q(j)$ find the closest black point w_j then measure $|pv_i| + \delta_{2,B}(v_i, w_j) + |w_jq|$. Return the minimum $\min_{i,j \in \{1,\ldots,13\}}\{|pv_i| + \delta_{2,B}(v_i, w_j) + |w_jq|\}$.

Lemma 2. *Given an acute isosceles triangle $\triangle ABC$ with $|AB| = |AC|$ being the long edges (i.e., $|AB| > |BC|$), and D being an point inside $\triangle ABC$, then when $\angle BAC = \frac{2\pi}{13}$ we have $|AD| + 1.998|CD| < 1.998|AB|$.*

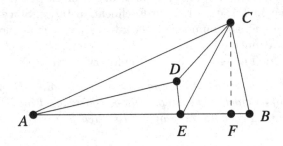

Fig. 4. Illustration of the proof of Lemma 2.

Proof. We refer to Fig. 4. First of all, we prove that D must lie on AB or AC, due to symmetry, say on AB, to achieve the maximum of $|AD| + 1.998|CD|$. Suppose it is not the case and D is properly inside $\triangle ABC$, we look at the isoceles triangle $\triangle ADE$, where E is on AB and $|AD| = |AE|$. We have $0 < \angle DAE$. Consequently, $\angle ADE < \pi/2$. Hence $\angle ADC + \angle CDE > 2\pi - \angle ADE > \frac{3}{2}\pi$, and $\angle CDE > \pi/2$ (as $\angle ADC < \pi$). As a matter of fact, $|CE| > |CD|$. Therefore, $|AD| + 1.998|CD| < |AE| + 1.998|CE|$.

When D lies on edge AB, we can see that $|AD|$ increases when D is moving from A to F (CF is perpendicular to AB), while $|CD|$ decreases. When D is moving from F to B, $|AD|$ keeps increasing and $|CD|$ also increases. Hence $|AD| + 1.998|CD|$ achieves the maximum value when D is at A (this solution is not relevant) or at B (that is the place for us to decide the maximum θ where $|AD| + 1.998|CD|$ achieves the maximum value of $1.998|AB|$).

Let θ denote $\angle BAC$ and let us first use a constant 2 instead of 1.998. With a simple calculation, when $\sin\frac{\theta}{2} = \frac{1}{4}$, $2|BC| = |AB|$. Hence, $\cos\theta = 1 - 2\sin^2\frac{\theta}{2} = \frac{7}{8}$. Consequently, in this case $\theta = \arccos\frac{7}{8} \approx 28.96°$.

When we select $\theta = \frac{2\pi}{13} \approx 27.6923°$, $\sin\frac{\theta}{2} \approx \sin 13.845° \approx 0.2393$ and D is at B, we have $1.998|CD| = 1.998|BC| = 1.998 \times 2 \times 0.2393|AB| = 0.9562|AB|$. In this case,

$$|AD| + 1.998|CD| = |AB| + 1.998|CD| = 1.9562|AB| < 1.998|AB|.$$

$\square$

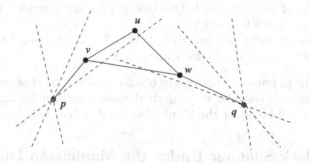

Fig. 5. Illustration of the proof of Theorem 2.

Theorem 2. *Given n red points R and $m(\geq 1)$ black points B in the plane, a red-black spanner T under the L_2 distance and of at most $13n + 3m$ edges can be constructed in $O((n+m)\log(n+m))$ time; moreover, the stretch factor of T is bounded by 1.998.*

Proof. We focus on the case to compute the shortest path between two red nodes p and q on the spanner T as the case between a red and a black node easily follows. We refer to Fig. 5. Let p and q be two red points and let the optimal solution of routing p to q be $OPT = \delta_2(p, q) = |pu| + |qu|$, where u is some black point in $S_p(i) \cap S_q(j), i, j \in \{1, 2, \ldots, 13\}$. According to our algorithm, the approximation solution might select a black point v in some of the 13 cones apexed at p; similarly, the algorithm might select a black point w in some cone at q, where u is in the intersection of these two cones. Hence,

$$
\begin{aligned}
d_T(p,q) &\leq |pv| + \delta_{2,B}(v,w) + |wq| \quad //\text{by the algorithm} \\
&\leq |pv| + \lambda_2|vw| + |wq| \quad //\text{by Delaunay stretch factor} \\
&\leq |pv| + 1.998(|uv| + |uw|) + |wq| \quad //\text{by triangle inequality} \\
&= (|pv| + 1.998|uv|) + (1.998|uw| + |wq|) \\
&\leq 1.998|pu| + 1.998|qu| \quad //\text{by Lemma 2} \\
&= 1.998(|pu| + |qu|) \\
&= 1.998 \cdot OPT
\end{aligned}
$$

$\square$

4.2 The Lower Bounds

In the introduction, we illustrate a lower bound in which to achieve the optimal stretch factor one must construct a spanner T of quadratic size. The same example can be slightly augmented to have the following lower bounds on the stretch factor $\alpha_2(T)$.

1. If T must be of linear size and the edges in T cannot intersect, then there is a lower bound 1.079 for $\alpha_2(T)$.
2. If T must be of linear size and the edges in T can intersect, then there is a lower bound $1 + \varepsilon$ for $\alpha_2(T)$, for some $\varepsilon > 0$.

Clearly, how to improve these lower bounds makes a good open problem. In the next section, we consider very much the same problem as in this section, except that we will be under the Manhattan (or L_1) distance.

5 Red-Black Spanner Under the Manhattan Distance in 2-D

The idea for constructing a red-black spanner T_1 under the Manhattan (L_1) distance is similar to the previous section. We will augment the Delaunay triangulation under the L_1 distance, which is known to have a tight stretch factor $\lambda_1 = \sqrt{4 + 2\sqrt{2}} \approx 2.613$ [2]. But the details and the analysis are different from the Euclidean case.

Given a set R of n red points and a set B of $m(\geq 1)$ black points, our algorithm for constructing a red-black spanner T and routing points on T is as follows.

1. Construct the L_1 Delaunay triangulation $DT_1(B)$ of black points. Let $\delta_{1,B}(u,v)$ be the shortest path between two black points u and v on $DT_1(B)$.

2. For each red point p, partition the plane into 8 sectors $S_p(i), i \in \{1, \ldots, 8\}$, around p, using lines through p and with a slope of $k\pi/4, k \in \{0, 1, 2, 3\}$. For convenience, we order these sectors in counterclockwise order, starting from the first one $S_p(1)$ corresponding to the range $(0, \pi/4]$. In each sector, find the closest black point under L_1, if exists, and add this red-black edge into $DT_1(B)$ to form the spanner T_1.

3. To find a route from a red point p to a black point u, in each sector $S_p(i)$ find the closest black point v_i, then measure $|pv_i| + \delta_{1,B}(u, v_i)$. Return the minimum $\min_{i \in \{1, \ldots, 8\}} \{|pv_i| + \delta_{1,B}(u, v_i)\}$.

4. To find a route from a red point p to a red point q, in each sector $S_p(i)$ find the closest black point v_i, and in each sector $S_q(j)$ find the closest black point w_j then measure $|pv_i| + \delta_{1,B}(v_i, w_j) + |w_j q|$. Return the minimum $\min_{i,j \in \{1, \ldots, 8\}} \{|pv_i| + \delta_{1,B}(v_i, w_j) + |w_j q|\}$.

We will make use of the stretch factor of a Delaunay triangulation under L_1, which is $\lambda_1 = \sqrt{4 + 2\sqrt{2}} \approx 2.613$ [2]. To prove a stretch factor of 3.613 for T_1, we first prove two lemmas.

Lemma 3. *Let p be a red point and the goal be routing p to a black point u. When the closest black point v (to p) and u are both in the first sector $S_p(1)$, the spanning ratio of routing p to u through T_1 is at most $1 + \lambda_1$.*

Proof. Due to space constraint, we leave the proof to the full version.   □

Lemma 4. *Let p be a red point and the goal be routing p to a black point u. When the closest black point v (to p) and u are both in the second sector $S_p(2)$, the spanning ratio of routing p to u through T_1 is at most $1 + \lambda_1$.*

Proof. Due to space constraint, we leave the proof to the full version.   □

We can symmetrically show that the stretch factor of routing a red point p to a black point u through T_1 remains $1 + \lambda_1$ when u, v are in other sectors of p. In fact, it is easy to show that the $1 + \lambda_1$ bound is tight for the algorithm using only three points p, v and u (where u, v are vertices in $DT_1(B)$). (We leave that as a simple exercise.) Next, we summarize with the following theorem.

Theorem 3. *Given n red points R and $m(\geq 1)$ black points B in the plane, a red-black spanner T_1 under the L_1 distance and of at most $8n + 3m$ edges can be constructed in $O((n + m) \log(n + m))$ time; moreover, the stretch factor of T_1 is bounded by $1 + \lambda_1 = 1 + \sqrt{4 + 2\sqrt{2}} \approx 3.613$.*

Proof. The running time for constructing $DT_1(B)$ and T_1 easily follows. To complete the proof of this theorem, we need to show the stretch factor of routing two red points p, q is also $1 + \lambda_1$. We use triangle inequality, similar to the proof of Theorem 2. Let u be the optimal solution to minimize $|pu|_1 + |uq|_1$, and let $u \in S_p(i) \cap S_q(j)$, for some $i, j \in \{1, \ldots, 8\}$; moreover, let v be the point in $S_p(i)$ which is the closest to p and w be the point in $S_q(j)$ which is the closest to q. Then

$$d_{T_1}(p,q) \leq |pv|_1 + \delta_{1,B}(v,w) + |wq|_1 \quad //\text{by the algorithm}$$
$$\leq |pv|_1 + \lambda_1|vw|_1 + |wq|_1 \quad //\text{by Delaunay stretch factor}$$
$$\leq |pv|_1 + \lambda_1(|uv|_1 + |uw|_1) + |wq|_1 \quad //\text{by triangle inequality}$$
$$= (|pv|_1 + \lambda_1|uv|_1) + (\lambda_1|uw|_1 + |wq|_1)$$
$$\leq (1+\lambda_1)|pu|_1 + (1+\lambda_1)|qu|_1 \quad //\text{by Lemmas 3, 4 and extensions}$$
$$= (1+\lambda_1) \cdot \delta_1(p,q)$$
$$\leq 3.613 \cdot \delta_1(p,q).$$

$\square$

6   Concluding Remarks

We investigate the red-black spanners for points in 1-d and 2-d in this paper. In 1-d we construct a linear-size red-black spanner under the 'no red-red edge' constraint that achieves the optimal stretch factor. In 2-d we construct linear-size spanners with constant stretch factors (1.998 and 3.613 respectively) by augmenting the corresponding Delaunay triangulations (under L_2 and L_1). This opens up with a lot of open questions in 2-d.

1. Our red-black spanners might involve edge intersections (crossings) between two red-black edges, and between a red-black edge and a black-black edge. Is it possible to have a red-black spanner with no edge crossings while keeping a small stretch factor? An instant try is to compute the Delaunay triangulation of $R \cup B$ and then remove all red-red edges. It certainly has no edge crossing but the stretch factor part falls off.
2. Due to the planarity of Delaunay triangulations, we could say that the average vertex degree in our spanners is a constant. Is it possible to have a red-black spanner with a constant maximum vertex degree, with no edge crossing, while still keeping a small stretch factor? Notice that starting with the bounded-degree spanner for black points, say using the one by Bose et al. [4], then use the same idea of augmentation, we could easily fulfill the constant maximum degree constraint. But it will not guarantee the 'no edge crossing' property. An alternative solution is to construct the bounded degree spanner of $R \cup B$ using [4] and then remove all red-red edges, but the proof of a bounded stretch factor becomes non-trivial.
3. In Sect. 2 we show that using an existing road network Q cannot help us obtain a good spanner. Can we augment Q, say, by adding a minimum number of edges to Q, to achieve a decent (bounded) stretch factor?

Acknowledgment. Part of this research was done while the first and last author visited University of Hyogo in late 2022.

References

1. Biniaz, A., Bose, P., Carufel, J.D., Gavoille, C., Maheshwari, A., Smid, M.H.M.: Towards plane spanners of degree 3. J. Comput. Geom. **8**(1), 11–31 (2017)
2. Bonichon, N., Gavoille, C., Hanusse, N., Perković, L.: The stretch factor of L 1- and L ∞-delaunay triangulations. In: Epstein, L., Ferragina, P. (eds.) ESA 2012. LNCS, vol. 7501, pp. 205–216. Springer, Heidelberg (2012). https://doi.org/10. 1007/978-3-642-33090-2_19
3. Bose, P., Smid, M.H.M.: On plane geometric spanners: a survey and open problems. Comput. Geom. **46**(7), 818–830 (2013)
4. Bose, P., Smid, M.H.M., Xu, D.: Delaunay and diamond triangulations contain spanners of bounded degree. Int. J. Comput. Geom. Appl. **19**(2), 119–140 (2009)
5. Chew, P.: There is a planar graph almost as good as the complete graph. In: Aggarwal, A. (ed.), Proceedings of the Second Annual ACM SIGACT/SIGGRAPH Symposium on Computational Geometry, Yorktown Heights, NY, USA, 2–4 June 1986, pp. 169–177. ACM (1986)
6. Dobkin, D.P., Friedman, S.J., Supowit, K.J.: Delaunay graphs are almost as good as complete graphs. In: 28th Annual Symposium on Foundations of Computer Science, Los Angeles, California, USA, 27–29 October 1987, pp. 20–26. IEEE Computer Society (1987)
7. Dumitrescu, A., Ghosh, A.: Lower bounds on the dilation of plane spanners. Int. J. Comput. Geom. Appl. **26**(2), 89–110 (2016)
8. Eppstein, D.: Spanning trees and spanners. In: Sack, J.-R., Urrutia, J. (eds.) Handbook of Computational Geometry, pp. 425–461. Elsevier Science Publishers B.V. North-Holland, Amsterdam (2000)
9. Kazemi, M.A., Sedighizadeh, M., Mirzaei, M.J., Homaee, O.: Optimal siting and sizing of distribution system operator owned EV parking lots. Appl. Energy **179**, 1176–1184 (2016)
10. Keil, M., Gutwin, C.A.: Classes of graphs which approximate the complete Euclidean graph. Discret. Comput. Geom. **7**, 13–28 (1992)
11. Liang, Y., Guo, C., Yang, J., Ding, Z.: Optimal planning of charging station based on discrete distribution of charging demand. IET Gener. Trans. Distrib. **14**(6), 965–974 (2020)
12. Narasimhan, G., Smid, M.H.M.: Geometric Spanner Networks. Cambridge University Press, Cambridge (2007)
13. Sadeghi-Barzani, P., Rajabi-Ghahnavieh, A., Kazemi-Karegar, H.: Optimal fast charging station placing and sizing. Appl. Energy **125**, 289–299 (2014)
14. Sperling, D.: New Transportation Fuels: A Strategic Approach to Technological Change. University of California Press, Berkeley (1990)
15. Wang, Y., Shi, J., Wang, R., Liu, Z., Wang, L.: Siting and sizing of fast charging stations in highway network with budget constraint. Appl. Energy **228**, 1255–1271 (2018)
16. Xia, G.: The stretch factor of the Delaunay triangulation is less than 1.998. SIAM J. Comput. **42**(4), 1620–1659 (2013)
17. Xia, G., Zhang, L.: Toward the tight bound of the stretch factor of delaunay triangulations. In: Proceedings of the 23rd Annual Canadian Conference on Computational Geometry, Toronto, Ontario, Canada, 10–12 August 2011 (2011)
18. Zhu, H., Pei, Z.: Data-driven layout design of regional battery swapping stations for electric bicycles. IFAC PapersOnLine **53**(5), 13–18 (2020)

Self-stabilizing $(\Delta + 1)$-Coloring in Sublinear (in Δ) Rounds via Locally-Iterative Algorithms

Xinyu Fu$^{(\boxtimes)}$, Yitong Yin, and Chaodong Zheng

State Key Laboratory for Novel Software Technology, Nanjing University, Nanjing, China
xyfu@smail.nju.edu.cn, {yinyt,chaodong}@nju.edu.cn

Abstract. Fault-tolerance is a central theme in distributed computing. Self-stabilization is a key property that guarantees a distributed system starting from an arbitrary state eventually converges to a desired behavior. Such strong level of fault-tolerance is often desirable due to the error-prone nature of distributed systems. Developing fast and robust coloring algorithms has been a central topic in the study of distributed graph algorithms. In this paper, we give a $(\Delta + 1)$-coloring algorithm with $O(\Delta^{3/4} \log \Delta) + \log^* n$ stabilization time, only using messages of size $O(\log n)$, on input graphs of n vertices and maximum degree Δ. This is the first self-stabilizing $(\Delta + 1)$-coloring algorithm with sublinear-in-Δ stabilization time. The key building block of our algorithm is a new locally-iterative $(\Delta + 1)$-coloring algorithm with $O(\Delta^{3/4} \log \Delta) + \log^* n$ runtime. To the best of our knowledge, this is the first locally-iterative $(\Delta+1)$-coloring algorithm with sublinear-in-Δ runtime. This answers an open question raised in [Barenboim, Elkin, and Goldberg, JACM '21].

1 Introduction

Distributed graph coloring is a fundamental and extensively studied problem in distributed computing [1,3–6,12,13,18,19,21,24]. As a locally checkable labeling problem, it is widely considered to be one of the benchmark problems for answering the fundamental question "what can be computed locally" [22]. This problem also has a wide range of applications in practice, including channel allocation, scheduling, and mutual exclusion [14,15].

In graph theory, for a graph $G = (V, E)$, a q-*coloring* is a mapping ϕ from V to a palette Q, where $|Q| = q$. A q-coloring is *proper* if $\phi(u) \neq \phi(v)$ for every edge $(u, v) \in E$. Distributed graph coloring is often studied in synchronous message-passing model [23]. In this model, a communication network is represented by an n-vertex graph $G = (V, E)$ with maximum degree Δ. Each vertex $v \in V$ hosts a processor and each edge $(u, v) \in E$ denotes a communication link between two vertices u and v. Each vertex $v \in V$ has a unique identifier $id(v)$ belonging to the set $[n] = \{0, 1, \cdots, n - 1\}$. In each synchronous *round*, vertices perform local computation and exchange messages with their neighbors. We restrict each

W. Wu and G. Tong (Eds.): COCOON 2023, LNCS 14422, pp. 232–243, 2024.
https://doi.org/10.1007/978-3-031-49190-0_17

message's size to $O(\log n)$ bits, as in the standard CONGEST model [23]. The time complexity of an algorithm is the maximum number of rounds required for all vertices to arrive at a solution for the considered problem.

Self-stabilization. Fault-tolerance is another central topic in the study of distributed computing. *Self-stabilization*, a concept proposed by Edsger W. Dijkstra [10], is a property that, roughly speaking, guarantees a distributed system starting from an arbitrary state eventually converges to a desired behavior. This concept is regarded as "a milestone in work on fault tolerance" by Leslie Lamport [16]. Indeed, over the last four decades, a vast collection of self-stabilizing distributed algorithms have been devised (see the classical monograph by Dolev [11] and the more recent one by Altisen et al. [2] for more detail), and several of them have seen practical applications [8,9].

More formally, we adopt the same self-stabilizing setting assumed by Barenboim, Elkin, and Goldenberg [5]. In this setting, the memory of each vertex consists of two parts: the immutable *read-only memory (ROM)* and the mutable *random access memory (RAM)*. The ROM part is faultless but cannot change during execution; and it may be used to store hard-wired data such as vertex identity and graph parameters, as well as the program code. The RAM part on the other hand, can change during algorithm execution; and it is for storing the internal states of the algorithm. The RAM part is subject to error and controlled by an *adversary* called Eve. At any moment during the execution, the adversary can examine the entire memory (including both ROM and RAM) of all vertices, and then make arbitrary changes to the RAM part of all vertices.

An algorithm is self-stabilizing if it can still compute a proper solution once the adversary stops disrupting its execution. Specifically, assume T_0 is the last round in which the adversary makes any changes to vertices' RAM areas, if it is always guaranteed that by the end of round $T_0 + T$ a desired solution is produced, then the algorithm is self-stabilizing with *stabilization time T*.

An Open Question. Interestingly, for the distributed $(\Delta + 1)$-coloring problem, there is a considerable efficiency gap between general algorithms and self-stabilizing ones. On the one hand, non-fault-tolerant distributed coloring algorithms with sublinear-in-Δ runtime were already known around 2016 [3,12]. On the other hand, once we desire self-stabilization, the state of the art which has linear-in-Δ stabilization time was only discovered in a recent breakthrough [5]. Though researchers have proposed generic techniques for converting general coloring algorithms into self-stabilizing ones [17], such approach usually results in large message size, rendering the converted algorithms unpractical. Therefore, an important open question is, can one compute a proper $(\Delta + 1)$-coloring with $o(\Delta) + \log^* n$ stabilization time, using only small messages?

1.1 Our Results

In this paper, we give affirmative answer to the above question:

Theorem 1 (Efficient Self-stabilizing Coloring Algorithm). *There exists a self-stabilizing coloring algorithm such that, for any input graph with n vertices* and maximum degree Δ, produces a proper $(\Delta+1)$-coloring with $O(\Delta^{3/4} \log \Delta) +$ $\log^* n$ stabilization time, using messages of $O(\log n)$ bits.

To the best of our knowledge, this is the first self-stabilizing algorithm for $(\Delta+1)$-coloring in the CONGEST model with sublinear-in-Δ stabilization time. The key building block of this algorithm is a new non-fault-tolerant *locally-iterative* coloring algorithm with sublinear-in-Δ runtime. Next, we will introduce the concept of locally-iterative algorithms, give an overview of the new locally-iterative coloring algorithm, and discuss the connection between locally-iterative algorithms and self-stabilizing algorithms.

New Locally-Iterative Coloring Algorithm. Locally-iterative coloring algorithms are introduced by Szegedy and Vishwanathan [24]. Throughout the execution of such algorithms, a proper coloring of the network graph is maintained and updated from round to round. Moreover, in each round, for each vertex $v \in V$, its next color is computed from its current color and the current colors of its neighbors $N(v) = \{u \in V \mid (u, v) \in E)\}$. More formally,

Definition 1. *In the synchronous message-passing model, an algorithm for graph coloring is said to be* locally-iterative *if it maintains a sequence of proper colorings ϕ_t of the input $G = (V, E)$ such that:*

- *The initial coloring ϕ_0 is constructed locally in the sense that, for every vertex v, its initial color $\phi_0(v)$ is computed locally from $id(v)$.*
- *In each round $t \geq 1$, every vertex v computes its next color $\phi_t(v)$ based only on its current color $\phi_{t-1}(v)$ along with the multiset of colors $\{\phi_{t-1}(u) \mid u \in N(v)\}$ appearing in v's neighborhood. Particularly, in each round $t \geq 1$, every vertex v only broadcasts $\phi_{t-1}(v)$ to its neighbors.*

Due to the simplicity and naturalness of its framework, locally-iterative algorithms often play the role of starting points for self-stabilizing algorithms. Indeed, currently the fastest known self-stabilizing $(\Delta + 1)$-coloring algorithm is converted from a simple yet elegant locally-iterative coloring algorithm [5].

In this paper, we also start with developing a non-fault-tolerant locally-iterative coloring algorithm. Nonetheless, to achieve sublinear-in-Δ time complexity, our locally-iterative algorithm adopts a more sophisticated three-phases framework used by several recent work [3,5]. In this framework, the first phase employs Linial's celebrated algorithm [18] (or some variant of it) and produces a proper α-coloring within $\log^* n + O(1)$ rounds, where $\alpha = O(\Delta^2)$. The second phase takes the proper $O(\Delta^2)$-coloring as input and reduces it into another proper β-coloring, where $\beta \in (\Delta+1, \alpha)$. In the last phase, a folklore reduce-one-color-per-round procedure is executed, hence within another $\beta - (\Delta+1)$ rounds, a proper $(\Delta + 1)$-coloring is obtained.

The key novelty of our new locally-iterative algorithm lies in a faster and stronger second phase: it only takes $O(\Delta^{3/4} \log \Delta)$ rounds and produces a proper $\Delta + O(\Delta^{3/4} \log \Delta)$ coloring. As a result, the total runtime of our locally-iterative algorithm is $O(\Delta^{3/4} \log \Delta) + \log^* n$. To sum up,

Theorem 2 (Efficient Locally-Iterative Coloring Algorithm). *There exists a locally-iterative coloring algorithm such that, for any input graph with n vertices and maximum degree Δ, produces a proper $(\Delta + 1)$-coloring within $O(\Delta^{3/4} \log \Delta) + \log^* n$ rounds, using messages of $O(\log n)$ bits.*

This is the first locally-iterative $(\Delta+1)$-coloring algorithm achieving a sublinear-in-Δ runtime while only using small messages. It is another major technical contribution of this paper. This algorithm also gives an affirmative answer to the main open question raised by previous best result [5].

Reconfigurable Locally-Iterative Coloring and Self-stabilization. In adopting the locally-iterative algorithm to the self-stabilizing setting, we cope with a strong level of "asynchrony" among vertices, as the adversary can manipulate vertices' states and put them in different stages of the algorithm. We have also crafted an error-correction procedure ensuring that once the adversary stops disrupting algorithm execution, any vertex with an "improper" state will be detected within one round and resets itself to some proper state.

Interestingly, we find that if a locally-iterative algorithm supports above "reconfiguration" (i.e., state resetting upon detecting illicit status), and if the algorithm's correctness is still enforced under such reconfiguration, then the locally-iterative algorithm in consideration can be modified into a self-stabilizing one with relative ease, with limited or no complexity overhead. Formally, we propose *reconfigurable locally-iterative coloring algorithms*. As an example, the locally-iterative coloring algorithm developed in [5] is reconfigurable out of the box, whereas converting ours into a self-stabilizing one needs more efforts.

Definition 2. *A locally-iterative coloring algorithm $\mathcal{A}$ is reconfigurable if it satisfies the following properties:*

- *The algorithm has a build-in function $\text{LEGIT}_{\mathcal{A}}$ that, upon inputting a vertex v's color and the colors of v's neighbors, outputs a single bit indicating whether v's status is legit or illicit.*
- *For each vertex v, its status after initialization is legit:*

$$\text{LEGIT}_{\mathcal{A}} \left(\phi_0(v), \{\phi_0(u) \mid u \in N(v)\} \right) = 1.$$

- *Normal execution maintains legitimacy. Moreover, when external interference occur, resetting illicit vertices to initial colors resumes legitimacy. Specifically, for any round $t \geq 1$, let $V_{\phi_{t-1}}$ denote the set of vertices that have illicit status in (a not necessarily proper) coloring ϕ_{t-1}, if every vertex updates its color using the following rule, then in ϕ_t all vertices' status are legit.*

$$\phi_t(v) \leftarrow \begin{cases} \phi_0(v) & \text{if } v \in V_{\phi_{t-1}}, \\ \text{UPDATE}_{\mathcal{A}} \left(\phi_{t-1}(v), \{\phi_{t-1}(u) \mid u \in N(v)\} \right) & \text{if } v \in V \setminus V_{\phi_{t-1}}. \end{cases}$$

2   The Locally-Iterative Coloring Algorithm

Our locally-iterative algorithm contains three phases, but vertices cannot depend on current round number to determine which phase they are in per Definition 1.

To solve this issue, we assign each phase an interval so that vertices running that phase will have colors in the corresponding interval. By assigning disjoint intervals to different phases, vertices can correctly determine its progress by observing its current color. More specifically, the intervals used by the three phases are I_1, I_2, and I_3, where

$$|I_1| = \ell_1, |I_2| = \ell_2, \text{ and } |I_3| = \ell_3,$$
$$I_1 \triangleq [\ell_3 + \ell_2, \ell_3 + \ell_2 + \ell_1), I_2 \triangleq [\ell_3, \ell_3 + \ell_2), \text{ and } I_3 \triangleq [0, \ell_3).$$

Next, we introduce each phase in more detail and give precise values for ℓ_1, ℓ_2, ℓ_3. Due to space constraints, complete description and analysis of our locally-iterative coloring algorithm are provided in the full version of the paper.[1]

First Phase: Linial Phase. The first phase runs a locally-iterative version of Linial's well-known coloring algorithm [18]. Let $n_0 = n$, and for $i \geq 1$, define

$$n_i = \begin{cases} 4(\Delta+1)^2 \log^2(n_{i-1}) & \text{if } n_{i-1} > 8(\Delta+1)^3, \\ 4(\Delta+1)^2 & \text{if } n_{i-1} \leq 8(\Delta+1)^3. \end{cases}$$

Let r^* be the smallest $r \geq 0$ such that $n_r \leq 4(\Delta+1)^2$. It has been shown $r^* \leq \log^* n + O(1)$. (See, e.g., Sect. 3.10 of [4].) During the Linial phase, vertices will reduce the number of colors used to n_i after i rounds, thus within $\log^* n + O(1)$ rounds the algorithm produces a proper $O(\Delta^2)$-coloring.

More specifically, we set interval length ℓ_1 and each vertex v's initial color in the following manner:

$$\phi_0(v) \leftarrow \ell_3 + \ell_2 + \sum_{i=1}^{r^*} n_i + id(v),$$

$$|I_1| = \ell_1 = \sum_{i=0}^{r^*} n_i = n + O(\log^* n \cdot \Delta^2 \cdot \log^2 n).$$

Furthermore, we partition I_1 into $r^* + 1$ sub-intervals $I_1^{(0)}, I_1^{(1)}, \cdots, I_1^{(r^*)}$, such that for each $0 \leq t \leq r^*$: $I_1^{(t)} \triangleq [\ell_3 + \ell_2 + \sum_{t+1 \leq i \leq r^*} n_i$, $\ell_3 + \ell_2 + \sum_{t \leq i \leq r^*} n_i)$. Notice that $I_1^{(r^*)} = [\ell_3 + \ell_2, \ell_3 + \ell_2 + n_{r^*})$. During the Linial phase, in each round, vertices will construct Δ-cover-free set families to compute their next colors. (Definition and relevant results regarding cover-free set systems are provided in the full paper.) This mechanism ensures after t rounds where $t \in [0, r^*]$, all vertices' colors are in interval $I_1^{(t)}$, and the coloring ϕ_t is proper. To sum up,

Lemma 1 (Linial Phase). *By the end of round $r^* = \log^* n + O(1)$, all vertices have completed the Linial phase, producing a proper coloring ϕ_{r^*} where $\phi_{r^*}(v) \in [\ell_3 + \ell_2, \ell_3 + \ell_2 + O(\Delta^2)) \subseteq I_1$ for every vertex v. Moreover, ϕ_t is proper for every round $t \in [1, r^*]$.*

---

[1] The full version of the paper is available at https://arxiv.org/abs/2207.14458.

Second Phase: Quadratic Reduction Phase. Before diving into the details, we introduce the concept of (arb)defective colorings as they are the key to understand the second phase of our algorithm. A coloring ϕ of an undirected graph $G = (V, E)$ is said to be: (1) *d-defective* if for every $v \in V$, the number of neighbors $u \in N(v)$ with $\phi(u) = \phi(v)$ is at most d; (2) *a-arbdefective* if we can define an orientation for each edge such that the out-degree of the oriented graph induced by each color class is at most a. Essentially, (arb)defective colorings are "improper colorings with bounded collisions".

The second phase is the most complex component of our algorithm, it is also the key for achieving sublinear-in-Δ runtime. Recall vertices use colors in $I_2 \triangleq [\ell_3, \ell_3 + \ell_2)$ during phase two, where $|I_2| = \ell_2$ and $|I_3| = \ell_3$. To give precise values for ℓ_2 and ℓ_3, we first define three integers and three prime numbers:

$$m_1 \triangleq 4\Delta^{3/2} \log^2(n_{r^*}), \quad m_2 \triangleq 4\sqrt{\Delta} \log^2(n_{r^*}), \quad \text{and} \quad m_3 \triangleq 16\sqrt{\Delta} \log^2(\lambda^2 m_2);$$

$$\lambda \in (\sqrt{m_1} + 1, 2(\sqrt{m_1} + 1)], \quad \mu \in (\sqrt{\Delta} + \sqrt{m_3}, 2(\sqrt{\Delta} + \sqrt{m_3})], \quad \text{and} \quad \tau \in (\sqrt{m_3}, 2\sqrt{m_3}].$$

Due to Bertrand-Chebyshev theorem [7], prime numbers λ, μ, τ exist. We set:

$$|I_2| = \ell_2 = 2\lambda^3(\mu + 1) \cdot m_3 = O(\Delta^{13/4} \log^5 \Delta),$$
$$|I_3| = \ell_3 = \Delta + (2\sqrt{m_3} + 1) \cdot \mu = \Delta + O(\Delta^{3/4} \log \Delta).$$

Since $\ell_2 = 2\lambda^3(\mu + 1) \cdot m_3$, for every color $(\ell_3 + i) \in I_2$ where $i \in [\ell_2]$, we can use a unique quadruple $\langle a, b, c, d \rangle$ to identify it, where:

$$a = \lfloor i / (2\lambda(\mu + 1)m_3) \rfloor,$$
$$b = \lfloor (i - a \cdot 2\lambda(\mu + 1)m_3) / (2\lambda(\mu + 1)) \rfloor,$$
$$c = \lfloor (i - a \cdot 2\lambda(\mu + 1)m_3 - b \cdot 2\lambda(\mu + 1)) / (\mu + 1) \rfloor,$$
$$d = i \bmod (\mu + 1).$$

In other words, $i = a \cdot 2\lambda(\mu + 1)m_3 + b \cdot 2\lambda(\mu + 1) + c \cdot (\mu + 1) + d$. Clearly, $a \in [\lambda^2]$, $b \in [m_3]$, $c \in [2\lambda]$, and $d \in [\mu + 1]$. Throughout the paper, for any round $t \geq r^* + 1$, for any vertex v, if $\phi_t(v) \in I_2$, we denote the values of $a(v), b(v), c(v), d(v)$ in $\phi_t(v)$ as $a_t(v), b_t(v), c_t(v), d_t(v)$.

We now introduce the second phase, which is further divided into three *stages*.

The first stage is the *transition-in stage*, which takes one round and transforms ϕ_{r^*} to a proper coloring with colors from interval I_2. (Recall that the Linial phase takes r^* rounds.) Specifically, we employ the defective coloring algorithm developed by Barenboim, Elkin, and Kuhn [6], with suitable parameters tailored for our purpose. As a result, by the end of round $r^* + 1$, for any vertex v, its $a(v)$ value may collide with up to $\Delta^{1/4}$ neighbors. Moreover, for neighbors with potentially colliding a value, we build $\Delta^{1/4}$-cover-free set families to assign distinct b values. In the end, ϕ_{r^*+1} still corresponds to a proper coloring. Lastly, we note that every vertex v initializes $c(v) = 0$ and $d(v) = \mu$ during the transition-in stage, though they are not used in this stage.

Once the transition-in stage is done, the a values of all vertices correspond to a $\Delta^{1/4}$-defective coloring, using a palette containing λ^2 colors, as $a \in [\lambda^2]$. The main objective of the second stage—which is called the *core stage*—is to start from this $\Delta^{1/4}$-defective λ^2-coloring to gradually obtain a $(2 \cdot \Delta^{1/4})$-arbdefective λ-coloring. Notice that this reduces the number of colors used—or more precisely, the range of the a values of all vertices—from $[\lambda^2]$ to $[\lambda]$. To achieve this quadratic reduction, for every vertex v, we interpret the first coordinate $a(v)$ of its color quadruple in the following manner:

$$a(v) = \hat{a}(v) \cdot \lambda + \tilde{a}(v), \text{ where } \hat{a}(v) = \lfloor a(v)/\lambda \rfloor \text{ and } \tilde{a}(v) = a(v) \bmod \lambda.$$

During the core stage, we run a locally-iterative arbdefective coloring algorithm inspired by [5] that makes a series of updates to $a(v)$ so that eventually $\hat{a}(v) = 0$, reducing $a(v)$ from $[\lambda^2]$ to $[\lambda]$. During this process, vertices also update their c values to implicitly define the orientations of edges: for neighbors u and v, vertex v points to vertex u if and only if $c(v) \geq c(u)$. By guaranteeing that the out-degree of the oriented graph induced by each a value is at most $2 \cdot \Delta^{1/4}$, the a and c values of vertices together constitute a $(2 \cdot \Delta^{1/4})$-arbdefective coloring during the core stage. On the other hand, similar to the first stage, we also use $(2 \cdot \Delta^{1/4})$-cover-free set families to assign distinct b values for neighbors with potentially colliding a value, hence enforcing the overall coloring is always proper. Lastly, we note that vertices may complete the core stage at different times: once a vertex v has $a_t(v) \in [\lambda]$ by the end of round t, its core stage is considered done.

The last stage of the second phase is called the *transition-out stage*, in which vertices produce a proper $(\Delta + O(\Delta^{3/4} \log \Delta))$-coloring using colors in interval I_3. The approach we took during the transition-out stage is inspired by the techniques developed by Barenboim [3]. Nonetheless, important adjustments are made on both the implementation and the analysis, as we are in the more restrictive locally-iterative setting, and have to take the "asynchrony" that vertices may start the transition-out stage in different rounds into consideration. We also note that the transition-out stage is the only stage where the d values of vertices' color quadruple are used.

We conclude this subsection by stating the key guarantees provided by the second phase. We also note that an alternative approach that phase two could take to achieve similar results is to adopt the techniques proposed by Maus [20].

Lemma 2 (Quadratic Reduction Phase). *By the end of round $r^* + 2 + 3\lambda$, all vertices have completed the quadratic reduction phase, producing a proper coloring $\phi_{r^*+2+3\lambda}$ where $\phi_{r^*+2+3\lambda}(v) \in I_3$ for every vertex v. Moreover, ϕ_t is proper for every round $t \in [r^* + 1, r^* + 2 + 3\lambda]$.*

Third Phase: Standard Reduction Phase. In the standard reduction phase, each vertex v maps color $\phi_{t_v^\#} \in I_3$ to another color in $[\Delta + 1] \subset I_3$, completing $(\Delta+1)$-coloring. Here, $t_v^\#$ denotes the smallest round number such that $\phi_{t_v^\#}(v) \in I_3$. Hence, $t_v^\# + 1$ is the first round in which v runs the standard reduction phase.

For each vertex v, for each round $t \geq t_v^\# + 1$, if every neighbor $u \in N(v)$ has also entered the standard reduction phase, and if v has the maximum color value in its one-hop neighborhood, then v will update its color to be the minimum value in $[\Delta + 1]$ that still has not been used by any of its neighbors. Clearly, such color must exist. In all other cases, v keeps its color unchanged in round t. Effectively, this procedure reduces the maximum color value used by any vertex by at least one in each round. Hence, within $\ell_3 - (\Delta + 1)$ rounds into the third phase, a proper $(\Delta + 1)$-coloring is obtained. As a result,

Lemma 3 (Standard Reduction Phase). *By the end of round $r^* + 1 + 3\lambda + (2\sqrt{m_3} + 1)\mu$, the coloring $\phi_{r^* + 1 + 3\lambda + (2\sqrt{m_3}+1)\mu}$ is a proper $(\Delta + 1)$-coloring. Moreover, ϕ_t is proper for every round $t \in [r^* + 3 + 3\lambda, \infty)$.*

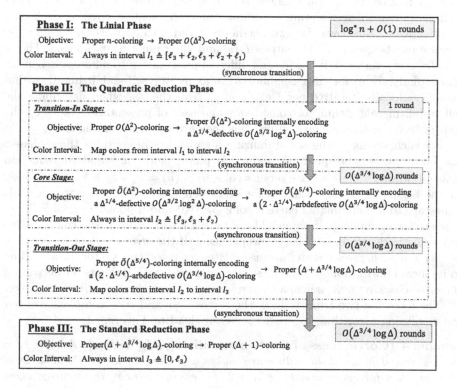

Fig. 1. Structure of the locally-iterative $(\Delta + 1)$-coloring algorithm.

Summary. We conclude this section by noting that Lemma 1, Lemma 2, and Lemma 3 together could easily lead to Theorem 2. Figure 1 provides a graphical overview of the locally-iterative coloring algorithm's structure.

3   The Self-stabilizing Coloring Algorithm

In this section, we discuss how to convert the above locally-iterative algorithm into a self-stabilizing one. To begin with, we specify what are stored in the ROM and the RAM areas of vertices: in the ROM area of a vertex v, we store its identity $id(v)$, graph parameters n and Δ, and the program code; in the RAM area of v, we store the colors of its local neighborhood, a boolean vector T_v of size Δ, and other variables that are used during execution.

The boolean vector T_v is used to determine the orientation of the edges incident to v, replacing the role of $c(v)$. Specifically, in the self-stabilizing algorithm, for each edge (v, u), vertex v maintains a bit in T_v denoted as $T_v[u]$, and we treat v points to u if and only if $T_v[u] = 1$. The reason that we replace $c(v)$ with bit vector T_v is that in the self-stabilizing setting, the adversary can employ a certain strategy to grow the c values indefinitely. However, a side effect of this replacement is that v must maintain a variable for *each* incident edge to determine its orientation. Moreover, for two neighbors v and u to correctly determine the orientation of edge (v, u), bit entries $T_v[u]$ and $T_u[v]$ must be exchanged. Therefore, for every vertex v, it has to send different information to different neighbors (particularly, $T_v[u]$ for each neighbor u), making our self-stabilizing algorithm no longer locally-iterative per Definition 1. Nonetheless, when describing the self-stabilizing algorithm, for consistency and ease of presentation, we keep the c entry in vertices' color quadruples, but they are not used throughout.

For each vertex v, the self-stabilizing algorithm still contains three phases: the Linial phase, the quadratic reduction phase, and the standard reduction phase. Initially, every vertex v sets its color to $\phi_0(v) = \ell_3 + \ell_2 + \sum_1^{r^*} n_i + id(v)$. At the beginning of each round t, for every neighbor $u \in N(v)$, vertex v sends a message to u including its current color $\phi_{t-1}(v)$ and a boolean variable $T_v[u]$. After receiving messages from neighbors, vertex v will perform an *error-checking* procedure to determine if it is in a proper state. If the error-checking passes then we say v is in a *proper* state, and v updates its color and vector T_v according to its local information and the messages received from neighbors. Otherwise, if the error-checking fails, v is in an *improper* state. In such case, v resets its color.

The following two lemmas show the correctness and time complexity of our self-stabilizing algorithm, together they easily lead to Theorem 1.

Lemma 4 (Correctness of the Self-stabilizing Algorithm). *Assume T_0 is the last round in which the adversary makes any changes to the RAM areas of vertices, then for every round $t \geq T_0 + 2$, for every vertex v, the error-checking procedure will not reset vertex v's color.*

Lemma 5 (Time Complexity of the Self-stabilizing Algorithm). *Assume T_0 is the last round in which the adversary makes any changes to the RAM areas of vertices, then for every round $t \geq T_0 + r^* + 4\lambda + 2 + 2(\sqrt{m_3} + 1)\mu$, every vertex v has $\phi_t(v) \in [\Delta + 1]$.*

We now introduce each phase in more detail. Complete description and analysis are provided in the full version of the paper due to space constraints.

The Linial Phase and the Transition-in Stage of the Quadratic Reduction Phase. At the beginning of a round t, if a vertex v finds its color $\phi_{t-1}(v)$ not in $I_2 \cup I_3$, it will do error-checking to see if any of the following conditions is satisfied: (1) its color collide with some neighbor; (2) its color is not in $(\bigcup_{i=1}^{r^*} I_1^{(i)}) \cup I_2 \cup I_3$, which implies v should be running the first iteration of the Linial phase but its current color is not $\ell_3 + \ell_2 + \sum_{i=1}^{r^*} n_i + id(v)$. If any of these conditions is satisfied, then vertex v treats itself in an improper state and resets its color to $\ell_3 + \ell_2 + \sum_{i=1}^{r^*} n_i + id(v)$. That is, it sets $\phi_t(v) = \ell_3 + \ell_2 + \sum_{i=1}^{r^*} n_i + id(v)$.

Otherwise, v is in a proper state with $\phi_{t-1}(v) \in I_1$. In such case, v first determines which interval $I_1^{(t')}$ it is in, and then runs either the Linial phase or the transition-in stage of the quadratic reduction phase, according to value of t'.

If $0 \le t' < r^*$, then vertex v computes its new color in a similar fashion as in the Linial phase of the locally-iterative algorithm.

If $t' = r^*$, then vertex v transforms its color from interval I_1 to I_2, effectively running the transition-in stage of the quadratic reduction phase. The transition-in stage of the self-stabilizing algorithm is similar to the one in the locally-iterative algorithm. The only difference is that vertices may end the Linial phase and start the transition-in stage in different rounds. This brings the side effect that the a values of all vertices are not necessarily $\Delta^{1/4}$-defective. Instead, we maintain a $\Delta^{1/4}$-arbdefective λ^2-coloring. Specifically, each vertex v still computes $a(v)$ based on its color and the colors of its neighbors using the defective coloring algorithm; moreover, v again uses $b(v)$ to differentiate itself from the neighbors with identical a value. On the other hand, v sets $T_v[u] = 1$ if $a(v)$ might collide with neighbor u, otherwise v sets $T_v[u] = 0$. (Recall that $T_v[u]$ and $T_u[v]$ determine the orientation of edge (u, v) in arbdefective colorings.)

The Core Stage of the Quadratic Reduction Phase. A vertex v with $\phi(v) \in I_2$ and $a(v) \ge \lambda$ should run the core stage. Nonetheless, before proceeding, it will do error-checking to see if any of the following conditions is satisfied: (1) there exists a neighbor u of v such that $a(v) = a(u)$ and $b(v) = b(u)$, effectively implying u and v have colliding color; (2) there exists a neighbor u of v such that $a(v) = a(u)$ yet $T_v[u] + T_u[v] = 0$, implying that the orientation of edge (u, v) is still undetermined when $a(v) = a(u)$; (3) the number of neighbors $u \in N(v)$ with $T_v[u] = 1$ is larger than $\Delta^{1/4}$, violating the bounded arboricity assumption during the core stage; (4) there exists a vertex $u \in N(v) \cup \{v\}$ with its color in I_2 and $a(u) \ge \lambda$, yet $b(u) \ge m_2$, violating the range requirement of b values during the core stage. If any of these conditions is satisfied, then vertex v resets its color. Otherwise, it executes the core stage of the quadratic reduction phase to reduce its a value from $[\lambda, \lambda^2)$ to $[0, \lambda)$.

The procedure we use in the self-stabilizing settings to transform a $\Delta^{1/4}$-arbdefective λ^2-coloring to a $(2 \cdot \Delta^{1/4})$-arbdefective λ-coloring is almost identical to the one we used in the locally-iterative setting. The only difference is that we have altered the definition of some variables to incorporate relevant bits in T_v. This is because, in the self-stabilizing setting, vertices start the core stage with an arbdefective coloring instead of a defective coloring.

Once the reduction of the a value occurs in some round t, vertex v obtains an $a_t(v) \in [\lambda]$, and updates $b_t(v)$ to differentiate itself from neighbors that may have colliding a value. It also sets $T_v[u] = 1$ for certain entries in T_v, recording the orientation of corresponding edges. Notice that T_v here is used to maintain the arboricity of a $(2 \cdot \Delta^{1/4})$-arbdefective λ-coloring for vertices with $a(v) < \lambda$, whereas in the transition-in stage, T_v is used to maintain the arboricity of a $\Delta^{1/4}$-arbdefective λ^2-coloring for vertices with $a(v) \geq \lambda$.

The Transition-Out Stage of the Quadratic Reduction Phase. At the beginning of a round t, if vertex v finds $\phi_{t-1}(v) \in I_2$ and $a(v) \in [\lambda]$, then it is in the transition-out stage. Once again, it does the following error-checking before proceeding: (1) there exists a neighbor u of v such that $a(v) = a(u)$ and $b(v) = b(u)$, effectively implying u and v have colliding color; (2) there exists a neighbor u of v such that $a(v) = a(u)$ yet $T_v[u] + T_u[v] = 0$, implying that the orientation of edge (u,v) is still undetermined when $a(v) = a(u)$; (3) the number of neighbors $u \in N(v)$ with $T_v[u] = 1$ is larger than $2 \cdot \Delta^{1/4}$, violating the bounded arboricity assumption during the transition-out stage. If any of these conditions is satisfied, then v is in an improper state and resets its color. Otherwise, it executes the transition-out stage to transform its color from I_2 to I_3.

For each vertex v, the transformation is similar to the transition-out stage of the locally-iterative algorithm, except that: (1) we replace the constraints on $c(v)$ with corresponding constraints on T_v; and (2) we add an error-checking mechanism for $d(v)$ as the adversary can arbitrarily manipulate it. If the error-checking for $d(v)$ fails, vertex v resets $d(v)$ to μ, so that later it can obtain a proper $d(v)$. We note that such resetting occurs at most once for each vertex once the adversary stops disrupting.

The Standard Reduction Phase. For a vertex v with its color in I_3, it considers itself in the standard reduction phase, whose error-checking procedure is simple: if its color collides with any neighbor, then it resets $\phi(v)$ to $\ell_3 + \ell_2 + \sum_1^{r^*} n_i + id(v)$. Otherwise, vertex v considers itself in a proper state, and runs exactly the same standard reduction procedure described in the locally-iterative setting.

Acknowledgements. This work is supported by the National Natural Science Foundation of China (NSFC) under Grant No. 62172207 and 62332009, and by the Department of Science and Technology of Jiangsu Province under Grant No. BK20211148.

References

1. Alon, N., Babai, L., Itai, A.: A fast and simple randomized parallel algorithm for the maximal independent set problem. J. Algorithms **7**(4), 567–583 (1986)
2. Altisen, K., Devismes, S., Dubois, S., Petit, F.: Introduction to Distributed Self-Stabilizing Algorithms. Morgan & Claypool, San Rafael (2019)
3. Barenboim, L.: Deterministic $(\Delta+1)$-coloring in sublinear (in Δ) time in static, dynamic, and faulty networks. J. ACM **63**(5), 1–22 (2016)
4. Barenboim, L., Elkin, M.: Distributed Graph Coloring: Fundamentals and Recent Developments. Morgan & Claypool Publishers, San Rafael (2013)
5. Barenboim, L., Elkin, M., Goldenberg, U.: Locally-iterative distributed $(\Delta+1)$-coloring and applications. J. ACM **69**(1), 1–26 (2021)

6. Barenboim, L., Elkin, M., Kuhn, F.: Distributed $(\Delta+1)$-coloring in linear (in Δ) time. SIAM J. Comput. **43**(1), 72–95 (2014)
7. Chebyshev, P.L.: Mémoire sur les nombres premiers. Journal de mathématiques pures et appliquées **1**, 366–390 (1852)
8. Chen, Y., Datta, A.K., Tixeuil, S.: Stabilizing inter-domain routing in the internet. J. High Speed Netw. **14**(1), 21–37 (2005)
9. Datta, A.K., Outley, E., Thiagarajan, V., Flatebo, M.: Stabilization of the x.25 connection management protocol. In: International Conference on Computing and Information, ICCI 1994, pp. 1637–1654 (1994)
10. Dijkstra, E.W.: Self-stabilizing systems in spite of distributed control. Commun. ACM **17**(11), 643–644 (1974)
11. Dolev, S.: Self-Stabilization. The MIT Press, Cambridge (2000)
12. Fraigniaud, P., Heinrich, M., Kosowski, A.: Local conflict coloring. In: Proceedings of the 2016 IEEE 57th Annual Symposium on Foundations of Computer Science, FOCS 2016, pp. 625–634. IEEE (2016)
13. Ghaffari, M., Kuhn, F.: Deterministic distributed vertex coloring: simpler, faster, and without network decomposition. In: Proceedings of the 62nd Annual Symposium on Foundations of Computer Science, FOCS 2022, pp. 1009–1020. IEEE (2022)
14. Guellati, N., Kheddouci, H.: A survey on self-stabilizing algorithms for independence, domination, coloring, and matching in graphs. J. Parallel Distrib. Comput. **70**(4), 406–415 (2010)
15. Kuhn, F.: Weak graph colorings: distributed algorithms and applications. In: Proceedings of the 21st Annual Symposium on Parallelism in Algorithms and Architectures, SPAA 2009, pp. 138–144. ACM (2009)
16. Lamport, L.: Solved problems, unsolved problems and non-problems in concurrency. ACM SIGOPS Oper. Syst. Rev. **19**(4), 34–44 (1985)
17. Lenzen, C., Suomela, J., Wattenhofer, R.: Local algorithms: self-stabilization on speed. In: Guerraoui, R., Petit, F. (eds.) SSS 2009. LNCS, vol. 5873, pp. 17–34. Springer, Heidelberg (2009). https://doi.org/10.1007/978-3-642-05118-0_2
18. Linial, N.: Distributive graph algorithms global solutions from local data. In: Proceedings of the 28th Annual Symposium on Foundations of Computer Science, FOCS 1987, pp. 331–335. IEEE (1987)
19. Luby, M.: A simple parallel algorithm for the maximal independent set problem. SIAM J. Comput. **15**(4), 1036–1053 (1986)
20. Maus, Y.: Distributed graph coloring made easy. In: Proceedings of the 33rd ACM Symposium on Parallelism in Algorithms and Architectures, SPAA 2021, pp. 362–372. ACM (2021)
21. Maus, Y., Tonoyan, T.: Local Conflict Coloring Revisited: Linial for Lists. In: Proceedings of the 34th International Symposium on Distributed Computing, DISC 2020, pp. 16:1–16:18. Schloss Dagstuhl-Leibniz-Zentrum für Informatik (2020)
22. Naor, M., Stockmeyer, L.: What can be computed locally? In: Proceedings of the 25th Annual ACM Symposium on Theory of Computing, STOC 1993, pp. 184–193. ACM (1993)
23. Peleg, D.: Distributed computing: a locality-sensitive approach. In: SIAM (2000)
24. Szegedy, M., Vishwanathan, S.: Locality based graph coloring. In: Proceedings of the 25th Annual ACM Symposium on Theory of Computing, STOC 1993, pp. 201–207. ACM (1993)

On Detecting Some Defective Items
in Group Testing

Nader H. Bshouty[iD] and Catherine A. Haddad-Zaknoon[(⊠)][iD]

Technion - Israel Institute of Technology, Haifa, Israel
{bshouty,catherine}@cs.technion.ac.il

Abstract. Group testing is an approach aimed at identifying up to d defective items among a total of n elements. This is accomplished by examining subsets to determine if at least one defective item is present. We focus on the problem of identifying a subset of $\ell < d$ defective items. We develop upper and lower bounds on the number of tests required to detect ℓ defective items in both the adaptive and non-adaptive settings while considering scenarios where no prior knowledge of d is available, and situations where some non-trivial estimate of d is at hand.

When no prior knowledge on d is available, we prove a lower bound of $\Omega(\frac{\ell \log^2 n}{\log \ell + \log \log n})$ tests in the randomized non-adaptive settings and an upper bound of $O(\ell \log^2 n)$ for the same settings. Furthermore, we demonstrate that any non-adaptive deterministic algorithm must ask $\Theta(n)$ tests, signifying a fundamental limitation in this scenario. For adaptive algorithms, we establish tight bounds in different scenarios. In the deterministic case, we prove a tight bound of $\Theta(\ell \log (n/\ell))$. Moreover, in the randomized settings, we derive a tight bound of $\Theta(\ell \log (n/d))$.

When d, or at least some non-trivial estimate of d, is known, we prove a tight bound of $\Theta(d \log(n/d))$ for the deterministic non-adaptive settings, and $\Theta(\ell \log(n/d))$ for the randomized non-adaptive settings. In the adaptive case, we present an upper bound of $O(\ell \log(n/\ell))$ for the deterministic settings, and a lower bound of $\Omega(\ell \log(n/d) + \log n)$. Additionally, we establish a tight bound of $\Theta(\ell \log(n/d))$ for the randomized adaptive settings.

Keywords: Group testing · Pooling design · Finding defectives partially

1 Introduction

Group testing is a technique for identifying a subset of items known as *defective items set* within a large amount of items using small number of tests called *group tests*. A group test is a subset of items, where the test result is *positive* if the subset contains at least one defective item and *negative* otherwise. Formally, let $X = \{1, 2, \ldots, n\}$ be a set of items, and $I \subseteq X$ is the set of defectives. A group test is set $Q \subseteq X$. The answer of the test Q with respect to the defective set I

© The Author(s), under exclusive license to Springer Nature Switzerland AG 2024
W. Wu and G. Tong (Eds.): COCOON 2023, LNCS 14422, pp. 244–271, 2024.
https://doi.org/10.1007/978-3-031-49190-0_18

is 1 if $Q \cap I \neq \emptyset$, and 0 otherwise. Throughout the paper, we denote the number of defective items by d and the number of items by $n := |X|$.

Group testing was formerly purposed by Robert Dorfman [12], for economizing mass blood testing during WWII. Since it was initially proposed, group testing methods have been utilized in a variety of applications including DNA library screening, product quality control and neural group testing for accelerating deep learning [17,18,27,29,36,38]. Among its recent applications, group testing has been advocated for accelerating mass testing for COVID-19 PCR-based tests around the world [8,20,23,24,30,35,41].

Several settings for group testing has been developed over the years. The distinction between *adaptive* and *non-adaptive* algorithms is widely considered. In *adaptive* algorithms, the tests can depend on the answers to the previous ones. In the non-adaptive algorithms, they are independent of the previous ones and; therefore, one can make all the tests in one parallel step.

Unlike conventional group testing, we consider the problem of finding only a subset of size $\ell < d$ from the d defective items. In [1], the authors solve this problem for $\ell = 1$ in the adaptive deterministic settings. They prove a tight bound of $\log n$ tests. For general ℓ, they prove an upper bound of $O(\ell \log n)$. Both results are derived under the assumption that d is known exactly to the algorithm. When no prior knowledge on d is available, Katona, [26], proves that for the deterministic non-adaptive settings, finding a single defective item requires n tests. For $\ell = 1$ in the adaptive deterministic settings, however, Katona proves a tight bound of $\Theta(\log n)$ tests.

The problem of detecting a subset of defective items with a predetermined size holds significant practical applications, especially when operational limitations necessitate the detection of a subset of a specific size ℓ. In Appendix B, we present some examples of real-world applications of this problem, including: (1) accelerating PCR-detectable syndromes and (2) abnormal event detection in surveillance camera videos using deep neural networks.

In this paper, we study the test complexity of this problem in adaptive, non-adaptive, randomized and deterministic settings. We establish lower and upper bounds for the scenario where no prior knowledge of the value of d is available. Moreover, we study the same problem when there is either an estimation or at least an upper bound on the value of d. In the literature, some results assume that d is known *exactly*, or some upper bound on the number of defective items is known in advance to the algorithm. In practice, only an estimate of d is known. In this paper, when we say that d is known in advance to the algorithm, we assume that some estimate D that satisfies $d/4 \leq D \leq 4d$ is known to the algorithm. All the results assume that $\ell \leq d/4$. If $\ell > d/4$, employ the algorithm that detects all defective items. Our results are summarized in the following subsection.

1.1   Detecting ℓ Defective Items from d Defective Items

The Table in Fig. 1 presents the results obtained by this work. Below, a concise explanation of the findings depicted in Fig. 1 is provided. The results indicated by $\star$ in Fig. 1 are the most challenging results of this paper.

		Adaptive			Non-Adaptive	
		Deterministic	Randomized		Deterministic	Randomized
d is known	Upper B.	(1) $\ell \log \frac{n}{\ell}$	(5) $\ell \log \frac{n}{d}$	Upper B.	(9)★ $d \log \frac{n}{d}$	(13) $\ell \log \frac{n}{d}$
	Lower B.	(2) $\ell \log \frac{n}{d} + \log n$	(6) $\ell \log \frac{n}{d}$	Lower B.	(10)★ $d \log \frac{n}{d}$	(14) $\ell \log \frac{n}{d}$
		Deterministic	Randomized		Deterministic	Randomized
d is unknown	Upper B.	(3) $\ell \log \frac{n}{\ell}$	(7) $\ell \log \frac{n}{d}$	Upper B.	(11) n	(15) $\ell \log^2 n$
	Lower B.	(4) $\ell \log \frac{n}{\ell}$	(8) $\ell \log \frac{n}{d}$	Lower B.	(12) n	(16)★ $\frac{\ell \log^2 n}{\log \ell + \log \log n}$

Fig. 1. Results of detecting ℓ from d defective items where $\ell \leq d/4$. In (1) and (2), the bounds are asymptotically tight when there is a constant $c < 1$, such that $d \leq n^c$. The results marked with ★ are the most difficult and challenging results of this paper.

- In (1) and (2), the algorithm is deterministic adaptive, and d is known in advance to the algorithm. The upper bound is $\ell \log(n/\ell) + O(\ell)$. The algorithm splits the items into equal sizes ℓ disjoint sets and uses binary search to detect ℓ defective items in the sets. The lower bound is $\max(\ell \log(n/d), \log n)$. In what follows, when we say input, we mean $I \subset [n]$, $|I| = d$ (the defective items)[1] and, when we say output, we mean a subset $L \subset I$ of size ℓ. A set of ℓ items can be an output of at most $\binom{n-\ell}{d}$ inputs. This gives a lower bound for the number of outputs of the algorithm, which, in turn, (its log) gives a lower bound $\ell \log(n/d) - \ell$ for the test complexity. For the lower bound $\log n$, we show that if the number of possible outputs of the algorithm is less than $n - d$, then one can construct a size d input I that contains no output. Therefore, the test complexity is at least $\log(n - d)$.
- In (3) and (4), the algorithm is deterministic adaptive, and d is unknown to the algorithm. The upper bound is $\ell \log(n/\ell) + O(\ell)$. The algorithm in (1) also works when d is unknown to the algorithm. The lower bound follows from (2) when we choose $d = 4\ell$.
- In (5) and (6), the algorithm is randomized adaptive, and d is known in advance to the algorithm. The upper bound is $\ell \log(n/d) + O(\ell)$. The algorithm uniformly at random chooses each element in X with probability $O(\ell/d)$ and puts the items in X'. We show that, with high probability, X' contains ℓ defective items and $|X'| = O(n\ell/d)$. Then, the algorithm deterministically detects ℓ defective items in X' using the algorithm of (3). This gives the result. For the lower bound, $\ell \log(n/d) - 1$, we use the same argument as in the proof of (2) in Fig. 1 with Yao's principle.

---

[1] A lower bound for the number of tests when the algorithm knows exactly d, is also a lower bound when the algorithm knows some estimate of d or does know d.

- In (7) and (8), the algorithm is randomized adaptive, and d is unknown. The upper bound is $\ell \log(n/\ell) + O(\ell + \log\log(\min(n/d, d))) = O(\ell \log(n/\ell))$. We develop a new algorithm for estimating d that uses $\log\log(\min(n/d, d))$ tests and then use the algorithm in (5). The lower bound follows from (4).
- In (9) and (10), the algorithm is deterministic non-adaptive, and d is known in advance to the algorithm. For the upper bound, we first define the $(2d, d+\ell)$-restricted weight one $t \times n$-matrix. This is a $t \times n$ 0-1-matrix such that any set of $2d$ columns contains at least $d + \ell$ distinct weight one vectors. Using this matrix, we show how to detect ℓ defective items with t tests. Then we show that there is such a matrix with $t = O(d \log(n/d))$ rows. We then give a tight lower bound. We show that such construction is not avoidable. From any non-adaptive algorithm, one can construct a $(2d, 1)$-restricted weight one $t \times n$-matrix. We then show that such a matrix must have at least $t = \Omega(d \log(n/d))$ rows.
- In (11) and (12), the algorithm is deterministic non-adaptive, and d is unknown. We show that any such algorithm must test all the items individually.
- In (13) and (14), the algorithm is randomized non-adaptive, and d is known in advance to the algorithm. The upper bound is $O(\ell \log(n/d))$. The algorithm runs $t = O(\ell)$ parallel iterations. At each iteration, it uniformly at random chooses each element in $X = [n]$ with probability $O(1/D)$ and puts it in X'. With constant probability, X' contains one defective item. Then it uses the algorithm that tests if X' contains exactly one defective item and detects it. The lower bound follows from (6).
- In (15) and (16), the algorithm is randomized non-adaptive, and d is unknown to the algorithm. The upper bound is $O(\ell \log^2 n)$. The algorithm runs the non-adaptive algorithm that gives an estimation $d/2 \le D \le 2d$ of d [4, 11,21], and, in parallel, it runs $\log n$ randomized non-adaptive algorithms that find ℓ defective items assuming $D = 2^i$ for all $i = 1, 2, \ldots, \log n$ (the algorithm in (13)). The lower bound is $\tilde{\Omega}(\ell \log^2 n)$. The idea of the proof is the following. Suppose a randomized non-adaptive algorithm exists that makes $\ell \log^2 n/(c \log R)$ tests where $R = \ell \log n$ and c is a large constant. We partition the internal $[0, n]$ of all the possible sizes of the tests $|Q|$ into $r = \Theta(\log n/\log R)$ disjoint sets $N_i = \{m|n/R^{8i+8} < m \le n/R^{8i}\}$, $i \in [r]$. We then show that, with high probability, there is an interval N_j where the algorithm makes at most $(\ell/c) \log n$ tests Q that satisfy $|Q| \in N_j$. We then show that if we choose uniformly at random a set of defective items I of size $d = (\ell \log n)^{8j+4}$, then with high probability, all the tests Q of size $|Q| < n/(\ell \log n)^{8j+8}$ give answers 0, and all the tests Q of size $|Q| > n/(\ell \log n)^{8j}$ give answers 1. So, the only useful tests are those in N_j, which, by the lower bound in (14) (and some manipulation), are insufficient for detecting ℓ defective items.

All the algorithms in this paper run in polynomial time, except for the algorithm in result (9), where we demonstrate its existence.

In Appendix A, we bring known results for detecting all the defective items.

2 Definitions and Preliminary Results

Let $X = [n] := \{1, 2, \ldots, n\}$ be a set of items that contains defective items $I \subseteq X$. In group testing, we test a subset $Q \subseteq X$ of items, and the answer to the test is $T_I(Q) := 1$ if $Q \cap I \neq \emptyset$, and $T_I(Q) = 0$ otherwise. The size of the set I is denoted by $d := |I|$. When we say that d is known in advance to the algorithm, we assume that some estimate D that satisfies $d/4 \leq D \leq 4d$ is known to the algorithm.

3 Non-adaptive Algorithms

In this section, we give lower and upper bounds on the test complexity for finding ℓ defective items in the non-adaptive settings.

3.1 Deterministic Algorithms

Definition 1. *A (r, s)-restricted weight one $t \times n$-matrix M is a $t \times n$ 0-1-matrix such that any r columns in M contains at least s distinct weight one vectors.*

That is, for every r distinct columns $j_1, j_2, \ldots, j_r \in [n]$ in M, there are s rows $i_1, i_2, \ldots, i_s \in [t]$ such that $\{(M_{i_k,j_1}, M_{i_k,j_2}, \ldots, M_{i_k,j_r})\}_{k=1,\ldots,s}$ are s distinct vectors of weight one.

We prove the following simple properties of such a matrix. The following lemma is obvious:

Lemma 1. *Let $\ell < s$. If M is (r, s)-restricted weight one $t \times n$-matrix, then it is $(r - \ell, s - \ell)$ and $(r, s - \ell)$-restricted weight one $t \times n$-matrix.*

Lemma 2. *Let $n > d > \ell > 0$. Let M be a $(2d, d + \ell)$-restricted weight one $t \times n$-matrix. Let $Q^{(i)} = \{j \mid M_{i,j} = 1\}$ for $i \in [t]$.*

1. *For every two sets $A \subset B \subseteq [n]$ where $|A| = d$ and $|B| = 2d$, there is $Q^{(i)}$ such that $Q^{(i)} \cap A = \emptyset$ and $Q^{(i)} \cap B \neq \emptyset$.*
2. *For every $C \subseteq E \subseteq [n]$ where $|C| = d$ and $|E| \leq 2d$ there are ℓ sets $Q^{(i_1)}, \ldots, Q^{(i_\ell)}$ such that $|Q^{(i_j)} \cap E| = |Q^{(i_j)} \cap C| = 1$ and for every $j_1 \neq j_2$, $Q^{(i_{j_1})} \cap C \neq Q^{(i_{j_2})} \cap C$.*

Proof. Consider the columns of M with the indices of A and B. There are $d + \ell$ distinct weight one vectors in the columns with indices B. Since $d + \ell > d$ and $|A| = d$, one of those vectors is zero in all the indices of A. Therefore, M contains a row i that is zero in the indices of A and of weight one on the indices of B. Thus, $Q^{(i)}$ satisfies $Q^{(i)} \cap A = \emptyset$ and $Q^{(i)} \cap B \neq \emptyset$. This proves 1.

Assume that $|E| = 2d$. Otherwise, add $2d - |E|$ new items to E. Consider the columns of M with the indices of E and $C \subseteq E$. There are $d + \ell$ distinct weight one vectors in the columns of M with indices of E. Since $C \subset E$ and $|E \backslash C| = d$, at least ℓ of those vectors are zero in the indices of $E \backslash C$ and weight one in the indices of C. Let $i_1, \ldots, i_\ell$ be the rows that correspond to those vectors. Then $|Q^{(i_j)} \cap E| = |Q^{(i_j)} \cap C| = 1$, and for every $j_1 \neq j_2$, $Q^{(i_{j_1})} \cap C \neq Q^{(i_{j_2})} \cap C$. □

The following lemma can be shown using standard probability techniques. Its proof is given in Appendix D.

Lemma 3. *Let $s \leq cr$ for some constant $1/2 < c < 1$. Let*

$$t = O\left(\frac{r\log(n/r) + \log(1/\delta)}{\log(1/c)}\right).$$

Consider a $t \times n$ 0-1-matrix M where $M_{i,j} = 1$ with probability $1/r$. Then, with probability at least $1 - \delta$, M is a (r,s)-restricted weight one $t \times n$-matrix. In particular, there is a (r,s)-restricted weight one $t \times n$-matrix with $t = O\left(\frac{r\log(n/r)}{\log(1/c)}\right)$.

We now show how to use the (r,s)-restricted weight one matrix for testing.

Lemma 4. *Let D be an integer. If there is a $t \times n$-matrix such that for every $D/4 \leq d' \leq 4D$, M is $(2d', d' + \ell)$-restricted weight one matrix, then there is a non-adaptive deterministic algorithm that, when $d/4 \leq D \leq 4d$ is known in advance to the algorithm, detects ℓ defective items and makes t tests.*

Proof. Since $d/4 \leq D \leq 4d$, we have $D/4 \leq d \leq 4D$, and therefore, the matrix M is $(2d, d + \ell)$-restricted weight one $t \times n$-matrix. The tests of the algorithm are $Q^{(i)} = \{j | M_{i,j} = 1\}$, $i \in [t]$. The algorithm is given in Fig. 2.

Let $X' = X$ after executing steps 3-4. We first show that $|X'| < 2d$ and $I \subset X'$, i.e., X' contains all the defective items. First, if $Answer_i = T_I(Q^{(i)}) = 0$, then $Q^{(i)} \cap I = \emptyset$, and therefore, $I \subset X \backslash Q^{(i)}$. Thus, by step 4, $I \subseteq X'$.

By step 4, it follows that if $T_I(Q^{(i)}) = 0$, then $X' \cap Q^{(i)} = \emptyset$. Now, assume to the contrary that $|X'| \geq 2d$. Consider any subset $X'' \subset X'$ of size $|X''| = 2d$. Since $|I| = d$, by Lemma 2, there is $Q^{(j)}$ such that $Q^{(j)} \cap I = \emptyset$ and $Q^{(j)} \cap X'' \neq \emptyset$. Therefore, $T_I(Q^{(j)}) = 0$ and $X' \cap Q^{(j)} \neq \emptyset$. A contradiction.

Now $I \subseteq X'$, $|I| = d$ and $|X'| \leq 2d$. By Lemma 2 it follows that there are ℓ sets $Q^{(i_1)}, \ldots, Q^{(i_\ell)}$ such that $|Q^{(i_j)} \cap I| = |Q^{(i_j)} \cap X'| = 1$, and for every $j_1 \neq j_2$, $Q^{(i_{j_1})} \cap I \neq Q^{(i_{j_2})} \cap I$. Therefore, step 6 detects at least ℓ defective items. □

1. Let $Answer_i = T_I(Q^{(i)})$.
2. Let $X = [n]$; $Y = \emptyset$.
3. For $i = 1$ to t
4.   If $Answer_i = 0$ then $X \leftarrow X \backslash Q^{(i)}$.
5. For $i = 1$ to t
6.   If $Answer_i = 1$ and $|Q^{(i)} \cap X| = 1$ then $Y \leftarrow Y \cup (Q^{(i)} \cap X)$.
7. Output Y.

Fig. 2. A non-adaptive deterministic algorithm for detecting ℓ defectives when $d/4 \leq D \leq 4d$ is known.

We are now ready to prove the upper bound. This proves (9) in Fig. 1.

Theorem 1. *Let $\ell \leq D/8$. There is a non-adaptive deterministic algorithm that, when $d/4 \leq D \leq 4d$ is known in advance to the algorithm, detects ℓ defective items and makes $O(d\log(n/d))$ tests.*

Proof. Since $d/4 \leq D \leq 4d$, we have $D/4 \leq d \leq 4D$. We construct a (r, s)-restricted weight one $t \times n$-matrix where $r = 8D$ and $s = 7\frac{3}{4}D + \ell$. Since $\ell \leq D/8$, we have $s/r \leq 0.985$, and by Lemma 3, there is a $(8D, 7\frac{3}{4}D + \ell)$-restricted weight one $t \times n$-matrix with $t = O\left(8D\log\frac{n}{8D}\right) = O\left(d\log\frac{n}{d}\right)$. Let $D/4 \leq d' \leq 4D$. By Lemma 1, M is also a $(8D, 7\frac{3}{4}D + \ell - (d' - D/4) = 8D - d' + \ell)$-restricted weight one $t \times n$-matrix and $(8D - (8D - 2d'), (8D - d' + \ell) - (8D - 2d')) = (2d', d' + \ell)$-restricted weight one $t \times n$-matrix. Then, by Lemma 4, the result follows. □

We now prove the lower bound. This proves (10) in Fig. 1.

Theorem 2. *Suppose some integer D is known in advance to the algorithm where $d/4 \leq D \leq 4d$. Any non-adaptive deterministic algorithm that detects one defective item must make at least $\Omega(d\log(n/d))$ tests.*

Proof. Consider any non-adaptive deterministic algorithm $\mathcal{A}$ that detects one defective item. Let M be a 0-1-matrix of size $t \times n$ that their rows are the 0-1-vectors that correspond to the tests of $\mathcal{A}$. That is, if $Q^{(i)}$ is the ith test of $\mathcal{A}$, then the ith row of M is $(M_{i,1}, \ldots, M_{i,n})$ when $M_{i,j} = 1$ if $j \in Q^{(i)}$ and $M_{i,j} = 0$ otherwise. Let $M^{(i)}$ be the ith column of M. Let $I = \{i_1, i_2, \ldots, i_w\} \subseteq [n]$ be any set of size $w \in \{d, d+1, \ldots, 2d\}$. If I is the set of defective items, then $\vee_{i \in I} M^{(i)}$ (bitwise or) is the vector of the answers of the tests of $\mathcal{A}$. Suppose that when I is the set of defective items, $\mathcal{A}$ outputs i_j. Consider the case when the set of defective items is $I' = I \backslash \{i_j\}$. Since the answer of $\mathcal{A}$ on I' is different from the answer on I, and $\mathcal{A}$ is deterministic, we must have $\vee_{i \in I} M^{(i)} \neq \vee_{i \in I'} M^{(i)}$. Therefore, columns I must contain a vector of weight one in M. So far, we proved that every $w \in \{d, d+1, \ldots, 2d\}$ columns in M contains a vector of weight one. This also implies that if $J \subset [n]$ and $|J| \in \{d, d+1, \ldots, 2d\}$, then $\oplus_{j \in J} M^{(j)} \neq 0$ (bitwise xor). This is because if $\oplus_{j \in J} M^{(j)} = 0$, then the columns J do not contain a vector of weight one.

Consider the maximum size subset $J_0 \subset [n]$, $|J_0| < d$ such that $\oplus_{j \in J_0} M^{(j)} = 0$. We claim that there is no set $J' \subset [n] \backslash J_0$, $|J'| \leq d$, such that $\oplus_{j \in J'} M^{(j)} = 0$. This is because if such J' exists, then $\oplus_{j \in J_0 \cup J'} M^{(j)} = 0$. Then if $|J_0 \cup J'| \in \{d, d+1, \ldots, 2d\}$, we get a contradiction, and if $|J_0 \cup J'| < d$, then $|J_0 \cup J'| > |J_0|$ and J_0 is not maximum, and again we get a contradiction. Therefore, no set $J' \subset [n] \backslash J_0$, $|J'| \leq d$ satisfies $\oplus_{j \in J'} M^{(j)} = 0$.

Consider the sub-matrix M' composed of the $n - |J_0| \geq n - d$ columns $[n] \backslash J_0$ of M. The above property shows that every $2d$ columns in M' are linearly independent over the field $GF(2)$. Then the result immediately follows from the bounds on the number of rows of the parity check matrix in coding theory [32]. We give the proof for completeness.

We now show that the xor of any d columns in M' is district from the xor of any other d columns. If there are two sets of d columns J_1 and J_2 that have the same xor, then $\oplus_{j \in J_1 \Delta J_2} M^{(j)} = 0$ and $|J_1 \Delta J_2| \leq 2d$. A contradiction. Therefore, by summing all the possible d columns of M', we get $\binom{n}{d}$ distinct vectors. Thus, the number of rows of M' is at least $t \geq \log\binom{n}{d} = \Omega\left(d\log\frac{n}{d}\right)$.   □

The following theorem summarizes the result on the lower bound when d is unknown to the algorithm (result (11) in Fig. 1). Result (12) follows from the algorithm that tests every item individually. Theorem 3 is proved in Appendix D.

Theorem 3. *If d is unknown, then any non-adaptive deterministic algorithm that detects one defective item must make at least $\Omega(n)$ tests.*

3.2   Randomized Algorithms

In this subsection, we summarize results (13)-(16) in Fig. 1. We give here a detailed proof of result (16). We start our discussion with Lemmas and Theorems that are used to show the results (13)-(15). Full proofs of these results are detailed in Appendix D. We start with result (13):

Theorem 4. *Suppose some integer D is known in advance to the algorithm where $d/4 \leq D \leq 4d$. There is a polynomial time non-adaptive randomized algorithm that makes $O(\ell\log(n/d)+\log(1/\delta)\log(n/d))$ tests and, with probability at least $1 - \delta$, detects ℓ defective items.*

Result (14) in Fig. 1 is captured in the following theorem:

Theorem 5. *Let $\ell \leq d \leq n/2$ and d be known in advance to the algorithm. Any non-adaptive randomized algorithm that, with probability at least $2/3$, detects ℓ defective items must make at least $\ell\log(n/d) - 1$ tests.*

The following Theorem gives the upper bound for non-adaptive randomized algorithms when d is unknown to the algorithm. This is result (15) in Fig. 1.

Theorem 6. *Let $c < 1$ be any constant, $\ell \leq n^c$, and d be unknown to the algorithm. There is a polynomial time non-adaptive randomized algorithm that makes $O(\ell\log^2 n+\log(1/\delta)\log^2 n)$ tests, and with probability at least $1-\delta$, detects ℓ defective items.*

We now prove the lower bound when d is unknown to the algorithm. This proves result (16) in Fig. 1.

Theorem 7. *Let $c < 1$ be any constant, $\ell \leq n^c$, and d be unknown to the algorithm. Any non-adaptive randomized algorithm that, with probability at least $3/4$, detects ℓ defectives must make at least $\Omega((\ell\log^2 n)/(\log\ell + \log\log n))$ tests.*

Proof. If $n^c \geq \ell \geq n^{1/32}$, then for $d = n^{(1+c)/2} > \ell$, by Theorem 5, the lower bound is $\Omega(\ell\log(n/d)) = \Omega((\ell\log^2 n)/(\log\ell + \log\log n))$. Hence, $\ell < n^{1/32}$ can be assumed. Suppose, to the contrary, there is a non-adaptive randomized algorithm $\mathcal{A}(s, I)$ that, with probability at least $3/4$, detects ℓ defective items and makes $t = (\ell\log^2 n)/(3072(\log\ell + \log\log n))$. Here s is the random seeds, and I is the set of defective items. Define the set of integers $N_i = \{k | n/(\ell \log n)^{8i+8} \leq k < n/(\ell \log n)^{8i}\}$ for $i = 1, 2, \ldots, r$ where $r = \log n/(32(\log \ell + \log \log n))$. Let t_i be a random variable representing the number of tests Q made by $\mathcal{A}(s, I)$ where $|Q| \in N_i$. Then $t \geq t_1 + t_2 + \cdots + t_r$ and $(\ell \log^2 n)/(3072(\log \ell + \log \log n)) = t = \mathbf{E}[t] \geq \sum_{i=1}^{r} \mathbf{E}[t_i]$. Therefore, there is $j \in [r]$ such that $\mathbf{E}[t_j] \leq \mathbf{E}[t]/r = (\ell \log n)/96$. By Markov's bound, with probability at least $1 - 4/96 = 1 - 1/24$ we have $t_j < (\ell \log n)/4$. Let $d = (\ell \log n)^{8j+4}$. Define the following sets random variables: M_1 the set of all tests Q that $\mathcal{A}(s, I)$ makes where $|Q| < n/(\ell \log n)^{8j+8}$, M_2 the set of all tests Q that $\mathcal{A}(s, I)$ makes where $|Q| \geq n/(\ell \log n)^{8j}$ and M_3 the set of tests Q that $\mathcal{A}(s, I)$ makes where $|Q| \in N_j = \{k | n/(\ell \log n)^{8j+8} \leq k < n/(\ell \log n)^{8j}\}$. For a set of defective items I, let $A_1(I)$ be the event that all the tests in M_1 give answers 0 and $A_2(I)$ the event that all the tests in M_2 give answers 1. Let $\mathcal{D}$ be the distribution over $I \subset [n]$, $|I| = d$, where the items of I are selected uniformly at random without replacement from $[n]$. Let $\mathcal{D}'$ be the distribution over $I = \{i_1, \ldots, i_d\} \subset [n]$, where the items of I are selected uniformly at random with replacement from $[n]$. Let B be the event that I, chosen according to $\mathcal{D}'$, has d items. Then, since $\ell < n^{1/32}$,

$$\Pr_{\mathcal{D}'}[\neg B] = 1 - \prod_{i=1}^{d-1}\left(1 - \frac{i}{n}\right) \leq \frac{d(d-1)}{2n} \leq \frac{(\ell \log n)^{16j+8}}{2n} \leq$$

$$\leq \frac{(\ell \log n)^{16r+8}}{2n} = \frac{(\ell \log n)^8}{2\sqrt{n}} = o(1).$$

We now have

$$\Pr_{I \sim \mathcal{D}}[\neg A_1(I)] \leq \Pr_{I \sim \mathcal{D}'}[(\exists Q \in M_1)Q \cap I \neq \emptyset] + \Pr_{I \sim \mathcal{D}'}[\neg B]$$

$$\leq t \Pr_{I \sim \mathcal{D}'}[Q \cap I \neq \emptyset | Q \in M_1] + o(1)$$

$$\leq \frac{\ell \log^2 n}{3072(\log \ell + \log \log n)}\left(1 - \left(1 - \frac{1}{(\ell \log n)^{8j+8}}\right)^d\right) + o(1)$$

$$\leq \frac{\ell \log^2 n}{3072(\log \ell + \log \log n)}\frac{d}{(\ell \log n)^{8j+8}} + o(1)$$

$$\leq \frac{1}{3072 \log \log n}\frac{1}{\ell^3 \log^2 n} + o(1) = o(1).$$

and

$$\Pr_{I \sim \mathcal{D}}[\neg A_2(I)] \leq \Pr_{I \sim \mathcal{D}'}[(\exists Q \in M_2)Q \cap I = \emptyset] + \Pr_{I \sim \mathcal{D}'}[\neg B]$$

$$\leq t \Pr_{I \sim \mathcal{D}'}[Q \cap I = \emptyset | Q \in M_2] + o(1)$$

$$\leq \frac{\ell \log^2 n}{3072(\log \ell + \log \log n)} \left(1 - \frac{1}{(\ell \log n)^{8j}}\right)^d + o(1)$$

$$\leq \frac{\ell \log^2 n}{3072(\log \ell + \log \log n)} e^{-\frac{d}{(\ell \log n)^{8j}}} + o(1)$$

$$\leq \frac{\ell \log^2 n}{3072(\log \ell + \log \log n)} e^{-\ell^4 \log^4 n} + o(1) = o(1).$$

We give a non-adaptive randomized algorithm that for $d = (\ell \log n)^{8j+4}$ makes $(\ell/4) \log n$ tests and with probability at least $2/3$ detects ℓ defective items. By Theorem 5, and since $\ell < n^{1/32}$ and $d = (\ell \log n)^{8j+4} \leq (\ell \log n)^{8r+4} = (\ell \log n)^4 n^{1/4} \leq n^{1/2}/2$, the test complexity is at least $\ell \log(n/d) - 1 \geq (1/2)\ell \log n$, and we get a contradiction.

The algorithm is the following. Choose uniformly at random a permutation $\phi : [n] \to [n]$. Consider the tests of algorithm $\mathcal{A}$. Let M_i, $i = 1, 2, 3$, be the sets defined above. Define $M_i' = \{(a_{\phi(1)}, \ldots, a_{\phi(n)}) | (a_1, \ldots, a_n) \in M_i\}$. Answer 0 for all the tests in M_1' and 1 for all the tests in M_2'. If $|M_3'| > (\ell/4) \log n$, then return FAIL. Otherwise, make all the tests in M_3'. Give the above answers of the tests to the algorithm $\mathcal{A}$ and let L be its output. Output $\phi^{-1}(L) = \{\phi^{-1}(i) | i \in L\}$.

Since ϕ is chosen uniformly at random, the new set of defective items $\phi(I)$ is distributed uniformly at random over all the subsets of $[n]$ of size d. The probability that the answers for tests in M_1' and M_2' are wrong is $o(1)$. The probability that $|M_3'| > (\ell/4) \log n$ is at most $1/24$. By the promises, the failure probability of $\mathcal{A}$ is at most $1/4$. Therefore, the probability that this algorithm fails is at most $1/4 + 1/24 + o(1) < 1/3$. This completes the proof. $\square$

4   Adaptive Algorithms

In this section, we study the test complexity of adaptive deterministic and adaptive randomized algorithms. We demonstrate the results (1)–(8) in Fig. 1. All proofs for the theorems in this section are detailed in Appendix E.

4.1   Deterministic Algorithms

The following Theorem summarizes the results (1) and (3) in Fig. 1. Here d can be known or unknown to the algorithm.

Theorem 8. *Let $d \geq \ell$. There is a polynomial time adaptive deterministic algorithm that detects ℓ defective items and makes at most $\ell \log(n/\ell) + 3\ell = O(\ell \log(n/\ell))$ tests.*

The following Theorem, summarizes the lower bound (2) in Fig. 1. We remind the reader that when we say that d is known in advance to the algorithm, we mean that an estimate D that satisfies $d/4 \leq D \leq 4d$ is known to the algorithm. The following lower bound holds even if the algorithm knows d exactly in advance.

Theorem 9. *Let $\ell \leq d \leq n/2$ and d be known in advance to the algorithm. Any adaptive deterministic algorithm that detects ℓ defective items must make at least $\max(\ell \log(n/d), \log n - 1) = \Omega(\ell \log(n/d) + \log n)$ tests.*

Note that the upper bound $O(\ell \log(n/\ell))$ in Theorem 8 asymptotically matches the lower bound $\Omega(\ell \log(n/d))$ in Theorem 9 when $d = n^{o(1)}$.

The following Theorem summarizes result (4) in Fig. 1.

Theorem 10. *Let $\ell \leq d \leq n/2$ and d be unknown to the algorithm. Any adaptive deterministic algorithm that detects ℓ defective items must make at least $\ell \log(n/\ell)$ tests.*

4.2   Randomized Algorithms

In this subsection, we demonstrate the results on the test complexity of adaptive randomized algorithms. The following theorem proves the upper bound when d is known in advance to the algorithm. This proves result (5) in Fig. 1.

Theorem 11. *Let $\ell \leq d/2$. Suppose some integer D is known in advance to the algorithm where $d/4 \leq D \leq 4d$. There is a polynomial time adaptive randomized algorithm that makes $\ell \log(n/d) + \ell \log \log(1/\delta) + O(\ell)$ tests and, with probability at least $1 - \delta$, detects ℓ defective items.*

The following theorem captures the lower bounds (6) and (8) in Fig. 1. These suit the case when d is known in advance.

Theorem 12. *Let $\ell \leq d \leq n/2$ and d be known in advance to the algorithm. Any adaptive randomized algorithm that, with probability at least $2/3$, detects ℓ defective items must make at least $\ell \log(n/d) - 1$ tests.*

In particular,

Theorem 13. *Let $\ell \leq d \leq n/2$ and d is unknown to the algorithm. Any adaptive randomized algorithm that, with probability at least $2/3$, detects ℓ defective items must make at least $\ell \log(n/d) - 1$ tests.*

The following theorem summarizes the upper bound when d is unknown to the algorithm. This proves result (7) in Fig. 1.

Theorem 14. *Let $\ell \leq d/2$ and d be unknown to the algorithm. There is a polynomial time adaptive randomized algorithm that detects ℓ defective items and makes $\ell \log(n/d) + \ell \log \log(1/\delta) + O(\ell + \log \log(\min(n/d, d)) + \log(1/\delta))$ tests.*

A   Known Results for Detecting All the Defective Items

The following results are known for detecting all the d defective items. See the Table in Fig. 3.

		Adaptive			Non-Adaptive	
		Deterministic	Randomized		Deterministic	Randomized
d is known	Upper B.	(1) $d\log\frac{n}{d}$	(5) $d\log\frac{n}{d}$	Upper B.	(9) $d^2\log n$	(13) $d\log\frac{n}{d}$
	Lower B.	(2) $d\log\frac{n}{d}$	(6) $d\log\frac{n}{d}$	Lower B.	(10) $\frac{d^2\log n}{\log d}$	(14) $d\log\frac{n}{d}$
		Deterministic	Randomized		Deterministic	Randomized
d is unknown	Upper B.	(3) $d\log\frac{n}{d}$	(7) $d\log\frac{n}{d}$	Upper B.	(11) n	(15) n
	Lower B.	(4) $d\log\frac{n}{d}$	(8) $d\log\frac{n}{d}$	Lower B.	(12) n	(16) n

Fig. 3. Results for the test complexity of detecting the d defective items. The lower bounds are in the Ω-symbol and the upper bound are in the O-symbol

- In (1) and (2) (in the table in Fig. 3), the algorithm is deterministic adaptive, and d is known in advance to the algorithm. The best lower bound is the information-theoretic lower bound $\log\binom{n}{d} \geq d\log(n/d)+\Omega(d)$. Hwang in [25] gives a generalized binary splitting algorithm that makes $\log\binom{n}{d} + d - 1 = d\log(n/d) + O(d)$ tests.
- In (3) and (4), the algorithm is deterministic adaptive, and d is unknown to the algorithm. The upper bound $d\log(n/d) + O(d)$ follows from [3,10, 14–16,34,39] and the best constant currently known in $O(d)$ is $5 - \log 5 \approx 2.678$ [39]. The lower bound follows from (2). In [5], Bshouty et al. show that estimating the number of defective items within a constant factor requires at least $\Omega(d\log(n/d))$ tests.
- In (5) and (6), the algorithm is randomized adaptive, and d is known in advance. The upper bound follows from (1). The lower bound follows from Yao's principle with the information-theoretic lower bound.
- In (7) and (8), the algorithm is randomized adaptive, and d is unknown to the algorithm. The upper bound $d\log(n/d) + O(d)$ follows from (3). The lower bound follows from (6).
- In (9) and (10), the algorithm is deterministic non-adaptive, and d is known in advance to the algorithm. The lower bound $\Omega(d^2\log n/\log d)$ is proved in [9,19,22,33]. A polynomial time algorithm that constructs an algorithm that makes $O(d^2\log n)$ tests was first given by Porat and Rothschild [31].

- In (11) and (12), the algorithm is deterministic non-adaptive and d is unknown to the algorithm. In [4], Bshouty shows that estimating the number of defective items within a constant factor requires at least $\Omega(n)$ tests. The upper bound is the trivial bound of testing all the items individually.
- In (13) and (14), the algorithm is randomized non-adaptive, and d is known in advance to the algorithm. The lower bound follows from (6). The upper bound is $O(d\log(n/d))$. The constant in the O-symbol was studied in [2,6,7,13] and referenced within. The best constant known in the O-symbol is $\log e \approx 1.443$ [7].
- In (15) and (16), the algorithm is randomized non-adaptive, and d is unknown to the algorithm. The lower bound $\Omega(n)$ follows from Yao's principle and the fact that, for a random uniform $i \in [n]$, to detect the defective items $[n]\backslash\{i\}$, we need at least $\Omega(n)$ tests. The upper bound is the trivial bound of testing all the items individually.

B   Applications

In many cases, the detection of a specific number of defective items, ℓ, is of utmost importance due to system limitations or operational requirements. For instance, in scenarios like blood tests or medical facilities with limited resources such as ventilators, doctors, beds, or medicine supply, it becomes crucial to employ algorithms that can precisely identify ℓ defectives instead of detecting all potential cases. This targeted approach offers significant advantages in terms of efficiency, as the time required to detect only ℓ defective items is generally much shorter than the time needed to identify all defects. By focusing on any subset of ℓ defectives, the algorithms proposed in this paper offer more efficient procedures.

B.1   Identifying a Subset of Samples that Exhibit a PCR-Detectable Syndrome

Polymerase Chain Reaction or *PCR* testing is a widely used laboratory technique in molecular biology. This technique is used to amplify specific segments of DNA or RNA in a sample, and therefore, allowing for detection, quantification and analyses of these specific genetic sequences [13,24,41]. PCR tests can be designed to identify various organisms, including pathogens such as viruses or bacteria (e.g. COVID-19), by targeting their unique DNA or RNA signatures. Although PCR tests are associated with high costs and time consumption, they are extensively utilized in a wide range of fields, including medical diagnostics, research laboratories, forensic analysis, and other applications that demand accurate and sensitive detection of genetic material. This popularity is primarily attributed to their exceptional accuracy. To enhance the efficiency and cost-effectiveness of PCR testing, group testing methodologies can be applied to PCR testing. Applying group testing to PCR involves combining multiple samples into a single test sample. The combined sample, also called the *group test*, is then examined. If the sample screening indicates an infectious sample, this implies that at least one of the original samples is infected. Conversely, if none of the samples in the combined sample exhibit signs of infection, then none of the individual samples are infected. Typically, PCR tests are conducted by specialized machines capable of simultaneously performing approximately 96 tests. Each test-run can span over several hours. Therefore, when applying group testing to accelerate PCR process, it is recommended to employ non-adaptive methodologies.

Assume that a scientific experiment need to be conducted over a group of study participants to examine the efficiency of a new drug developed for medicating some disease related to bacterial or virus infection. Suppose that a PCR test is required to check whether a participant is affected by the disease or not. Moreover, assume that the number of the participants that volunteered for the experiment is n and the incidence rate of the infection among them is known in advance, denote that by p. Therefore, an approximation of the number of infected participants can be derived from n and p, denote that value by d. In situations where logistic constraints necessitate selecting a limited number of infected individuals, specifically $\ell \leq d$, to participate in an experiment, a non-adaptive group testing algorithm for identifying ℓ defectives (virus carriers) from n samples when d is known can be employed.

B.2   Abnormal Event Detection in Surveillance Camera Videos

Efficiently detecting abnormal behavior in surveillance camera videos plays a vital role in combating crimes. These videos are comprised of a sequence of continuous images, often referred to as *frames*. The task of identifying suspicious behavior within a video is equivalent to searching for abnormal behavior within a collection of frames. Training *deep neural networks* (shortly, DNN) for automating suspicious image recognition is currently a widely adopted approach for the task [28,37,40]. By utilizing the trained DNN, it becomes possible to classify a new image and determine whether it exhibits suspicious characteristics or not. However, once the training process is complete, there are further challenges to address, specially when dealing with substantial amount of images that need to be classified via the trained network. In this context, *inference* is the process of utilizing the trained model to make predictions on new data that was not part of the training phase. Due to the complexity of the DNN, inference time of images can cost hundreds of seconds of GPU time for a single image. Long inference time poses challenges in scenarios where real-time or near-real-time processing is required, prompting the need for optimizing and accelerating the inference process.

The detection of abnormal behavior in surveillance camera videos is often characterized by an imbalanced distribution of frames portraying abnormal behavior, also called *abnormal frames*, in relation to the total number of frames within the video. Denote the total number of frames in a video by n and the number of abnormal frames by d. To identify suspicious behavior in a video, the goal is to find at least one abnormal frame among the d frames. In most cases, we cannot assume any non-trivial upper bound or estimation of any kind for

d. Therefore, applying non-adaptive group testing algorithms for finding $\ell < d$ defectives when d is unknown best suits this task.

It is unclear, however, how group testing can be applied to instances like images. Liang and Zou, [29], proposed three different methods for pooling image instances: 1) merging samples in the pixel space, 2) merging samples in the feature space, and 3) merging samples hierarchically and recursively at different levels of the network. For each grouping method, they provide network enhancements that ensure that the group testing paradigm continues to hold. This means that a positive prediction is inferred on a group if and only if it contains at least one positive image (abnormal frame).

C   Useful Lemmas

We will use the following version of Chernoff's bound.

Lemma 5. Chernoff's Bound. *Let* $X_1, \ldots, X_m$ *be independent random variables taking values in* $\{0,1\}$. *Let* $X = \sum_{i=1}^{m} X_i$ *denotes their sum, and* $\mu = \mathbf{E}[X]$ *denotes the sum's expected value. Then*

$$\Pr[X > (1+\lambda)\mu] \leq \left(\frac{e^\lambda}{(1+\lambda)^{(1+\lambda)}}\right)^\mu \leq e^{-\frac{\lambda^2\mu}{2+\lambda}} \leq \begin{cases} e^{-\frac{\lambda^2\mu}{3}} & \text{if } 0 < \lambda \leq 1 \\ e^{-\frac{\lambda\mu}{3}} & \text{if } \lambda > 1. \end{cases} \quad (1)$$

In particular,

$$\Pr[X > \Lambda] \leq \left(\frac{e\mu}{\Lambda}\right)^\Lambda. \quad (2)$$

For $0 \leq \lambda \leq 1$ *we have*

$$\Pr[X < (1-\lambda)\mu] \leq \left(\frac{e^{-\lambda}}{(1-\lambda)^{(1-\lambda)}}\right)^\mu \leq e^{-\frac{\lambda^2\mu}{2}}. \quad (3)$$

D   Proofs for Non-adaptive Settings

In this section, we give all the proofs that were stated in Sect. 3. We restate the Theorems and Lemmas for convenience.

D.1   Deterministic Algorithms

In this subsection, we prove Theorem 3. This proves result (12) in Fig. 1. Result (11) follows from the algorithm that tests every item individually. Moreover, we give a detailed proof for Lemma 3.

Lemma 3. *Let* $s \leq cr$ *for some constant* $1/2 < c < 1$. *Let*

$$t = O\left(\frac{r\log(n/r) + \log(1/\delta)}{\log(1/c)}\right).$$

Consider a $t \times n$ *0-1-matrix* M *where* $M_{i,j} = 1$ *with probability* $1/r$. *Then, with probability at least* $1 - \delta$, M *is a* (r,s)-*restricted weight one* $t \times n$-*matrix. In particular, there is a* (r,s)-*restricted weight one* $t \times n$-*matrix with* $t = O\left(\frac{r\log(n/r)}{\log(1/c)}\right)$.

Proof. Consider any r columns $J = \{j_1, \ldots, j_r\}$ in M. Let A_J be the event that columns J in M do not contain at least s distinct weight one vectors. For every $i \in [t]$, the probability that $(M_{i,j_1}, \ldots, M_{i,j_r})$ is of weight 1 is $\binom{r}{1}(1/r)(1 - 1/r)^{r-1} \geq 1/2$. In every such row, the entry that is equal to 1 is distributed uniformly at random over J. Let m_J be the number of such rows. The probability that columns J in M do not contain at least s distinct weight one vectors is at most $\Pr[A_J | m_J = m] \leq \binom{r}{s-1}\left(\frac{s-1}{r}\right)^m \leq 2^r c^m$. Since $\mathbf{E}[m_J] \geq t/2$, by Chernoff's bound (Lemma 5), $\Pr\left[m_J < \frac{t}{4}\right] \leq 2^{-t/16}$. Therefore, the probability that M is not (r, s)-restricted weight one $t \times n$-matrix is at most

$$\Pr[(\exists J \subset [n], |J| = r)A_J] \leq \binom{n}{r}\Pr[A_J] \leq \binom{n}{r}\left(\Pr[A_J | m_J \geq \frac{t}{4}] + \Pr[m_J < \frac{t}{4}]\right)$$
$$\leq \binom{n}{r}\left(2^r c^{t/4} + 2^{-t/16}\right) \leq \binom{n}{r}2^{r+1}c^{t/16} \leq \delta.$$

$\square$

Theorem 3. *If d is unknown, then any non-adaptive deterministic algorithm that detects one defective item must make at least $\Omega(n)$ tests.*

Proof. Consider any non-adaptive deterministic algorithm $\mathcal{A}$ that detects one defective item. Let M be a 0-1-matrix of size $t \times n$ whose rows correspond to the tests of $\mathcal{A}$. Suppose for the set of defective items $I_0 = [n]$ the algorithm outputs i_1, for the set $I_1 = [n]\backslash\{i_1\}$ outputs i_2, for $I_2 = [n]\backslash\{i_1, i_2\}$ outputs i_3, etc. Obviously, $\{i_1, \ldots, i_n\} = [n]$. Now, since the output for I_0 is distinct from the output for I_1, we must have a row in M that is equal to 1 in entry i_1 and zero elsewhere. Since the output for I_1 is distinct from the output for I_2, we must have a row in M that is equal to 1 in entry i_2 and zero in entries $[n]\backslash\{i_1, i_2\}$. Etc. Therefore, M must have at least n rows. $\square$

D.2 Random Algorithms

In this subsection, we give detailed proofs for results (13)–(15) in Fig. 1, that are summarized in Subsect. 3.2. We start with proving the following lemma:

Lemma 6. *There is a non-adaptive deterministic algorithm that makes $t = \log n + 0.5 \log \log n + O(1)$ tests and decides whether $d \leq 1$ and if $d = 1$ detects the defective item.*

Proof. We define a 0-1-matrix M, where the rows of the matrix correspond to the tests of the algorithm. The size of the matrix is $t \times n$, where t is the smallest integer such that $n \leq \binom{t}{\lfloor t/2 \rfloor}$ and its columns contain distinct Boolean vectors of weight $\lfloor t/2 \rfloor$. Therefore $t = \log n + 0.5 \log \log n + O(1)$.

Now, if there are no defective items, we get 0 in all the answers of the tests. If there is only one defective item, and it is $i \in [n]$, then the vector of answers to the tests is equal to the column i of M. If there is more than one defective item, then the weight of the vector of the answers is greater than $\lfloor t/2 \rfloor$. $\square$

Theorem 4 shows result (13) in Fig. 1:

Theorem 4. *Suppose some integer D is known in advance to the algorithm where $d/4 \leq D \leq 4d$. There is a polynomial time non-adaptive randomized algorithm that makes $O(\ell \log(n/d) + \log(1/\delta) \log(n/d))$ tests and, with probability at least $1 - \delta$, detects ℓ defective items.*

Proof. If $\ell \geq D/32$, then the non-adaptive randomized algorithm that finds all the defective items makes $O(d \log(n/d)) = O(\ell \log(n/d))$ tests. So, we may assume that $\ell < D/32 \leq d/8$.

Let $\ell \leq d/8$. The algorithm runs $t = O(\ell + \log(1/\delta))$ iterations. At each iteration, it uniformly at random chooses each element in $X = [n]$ with probability $1/(2D)$ and puts it in X'. If $|X'| > 4n/D$, then it continues to the next iteration. If $|X'| \leq 4n/D$, then it uses the algorithm in Lemma 6 to detect if X' contains one defective item, and if it does, it detects the item. If X' contains no defective item or more than one item, it continues to the next iteration.

Although the presentation of the above algorithm is adaptive, it is clear that all the iterations can be run non-adaptively.

Let A be the event that X' contains exactly one defective item. The probability of A is

$$\Pr[A] = \binom{d}{1} \frac{1}{2D} \left(1 - \frac{1}{2D}\right)^{d-1} \geq \frac{1}{10}.$$

Since $\mathbf{E}[|X'|] = n/(2D)$, by Chernoff's bound (Lemma 5)

$$\Pr[|X'| > 4n/D] \leq \left(\frac{e^7}{8^8}\right)^{n/(2D)} \leq \frac{1}{20}.$$

Therefore,

$$\Pr[A \text{ and } |X'| \leq 4n/D] \geq \frac{1}{20}.$$

Now, assuming A occurs, the defective in X' is distributed uniformly at random over the d defective items. Since $\ell < d/8$, at each iteration, as long as the algorithm does not get ℓ defective items, the probability of getting a new defective item in the next iteration is at least $7/8$. Let B_i be the event that, in iteration i, the algorithm gets a new defective item. Then

$$\Pr[B_i] = \frac{7}{8} \Pr[A \text{ and } |X'| \leq 4n/D] \geq \frac{7}{160}.$$

By Chernoff's bound (Lemma 5), after $O(\ell + \log(1/\delta))$ iterations, with probability at least $1 - \delta$, the algorithm detects ℓ defective items.

Therefore, by Lemma 6, the test complexity of the algorithm is

$$O((\ell + \log(1/\delta)) \log |X'|) = O\left(\ell \log \frac{n}{d} + \log(1/\delta) \log \frac{n}{d}\right).$$

$\square$

The following lower bound follows from Theorem 12 from Sect. 4. This is result (14) in Fig. 1.

Theorem 5. *Let $\ell \leq d \leq n/2$ and d be known in advance to the algorithm. Any non-adaptive randomized algorithm that, with probability at least $2/3$, detects ℓ defective items must make at least $\ell \log(n/d) - 1$ tests.*
In [4,11,21], the following is proved

Lemma 7. *There is a polynomial time non-adaptive randomized algorithm that makes $O(\log(1/\delta) \log n)$ tests and, with probability at least $1 - \delta$, finds an integer D that satisfies $d/2 < D < 2d$.*

Result (15) in Fig. 1 is summarized in Theorem 6.

Theorem 6. *Let $c < 1$ be any constant, $\ell \leq n^c$, and d be unknown to the algorithm. There is a polynomial time non-adaptive randomized algorithm that makes $O(\ell \log^2 n + \log(1/\delta) \log^2 n)$ tests, and with probability at least $1 - \delta$, detects ℓ defective items.*

Proof. We make all the tests of the non-adaptive algorithm that, with probability at least $1 - \delta/2$, $1/4$-estimate d, i.e., finds an integer D such that $d/4 < D < 4d$. By Lemma 7, this can be done with $O(\log(1/\delta) \log n)$ tests.
We also make all the tests of the non-adaptive algorithms that, with probability at least $1 - \delta/2$, detects ℓ defective items for all $d = 2^i \ell$, $i = 1, 2, \ldots, \log(n/\ell)$. By Theorem 4, this can be done with

$$O\left(\sum_{i=1}^{\log(n/\ell)} \ell \log \frac{n}{2^i \ell} + \log \frac{2}{\delta} \log \frac{n}{2^i \ell} \right) = O((\ell + \log(1/\delta)) \log^2 n)$$

tests. □

E   Proofs for Adaptive Settings

In this section, we bring all the proofs of the Theorems that appeared in Sect. 4. We restate the Theorems for convenience.

E.1   Deterministic Algorithms

In the following, we bring proofs for all the Theorems in Sect. 4.1.

Theorem 8. *Let $d \geq \ell$. There is a polynomial time adaptive deterministic algorithm that detects ℓ defective items and makes at most $\ell \log(n/\ell) + 3\ell = O(\ell \log(n/\ell))$ tests.*

Proof. We first split the items $X = [n]$ to ℓ disjoint sets $X_1, \ldots, X_\ell$ of (almost) equal sizes (each of size $\lfloor n/\ell \rfloor$ or $\lceil n/\ell \rceil$). Then we use the binary search algorithm (binary splitting algorithm) for each i to detect all the defective items in X_i until we get ℓ defective items.

Each binary search takes at most $\lceil \log(n/\ell) \rceil + 1$ tests, and testing all X_i takes at most ℓ tests.   □

The following Theorem, summarizes the lower bound (2) in Fig. 1. We remind the reader that when we say that d is known in advance to the algorithm, we mean that an estimate D that satisfies $d/4 \le D \le 4d$ is known to the algorithm. The following lower bound holds even if the algorithm knows d exactly in advance.

Theorem 9. *Let $\ell \le d \le n/2$ and d be known in advance to the algorithm. Any adaptive deterministic algorithm that detects ℓ defective items must make at least $\max(\ell \log(n/d), \log n - 1) = \Omega(\ell \log(n/d) + \log n)$ tests.*

Proof. Let A be an adaptive deterministic algorithm that detects ℓ defective items. Let $L_1, \ldots, L_t$ be all the possible ℓ-subsets of X that A outputs. Since the algorithm is deterministic, the test complexity of A is at least $\log t$. Since $L_i \subseteq I$ (the set of d defective items), each L_i can be an output of at most $\binom{n-\ell}{d-\ell}$ sets I. Since the number of possible sets of defective items I is $\binom{n}{d}$, we have

$$ t \ge \frac{\binom{n}{d}}{\binom{n-\ell}{d-\ell}} \ge \frac{n(n-1)\cdots(n-\ell+1)}{d(d-1)\cdots(d-\ell+1)} \ge \left(\frac{n}{d}\right)^{\ell}. $$

Therefore the test complexity of A is at least $\log t \ge \ell \log(n/d)$.

We now show that $t > n - d$. Now suppose, to the contrary, that $t \le n - d$. Choose any $x_i \in L_i$ and consider any $S \subseteq X \backslash \{x_i | i \in [t]\}$ of size d. For the set of defective items $I = S$, the algorithm outputs some L_i, $i \in [t]$. Since $L_i \not\subseteq S$, we get a contradiction. Therefore, $t > n - d$ and $\log t > \log(n-d) \ge \log(n/2) = \log n - 1$.   □

Note that the upper bound $O(\ell \log(n/\ell))$ in Theorem 8 asymptotically matches the lower bound $\Omega(\ell \log(n/d))$ in Theorem 9 when $d = n^{o(1)}$.

The following Theorem proves result (4) in Fig. 3.

Theorem 10. *Let $\ell \le d \le n/2$ and d be unknown to the algorithm. Any adaptive deterministic algorithm that detects ℓ defective items must make at least $\ell \log(n/\ell)$ tests.*

Proof. Since the algorithm works for any d, we let $d = 4\ell$. Then by the first bound in Theorem 9, the result follows.   □

E.2   Random Algorithms

In this subsection, we demonstrate the results on the test complexity of adaptive randomized algorithms. The following theorem proves the upper bound when d is known in advance to the algorithm. This proves result (5) in Fig. 1.

Theorem 11. *Let $\ell \leq d/2$. Suppose some integer D is known in advance to the algorithm where $d/4 \leq D \leq 4d$. There is a polynomial time adaptive randomized algorithm that makes $\ell \log(n/d) + \ell \log \log(1/\delta) + O(\ell)$ tests and, with probability at least $1 - \delta$, detects ℓ defective items.*

Proof. Let $c = 32 \log(2/\delta)$. If $D < c\ell$, we can use the deterministic algorithm in Theorem 8. The test complexity is $\ell \log(n/\ell) + 2\ell = \ell \log(cn/D) + 2\ell = \ell \log(n/d) + \ell \log \log(1/\delta) + O(\ell)$.

If $D > c\ell$, then the algorithm uniformly at random chooses each element in X with probability $c\ell/D < 1$ and puts the items in X'. If $|X'| \leq 3c\ell n/D$, then deterministically detects ℓ defective items in X' using Theorem 8.

Let Y_i be an indicator random variable that is 1 if the ith defective item is in X' and 0 otherwise. Then $\mathbf{E}[Y_i] = c\ell/D$. The number of defective items in X' is $Y = Y_1 + \cdots + Y_d$ and $\mu := \mathbf{E}[Y] = cd\ell/D \geq c\ell/4$. By Chernoff's bound (Lemma 5), we have $\Pr[Y < \ell] \leq e^{-(1-4/c)^2 c\ell/8} < e^{-c\ell/32} \leq \delta/2$. Also, $\mathbf{E}[|X'|] = c\ell n/D$, and by Chernoff's bound (Lemma 5), $\Pr[|X'| > 3c\ell n/D] \leq (e/3)^{3c\ell n/D} \leq \delta/2$. Therefore, with probability at least $1 - \delta$, the number of defective items in X' is at least ℓ and $|X'| \leq 3c\ell n/D$. Therefore, with probability at least $1 - \delta$, the algorithm detects ℓ defective items.

Since $|X'| \leq 3c\ell n/D \leq 12c\ell n/d$, by Theorem 8, the test complexity is at most $\ell \log(|X'|/\ell) + 2\ell = \ell \log(n/d) + \ell \log \log(1/\delta) + O(\ell)$. □

We now prove the lower bound when d is known in advance to the algorithm. This proves results (6) and (8) in Fig. 1. These are summarized in Theorem 12.

Theorem 12. *Let $\ell \leq d \leq n/2$ and d be known in advance to the algorithm. Any adaptive randomized algorithm that, with probability at least $2/3$, detects ℓ defective items must make at least $\ell \log(n/d) - 1$ tests.*

Proof. We use Yao's principle in the standard way. Let $A(s, I)$ be any adaptive randomized algorithm that, with probability at least $2/3$, detects ℓ defective items. Here s is the random seeds, and I is the set of defective items. Let $X(I, s)$ be an indicator random variable that is equal 1 if $A(s, I)$ returns a subset $L \subset I$ of size ℓ and 0 otherwise. Then for every I, $\mathbf{E}_s[X(s, I)] \geq 2/3$. Therefore, $\mathbf{E}_s[\mathbf{E}_I[X(s, I)]] = \mathbf{E}_I[\mathbf{E}_s[X(s, I)]] \geq 2/3$, where the distribution in $\mathbf{E}_I$ is the uniform distribution. Thus, there is a seed s_0 such that $\mathbf{E}_I[X(s_0, I)] \geq 2/3$. That is, for at least $2\binom{n}{d}/3$ sets of defective items I, the *deterministic* algorithm $A(s_0, I)$ returns $L \subseteq I$ of size ℓ. Now, similar to the proof of Theorem 9, the algorithm $A(s_0, I)$ makes at least

$$\log \frac{\frac{2}{3}\binom{n}{d}}{\binom{n-\ell}{d-\ell}} \geq \ell \log(n/d) - 1.$$

□

In particular,

Theorem 13. *Let $\ell \leq d \leq n/2$ and d is unknown to the algorithm. Any adaptive randomized algorithm that, with probability at least $2/3$, detects ℓ defective items must make at least $\ell \log(n/d) - 1$ tests.*

We now prove the upper bound when d is unknown to the algorithm. This proves result (7) in Fig. 1.

Theorem 14. *Let $\ell \leq d/2$ and d be unknown to the algorithm. There is a polynomial time adaptive randomized algorithm that detects ℓ defective items and makes $\ell \log(n/d) + \ell \log \log(1/\delta) + O(\ell + \log \log(\min(n/d, d)) + \log(1/\delta))$ tests.*

Proof. We first estimate d within a factor of 2 and probability at least $1 - \delta/2$. By Lemma 9, this can be done in $2 \log \log(n/d) + O(\log(1/\delta))$. Then, by Theorem 11, the result follows. $\quad\square$

F   Estimating d

The following lemma follows from [5,21].

Lemma 8. *Let $\epsilon < 1$ be any positive constant. There is a polynomial time adaptive algorithm that makes $O(\log \log d + \log(1/\delta))$ expected number of tests and with probability at least $1 - \delta$ outputs D such that $(1 - \epsilon)d \leq D \leq (1 + \epsilon)d$.*

In Appendix F, we use a similar technique to prove:

Lemma 9. *Let $\epsilon < 1$ be any positive constant. There is a polynomial time adaptive algorithm that makes $O(\log \log(\min(d, n/d)) + \log(1/\delta))$ expected number of tests and with probability at least $1 - \delta$ outputs D such that $(1 - \epsilon)d \leq D \leq (1 + \epsilon)d$.*

To prove Lemma 9, we first prove:

Lemma 10. *Let $\epsilon < 1$ be any positive constant. There is a polynomial time adaptive algorithm that makes $O(\log \log(n/d) + \log(1/\delta))$ expected number of tests and with probability at least $1 - \delta$ outputs D such that $(1 - \epsilon)d \leq D \leq (1 + \epsilon)d$.*

We first give an algorithm that makes $O(\log \log(n/d))$ expected number of tests and outputs D that with probability at least $1 - \delta$ satisfies

$$\frac{\delta d^2}{4n \log^2(2/\delta)} \leq D \leq d. \tag{4}$$

The algorithm is

1. $\lambda = 2$.
2. Let each $x \in [n]$ be chosen to be in the test Q with probability $1 - 2^{-\lambda/n}$.
3. If $T_I(Q) = 0$ then $\lambda \leftarrow \lambda^2$; Return to step 2.
4. $D = \delta n/(4\lambda)$.
5. Output D.

We now prove

Lemma 11. *We have*

$$\Pr\left[\frac{\delta d^2}{4n\log^2(2/\delta)} \le D \le d\right] \ge 1 - \delta.$$

Proof. Let $\lambda_i = 2^{2^i}$ and Q_i be a set where each $x \in [n]$ is chosen to be in $Q_i \subseteq [n]$ with probability $1 - 2^{-\lambda_i/n}$, $i = 0, 1, \cdots$. Let i' be such that $\lambda_{i'} < \delta n/(4d)$ and $\lambda_{i'+1} \ge \delta n/(4d)$. Let $D = \delta n/(4\lambda_j)$ be the output of the algorithm. Then, since $\lambda_i \le \lambda_{i+1}/2$, we have $\lambda_{i'-t} < \delta n/(2^{t+2}d)$ and

$$\Pr[D > d] = \Pr[\delta n/(4\lambda_j) > d] = \Pr[\lambda_j < \delta n/(4d)] = \Pr[j \in \{0, 1, \dots, i'\}]$$

$$= \Pr[T_I(Q_0) = 1 \vee T_I(Q_1) = 1 \vee \cdots \vee T_I(Q_{i'}) = 1] \le \sum_{i=0}^{i'} \Pr[T_I(Q_i) = 1]$$

$$= \sum_{i=0}^{i'}(1 - 2^{-d\lambda_i/n}) \le \sum_{i=0}^{i'}\frac{d\lambda_i}{n} \le \cdots + \frac{\delta}{8} + \frac{\delta}{4} \le \frac{\delta}{2}.$$

Also, since $\lambda_j > a$ implies $\lambda_{j-1} > \sqrt{a}$,

$$\Pr\left[D < \frac{\delta d^2}{4n\log^2(2/\delta)}\right] = \Pr\left[\lambda_j \ge \frac{n^2}{d^2}\log^2\frac{2}{\delta}\right]$$

$$= \Pr\left[T_I(Q_{j-1}) = 0 \ \wedge \ \lambda_j \ge \frac{n^2}{d^2}\log^2\frac{2}{\delta}\right]$$

$$\le 2^{-d\lambda_{j-1}/n} \le 2^{-\log(2/\delta)} = \frac{\delta}{2}.$$

This completes the proof. □

Lemma 12. *The expected number of tests of the algorithm is* $\log\log(n/d) + O(1)$.

Proof. For $(n/d)^2 > \lambda_k \ge (n/d)$, the probability that the algorithm makes $k + t + 1$ tests is less than

$$2^{-d\lambda_{k+t}/n} = 2^{-d\lambda_k^{2^t}/n} \le 2^{-(n/d)^{2^t-1}}.$$

Therefore the expected number of tests of the algorithm is at most $k + O(1)$. Since $\lambda_k = 2^{2^k} < (n/d)^2$, we have $k = \log\log(n/d) + O(1)$. □

We now give another adaptive algorithm that, given that (4) holds, it makes $\log\log(n/d) + O(\log\log(1/\delta))$ tests and with probability at least $1 - \delta$ outputs D' that satisfies $d\delta/8 \le D' \le 8d/\delta$.

By (4), we have

$$1 \le \frac{d}{D} \le H := \sqrt{\frac{4\log^2(2/\delta)}{\delta}\frac{n}{D}}$$

Let $\tau = \lceil \log(1 + \log H) \rceil$. Then $1 \leq d/D \leq 2^{2^\tau - 1}$ and $0 \leq \log(d/D) \leq 2^\tau - 1$.

Consider an algorithm that, given a hidden number $0 \leq i \leq 2^\tau - 1$, binary searches for i with queries of the form "Is $i > m$". Consider the tree $T(\tau)$ that represents all the possible runs of this algorithm, with nodes labeled with m. See, for example, the tree $T(4)$ in Fig. 4.

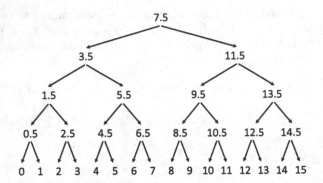

Fig. 4. The tree $T(4)$, which is all the runs of the binary search algorithm for $0 \leq i \leq 15$. Suppose we search for the hidden number $i = 9$. We start from the tree's root, and the first query is "Is $i > 7.5$". The answer is yes, and we move to the right son of the root. The following query is "Is $i > 11.5$" the answer is no, and we move to the left son. Etc.

We will do a binary search for an integer close to $\log(d/D)$ in the tree $T(\tau)$. The algorithm is the following

1. Let $\ell = 0; r = 2^\tau - 1$;
2. While $\ell \neq r$ do
3. Let $m = (\ell + r)/2$
4. Let each $x \in [n]$ be chosen to be in the test Q with probability $1 - 2^{-1/(2^m D)}$.
5. If $T_I(Q) = 1$ then $\ell = \lceil m \rceil$ else $r = \lfloor m \rfloor$.
6. Output $D' := D2^\ell$.

We first prove

Lemma 13. *Consider $T(\tau)$ for some integer τ. Consider an integer $0 \leq i \leq 2^\tau - 1$ and the path P_i in $T(\tau)$ from the root to the leave i. Then*

1. *P_i passes through a node labeled with $i - 1/2$, and the next node in P_i is its right son.*
2. *P_i passes through a node labeled with $i + 1/2$, and the next node in P_i is its left son.*

Proof. If the path does not go through the node labeled with $i - 1/2$ (resp. $i + 1/2$), then, in the search, we cannot distinguish between i and $i - 1$ (resp. $i + 1$). Obviously, if we search for i and reach the node labeled with $i - 1/2$, the next node in the binary search is the right son. $\qquad\square$

Now, by Lemma 13, if the algorithm outputs ℓ', then there is a node labeled with $m = \ell' - 1/2$ that the algorithm went through, and the answer to the test was 1. That is, the algorithm continues to the right node.

$$\Pr\left[D' > \frac{8d}{\delta}\right] = \Pr\left[D2^\ell > \frac{8d}{\delta}\right] = \Pr\left[\ell > \log\frac{d}{D} + \log\frac{8}{\delta}\right]$$

$$= \sum_{\ell'=\lceil\log(d/D)+\log(8/\delta)\rceil}^{2^\tau-1} \Pr[\ell = \ell']$$

$$= \sum_{\ell'=\lceil\log(d/D)+\log(8/\delta)\rceil}^{2^\tau-1} \Pr[\text{Answer in node labeled with } m = \ell' - 1/2 \text{ is 1}]$$

$$= \sum_{\ell'=\lceil\log(d/D)+\log(8/\delta)\rceil}^{2^\tau-1} 1 - 2^{-d/(2^{\ell'-1/2}D)}$$

$$\leq \sum_{\ell'=\lceil\log(d/D)+\log(8/\delta)\rceil}^{2^\tau-1} \frac{d}{D2^{\ell'-1/2}} \leq \frac{\delta}{4} + \frac{\delta}{8} \cdots \leq \frac{\delta}{2}.$$

By Lemma 13, if the algorithm outputs ℓ', then there is a node labeled with $m = \ell' + 1/2$ that the algorithm went through, and the answer to the test was 0.

$$\Pr\left[D' < \frac{\delta d}{8}\right] = \Pr\left[D2^\ell < \frac{\delta d}{8}\right] = \Pr\left[\ell < \log\frac{d}{D} - \log\frac{8}{\delta}\right]$$

$$= \sum_{\ell'=0}^{\lfloor\log(d/D)-\log(8/\delta)\rfloor} \Pr[\ell = \ell']$$

$$= \sum_{\ell'=0}^{\lfloor\log(d/D)-\log(8/\delta)\rfloor} \Pr[\text{Answer in node labeled with } m = \ell' + 1/2 \text{ is 0}]$$

$$= \sum_{\ell'=0}^{\lfloor\log(d/D)-\log(8/\delta)\rfloor} 2^{-d/(D2^{\ell'+1/2})}$$

$$\leq 2^{-4/\delta} + 2^{-8/\delta} + 2^{-16/\delta} + \cdots \leq \frac{\delta}{4} + \frac{\delta}{8} + \cdots = \frac{\delta}{2}.$$

Therefore, with probability at least $1 - \delta$, D' satisfies $d\delta/8 \leq D' \leq 8d/\delta$.

Lemma 14. *The number of tests of the algorithm is* $\log\log(n/d) + O(\log\log(1/\delta))$.

Proof. Since, by (4), $\delta d^2/(4n \log^2(2/\delta)) \leq D$, the number of tests is

$$\tau + 1 \leq \log\log H + 3$$

$$\leq 3 + \log\log \sqrt{\frac{4\log^2(2/\delta)}{\delta} \frac{n}{D}}$$

$$\leq 3 + \log\log \left(\frac{4\log^2 \frac{2}{\delta}}{\delta} \cdot \frac{n}{d}\right) = \log\log \frac{n}{d} + O\left(\log\log \frac{1}{\delta}\right).$$

$\square$

Finally, given D' that satisfies $d\delta/8 \leq D' \leq 8d/\delta$, Falahatgar et al. [21] presented an algorithm that, for any constant $\epsilon > 0$, makes $O(\log(1/\delta))$ queries and, with probability at least $1-\delta$, returns an integer D'' that satisfies $(1-\epsilon)d \leq D'' \leq (1+\epsilon)d$.

By Lemma 12 and 14, Lemma 10 follows.

One way to prove Lemma 9 is by running the algorithms in Lemma 8 and Lemma 10 in parallel, one step in each algorithm, and halt when one of them halts. Another way is by using the following result.

Lemma 15. *Let d and m be integers, and $\epsilon \leq 1$ be any real number. There is a non-adaptive randomized algorithm that makes $O((1/\epsilon^2)\log(1/\delta))$ tests and*

- *If $d < m$ then, with probability at least $1 - \delta$, the algorithm returns 0.*
- *If $d > (1+\epsilon)m$, then, with probability at least $1 - \delta$, the algorithm returns 1.*
- *If $m \leq d \leq (1+\epsilon)m$ then, the algorithm returns 0 or 1.*

Proof. Consider a random test $Q \subseteq X$ where each $x \in X$ is chosen to be in Q with probability $1 - (1+\epsilon)^{-1/(m\epsilon)}$. The probability that $T_I(Q) = 0$ is $(1+\epsilon)^{-d/(m\epsilon)}$. Since

$$\Pr[T_I(Q) = 0|d < m] - \Pr[T_I(Q) = 0|d > (1+\epsilon)m] \geq (1+\epsilon)^{-1/\epsilon} - (1+\epsilon)^{-(1+\epsilon)/\epsilon}$$

$$= (1+\epsilon)^{-1/\epsilon}\frac{\epsilon}{1+\epsilon}$$

$$\geq \frac{\epsilon}{2e}.$$

By Chernoff's bound (Lemma 5), we can, with probability at least $1-\delta$, estimate $\Pr[T_I(Q) = 0]$ up to an additive error of $\epsilon/(8e)$ using $O((1/\epsilon^2)\log(1/\delta))$ tests. If the estimation is less than $(1+\epsilon)^{-(1+\epsilon)/\epsilon} + \epsilon/(4e)$ we output 0. Otherwise, we output 1. This implies the result. $\square$

Now, to prove Lemma 9, we first run the algorithm in Lemma 15 with $m = \sqrt{n}$ and $\epsilon = 1$. If the output is 0 ($d < 2\sqrt{n}$), then we run the algorithm in Lemma 8. Otherwise, we run the algorithm in Lemma 10.

References

1. Ahlswede, R., Deppe, C., Lebedev, V.: Finding one of d defective elements in some group testing models. Probl. Inf. Transm. **48**, 04 (2012)
2. Balding, D.J., Bruno, W.J., Torney, D., Knill, E.: A comparative survey of non-adaptive pooling designs. In: Speed, T., Waterman, M.S. (eds.) Genetic Mapping and DNA Sequencing. The IMA Volumes in Mathematics and its Applications, vol. 81, pp. 133–154. Springer, New York, NY (1996). https://doi.org/10.1007/978-1-4612-0751-1_8
3. Bar-Noy, A., Hwang, F.K., Kessler, I., Kutten, S.: A new competitive algorithm for group testing. Discret. Appl. Math. **52**(1), 29–38 (1994)
4. Bshouty, N.H.: Lower bound for non-adaptive estimation of the number of defective items. In: 30th International Symposium on Algorithms and Computation, ISAAC 2019, December 8–11, 2019, Shanghai University of Finance and Economics, Shanghai, China, pp. 2:1–2:9 (2019)
5. Bshouty, N.H., Bshouty-Hurani, V.E., Haddad, G., Hashem, T., Khoury, F., Sharafy, O.: Adaptive group testing algorithms to estimate the number of defectives. In: Algorithmic Learning Theory, ALT 2018, 7–9 April 2018, Lanzarote, Canary Islands, Spain, pp. 93–110 (2018)
6. Bshouty, N.H., Diab, N., Kawar, S.R., Shahla, R.J.: Non-adaptive randomized algorithm for group testing. In: International Conference on Algorithmic Learning Theory, ALT 2017, 15–17 October 2017, Kyoto University, Kyoto, Japan, pp. 109–128 (2017)
7. Bshouty, N.H., Haddad, G., Haddad-Zaknoon, C.A.: Bounds for the number of tests in non-adaptive randomized algorithms for group testing. In: Chatzigeorgiou, A., et al. (eds.) SOFSEM 2020. LNCS, vol. 12011, pp. 101–112. Springer, Cham (2020). https://doi.org/10.1007/978-3-030-38919-2_9
8. Cabrera Alvargonzalez, J.J., et al.: Pooling for SARs-COV-2 control in care institutions. BMC Inf. Dis. **20**(1), 1–6 (2020)
9. Chen, H., Hwang, F.K.: Exploring the missing link among d-separable, d⁻-separable and d-disjunct matrices. Discret. Appl. Math. **155**(5), 662–664 (2007)
10. Cheng, Y., Du, D., Xu, Y.: A zig-zag approach for competitive group testing. INFORMS J. Comput. **26**(4), 677–689 (2014)
11. Damaschke, P., Muhammad, A.S.: Randomized group testing both query-optimal and minimal adaptive. In: Bieliková, M., Friedrich, G., Gottlob, G., Katzenbeisser, S., Turán, G. (eds.) SOFSEM 2012. LNCS, vol. 7147, pp. 214–225. Springer, Heidelberg (2012). https://doi.org/10.1007/978-3-642-27660-6_18
12. Dorfman, R.: The detection of defective members of large populations. Ann. Math. Stat. **14**(4), 436–440 (1943)
13. Du, D., Hwang, F.K.: Pooling Design and Nonadaptive Group Testing: Important Tools for DNA Sequencing. World Scientific Publishing Company, Singapore (2006)
14. Du, D., Hwang, F.K.: Competitive group testing. Discret. Appl. Math. **45**(3), 221–232 (1993)
15. Du, D., Park, H.: On competitive group testing. SIAM J. Comput. **23**(5), 1019–1025 (1994)
16. Du, D., Xue, G., Sun, S., Cheng, S.: Modifications of competitive group testing. SIAM J. Comput. **23**(1), 82–96 (1994)
17. Du, D.-Z., Hwang, F.K.: Combinatorial Group Testing and Its Applications. World Scientfic Publishing, Singapore (1993)

18. D.-Z. Du and F. K. Hwang. Pooling Designs And Nonadaptive Group Testing: Important Tools For DNA Sequencing. World Scientfic Publishing, Singapore (2006)

19. D'yachkov, A.G., Rykov, V.V.: Bounds on the length of disjunctive codes. Probl. peredachi Inf. **18**(3), 7–13 (1982)

20. Eis-Hübinger, A.M.: Ad hoc laboratory-based surveillance of SARS-CoV-2 by real-time rt-RT-PCR using minipools of RNA prepared from routine respiratory samples. J. Clin. Virol. **127**, 104381 (2020)

21. Falahatgar, M., Jafarpour, A., Orlitsky, A., Pichapati, V., Suresh, A.T.: Estimating the number of defectives with group testing. In: IEEE International Symposium on Information Theory, ISIT 2016, Barcelona, Spain, 10–15 July 2016, pp. 1376–1380. IEEE (2016)

22. Füredi, Z.: On r-cover-free families. J. Comb. Theory, Ser. A **73**(1), 172–173 (1996)

23. Gollier, C., Gossner, O.: Group testing against covid-19. Covid Economics, pp. 32–42, April 2020

24. Haddad-Zaknoon, C.A.: Heuristic random designs for exact identification of defectives using single round non-adaptive group testing and compressed sensing. In: The Fourteenth International Conference on Bioinformatics, Biocomputational Systems and Biotechnologies, BIOTECHNO 2022 (2022)

25. Hwang, F.K.: A method for detecting all defective members in a population by group testing. J. Amer. Stat. Assoc. **67**, 605–608 (1972)

26. Katona, G.O.: Finding at least one excellent element in two rounds. J. Stat. Planning Inf. **141**(8), 2946–2952 (2011)

27. Kautz, W., Singleton, R.: Nonrandom binary superimposed codes. IEEE Trans. Inf. Theory **10**(4), 363–377 (1964)

28. Kuppusamy, P., Bharathi, V.: Human abnormal behavior detection using CNNs in crowded and uncrowded surveillance - a survey. Meas. Sens. **24**, 100510 (2022)

29. Liang, W., Zou, J.: Neural group testing to accelerate deep learning. In: IEEE International Symposium on Information Theory, ISIT 2021. IEEE (2021)

30. Mentus, C., Romeo, M., DiPaola, C.: Analysis and applications of adaptive group testing methods for covid-19. medRxiv (2020)

31. Porat, E., Rothschild, A.: Explicit nonadaptive combinatorial group testing schemes. IEEE Trans. Inf. Theory **57**(12), 7982–7989 (2011)

32. Roth, R.M.: Introduction to Coding Theory. Cambridge University Press, Cambridge (2006)

33. Ruszinkó, M.: On the upper bound of the size of the r-cover-free families. J. Comb. Theory Ser. A **66**(2), 302–310 (1994)

34. Schlaghoff, J., Triesch, E.: Improved results for competitive group testing. Comb. Probab. Comput. **14**(1–2), 191–202 (2005)

35. Shani-Narkiss, H., Gilday, O.D., Yayon, N., Landau, I.D.: Efficient and practical sample pooling for high-throughput PCR diagnosis of covid-19. medRxiv (2020)

36. Sobel, M., Groll, P.A.: Group testing to eliminate efficiently all defectives in a binomial sample. Bell Syst. Tech. J. **38**, 1179–1252 (1959)

37. Wang, W., Siau, K.: Artificial intelligence, machine learning, automation, robotics, future of work and future of humanity: a review and research agenda. J. Database Manage. (JDM) **30**(1), 61–79 (2019)

38. Wolf, J.: Born again group testing: multiaccess communications. IEEE Trans. Inf. Theory **31**(2), 185–191 (1985)

39. Wu, J., Cheng, Y., Du, D.: An improved zig zag approach for competitive group testing. Discret. Optim. **43**, 100687 (2022)

40. Xie, S., Girshick, R., Dollár, P., Tu, Z., He, K.: Aggregated residual transformations for deep neural networks. In: Proceedings of the IEEE Conference on Computer Vision and Pattern Recognition, pp. 1492–1500 (2017)
41. Yelin, I., et al.: Evaluation of covid-19 rt-qPCR test in multi-sample pools. medRxiv (2020)

An Efficient Data Analysis Method for Big Data Using Multiple-Model Linear Regression

Bohan Lyu[1,2]($\boxtimes$) and Jianzhong Li[1,2]

[1] Harbin Institute of Technology, Harbin, Heilongjiang, China
18b903024@stu.hit.edu.cn, lijzh@hit.edu.cn
[2] Shenzhen Institute of Advanced Technology, Chinese Academy of Sciences,
Shenzhen, China

Abstract. This paper introduces a new data analysis method for big data using a newly defined regression model named multiple model linear regression(MMLR), which separates input datasets into subsets and construct local linear regression models of them. The proposed data analysis method is shown to be more efficient and flexible than other regression based methods. This paper also proposes an approximate algorithm to construct MMLR models based on (ϵ, δ)-estimator, and gives mathematical proofs of the correctness and efficiency of MMLR algorithm, of which the time complexity is linear with respect to the size of input datasets. This paper also empirically implements the method on both synthetic and real-world datasets, the algorithm shows to have comparable performance to existing regression methods in many cases, while it takes almost the shortest time to provide a high prediction accuracy.

Keywords: Data analysis · Big data · Linear regression · Segmented regression · Machine learning

1 Introduction

Data analysis plays an important role in various aspects, because it tells the features of data and helps predicting tasks. Regression with parametric models, especially linear models, is a typical data analysis method.

Let $\boldsymbol{DS} = \{(y, x_1, x_2, \cdots, x_k)\}$ be a $k + 1$ dimensional dataset with n elements, where y is called as response variable, x_i is called as explanatory variable for reach $1 \le i \le k$. The task of regression is to determine a function $\hat{y} = f(x_1, x_2, \cdots, x_k)$ using $\boldsymbol{DS}$, minimizing $\mathbb{E}(y - \hat{y})^2$. As for linear regression, $f(x_1, \cdots, x_k)$ is a linear function of x_is. And there's the assumption that

This work was supported by the National Natural Science Foundation of China under grants 61832003, Shenzhen Science and Technology Program (JCYJ202208181002205012) and Shenzhen Key Laboratory of Intelligent Bioinformatics (ZDSYS20220422103800001).

$y = \beta_0 + \beta_1 x_1 + \beta_2 x_2 + \cdots + \beta_k x_k + \varepsilon$ for each $(y, x_1, x_2, \cdots, x_k) \in \boldsymbol{DS}$, where ε is a random noise obeying normal distribution, and β_i $(1 \leq i \leq k)$ are constants.

Under such assumptions, linear regression model has statistical advantages and high interpretability. The numeric value of parameters can show the importance of variables, and the belonging information about the confidence coefficients and intervals make the model more credible in practice [9]. Therefore, linear regression is widely used in research areas requiring high interpretability, such as financial prediction, investment forecasting, biological and medical modelling, etc. Most machine learning and deep learning models might be more precise in predicting tasks, but the black-box feature limits their ranges of application.

However, linear regression still faces challenges in case of big data. Because big data has a feature that different subsets of a dataset fitting highly different regression models, which is described as *diverse predictor-response variable relationships*(DPRVR) in [5]. An example of real-world data is TBI dataset in [6], which is used to predict traumatic brain injury patients' response with sixteen explanatory variables. The researchers in [5] have shown that the *root mean square error* $(RMSE)$ is 10.45 when one linear regression model is used to model the whole TBI. While TBI is divided into 7 subsets and 7 different linear regression models are used individually, the $RMSE$ is reduced to 3.51. TBI shows that the DPRVR commonly appears in real-world data and big datasets. This feature indicates that it is much better to use multiple linear models rather than only one to model a big dataset.

Nevertheless, there is no efficient multi-model based regression algorithms for big datasets till now since the time complexities of existing multi-model based regression algorithms are too high to model big datasets. Piecewise Linear Regression, or segmented regression [1, 2, 7, 10, 11] are the only kinds so far. They divide the input datasets into several connected areas, and then construct local models using data points in each connected area, which is similar to the example in Fig. 1 from Appendix A of [8]. These kinds of regression models has high prediction accuracy because it considers DPRVR, and has high interpretability since the local models could have explicit expressions.

But there are still shortcomings of the existing multiple models based regression methods, which are shown as follows.

1. The time complexity of the methods is high. The state-of-art algorithm, PLR, has the time complexity of $O(k^2 n^5)$ [10].
2. The subsets being used to construct multiple regression models must be hyper cubes [4] or generated by partition the given dataset by hyperplanes [7]. Thus, the accuracy of the methods is lower when the subsets of a given dataset are not hyper cubes or can not be generated by hyperplanes.
3. Some methods need apriori knowledge that is difficult to get [1].

To overcome the three disadvantages above, this paper proposes a new multi-model based linear regression method named as MMLR.

Specifically, MMLR algorithm is outlined in Algorithm 1. Noticing that every $d \in \boldsymbol{DS}$ has the form $d = (y, x_1, x_2, \cdots, x_k)$ and the output of the MMLR algorithm is $M = \{(f_i, S_i)\}$, where S_i is a subset of DS and f_i is a linear regression model fits S_i.

Algorithm 1: MMLR

Input: $(k+1)$-dimensional Dataset DS with n data points;
Output: regression model $M = \{(f_i, S_i)\}$
1 $i = 0, M = \emptyset, WS = DS$;
2 **while** $WS \neq \emptyset$ **do**
3      $i = i + 1$;
4   Select a hypercube H_i with small size in WS;
5   Construct a linear regression model f_i of H_i ;
6   Compute the error bound eb_i using f_i to predict H_i ;
7      $S_i = H_i$;
8   **for** *each* $(y, x_1, ..., x_k)$ *in* $WS - H_i$ **do**
9   **if** $|f_i(x_1, ..., x_k) - y| < eb_i$ **then**
10          $S_i = S_i \cup (y, x_1, ..., x_k)$;
11     $WS = WS - S_i$, $M = M \cup \{(f_i, S_i)\}$;
12 **return** M

This paper has proved that the time for constructing every f_i is $O(k^2/\varepsilon^2)$, where ε is a user-given upper bound of the max error of all f_i's parameters. It's been proved that the time cost of MMLR algorithm is $O(m(n + (k/\varepsilon)^2 + k^3))$ in Sect. 4, where m is the number of models. The time complexity of MMLR is much lower than $O(k^2 n^5)$, since $(k/\varepsilon)^2$ is far more less than n^5. Therefore, the disadvantage 1 above is overcome.

From steps 4–10 of the MMLR algorithm, every subset S_i can be any shape rather than a hypercube or a subset generated by partitioning DS using only hyper-planes. Thus, the disadvantage 2 above is overcome also.

The MMLR algorithm iteratively increase the number of regression models so that it could always find the suitable number of models to optimize the prediction accuracy without knowing the number of models as apriori knowledge. In fact, MMLR only take DS as input, which overcomes the disadvantage 3 above.

The major contributions of this paper are as follows.

- The problem of constructing the optimized multiple linear regression models for a given dataset is formally defined and analyzed.
- A heuristic MMLR algorithm is designed for solving the problem above. MMLR algorithm can deal with the DPRVR of big datasets, and overcomes the disadvantages of the existing multi-model based linear regression methods.
- The time complexity of MMLR algorithm is analyzed, which is lower than the state-of-art algorithm. The accuracy of MMLR algorithm and related mathematical conclusions are proved.

The rest of this paper is organized as follows. Section 2 gives the formal definition of the problem. Section 3 proves the necessary mathematical theorems. Section 4 gives the design and analysis of the algorithm. Finally, Sect. 5 concludes the paper.

2 Preliminaries and Problem Definition

2.1 Regression and Linear Regression

The definition of traditional linear regression problem is given as follows.

Definition 1 (Linear Regression Problem).
   Input: A numerical dataset $DS = \{(\boldsymbol{x}_i, y_i) \,|\, 1 \le i \le n\}$, where $\boldsymbol{x}_i \in \mathbb{R}^k, y_i \in \mathbb{R}$, $y_i = f(\boldsymbol{x}_i) + \varepsilon_i$ for some function f, $\varepsilon_i \sim N(0, \sigma^2)$, and all ε_is are independent.
   Output: A function $\hat{f}(\boldsymbol{x}) = \boldsymbol{\beta} \cdot (1, \boldsymbol{x}) = \beta_0 + \beta_1 x_1 + \cdots + \beta_k x_k$ such that $\mathbb{E}[(y - \hat{f}(\boldsymbol{x}))^2]$ is minimized for $\forall (\boldsymbol{x}, y) \in DS$.

Douglas C. Montgomery, Elizabeth A. Peck and G. Geoffrey Vining have proved that minimize the $\mathbb{E}[(y - \hat{f}(\boldsymbol{x}))^2]$ is equivalent to minimizing $\sum (y_i - \hat{f}(\boldsymbol{x}_i))^2$ on DS in the case of linear regression [9]. Besides, it's trivial that a k-dimensional linear function can perfectly fit any n data points when $n \le k+1$. It is said DS to be centralized, if using $x_{ij} - \overline{x}_j$ to substitute x_{ij}, using $y_i - \overline{y}$ to substitute y_i, for $i = 1, 2, \cdots, n, j = 1, 2, \cdots, k$, where $\overline{x}_j = (x_{1j} + \cdots + x_{nj})/n, \overline{y} = (y_1 + \cdots + y_n)/n$. Obviously, when DS is centralized, there's always $\beta_0 = 0$ so that $\boldsymbol{\beta} = (\beta_1, \cdots, \beta_k)$. Therefore, this paper always assumes that $n > k + 1$, DS is centralized and all $\boldsymbol{x}_i$ are i.i.d. and uniformly drawn from the value domain of $\boldsymbol{x}$ for convenience of analysis.

The simplest method to construct linear regression model is *pseudo-inverse matrix method*. It transforms the given DS into a vector $\boldsymbol{y} = (y_1, y_2, \cdots, y_n)$ and an $n \times (k+1)$ matrix $\boldsymbol{X}$:

$$\boldsymbol{X} = \begin{pmatrix} 1 & x_{11} & x_{12} & \cdots & x_{1k} \\ 1 & x_{21} & x_{22} & \cdots & x_{2k} \\ \vdots & \vdots & \vdots & & \vdots \\ 1 & x_{n1} & x_{n2} & \cdots & x_{nk} \end{pmatrix}.$$

The x_{ij} in $\boldsymbol{X}$ equals to the value of the i-th data point's j-th dimension in DS. It is called the *data matrix* of DS. Then, using the formula $\hat{\boldsymbol{\beta}} = (\boldsymbol{X}'\boldsymbol{X})^{-1}\boldsymbol{X}'\boldsymbol{y}$, the linear regression model $\hat{f}$ could be constructed. The time complexity of pseudo-inverse matrix method is $O(k^2 n + k^3)$ [9]. When k is big enough, gradient methods is more efficient than pseudo-inverse matrix method. Generally, every method's complexity has the bound $O(k^2 n + k^3)$, so this paper use it as the time complexity of linear regression in common.

To judge the goodness of a linear regression model $\hat{f}$, the p-value of F-test is a convincing criterion, which denotes as $p_F(\hat{f})$ in this paper. Generally, linear regression model $\hat{f} \cong f$ when $p_F(f) < 0.05$, and $p_F(f)$ can be calculated in $O(n)$ time [9].

Besides, this paper uses the $(\epsilon, \delta) - estimator$ to analyze the performance of linear regression models constructing by subsets of DS. The formal definition of it is as follows.

Definition 2 ((ϵ, δ)-estimator).
$\hat{I}$ is an $(\epsilon, \delta) - estimator$ of I if $\Pr\{|\hat{I} - I| \geq \epsilon\} \leq \delta$ for any $\epsilon \geq 0$ and $0 \leq \delta \leq 1$, where $\Pr(X)$ is the probability of random event X.

Intuitively, $(\epsilon, \delta) - estimator$ $\hat{I}$ of I has high possibility to be very close to I's real value. By controlling some parameters for the calculation of $\hat{I}$, one can get an I's arbitrarily precise estimator.

2.2 Multiple-Model Linear Regression

In rest of the paper, the multiple-model linear regression is denoted as MMLR. The MMLR problem is defined as follows.

Definition 3 (Optimal MMLR problem).
  Input: A numerical dataset $DS = \{(\boldsymbol{x}_i, y_i) \,|1 \leq i \leq n\}$, where $\boldsymbol{x}_i \in \mathbb{R}^k$, $y_i \in \mathbb{R}$, $y_i = f(\boldsymbol{x}_i) + \varepsilon_i$ for a function f, $\varepsilon_i \sim N(0, \sigma^2)$ and all ε_is are independent, maximal model number M_0, smallest volume n_0.
  Output: $M_{opt} = \{(\hat{g}_i, S_i)|1 \leq i \leq m\}$ such that $MSE(M) = \sum_{1 \leq i \leq m} \sum_{(\boldsymbol{x}_j, y_j) \in S_i} (y_j - \hat{g}_i(\boldsymbol{x}_j))^2$ is minimized, where $\hat{g}_i$ is a linear regressoin models of $S_i \subseteq DS$, $S_1 \cup S_2 \cup \cdots \cup S_m = DS$, $S_i \cap S_j = \emptyset$ for $i \neq k$, and $|S_i| \geq n_0, m \leq M_0$.

The optimal MMLR problem is expensive to solve since it has a similar to Piecewise Linear Regression problem [10]. By far, the best algorithm to solve piecewise linear regression problem without giving the number of pieces beforehand is $O(k^2 n^5)$ [10]. Therefore, this paper focuses on the approximate optimal solution of MMLR problem. To bound the error of a linear function, this paper set $|f_1 - f_2| = \|\boldsymbol{\beta}_1 - \boldsymbol{\beta}_2\|_\infty = \max_{1 \leq i \leq k} |b_{1i} - b_{2i}|$, where $f_j(\boldsymbol{x}) = \boldsymbol{\beta}_j \boldsymbol{x} = b_{j1}x_1 + \cdots b_{jk}x_k, j = 1, 2$. The definition of Approximately Optimal MMLR problem is as follows.

Definition 4 (Approximately Optimal MMLR problem).
  Input: A numerical dataset $DS = \{(\boldsymbol{x}_i, y_i) \,|1 \leq i \leq n\}$, where $\boldsymbol{x}_i \in \mathbb{R}^k$, $y_i \in \mathbb{R}$, $y_i = f(\boldsymbol{x}_i) + \varepsilon_i$ for a function f, $\varepsilon_i \sim N(0, \sigma^2)$ and all ε_is are independent, $\epsilon > 0, 0 \leq \delta \leq 1$, maximal model number M_0, smallest volume n_0.
  Output: $M = \{(\hat{f}_i, S_i)|1 \leq i \leq m\}$ such that $\Pr\{\max_i |\hat{f}_i - \hat{g}_i| \geq \epsilon\} \leq \delta$, $\hat{f}_i$ is a linear regression models of $S_i \subseteq DS$, $S_1 \cup S_2 \cup \cdots \cup S_m = DS$ and $S_i \cap S_j = \emptyset$ for $i \neq k$, and $|S_i| \geq n_0, m \leq M_0$, $M_{opt} = \{(\hat{g}_i, S_i)|1 \leq i \leq m\}$.

In the end of this section, necessary denotations are given as follows. Intuitively, DS is the input dataset and $|DS| = n$, $(\boldsymbol{x}_i, y_i) \in DS$ denotes the i-th data points, $i = 1, 2, \cdots, n$, $\boldsymbol{x}_i = (x_{i1}, x_{i2}, \cdots, x_{ik}) \in \mathbb{R}^k, y_i \in \mathbb{R}$. The

j-th linear function is denoted as $f_j(x) = \beta_{j0} + \beta_{j1}x_1 + \cdots + \beta_{jk}x_k$, and $\beta_j = (\beta_{j0}, \beta_{j1}, \cdots, \beta_{jk})$ is the coefficient vector. $X_{n \times (k+1)}$ is the data matrix of DS, of which $X_{ij} = x_{ij}$. Finally, every estimator of a value, vector or function I is $\hat{I}$.

3 Mathematical Foundations

This section gives proofs of the necessary mathematical theorems used for the MMLR algorithm. We first discuss the settings and an existed mathematical result used in this section.

Suppose that D is a centralized dataset of size n, it's reasonable to set $y_i = f(x_i) = \beta x = \beta_1 x_{i1} + \cdots + \beta_k x_{ik} + \varepsilon_i$ for each $(x_{i1}, x_{i2}, \cdots, x_{ik}, y_i) \in D$, $\varepsilon_i \sim N(0, \sigma^2)$, from the discussion in Sect. 2.1. Besides, $\hat{f}(x) = \hat{\beta}x$ is the linear regression model of D constructed by *least square criterion*. Thus there's the result in the following *Lemma 1*.

Lemma 1. *Let $X_{n \times k}$ be the data matrix of D, $n \geq k + 2$, and $\hat{\beta}$ be the least square estimator of β, then $\hat{\beta}$ is unbiased, and the covariance matrix of $\hat{\beta}$ is:*

$$\mathrm{Var}(\hat{\beta}) = \sigma^2 (X'X)^{-1}.$$

3.1 Theorems Related to Sampling

MMLR separates an input dataset into several disjoint subsets and construct local models for them. This subsection discusses how to construct local linear regression model on one subset efficiently.

The main process of constructing $\hat{f}$ using subset $PS \subseteq D$, *Cons-$\hat{f}$* for short, is given as follows.

Step 1: Independently sample $PS_1, PS_2, \cdots, PS_t$ without replacement from D, where $|PS_i| = p$ for $1 \leq i \leq t$.

Step 2: Use least square method to construct linear regression model $\hat{f}^{(i)} = \sum_{j=1}^{k} \hat{\beta}^{(i)}x$ for each PS_i.

Step 3: Let $\hat{\beta}$ be the average of all $\hat{\beta}^{(i)}$, where

$$\hat{\beta}_j = \frac{1}{t} \sum_{i=1}^{t} \hat{\beta}_j^{(i)}, j = 1, \cdots, k.$$

Let $PS = PS_1 \cup PS_2 \cup \cdots \cup PS_t$ and $|PS| = n_s$, then $\hat{f}$ is a linear regression model constructed from PS. *Theorem 1* given in this section shows that $\hat{f}$ satisfies

$$\Pr\left\{ \max_j |\hat{\beta}_j - \beta_j| \geq \epsilon \right\} \leq \delta \tag{1}$$

for given $\epsilon \geq 0$ and $0 \leq \delta \leq 1$ if $|PS|$ is big enough.

By the steps of *Cons-$\hat{f}$*, all $\hat{\beta}^{(i)}$ obey the same distribution since every PS_i is sampled from D by the same way independently. Noticing that $\hat{\beta}_j^{(i)}$ is a least-square estimator of β constructed using PS_i. *Lemma 1* shows that $\hat{\beta}$ satisfies $\mathbb{E}(\hat{\beta}) = \beta$ and $\text{Var}(\hat{\beta}) = \sigma^2(X'X)^{-1}$ when $n \geq k + 2$. So let $p = k + 2$ and $n_s = pt = (k + 2)t$, the minimum n_s such that $\Pr\{\max |\hat{\beta}_j - \beta_j| \geq \epsilon\} \leq \delta$ only depends on t. Furthermore, we can give the following *Lemma 2*, and then prove *Theorem 1*.

Lemma 2. *Letting $\hat{\beta} = (\hat{\beta}_1, \cdots, \hat{\beta}_j)$ be constructed by the procedure of Cons-$\hat{f}$, and X_i be the data matrix of the PS_i, for $1 \leq i \leq t$, then $\hat{\beta}$ is an unbiased estimator of β and the covariance matrix of $\hat{\beta}$ is:*

$$\text{Var}(\hat{\beta}) = \frac{\sigma^2}{t^2} \sum_{i=1}^{t} (X_i'X_i)^{-1}.$$

Proof. By the linearity of mathematical expectation and sum of mutually independent variables' variance, the conclusion of this lemma is obvious [3].   □

Theorem 1. *Let $\hat{\beta} = (\hat{\beta}_0, \hat{\beta}_1, \cdots, \hat{\beta}_j)$ be constructed by the procedure of Cons-$\hat{f}$. If $\Phi(\frac{\epsilon\sqrt{t}}{\nu}) \geq \frac{2-\delta}{2}$, then*

$$\Pr\Big\{ \max_j |\hat{\beta}_j - \beta_j| \geq \epsilon \Big\} \leq \delta,$$

for any $\epsilon > 0$ and $0 \leq \delta \leq 1$, where $\Phi(x)$ is the distribution function of standard normal distribution, and ν is the biggest standard deviation of all $\hat{\beta}_j$s.

Proof. The proof of *Theorem 1* is shown in Appendix B of [8].   □

Theorem 1 can be used to decide the needed n_s to satisfy inequality (1) by the following steps.

1. Check the table of normal distribution function for $\Phi(\frac{\epsilon\sqrt{t}}{\nu}) \geq \frac{2-\delta}{2}$, and $\nu = \max \nu_j = \max \frac{\sigma e_{jj}}{t} = \frac{\sigma \max e_{jj}}{t}$, get

$$t > (\frac{1}{\epsilon}\Phi^{-1}(\frac{2-\delta}{2})\sigma \max \sqrt{e_{jj}})^{2/3};$$

2. Let $t > (\frac{1}{\epsilon}\Phi^{-1}(\frac{2-\delta}{2})\sigma)^{2/3}$ since e_{jj} is always far more less than 1;
3. Let $n_s = p(\frac{1}{\epsilon}\Phi^{-1}(\frac{2-\delta}{2})\sigma)^{2/3}$.

Besides, it doesn't necessary to carry out the three steps of *cons-$\hat{f}$* in practice. In fact, we can directly construct $\hat{f}$ on the sample PS by using least square method to satisfy inequation (1). The following *Theorem 2* shows the correctness.

Theorem 2. *Suppose $PS, t, n_s, \hat{\beta} = (\hat{\beta}_1, \cdots, \hat{\beta}_k)$ are defined as in previous part, $\tilde{\beta} = (\tilde{\beta}_1, \cdots, \tilde{\beta}_k)$ is the least square estimator of PS's β, then if $\hat{\beta}$ satisfies inequality (1), $\tilde{\beta}$ also satisfies inequality (1).*

Proof. The proof of *Theorem* 2 is shown in Appendix C of [8]. □

From *Theorem* 2, the *Cons-$\hat{f}$* can be simplified to the following *Cons-$\hat{f}$-New*, which is used in *Algorithm* 3 in Sect. 4.

Step 1: Sample PS from $\boldsymbol{D}$, where $|PS| = n_s$.

Step 2: Use least square method to construct the parameters of linear regression model $\hat{\boldsymbol{\beta}} = (\hat{\beta}_1, \hat{\beta}_2, \cdots, \hat{\beta}_k)$.

Finally, the following *Theorem* 3 shows that n_s is not necessary to be large.

Theorem 3. *There exists an $n_s = O(\frac{1}{\epsilon^2})$ such that*

$$\Pr\left\{ \max_j |\hat{\beta}_j - \beta_j| \geq \epsilon \right\} < 10^{-6}.$$

Proof. The proof of *Theorem* 3 is shown in Appendix D of [8]. □

3.2 Theorem Related to the Measures of Subsets

By *Algorithm* 1, MMLR constructs $\hat{f}_i$ using subsets $H_i \cap \boldsymbol{DS} \subseteq \boldsymbol{DS}$ initially, where H_i is a hypercube whose edges are parallel to coordinate axis of $\mathbb{R}^k$. Subsection 3.1 shows that using a subset PS randomly sampled from $H_i \cap \boldsymbol{DS}$ to construct $\hat{f}_i$ can be very accurate. However, the measures of H_i is also a key factor influencing the accuracy of $\hat{f}_i$. Intuitively, the larger H_i is, the more accurate $\hat{f}_i$ is. However, the H_i is not necessary to be very large when ϵ and δ are given. This subsection discusses the necessary measure of H_i to satisfy inequality (1). The following subsection has the same mathematical assumptions as Sect. 3.1
.

Limited by the length of the paper, we can only show the following necessary conclusions, the proofs of them and other intermediate results *Lemma 3.1–3.4* are shown in Appendix E of [8].

Lemma 3. *Given $\epsilon > 0, 0 < \delta < 1$, then for any $\epsilon' > 0$, there exists an n_0 such that when $L \geq \frac{4\sqrt{3}\sigma}{\epsilon\sqrt{n\delta}}$ and $n > n_0$, $\Pr\{|\hat{\beta}_j - \beta_j| \geq \epsilon\} \leq \delta$ holds with possibility no less than $1 - \epsilon'$. Further, $\mathbb{E}(|\hat{\beta}_j - \beta_j|)$ is monotonically decreasing at nL^2.*

The Lemmas in this section show that for any $j = 1, 2, \cdots, k$, the value range of the j-th dimension of $\boldsymbol{D}$ influences the error of $\hat{\beta}_j$, and $\Pr\{|\hat{\beta}_j - \beta_j| \geq \epsilon\}$ is in inverse proportion to L^2. It means that when n is fixed, one can sample from larger value range of $\boldsymbol{D}$ to get more precise $\hat{\beta}_j$.

Besides, the sample size n has no need to be very big. The $\Pr\{\boldsymbol{x}'\boldsymbol{Ax} \geq \frac{1}{2}\boldsymbol{x}'\boldsymbol{x}\}$ in *Lemma 3.3* has a very fast convergence speed. When $n > 3k$, $\Pr\{\boldsymbol{x}'\boldsymbol{Ax} \geq \frac{1}{2}\boldsymbol{x}'\boldsymbol{x}\}$ has already larger than 0.99. From *Lemma 3.2* we could know that when $n > 65$, $\Pr\{\boldsymbol{x}'\boldsymbol{x} \geq \frac{nL^2}{24}\} \geq 0.95$. Since $\Pr\{|\hat{\beta}_j - \beta_j| \geq \epsilon\} \leq \delta$ requires both $\boldsymbol{x}'\boldsymbol{x} \geq \frac{nL^2}{24}$ and $\boldsymbol{x}'\boldsymbol{Ax} \geq \frac{1}{2}\boldsymbol{x}'\boldsymbol{x}$, the inequality holds with possibility larger than 0.95 when $n > \max\{65, 3k\}$.

According to the discuss above, we propose the following process to construct $\hat{f}$ using subset $HS \subset \boldsymbol{D}$ with smallest $|HS|$, *Cons $- H\hat{f}$* for short.

$Cons - H\hat{f}$:

Step 1: Given $n_s > \max\{65, 3k\}$, calculate $L = \frac{4\sqrt{3}\sigma}{\epsilon\sqrt{n_s}\delta}$.

Step 2: Randomly choose a data point $d \in D$, which satisfies $\min x_i + \frac{L}{2} \leq d_i \leq \max x_i - \frac{L}{2}$ for $i = 1, 2, \cdots, k$. Let $H = [d_1 - \frac{L}{2}, d_1 + \frac{L}{2}] \times \cdots \times [d_k - \frac{L}{2}, d_k + \frac{L}{2}]$

Step 3: Let $HS = D \cap H$. If $|HS| \geq n_s$, use least square method to construct $\hat{f}$ on HS; else, increase H till $|HS| \geq n_s$ or $|HS| = |D|$, use least square method to construct $\hat{f}$ on HS.

In conclusion, the following *Theorem 4* is correct by the discussion above.

Theorem 4. *If* $|DS| \geq n_s$ *and* $\max x_i - \min x_i \geq L$ *for* $i = 1, \cdots, k$, *the least square estimator* $\hat{f}$ *constructing by* $Cons - H\hat{f}$ *satisfies* $\Pr\left\{\max_j |\hat{\beta}_j - \beta_j| \geq \epsilon\right\} \leq \delta$.

4   Algorithm and Analysis

This section shows the pseudo-code of MMLR algorithm, as well as the details of illustrations and analysis.

4.1   Algorithm

Firstly, the general idea of MMLR has been shown in Sect. 1. The detail of MMLR algorithm is shown in *Algorithm* 2, the invoked algorithm *Subset* is shown in *Algorithm* 3. Specifically, MMLR uses pre-processing, pre-modelling, examine, grouping these four important phases to iteratively solving the problem.

Line 1–8 are the pre-processing phase of MMLR. Line 1–5 construct a linear regression model on the whole dataset DS. If the regression model is precise enough, there's no need to use multiple models fitting DS. MMLR uses the index p_F to determine whether one linear model is enough. If not, MMLR would begin to construct multiple model M. In line 7–8, MMLR firstly calculate the smallest sample size using an estimate of σ^2. There are already methods to precisely get σ^2's estimate [9]. Suppose that algorithm $estimate(DS)$ take a dataset DS as input and output σ^2, MMLR can use anyone of them, which is shown in Line 7. Thus MMLR could calculate n_s and L for further work in Line 8.

Line 9–19 is an iteration to construct every $\hat{f}_i$ and related S_i. Generally, MMLR samples small subsets to construct local models then finds the data points fitting them. When every iteration ends, MMLR abandons those data points from DS. The terminal condition is that when $|DS| \leq n_0$ or current $m = M_0 - 1$. At this time, the data points left would be marked as S_{m+1}.

Line 10–14 is pre-modelling phase. In this part, MMLR firstly prudently chooses a small area of the whole value range and samples from the data points in this area invoking $Subset(DS, n_s, L)$ in Line 10, 13. S is the sampled subset, and D denotes the rest part. After that MMLR construct a regression model f and get its statistical characteristic. By *Theorem* 2 and 4, f is a highly precise

---

Algorithm 2: MMLR($DS, \epsilon, \delta, M_0, N_0$)

Input: A k-dimensional dataset DS with N data points, error bounds
$\epsilon > 0, 0 < \delta < 1$, least subset volume M_0 and largest model number N_0;

Output: M, i.e. the approximate set of linear regression models and subsets of
DS fitting them

1 $M \leftarrow \emptyset$, $m \leftarrow 0$, $S \leftarrow DS$;
2 $f \leftarrow$ the linear regression model of S, $p_F \leftarrow$ the F-test's p-value of f on S ;
3 **if** $p_F < 0.05$ **then**
4 | $M \leftarrow \{(f, S)\}$;
5 | **return** M ;
6 **else**
7 | $n \leftarrow |DS|$, $D \leftarrow DS$, $\sigma^2 \leftarrow estimate(DS)$;
8 | $n_s \leftarrow (k+1)(\frac{1}{\epsilon}\Phi^{-1}(\frac{2-\delta}{2})\sigma)^{2/3}$, $L \leftarrow \dfrac{4\sqrt{3}\sigma}{\epsilon\sqrt{\max\{65, 3k\}\delta}}$;
9 | **While** ($n > N_0$ and $m < M_0 - 1$) **Do**
10 | | $S \leftarrow$ Subset(DS, n_s, L), $D \leftarrow D - S$;
11 | | $f \leftarrow$ linear regression model of S, $p_F \leftarrow$ the F-test's p-value of f on S ;
12 | | **While** ($p_F \geq 0.05$ and $D \neq \emptyset$) **Do**
13 | | | $S \leftarrow$ Subset(DS, n_s, L);
14 | | | $f \leftarrow$ linear regression model of S, $p_F \leftarrow$ the F-test's p-value of f on S ;
15 | | $\tilde{\sigma}_f^2 = \dfrac{(y - X\hat{\beta})'(y - X\hat{\beta})}{n}$, where β is the coefficients of f, y and X are response variables and data matrix of S respectively, $b_f \leftarrow 3\tilde{\sigma}_f$;
16 | | **for** each $d = (x, y)$ in $DS - S$ **do**
17 | | | **if** $|f(x) - y| \leq b_f$ **then**
18 | | | | $S \leftarrow S \cup \{d\}$;
19 | | $M \leftarrow M \cup \{(f, S)\}$, $DS \leftarrow DS - S$, $n \leftarrow |DS|$, $m \leftarrow m + 1$;
20 | $S \leftarrow DS$, $f \leftarrow$ linear regression model of S ;
21 | $M \leftarrow M \cup \{(f, S)\}$;
22 **return** M;

---

model, of which p_F should be small enough. If not, S does not fit one linear model, which means MMLR should sample again in Line 12–14.

Line 15–18 is the examine phase. In Line 15, MMLR firstly gives the fitting bound b_f of f. The fitting bound means that the max prediction error of a data point (x, y) if it fits f. So MMLR can figure out (x, y) fitting f if $|f(x) - y| \leq b_f$. If a data point (x, y) fit f, $y - f(x) \sim N(0, \sigma_f^2)$. As shown in [9], $\tilde{\sigma}_f^2 = \frac{(y - X\hat{\beta})'(y - X\hat{\beta})}{n}$ is a unbiased estimator of σ_f^2. According to the characteristics of normal distribution, MMLR chooses $3\tilde{\sigma}_f$ as the fitting bound of f, since $\Pr\{|\xi - 0| \leq 3\sigma\} \approx 0.99$ when $\xi \sim N(0, \sigma^2)$. MMLR tests all data points that are not assigned into existing model's acting scope by checking whether $f(x) - y \leq 3\tilde{\sigma}_f$. After getting all data points that belongs to this one linear model, MMLR updates M, and DS by deleting those points in Line 19.

Algorithm 3: Sample(DS, n_s, L)

Input: A k-dimensional dataset DS with N data points, smallest sample size
n_s, value range L;

Output: subset $S \subset DS$

1 $D \leftarrow$ subset of DS that every (x, y) satisfies
   $\min x_i + \frac{L}{2} \le d_i \le \max x_i - \frac{L}{2}, i = 1, 2, \cdots, k$;

2 $d(d_1, \cdots, d_k, y_d) \leftarrow$ randomly choose a data point from D ;

3 $H \leftarrow [d_1 - \frac{L}{2}, d_1 + \frac{L}{2}] \times \cdots \times [d_k - \frac{L}{2}, d_k + \frac{L}{2}]$, $S \leftarrow DS \cap H$;

4 **if** $|S| \ge n_s$ **then**

5 $\quad$ $S \leftarrow$ uniformly randomly choose $\min\{n_s, \max\{65, 3k\}\}$ from S ;

6 $\quad$ **return** S ;

7 **else if** $|S| < \max\{65, 3k\}$ **then**

8 $\quad$ **While** $(|S| < \max\{65, 3k\})$ **Do**

9 $\quad\quad$ $j \leftarrow$ randomly choose from $1, 2, \cdots, k$;

10 $\quad\quad$ $L_j \leftarrow \frac{\max\{65, 3k\}}{|S|} L$;

11 $\quad\quad$ **if** $d_j + \frac{L_j}{2} > \max x_j$ **then**

12 $\quad\quad\quad$ $H_j \leftarrow [d_j - \frac{L_j}{2}, d_j + \frac{L}{2}]$;

13 $\quad\quad$ **else**

14 $\quad\quad\quad$ $H_j \leftarrow [d_j - \frac{L}{2}, d_j + \frac{L_j}{2}]$;

15 $\quad\quad$ $S \leftarrow DS \cap H$;

16 **else**

17 $\quad$ **return** S ;

MMLR iteratively carries out the pre-modelling phase and examine phase until $|DS|$ is small enough. When $|DS| < N_0$ or the number of models $m = M_0 - 1$, the iteration stops. The current DS and the linear regression model of it will be settled as (f_m, S_m) and added into M. So far, MMLR gets a solution of the Approximately Optimal MMLR Problem.

4.2   Analysis

By the steps of *Algorithm 2*, *Theorem 2* and *4*, the correctness of *Algorithm 2* is given as the following theorem.

Theorem 5. *The* $M = \{(\hat{f}_i, S_i) | 1 \le i \le m\}$ *constructed by* Algorithm 2 *satisfies* $\Pr\{\max_i |\hat{f}_i - \hat{g}_i| \ge \epsilon\} \le \delta$, *where* $M_{opt} = \{(\hat{g}_i, S_i) | 1 \le i \le m\}$.

At last, when the input dataset satisfies some universal assumptions, the following theorem shows the time complexity of MMLR Algorithm. The proof is shown in Appendix F of [8].

Theorem 6 (Time complexity of MMLR). *Suppose that* DS *is uniformly distributed in a big enough value range, the value range of* DS *could be divided into* $m \le M_0$ *continuous areas* $A_1, \cdots, A_m$, $S_i = DS \cap A_i$, $|S_i| \ge n_0$ *and* S_i *can* *be fitted by a linear function, then the expected time complexity of* Algorithm *2 is* $O(M_0(N + k^3 + \frac{k^2}{\epsilon^2}))$.

Such assumptions are common in low dimension situations [1,4,7], such as $k \le 7$. Besides, for many datasets, one can control the value of σ and L under prudent normalization of DS, so as to make DS satisfy the assumptions. Several experiment results on both synthetic and real-world datasets are shown in Appendix G of [8].

5 Conclusion and Future Work

This paper introduces a new data analysis method using multiple-model linear regression, called MMLR. This paper gives the approximate MMLR algorithm and related mathematical proofs. MMLR has the advantages of high interpretability, high predicting precision and high efficiency of model constructing. It can deal with $DPRVR$ of big datasets, and the expected time complexity under some universal assumptions is $O(M_0(N + k^3 + \frac{k^2}{\epsilon^2}))$, which is lower than the existing segmented regression methods.

However, there are still challenges and future work of multiple-model regression. Firstly, the linear model could be replaced by any parametric models. Since they also have high interpretability and low time cost to construct. The conclusions of smallest sample size and measures should be calculated in another way, which is a challenge of mathematical reasoning. Secondly, a more versatile algorithm of choosing subsets is required. Since several datasets might not satisfy the assumption of *Theorem* 6, and normalization is not enough to make the Algorithms work efficiently, a more flexibly sampling method might help. Lastly, when the dimension of DS is too high($k > 10$), MMLR algorithm is not suitable since there's always no enough data points in H. Some dimensional reduction methods might mitigate the problem.

In conclusion,multiple model regression methodology has the potential to make great contribution to data analysis, and need more attention on its corresponding problems.

References

1. Arumugam, M.: EMPRR: a high-dimensional EM-based piecewise regression algorithm. Ph.D. thesis, University of Nebraska-Lincoln (2003)
2. Bemporad, A.: A piecewise linear regression and classification algorithm with application to learning and model predictive control of hybrid systems. IEEE Trans. Autom. Control **68**, 3194–3209 (2022)
3. Bertsekas, D., Tsitsiklis, J.N.: Introduction to Probability, vol. 1. Athena Scientific, Nashua (2008)
4. Diakonikolas, I., Li, J., Voloshinov, A.: Efficient algorithms for multidimensional segmented regression. arXiv preprint arXiv:2003.11086 (2020)
5. Dong, G., Taslimitehrani, V.: Pattern-aided regression modeling and prediction model analysis. IEEE Trans. Knowl. Data Eng. **27**(9), 2452–2465 (2015)

6. Hukkelhoven, C.W., et al.: Predicting outcome after traumatic brain injury: development and validation of a prognostic score based on admission characteristics. J. Neurotrauma **22**(10), 1025–1039 (2005)

7. Lokshtanov, D., Suri, S., Xue, J.: Efficient algorithms for least square piecewise polynomial regression. In: ESA21: Proceedings of European Symposium on Algorithms (2021)

8. Lyu, B., Li, J.: An efficient data analysis method for big data using multiple-model linear regression (2023)

9. Montgomery, D.C., Peck, E.A., Vining, G.G.: Introduction to linear regression analysis. Technical report, Wiley (2021)

10. Siahkamari, A., Gangrade, A., Kulis, B., Saligrama, V.: Piecewise linear regression via a difference of convex functions. In: International Conference on Machine Learning, pp. 8895–8904. PMLR (2020)

11. Wang, Y., Witten, I.H.: Induction of model trees for predicting continuous classes (1996)

Multi-Load Agent Path Finding
for Online Pickup and Delivery Problem

Yifei Li[ID], Hao Ye, Ruixi Huang, Hejiao Huang[✉][ID], and Hongwei Du

School of Computer Science and Technology, Harbin Institute of Technology
(Shenzhen), Shenzhen 518055, China
{liyifei,yehao,huangruixi}@stu.hit.edu.cn, {huanghejiao,hwdu}@hit.edu.cn

Abstract. The Multi-Agent Pickup and Delivery (MAPD) problem, a variant of the lifelong Multi-Agent Path Finding (MAPF) problem, allows agents to move from their initial locations via the pickup locations of tasks to the delivery locations. In general MAPD problem, the agent is single-load and completes only one task at a time. However, many commercial platforms, e.g., Amazon and JD, have recently deployed multi-load agents to improve efficiency in their automated warehouses. As the multi-load agents can complete multiple tasks at once instead of just one, existing solutions for the general MAPD are unsuitable for the multi-load agent scenario. Meanwhile, a few works focus on the schedule of multi-load agents because it is hard to assign tasks to suitable multi-load agents and find conflict-free paths in real-time for multi-load agents. Therefore, in this paper, we formally define the Multi-Load Agent Pickup and Delivery (MLAPD) problem, in which the multi-load agents complete the real-time pickup-and-delivery tasks and avoid conflicts with each other to minimize the sum of the travel costs and the delay costs. For solving the MLAPD problem, we propose an efficient task assignment algorithm and a novel dynamic multi-agent path finding algorithm. Extensive experiments show that compared with the state-of-the-art, our solution can complete an additional 4.31%~138.33% of tasks and save 0.38%~12.41% of total costs while meeting real-time requirements.

Keywords: Multi-load agent · online task planning · optimization

1 Introduction

Recently, many commercial platforms, e.g., Amazon [1], Cainiao [2], and JD [9], are deploying multi-load agents to improve the operation efficiency of their automated warehouses. Therefore, the Multi-Agent Pickup and Delivery (MAPD) problem receives significant attention [4,5,12,13] because it is considered the core problem in improving efficiency and reducing costs in industrial scenarios. The existing works studying the MAPD problem mainly focus on two phases: task assignment and path finding. The task assignment phase refers to assigning tasks to suitable agents. For example, Dang et al. [6] propose an adaptive large neighborhood search algorithm to dynamically select the task assignment strategy for each

W. Wu and G. Tong (Eds.): COCOON 2023, LNCS 14422, pp. 285–296, 2024.
https://doi.org/10.1007/978-3-031-49190-0_20

agent. Liu et al. [11] assign tasks to agents by solving a special traveling salesman problem. Hu et al. [8] and Shi et al. [15] use learning-based methods to adaptively assign tasks to the most suitable agents. Moreover, many works focus on the path finding phase, i.e., the Multi-Agent Path Finding (MAPF) problem. Some previous works [3, 7] use the space-time A* algorithm or its variants to solve the MAPF problem. These solutions are efficient but the new planned path in the space-time A* algorithm may take a detour due to the previous paths occupying some grids on the shortest path of the new planned path. To address this drawback, recent works [10, 14] use a new algorithm, namely the Conflict-Based Search (CBS) algorithm, to find the optimal result of multi-agent path finding. Although these works can find satisfactory task assignments or efficiently find conflict-free paths to solve the MAPD problem, there are still two issues.

- First, most existing works [10,12,14,15] solving the MAPD problem only consider the single-load agent and ignore the multi-load agent. The single-load agent can complete one task at once, i.e., it can complete a new task if and only if the currently executing task is delivered. However, the schedule of the multi-load agent is more complex. Specifically, when we assign tasks to single-load agents, we only schedule the paths for agents to complete one task. Nevertheless, if the agent is multi-load, there may be a set of paths that need to visit in the schedule of the agent, so we not only need to consider the paths for the agent to complete the assigned task but also need to consider the existing paths in the schedule of the agent. Therefore, these solutions are unsuitable for the multi-load agent scenario.
- Second, most existing solutions [3,6,7,11] only solve the task assignment or the path finding. They ignore that the result of the task assignment can affect the quality of the path finding and vice versa. Accordingly, their solutions are ineffective in practice. Recently, a few works [5,12] realize that task assignment and path finding should be solved simultaneously. They first use the greedy approach to assign tasks to the nearest agents, and then simplify the path finding by prohibiting conflicts between subsequent paths and previous paths. Note that they use the space-time A* algorithm to find the conflict-free paths, so the disadvantage of A* yet exists in their solutions. Thus, their solutions also cannot cope with the MAPD problem well.

To address these issues, we focus on the pickup and delivery problem for multi-load agents. For solving this problem, we need a task assignment algorithm to efficiently assign tasks to the multi-load agent while considering that the multi-load agent already has multiple tasks to complete. Meanwhile, we also require a multi-agent path finding algorithm, which finds conflict-free paths in real time for multi-load agents, to overcome the shortcomings of the space-time A* algorithm.

Our Contribution: In this paper, we first formally define the *Multi-Load Agent Pickup and Delivery (MLAPD)* problem and analyze its competitive hardness. Then, we not only propose an efficient task assignment algorithm and a novel dynamic multi-agent path finding algorithm to solve the MLAPD problem but also analyze the complexities of the proposed algorithms in detail. Moreover, we use a heuristic method to improve the efficiency of the path finding algorithm in practice. Finally, we conduct extensive experiments, which show that our solution can complete an additional 4.31%~138.33% of tasks and save 0.38%~12.41% of total costs while meeting real-time requirements.

We organize the rest of this paper as follows. We first present the problem formulation in Sect. 2. In Sect. 3, we propose two algorithms for the MLAPD problem and give detailed analyses. After conducting the empirical study in Sect. 4, we conclude this paper and show the future work in Sect. 5.

2 Problem Formulation

In this paper, the warehouse provides a centralized service. Briefly, the warehouse collects task information and assigns tasks to suitable multi-load agents in time. Based on this fashion, we first give the definitions of task, agent, and cost. Then, we formally define the MLAPD problem and analyze its hardness.

Definition 1 (Task). *A task $\tau \in \Gamma$ is represented as a four-entry tuple $\tau = \langle l_\tau^p, l_\tau^d, t_\tau, w_\tau \rangle$, where l_τ^p and l_τ^d are the pickup location and delivery location, respectively, t_τ is the deadline, and w_τ indicates the weight.*

Definition 2 (Agent). *An agent $v \in V$ is denoted as a four-entry tuple $v = \langle l_v, c_v, \Gamma_v, S_v \rangle$, where l_v is the location, c_v shows the maximum capacity, Γ_v is the set of tasks to be completed, and S_v is the schedule of Γ_v.*

The schedule $S_v = \langle l_1, l_2, ..., l_n \rangle$, where l_i is a pickup or delivery location of an assigned task. Note that there are several constraints: (i) a task only can be assigned to an agent; (ii) all tasks only can be delivered after picking up; (iii) whenever the unoccupied capacity $\overline{c_v} = c_v - \sum_{\tau \in \Gamma_v} w_\tau \geq 0$.

Definition 3 (Cost). *The cost of v completing τ, which contains travel cost and delay cost, is denoted as follows,*

$$c(v, \tau) = \alpha \cdot t(v, \tau) + (1 - \alpha) \cdot d(v, \tau) \tag{1}$$

where $\alpha \in (0, 1)$ is a factor to balance the travel cost and delay cost. $t(v, \tau)$ is the travel distance, and delay cost $d(v, \tau) = \max(t_\tau^v - t_\tau, 0)$, where t_τ^v is the actual completion time of v completing τ. For simplicity, we assume that the unit travel distance cost and unit delay cost are both 1. Then, we formally define the MLAPD problem as follows.

Definition 4 (MLAPD problem). *Given a set of agents and a stream of tasks, the MLAPD problem aims to assign tasks to multi-load agents and plan conflict-free paths to minimize the total cost $C(V, \Gamma) = \sum_{v \in V, \tau \in \Gamma} c(v, \tau)$.*

The MLAPD problem is NP-hard because we can easily construct an instance of the MLAPD problem and map it to a classic NP-hard problem named the Vehicle Routing Problem. We further prove that the competitive ratio of the MLAPD problem cannot be guaranteed by online deterministic algorithms.

Theorem 1. *The competitive ratio of the MLAPD problem cannot be guaranteed by online deterministic algorithms.*

Proof. We assume that there is an online deterministic algorithm ALG that can minimize the total cost while guaranteeing a constant competitive ratio c in the MLAPD problem. The total cost of ALG is no more than c times of an oblivious adversary OPA, who knows all release information of tasks before the entire process. In other words, if we find an instance where the ratio of the total cost of ALG to OPA is more than c, no $c-$competitive algorithm exists.

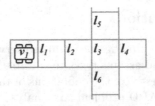

Fig. 1. An instance of an agent completing a task

We construct an instance as shown in Fig. 1. There is an agent v_1 and grids $l_1, l_2, ..., l_6$. A task τ appears at time 1 with the deadline 4. The pickup location l_τ^p and the delivery location l_τ^d are l_1 and l_4, respectively. Meanwhile, there will be n agents continuously passing through l_3 at time 4 following the path from l_6 to l_5. The tasks assigned to these n agents can be completed exactly before the deadlines. OPA knows the release information of τ, so v_1 in OPA can move to l_1 and pick τ up at time 1 immediately. The total cost of OPA is $\alpha \cdot 3 + (1-\alpha) \cdot 0 = 3\alpha$ because v_1 does not have conflicts with other agents at l_3 when the time is 3. However, for ALG, v_1 departs at time 1 because ALG assigns τ to v_1 after τ is released. Therefore, v_1 in ALG will arrive at l_3 when time is 4, which conflicts with other n agents, i.e., the total cost of ALG is $\alpha \cdot 3 + (1-\alpha) \cdot n = n + (3-n)\alpha$. Then, the competitive ratio of ALG to OPA is $c = \frac{n+(3-n)\alpha}{3\alpha}$. In conclusion, $c \to \infty$ if n is large enough. Thus, no $c-$competitive algorithms exist and this theorem is correct.

3   Solution

To solve the MLAPD problem, in this section, we first propose an efficient task assignment algorithm to quickly assign tasks to suitable multi-load agents and then propose a novel dynamic multi-agent path finding algorithm to find conflict-free paths in real-time scenarios.

3.1   Cost-Based Task Assignment Algorithm

Task assignment is one of the core components of the MLAPD problem. Existing works [5,12] assign the task to the nearest agent and ignore other more suitable agents so that they cannot find the most suitable agent. To address this draw-back, we formulate the task assignment as a min-cost max-flow problem and propose a *Cost-Based Task Assignment (CBTA)* algorithm. In what follows, we first introduce the mathematical model. We denote the directed agent-task graph as G, where N is the node set, and A is the arc set, i.e., $G = \langle N, A \rangle$. Except for the source and the sink, each node $i \in N$ can be an agent or a task. The details of the capacity u_{ij} and the cost c_{ij} of each arc $(i, j) \in A$ are shown in Algorithm 1 in detail. Briefly, the mathematical model of task assignment in the MLAPD problem is shown below,

$$Minimize \ C(V, \Gamma) = \sum_{(i,j) \in A} c_{ij} x_{ij} \qquad (2)$$

subject to

$$\sum_{j:(i,j) \in A} x_{ij} - \sum_{j:(i,j) \in A} x_{ji} = b(i), \qquad for \ all \ i \in N \qquad (3)$$

$$0 \le x_{ij} \le c_v, \qquad for \ all \ (i, j) \in A \qquad (4)$$

where x_{ij} denotes the flow from the node i to the node j and $b(i)$ is the supply or the demand of each node. Note that $b(i)$s of all agent or task nodes are 0 because they are intermediate nodes.

Following this model, the CBTA considers the costs between all released tasks and all agents with unoccupied capacity and quickly assigns each task to the most suitable agent to minimize the total costs in each batch. More specifically, we first construct an agent-task graph, which contains a source, a sink, agents, and tasks. Then, we run the *Successive Shortest Path (SSP)* algorithm, an efficient min-cost max-flow algorithm, to find the minimum cost maximum flow on this graph, which means finding the best assignment between agents and tasks. Next, we show the pseudo-code of CBTA in Algorithm 1 and analyze the time complexity.

Algorithm Detail. The pseudo-code is shown in Algorithm 1, where the input is an agent set V and a task set Γ and the output is an agent-task assignment $\mathcal{M}$. In the beginning, we initialize an assignment $\mathcal{M}$ and an agent-task graph $\mathcal{G}$ by $\emptyset$. We also set a source S and a sink T (line 1). After that, we connect the source S with each agent v. $\langle S, v, \overline{c_v}, 0 \rangle$ means the capacity of arc (S, v) is the unoccupied capacities $\overline{c_v}$ and the cost is 0 (lines 2–3). Meanwhile, we connect each task with the sink and set the arc capacity to 1, which means each task can be assigned once (lines 4–5). After that, we first get the insert location set $\mathcal{I}'$, which is decided by the deadline constraint (line 10). Then, we calculate the costs between tasks and agents by trying to insert the task into each adjacent location pair in the location set $\mathcal{I}'$ and add edges with costs into $\mathcal{G}$ (lines 6–16). Note that the edge costs between source and agents or tasks and sink are all 0 but the edge costs between agents and tasks are costs we have calculated. After constructing the agent-task graph, we call the SSP algorithm to find the minimum cost flow

---

Algorithm 1: CBTA Algorithm

Input: *An agent set V, a task set Γ*
Output: *An agent-task assignment M*

1 Initialize an assignment $\mathcal{M}$, an agent-task graph $\mathcal{G}$, a source point S, a sink point T;
2 **foreach** $v \in V$ **do**
3 | Add $\langle S, v, \overline{c_v}, 0 \rangle$ into $\mathcal{G}$;
4 **foreach** $\tau \in \Gamma$ **do**
5 | Add $\langle \tau, T, 1, 0 \rangle$ into $\mathcal{G}$;
6 **foreach** $(v, \tau) \in V \times \Gamma$ **do**
7 | Initialize an insert result set $\mathcal{I}$;
8 | $c(v, \tau) = \infty$;
9 | **if** $c_v > 0$ **then**
10 | Find the insert location set $\mathcal{I}'$ from the schedule of agent v;
11 | **foreach** *adjacent location pair* $(l_i, l_{i+1}) \in \mathcal{I}'$ **do**
12 | Insert l_τ^s, l_τ^d between l_i and l_{i+1}, respectively, and calculate path lengths by calling A* algorithm;
13 | Add path lengths into $\mathcal{I}$;
14 | Get the minimum insert result and calculate $c(v, \tau)$;
15 | **if** $c(v, \tau)$ *is not* ∞ **then**
16 | Add $\langle v, \tau, 1, c(v, \tau) \rangle$ into $\mathcal{G}$;
17 $\mathcal{M} \leftarrow$ call SSP algorithm based on $\mathcal{G}$;
18 **while** ∃ *tasks can be packaged* **do**
19 | Compare costs between packaged or not and packaged these tasks if can improve the result;
20 | Update $\mathcal{M}$;
21 **return** $\mathcal{M}$;

---

on $\mathcal{G}$ (line 17). Finally, if any two tasks are assigned to the same agent and have the same insert location, we try to package them. If packaged tasks can reduce the total costs of the assignment, we update $\mathcal{M}$ and return $\mathcal{M}$ (lines 18–21).

Theorem 2. *The time complexity of Algorithm 1 is $O(|V||\Gamma|(k + k\log k) + n^2\log n)$, where $|V|$ and $|\Gamma|$ are the number of agents and the number of tasks, k is the number of grids in the warehouse, and n is the number of edges in the agent-task graph.*

Proof. $O(|V|)$ is the time complexity of connecting source and agents, where $|V|$ is the number of agents (lines 2–3). $O(|\Gamma|)$ is the time complexity of connecting tasks and sink, and $|\Gamma|$ is the number of agents (lines 4–5). When we connect agents and tasks, the time complexity is $O(|V||\Gamma|(c_v + |\mathcal{I}'|(k + k\log k)))$, where k is the number of grids in the warehouse, c_v is the maximum capacity of the agent, and $|\mathcal{I}'|$ is the maximum length of the potential insert location set. Note that $O(|V||\Gamma|(c_v + |\mathcal{I}'|(k + k\log k)))$ can be simplified to $O(|V||\Gamma|(k + k\log k))$ because c_v is a constant and $|\mathcal{I}'| \leq 2c_v$ (lines 6–16). The time complexity of the SSP algorithm is $O(mnc_v + n^2 c_v \log n)$, where m and n are the numbers of nodes and edges in the network, i.e., constructed agent-task graph. Specifically, nc_v is the maximum supply in the network. $O(m + n\log n)$ is the time complexity of solving the shortest path problem with nonnegative arc lengths in the network (line 17). In addition, $O(|\Gamma|)$ is the time complexity of packaging tasks (lines 18–20). Therefore, the time complexity of Algorithm 1 is $O(|V| + 2|\Gamma| + |V||\Gamma|(k + k\log k) + mnc_v + n^2 c_v \log n)$, which can be simplified to $O(|V||\Gamma|(k + k\log k) + n^2 \log n)$. This completes the theorem.

3.2   Dynamic Conflict-Based Search Algorithm

Path finding is considered another core component to solving the MLAPD problem. Previous works [4,5] use the space-time A* algorithm to find conflict-free paths for agents. Although the space-time A* algorithm is highly efficient, their solutions plan new paths while avoiding conflict with existing paths, which causes existing paths to affect the qualities of the new paths. To solve this issue, we propose the *Dynamic Conflict-Based Search (DCBS)* algorithm, which plans the conflict-free paths for agents dynamically. Briefly, the DCBS only finds the conflict-free paths in the current batch for all agents and ignores subsequent conflicts. Therefore, DCBS has two advantages. One is that compared with the CBS algorithm which finds entire conflict-free paths, DCBS has better efficiency because it focuses on solving a few conflicts in one batch. The other is that DCBS allows existing paths to be changed when planning new paths, so it overcomes the disadvantage of the space-time A* algorithm. Moreover, we also use a heuristic method to further improve the efficiency of DCBS to satisfy real-time requirements. In what follows, we present the DCBS algorithm in detail.

Algorithm Detail. We present the pseudo-code of the DCBS algorithm in Algorithm 2, where the input is an agent set V and a batch size b and the output is a planning path set of current batch $\mathcal{P}$. First, we initialize $\mathcal{P}$, an empty node set $\mathcal{N}$, and a root node n_{root} (line 1). Then, we calculate the paths in the current batch for all agents and add these paths into $\mathcal{P}$ (lines 2–4). For n_{root}, we update its solution, constraint, and cost by $\mathcal{P}$, $\emptyset$, and the sum of all paths' lengths in $\mathcal{P}$, respectively (lines 5–7). Then, we add n_{root} into $\mathcal{N}$ and select the node with minimum cost in $\mathcal{N}$ as the target node n_t (lines 8–9). Note that n_{root} is the only node in $\mathcal{N}$ currently, so n_{root} is the current target node n_t. If there is no conflict in $n_t.solution$, we return $\mathcal{P}$. Otherwise, we create two child nodes n_l and n_r, and add the conflict into these two nodes' constraints (lines 10–12). Note that the solutions and costs of child nodes are the same as n_t. After that, we use the *UpdatePaths(·)* to update the solutions of child nodes (line 13). Specifically, *UpdatePaths(·)* uses the space-time A* algorithm to find the new path (the length equals b) for one agent while prohibiting the other agent from passing the conflict grid at the conflict time based on the constraint. Moreover, we use the sum of all paths' lengths and conflicts' number of each child node as its cost and add child nodes into the node set $\mathcal{N}$ (lines 14–15). Then, we find the target node from $\mathcal{N}$ as above (line 16) and repeat the loop until no conflicts exist. Finally, we return the corresponding planning path set $\mathcal{P}$ (lines 17–18).

Algorithm 2: DCBS Algorithm

    Input: *An agent set V, a batch size b*
    Output: *A planning path set $\mathcal{P}$*
1 Initialize a planning path set $\mathcal{P}$, a node set $\mathcal{N}$, a root node n_{root};
2 **foreach** $v \in V$ **do**
3   Calculate the path p_v^b of v in the current batch b;
4   Add p_v^b into $\mathcal{P}$;
5 $n_{root}.solution \leftarrow \mathcal{P}$;
6 $n_{root}.constraint \leftarrow \emptyset$;
7 $n_{root}.cost \leftarrow$ the sum of all paths' lengths in $\mathcal{P}$;
8 Add n_{root} into $\mathcal{N}$;
9 Select the node with minimum cost in $\mathcal{N}$ as the target node n_t;
10 **while** $\exists$ *conflict in* $n_t.solution$ **do**
11     $\mathcal{N} \leftarrow$ remove n_t from $\mathcal{N}$;
12   Create two child node n_l and n_r for conflicted agents and add
       corresponding conflict into $n_l.constraint$ and $n_r.constraint$;
13   Update $n_l.solution$ and $n_r.solution \leftarrow UpdatePaths(\cdot)$;
14   Update $n_l.cost$ and $n_r.cost$ based on the sum of all paths' lengths and
       conflicts' number in their solution;
15   Add n_l and n_r into $\mathcal{N}$;
16   Select the node with minimum cost in $\mathcal{N}$ as the target node n_t;
17 $\mathcal{P} \leftarrow n_t.solution$;
18 **return** $\mathcal{P}$;

Analysis. If the total number of conflicts $|C|$ is known in Algorithm 2, the time complexity can be computed as $O((|V| + |C|)(k + klogk))$. $O(|V|(k + klogk))$ is the time complexity of calculating the solution of the root node (lines 2–4). $O(|C|(k + klogk))$ is the time complexity of searching the node with a conflict-free solution (lines 10–16). Note that the time complexity of computing conflicts' number in the solutions of nodes is $O(|C|b)$, which can be ignored because batch size b is a given constant (lines 14). Nevertheless, the conflicts' number $|C|$ is dynamically changed when choosing different target nodes to split, and the number of conflicts shows an exponential explosive growth trend with the number of agents. Therefore, it is difficult to determine the number of conflicts, i.e., it is hard to compute the time complexity of the DCBS algorithm.

To improve the efficiency of the DCBS algorithm in practice as possible, following existing works [10,14], we use a heuristic method to accelerate the search for the most likely conflict-free node. Briefly, as shown in Algorithm 2 (line 14), we count the conflicts number in the solution of each node and add this value into the cost of the node. After that, the minimum cost node means not only fewer costs but also fewer potential conflicts. In other words, we are more likely to find the conflict-free paths set when selecting the minimum cost node.

4   Empirical Study

In this section, we first show the experimental settings containing the map, parameters, compared algorithms, and metrics. Then, we present the experimental results and analyses.

4.1   Experimental Settings

We perform all evaluations in a 21×35 pickup and delivery warehouse map, which is widely used in many related works [4,5,12]. The warehouse is shown in Fig. 2, where blue grids are pickup and delivery points, black grids are obstacles or walls, white grids are corridors, and orange grids represent the current locations of agents.

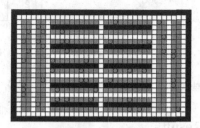

Fig. 2. A warehouse map with 21×35 grids

Moreover, all tasks are released in one hour with weight 1, and are executed in an online fashion. We set α to 0.5 to balance the travel cost and delay cost. All parameter settings are listed in Table 1, and the default values are bold.

Table 1. Parameter settings

Parameter	Value
# of Tasks $\|\Gamma\|$	2000, 4000, **6000**, 8000, 10000
# of Agents $\|V\|$	10, 20, **30**, 40, 50
Maximum capacity c	1, 2, **3**, 4 , 5
Batch size b	5 s, 10 s, **15 s**, 20 s,3 0 s

We compare our algorithm CBDC which first uses the CBTA algorithm (see Sect. 3.1) to assign tasks and then uses the DCBS algorithm (see Sect. 3.2) to find paths in each batch for agents, the TPMC [5] that assigns tasks to the nearest agents and uses the space-time A* algorithm to plan paths, and the RMCA [4] that assigns tasks to the agents with the minimum cost and also plans paths by the space-time A* algorithm. More specifically, we evaluate the effectiveness of all algorithms by completion ratio (CR) and total cost (TC) and test the efficiency by the average processing time of each batch (PT).

294   Y. Li et al.

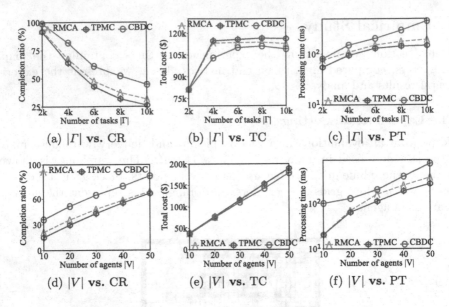

Fig. 3. Performance of $|\Gamma|$, $|V|$ **vs. CR, TC, and PT**

4.2 Experimental Results

Effect of the Task Number $|\Gamma|$. We show the effect of task number $|\Gamma|$ in Figs. 3(a), 3(b), and 3(c). When increasing $|\Gamma|$, as the number of agents and the maximum capacity are constants, i.e., the maximum transport capability is limited, the CRs of all algorithms decrease. Among them, CBDC performs best, its CR dominates TPMC and RMCA and the TC is also better than them, e.g., when $|\Gamma| = 2k$, the TC of RMCA is close to CBDC but still 0.38% higher. When $|\Gamma| > 6k$, the TC of CBDC does not increase showing that it cannot serve more tasks. However, more tasks mean CBDC can assign more proper tasks to agents, so the total cost has a slight decrease. Moreover, PTs of all algorithms raise when extending the task number.

Effect of the Agent Number $|V|$. The effect of the agent number $|V|$ is plotted in Figs. 3(d), 3(e), and 3(f). When we increase $|V|$, more agents can serve more tasks so all CRs raise. CBDC has the best CR. Note that the CR of CBDC is even 138.33% higher than that of TPMC when $|V| = 10$. As RMCA considers the task assignment between all agents and tasks in each batch, RMCA performs similarly to TPMC and has a little improvement in CR. In addition, CBDC plans conflict-free paths dynamically. Therefore, compared with TPMC and RMCA, CBDC has a similar TC with serving more tasks simultaneously. TPMC has the best PT as it only tries to assign each task to the nearest agent with unoccupied capacities. Overall, CBDC can serve more tasks than other algorithms but reduce TC while satisfying real-time requirements.

Effect of the Maximum Capacity c. Increasing the maximum capacity of each agent also means increasing the maximum transport capability of the entire

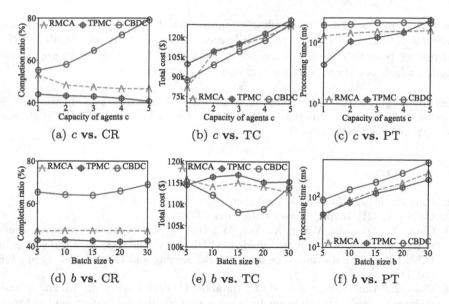

Fig. 4. Performance of c, b, vs. **CR, TC, and PT**

warehouse. Therefore, we can observe that the CR of CBDC raises with extending c and the TC also increases because agents complete more tasks. When the capacity is 1, the CR of CBDC is 4.31% higher than RMCA and its TC is 12.41% lower than TPMC. It is worth noting that TPMC and RMCA do not have an expected increase in CR. The main reason is that they use A* to plan conflict-free paths. When we extend c, more previous paths mean the lengths of new paths increase, so the TCs increases but the CRs even decrease. It is worth noting that when $c = 5$, CBDC has a similar PT with TPMC but serves more than twice the number of tasks than TPMC.

Effect of the Batch Size b. As the increase of batch size b, each batch contains more release tasks. However, the fixed number of agents and capacities limit the maximum transport capability, so CRs of TPMC and RMCA are almost no changes. It is inserting that CBDC can find better task assignment when we extend b because it considers the assignments between all agents and release tasks in each batch. Therefore, the TC of CBDC decreases when $b \leq 15$ because of better assignments and increases when $b > 15$ because of serving more tasks.

5   Conclusion

In this paper, we formally define the MLAPD problem and analyze its hardness. Existing works cannot cope with the MLAPD problem well. In response, we propose an efficient task assignment algorithm to assign tasks to the most suitable multi-load agents and a novel dynamic multi-agent path finding algorithm to find conflict-free paths for agents to complete assigned tasks. Extensive experiments show that our solution can complete an additional 4.31%∼138.33% of tasks and save 0.38%∼12.41% of total costs while meeting real-time requirements. In the future, we plan to extend our proposed algorithms to schedule heterogeneous agents, which is a more complex but realistic scenario.

Acknowledgements. This work is financially supported by Shenzhen Science and Technology Program under Grant No.GXWD20220817124827001 and No.JCYJ20210324132406016.

References

1. Amazon. https://logistics.amazon.com/marketing
2. Cainiao (2021). https://www.cainiao.com
3. Chen, X., Feng, L., Wang, X., Wu, W., Hu, R.: A two-stage congestion-aware routing method for automated guided vehicles in warehouses. In: Proceedings of IEEE International Conference on Network, Sensing and Control, vol. 1, pp. 1–6 (2021)
4. Chen, Z., Alonso-Mora, J., Bai, X., Harabor, D.D., Stuckey, P.J.: Integrated task assignment and path planning for capacitated multi-agent pickup and delivery. IEEE Rob. Autom. Lett. **6**(3), 5816–5823 (2021)
5. Çilden, E., Polat, F.: Multiagent pickup and delivery for capacitated agents. In: Proceedings of PAAMS International Conference, pp. 76–87 (2022)
6. Dang, Q., Singh, N., Adan, I., Martagan, T., van de Sande, D.: Scheduling heterogeneous multi-load AGVS with battery constraints. Comput. Oper. Res. **136**, 105517 (2021)
7. Goldenberg, M., et al.: Enhanced partial expansion A. Artif. Intell. Res. **50**, 141–187 (2014)
8. Hu, H., Jia, X., He, Q., Fu, S., Liu, K.: Deep reinforcement learning based AGVS real-time scheduling with mixed rule for flexible shop floor in industry 4.0. Comput. Ind. Eng. **149**, 106749 (2020)
9. JD. https://www.jdl.cn
10. Li, J., Ruml, W., Koenig, S.: Eecbs: a bounded-suboptimal search for multi-agent path finding. In: Proceedings of the AAAI Conference on Artificial Intelligence, vol. 35, pp. 12353–12362 (2021)
11. Liu, M., Ma, H., Li, J., Koenig, S.: Task and path planning for multi-agent pickup and delivery. In: Proceedings of the International Joint Conference on Autonomous Agents and Multiagent Systems (2019)
12. Ma, H., Li, J., Kumar, T.K.S., Koenig, S.: Lifelong multi-agent path finding for online pickup and delivery tasks. In: Proceedings of the International Conference on Autonomous Agents Multiagent Systems, pp. 837–845 (2017)
13. Salzman, O., Stern, R.: Research challenges and opportunities in multi-agent path finding and multi-agent pickup and delivery problems. In: Proceedings of the International Conference on Autonomous Agents and MultiAgent Systems, pp. 1711–1715 (2020)
14. Sharon, G., Stern, R., Felner, A., Sturtevant, N.R.: Conflict-based search for optimal multi-agent pathfinding. Artif. Intell. **219**, 40–66 (2015)
15. Shi, D., Tong, Y., Zhou, Z., Xu, K., Tan, W., Li, H.: Adaptive task planning for large-scale robotized warehouses. In: Proceedings of the IEEE International Conference on Data Engineering, pp. 3327–3339 (2022)

Improved Sourcewise Roundtrip Spanners with Constant Stretch

Eli Stafford and Chunjiang Zhu[✉]

Department of Computer Science, UNC Greensboro, Greensboro, NC, USA
chunjiang.zhu@uncg.edu

Abstract. Graph spanners are a sparse subgraph of a graph such that shortest-path distances for all pairs of vertices are approximately preserved with a factor called stretch, and roundtrip-spanners are defined for directed graphs to preserve roundtrip distances instead of one-way distances. Sourcewise roundtrip-spanners can approximate roundtrip distances for only some pairs of vertices $S \times V$ for source vertices $S \subseteq V$ and are more generalized than traditional all-pairs roundtrip-spanners. While general roundtrip-spanners have made progress in the realm of constant stretch, it is unknown whether constant stretch (with small number of edges dependent on $|S|$) can be achieved in the sourcewise setting. In this paper, we provide an algorithm that, for a weighted, directed graph with n vertices, m edges G and a set of sources S of size s, constructs a sourcewise roundtrip-spanner with stretch 3 and $\tilde{O}(n\sqrt{s})$ expected edges in $\tilde{O}(ms)$ time. Moreover, we develop a faster $\tilde{O}(m\sqrt{n}/\epsilon^2)$-time algorithm with stretch $(5+\epsilon)$ and $\tilde{O}(n\sqrt{s}/\epsilon^2)$ edges when S is randomly picked with size $s = \Omega(\sqrt{n})$. Our algorithms combine ideas from [RTZ08, RTZ05] and adapt the algorithm of [DW20] to the sourcewise case.

Keywords: Graph Spanners · Roundtrip Spanners · Graph Algorithms · Randomized Algorithms

1 Introduction

Graph spanners are sparse graph structures that approximate pairwise shortest-path distances in graphs [PS89]. In an undirected graph $G(V, E)$, a spanner of *stretch* α, or called α-spanner, is a subgraph $H(V, E' \subseteq E)$ of G such that for every $u, v \in V$, their distance in H is at most α times of their original distance in G. It was well-established that for an integer $k > 1$, every undirected graph on n vertices has a $(2k - 1)$-spanner of size (number of edges) $O(n^{1+1/k})$ [ADD+93, TZ05]. The definition of spanners can be easily extended to directed graphs but it becomes trivial in this setting because of the well-known lower bound $\Omega(n^2)$ on the size. Research efforts were then devoted to roundtrip-spanners: a k-roundtrip-spanner is a subgraph that preserves all-pairs *roundtrip distances*

Chunjiang Zhu is supported by UNC Greensboro Start-up Funds and Faculty First Award.

up to a factor of k [RTZ08], where the roundtrip distance between $u, v \in G$ is the distance from u to v plus the distance from v to u in G. Chechik *et al.* [CLRS20] proposed algorithms that construct a roundtrip-spanner of stretch $O(k \log \log n)$ (or $O(k \log k)$) and size $\tilde{O}(n^{1+1/k})$ in time $\tilde{O}(m^{1+1/k})$ with high probability (whp), where m is the number of edges and $\tilde{O}(\cdot)$ hides polylog terms. Moreover, they developed an algorithm that can compute a *constant* stretch 8-roundtrip-spanner of $\tilde{O}(n^{1.5})$ edges in $\tilde{O}(m\sqrt{n})$ time whp. Later in [DW20], the stretch was improved to $(5 + \epsilon)$ while keeping other parameters asymptotically the same. Chechik and Lifshitz [CL21] developed an algorithm that can compute a 4-roundtrip-spanner of size $O(n^{1.5})$ in time $\tilde{O}(n^2)$ w.h.p, which is optimal for dense graphs of $m = \Theta(n^2)$.

Recently, *sourcewise* roundtrip-spanners have been studied in [ZL17, ZHL21] as a more general object of the all-pairs roundtrip-spanners. Given any source vertex set $S \subseteq V$ in a digraph $G(V, E)$, a *k-sourcewise-roundtrip-spanner* is a subgraph $H(V, E' \subseteq E)$ such that, for every $u, v \in S \times V$, the roundtrip distance between u and v in H is at most k times their original roundtrip distance in G. Traditional roundtrip-spanners (ones that are concerned with all pairs $V \times V$) have found concrete applications in networking and routing problems [SS10, RTZ08], and are closely related to cycle and girth approximation [PRS+18, DW20, CL21]. Being generalized versions of these traditional structures, it is believed that sourcewise roundtrip-spanners can have wide algorithmic applications, including sourcewise roundtrip distance oracles, roundtrip compact routing schemes, and low distortion embeddings [ABS+20].

Sourcewise roundtrip-spanners concern themselves with a subset of possible pairs ($S \times V \subseteq V \times V$), so for any source set a traditional k-roundtrip-spanner is by definition also a valid k-sourcewise-roundtrip-spanner, though clearly larger than optimal. The overarching goal of a sourcewise roundtrip-spanner algorithm is to provide algorithms that can produce sourcewise roundtrip-spanners in *faster* runtime and/or with *fewer* edges than these traditional algorithms when $S \subset V$. Here we also try and match the performance of previous algorithms when $S = V$, thereby making our algorithms true generalizations. Previous sourcewise roundtrip-spanner results include the algorithm of Zhu and Lam [ZL17] that attained stretch $(2k + \epsilon)$ and size $\tilde{O}(ns^{1/k})$ in time $\tilde{O}(ms)$, where s is the size of S. Zhu *et al.* [ZHL21] then improved the running time to nearly linear time of $\tilde{O}(ms^{1/k})$ while having $O(k \log n)$ stretch for $\tilde{O}(ns^{1/k})$ edges. These results outperform general roundtrip-spanners when s is small. While general roundtrip-spanners have made progress in the realm of constant stretch, it is still open whether constant stretch with small number of edges dependent on $|S|$ can be achieved in sourcewise roundtrip-spanners.

Our Contributions. In this paper, we answer the question affirmatively by constructing sourcewise roundtrip-spanners of stretch 3 and $5 + \epsilon$. The main results are formally provided in Theorems 1 and 2. Our results essentially produce a 3-sourcewise-roundtrip-spanner with $\tilde{O}(n\sqrt{s})$ edges in $\tilde{O}(ms)$ time, and when $s = \Omega(\sqrt{n})$, a $(5 + \epsilon)$-sourcewise-roundtrip-spanner with asymptotically the same size in $\tilde{O}(m\sqrt{n}/\epsilon^2)$ time. One might enforce the stretch $(2k + \epsilon)$ of

[ZL17] (or $O(k \log n)$ in [ZHL21]) to match our stretch 3, but the size becomes $\Omega(ns^{2/3})$, much larger than our size $\tilde{O}(n\sqrt{s})$. In addition, our size matches the size of 3-sourcewise spanner in undirected graphs (up to polylog factor) [RTZ05]. One may also set the spanner size of [ZL17,ZHL21] to $\tilde{O}(n\sqrt{s})$, but their stretch becomes $(4+\epsilon)$ and $O(\log n)$, much larger than our stretch 3. That is, our algorithm exhibits a better stretch-size tradeoff than the arts in source-wise roundtrip-spanners. But the running time $\tilde{O}(ms)$ is large especially when s is a large integer. When S is a uniform random subset of V of size $s = \Omega(\sqrt{n})$, we develop the second algorithm to improve the running time to $\tilde{O}(m\sqrt{n}/\epsilon^2)$ at the expense of a larger stretch $(5 + \epsilon)$. Compared with the all-pairs roundtrip-spanners [DW20], we reduce the number of edges from $\tilde{O}(n^{1.5} \cdot \log^2(M)/\epsilon^2)$ to $\tilde{O}(n\sqrt{s} \cdot \log^2(M)/\epsilon^2)$.

Theorem 1 (Stretch 3). *For a graph $G(V,E)$ with n vertices and m edges, and a source vertex set $S \subseteq V$ of size s, there exists an algorithm that constructs a 3-sourcewise-roundtrip-spanner with $\tilde{O}(n\sqrt{s})$ expected edges in $\tilde{O}(ms)$ time.*

Theorem 2 (Stretch $5 + \epsilon$). *For a graph $G(V,E)$ with edge weights in $\{1,...,M\}$, randomly sampled source vertex set $S \subseteq V$ of size $s = \Omega(\sqrt{n})$ and a parameter $\epsilon > 0$, there is an $\tilde{O}(m\sqrt{n}\log^2(M)/\epsilon^2)$-time algorithm that whp constructs a $(5+\epsilon)$-sourcewise-roundtrip-spanner of $\tilde{O}(n\sqrt{s}\log^2(M)/\epsilon^2)$ edges.*

Related Work. Graph spanners in undirected graphs have received extensive studies because of the power in approximating shortest-path distances. Distances can be preserved in different formats, e.g., multiplicative stretch αd [ADD+93,TZ05], additive surplus $d + \beta$ [EP04,Che13,Woo10] or mixed form $\alpha d + \beta$ [Pet09,BKMP10], where d is the original distance. For more details and variants of spanners, readers are referred to the good survey [ABS+20]. In digraphs, there have been efforts for improving the stretch-size-runtime tradeoff of roundtrip-spanners. Cen et al. [CDG20] proposed an algorithm that can construct a $(2k-1)$-roundtrip-spanner of size $O(kn^{1+1/k}\log n)$ in $\tilde{O}(kmn\log M)$ time, improving the stretch factor in a prior work [ZL18]. Pachocki et al. [PRS+18] devised a fast algorithm that constructs an $O(k\log n)$-roundtrip-spanner of size $\tilde{O}(n^{1+1/k})$ in time $\tilde{O}(mn^{1/k})$ whp. Chechik et al. [CLRS20] improved the stretch to $O(k\log\log n)$ and $O(k\log k)$ respectively, while using the same runtime as the algorithm in [PRS+18]. In addition, they developed an $\tilde{O}(m\sqrt{n})$ time algorithm that computes an 8-roundtrip-spanner of expected $\tilde{O}(n^{1.5})$ edges. Later in [DW20], the stretch was improved to $(5+\epsilon)$ for the same runtime. Recently, the stretch was further reduced to 4 with runtime $\tilde{O}(n^2)$ and the polylog factor in the size was also peeled off [CL21].

Another thread of works focuses on approximating distances for *some* pairs of vertices, instead of all pairs. Coppersmith and Elkin [CE06] studied exact sourcewise and pairwise preservers that preserve distances in $S \times V$ with sources $S \subseteq V$ and $P \subseteq V \times V$, respectively. They showed that for a set of sources S of size s and a set of vertex pairs P of size p, every graph G contains a sourcewise preserver of size $O(\min\{n^{1/2}s^2, ns\})$ and a pairwise preserver of size

$O(\min\{np^{1/2}, n^{1/2}p\})$. Recently, Bodwin [Bod17] improved the size of the pair-wise preserver to $O(n + n^{2/3}p)$, even in the case of digraphs. Roditty et al. [RTZ05] proposed an algorithm that constructs a sourcewise $(2k-1)$-spanner of size $O(kns^{1/k})$ in expected $\tilde{O}(kms^{1/k})$ time. Sourcewise roundtrip-spanners were firstly studied in [ZL17] that constructs $(2k+\epsilon)$-sourcewise-roundtrip-spanners of size $\tilde{O}(ns^{1/k})$ in time $\tilde{O}(ms)$. Recently, a faster algorithm [ZHL21] can construct, in nearly linear time $\tilde{O}(ms^{1/k})$, $O(k \log n)$-sourcewise-roundtrip-spanners of the same size.

2   Definitions and Notations

A directed graph $G(V, E, W)$ consists of a vertex set V, an edge set E and a weight function W. W assigns weights to each edge $e \in E$ and can be omitted from the presentation if it is clear from the context. The maximum edge weight amongst all edges is denoted by M. A *(one-way) shortest path* from u to v in G is a path with the minimum distance amongst all paths from u to v in G. This can be represented in shorthand as $u \to v$. Its distance is called the *(one-way) distance*, $d_G(u, v)$ or $d_G(u \to v)$, from u to v in G and the subscript G can be ignored when clear. A *roundtrip shortest path* between u and v in G is the concatenation of a shortest path from u to v and a shortest path from v to u in G. Its distance is called the *roundtrip distance* $d(u \leftrightarrows v)$ between u and v in G.

An *(out-)shortest path tree* from a vertex v to an arbitrary vertex set U in G is a subgraph that is a tree T of G, such that for every $u \in U$, $d(v \to u)$ in T is $d(v \to u)$ in G. This tree is said to be *centered* at u. Analogously for in-shortest path trees. We follow a common assumption that all vertex degrees are at most $\lceil m/n \rceil$ since otherwise, we can construct such a graph with the same number of vertices and edges as the original graph using the technique in [CLRS20].

3   3-Sourcewise-Roundtrip-Spanner

In this section, we present an algorithm for constructing 3-sourcewise-roundtrip-spanners. We combine ideas from Roddity et al. [RTZ08,RTZ05] and provide a generalization of a result from [RTZ08] which presented a 3-roundtrip-spanner in the form of a routing scheme.

For graph $G(V, E)$, their premise is to sample a set A from V at a probability of $1/\sqrt{n}$. The shortest paths from all vertices $v \in V$ to and from every vertex $a \in A$ are recorded in the spanner H. Then for all vertices $v \in V$, they define the center $cent(v)$, which is the vertex in A with the minimum roundtrip distance to v. They can then conclude that there are $\tilde{O}(\sqrt{n})$ vertices with roundtrip distance to v no larger than[1] to their center whp, or called bunch $B(v)$[2]. Any vertex in this "short" set has its shortest path directly recorded. They then prove any

---

[1] When roundtrip distances are the same, use one-way distances and vertex identifiers for comparison. See their definition of roundtrip ordering which we do not need.

[2] Called clusters in Sect. 5 of [RTZ08].

vertex outside of this set $B(v)$ has a bounded stretch via the triangle inequality. To generalize this approach to the sourcewise case there were two main obstacles. Firstly, it was unclear that how to define bunches and $cent()$ in the sourcewise setting. Secondly, their approach describes a single direction. When $S = V$, one only has to approximate one-way paths, as all these one-way paths together create a roundtrip-spanner. In contrast, we need to approximate all paths $v \to u$ but also all paths $u \to v$, where $u \in S$ and $v \in V$.

For a source vertex set S of size s, we first compute a subset $A \subseteq S$ of expected size $O(\sqrt{s} \log s)$ by independently sampling each source in S with probability of $s^{-0.5}$. For each $u \in S$, let its *center* be the sampled source with the minimum roundtrip distance with u and ties are broken arbitrarily. That is,

$$cent(u) \in \{a \in A \mid d(a \leftrightarrows u) \leq d(q \leftrightarrows u), \forall q \in A\}.$$

We then apply Dijkstra's search from each vertex $a \in A$ in both forward and backward directions and add the edges of the in- and out-shortest path trees into the spanner H. Now the number of edges in H is $\tilde{O}(n\sqrt{s})$. For each source $u \in S$ and every $v \in V$, we will bound the distance approximation in the $v \to u$ direction as well as the $u \to v$ direction. We are able to identify the following important subsets of sources S called bunches and bound their expected size.

Lemma 1. *Using the sample set A and the definition of $cent()$, we can define, for each $v \in V$, the set $B(v) = \{u \in S \mid d(v \leftrightarrows u) \leq d(cent(u) \leftrightarrows u)\}$ [3] such that $E[\|B(v)\|] \leq 2\sqrt{s}$.*

Proof. Suppose for contradiction that $E[\|B(v)\|] > 2\sqrt{s}$. Because $B(v) \subseteq S$ and each vertex in S is sampled into A with probability $1/\sqrt{s}$, whp there exists at least one vertex of $B(v)$ that is sampled and included in A. Let $w \in B(v) \cap A$ be such a vertex without loss of generality. Since $w = cent(w)$, we have $d(v \leftrightarrows w) > d(w \leftrightarrows cent(w)) = 0$. This contradicts with $w \in B(v)$, completing the proof. $\square$

Next, we prove the following lemmas that are important to bound the number of edges in the generated spanner. Lemma 2 is for controlling the number of edges in the shortest paths from $v \in V$ to sources $B(v)$ while Lemma 3 is for the shortest paths from sources $B(v)$ to v.

Lemma 2. *For any shortest path $v \to u$, where $v \in V$ and $u \in B(v)$, for any vertex w on that path, $u \in B(w)$.*

Proof. By definition of $B(v)$, we have that $d(v \leftrightarrows u) \leq d(cent(u) \leftrightarrows u)$. Since w is on the shortest path from v to u, we know that the roundtrip shortest path $v \leftrightarrows u$ is a valid roundtrip path between u and w. Thus the distance of the shortest roundtrip path $v \leftrightarrows u$ upper bounds the roundtrip shortest distance $d(w \leftrightarrows u)$, i.e., $d(w \leftrightarrows u) \leq d(v \leftrightarrows u)$. Combining the two inequalities, we get that $d(w \leftrightarrows u) \leq d(cent(u) \leftrightarrows u)$ and thus $u \in B(w)$. $\square$

---

[3] Our definition of bunches is not the same as the classic definition in [RTZ05,RTZ08], though they have the same name. Firstly, our bunches are only from S; Secondly, the distance upper bound is $d(cent(u) \leftrightarrows u)$, which is based on u instead of v.

Algorithm 1. 3-Sourcewise-Roundtrip-Spanner

Input: $G(V, E), S \subseteq V$
Output: 3-sourcewise-roundtrip-spanner H
1: $H \leftarrow \emptyset$
2: $A \leftarrow$ Samples of each $u \in S$ with probability $s^{-0.5}$
3: Run Dijsktra's algorithm centered at each $u \in S$, creating in/out trees $E_{in}(u)$ and $E_{out}(u)$ to/from all $v \in V$, recording each $d(u \rightarrow v)$, $d(v \rightarrow u)$, the edge $f(v, u)$ (the first edge on the path $v \rightarrow u$) and $l(u, v)$ (the last edge of $u \rightarrow v$).
4: **for all** $a \in A$ **do**
5:     $H \leftarrow H \cup E_{in}(a) \cup E_{out}(a)$
6: **for all** $u \in S$ **do**
7:     $C(u) \leftarrow \{v \in V \mid d(cent(u) \leftrightarrows u) \leq d(cent(u) \leftrightarrows u)\}$ $\{u \in B(v)$ iff $v \in C(u)\}$
8:     **for all** $v \in C(u)$ **do**
9:         $H \leftarrow H \cup \{f(v, u), l(u, v)\}$
10: **return** H

Lemma 3. *For any path $u \rightarrow v$ where $v \in V$ and $u \in B(v)$, for any vertex w on this path, $u \in B(w)$.*

Proof. By definition of $B(v)$, we have $d(v \leftrightarrows u) \leq d(cent(u) \leftrightarrows u)$. Because w is on the shortest path from u to v, we get that the roundtrip distance $d(v \leftrightarrows u)$ is an upper bound on the roundtrip distance $d(w \leftrightarrows u)$, i.e., $d(w \leftrightarrows u) \leq d(v \leftrightarrows u)$. Therefore, we have $d(w \leftrightarrows u) \leq d(cent(u) \leftrightarrows u)$ and thus $u \in B(w)$.   □

Algorithm 1 presents the algorithm for constructing 3-sourcewise-roundtrip-spanners as in Theorem 1. We prove Theorem 1 through the following two lemmas, where H_1 and H_2 are a subgraph of the spanner H constructed in Algorithm 1 and referred to only for the purpose of the proof.

Lemma 4. *For a graph $G(V, E)$ on n vertices and a source vertex set $S \subseteq V$ of size s, we can create a spanner H_1 such that for each $v \in V$ and each $u \in S$, there exists a path $v \rightarrow u$ such that $d_{H_1}(v \rightarrow u) < d(v \rightarrow u) + d(v \leftrightarrows u)$. The expected number of edges in H_1 is $\tilde{O}(n\sqrt{s})$.*

Proof. We construct the spanner H_1 by first including the edges of the in- and out-shortest path trees from all $a \in A$. Then for $v \in V$ and $u \in B(v)$, we add the first edge $f(v, u)$ in the shortest path from v to u. Based on Lemmas 1 and 2, the expected number of edges in H_1 is $\tilde{O}(n\sqrt{s})$.

By construction, for $v \in V$ and $u \in B(v)$, all edges in the shortest path from v to u are included into H_1 and thus $d_{H_1}(v \rightarrow u) = d_G(v \rightarrow u)$. For $u \in S \setminus B(v)$, the distance is approximated by the path $v \rightarrow cent(u)$ and $cent(u) \rightarrow u$. The following inequality describes how the distance of this path is bounded:

$$
\begin{aligned}
d_{H_1}(v \rightarrow u) &= d(v \rightarrow cent(u)) + d(cent(u) \rightarrow u) \\
&\leq d(v \rightarrow u) + d(u \rightarrow cent(u)) + d(cent(u) \rightarrow u) \\
&= d(v \rightarrow u) + d(cent(u) \leftrightarrows u) \\
&< d(v \rightarrow u) + d(v \leftrightarrows u).
\end{aligned}
$$

The first inequality follows from the triangle inequality, substituting $d(v \to cent(u))$ with $d(v \to u) + d(u \to cent(u))$, and the last inequality derives from the fact that $u \notin B(v)$. $\square$

Analogously, we can prove the following corollary for the $u \to v$ direction based on Lemmas 1 and 3.

Corollary 1. *For a graph $G(V, E)$ on n vertices, and a source vertex set $S \subseteq V$ of size s, we can create a spanner H_2 such that for each $v \in V$ and each $u \in S$, there exists a path $u \to v$ such that $d_{H_2}(u \to v) < d(u \to v) + d(u \leftrightarrows v)$. The expected number of edges in H_2 is $\tilde{O}(n\sqrt{s})$.*

By combining Lemma 4 and Corollary 1, we get the stretch factor of 3 in Theorem 1. For the running time, the computation of the sets $B(v)$ for every $v \in V$ can be obtained by the "dual", the clusters $C(u)$ for every $u \in S$, where $u \in B(v)$ iff $v \in C(u)$, and this incurs $\tilde{O}(ms)$ time. It should be noted that in undirected graphs, one can compute a similar set of clusters $C(v)$ in expected time $\tilde{O}(m\sqrt{s})$ by growing a shortest path tree from v and stopping when the distance is large enough [RTZ05]. But the technique cannot be easily adapted to digraphs since the growing in either forward or backward direction may include more vertices than those in $C(v)$. Thus the total running time is $\tilde{O}(ms)$.

4 $(5 + \epsilon)$-Sourcewise-Roundtrip-Spanner

To improve the running time $\tilde{O}(ms)$ for $s = \Omega(\sqrt{n})$, in this section, we present an $\tilde{O}(m\sqrt{n}/\epsilon^2)$-time algorithm that whp constructs a $(5+\epsilon)$-sourcewise-roundtrip-spanner of size $\tilde{O}(n\sqrt{s}/\epsilon^2)$, while adding the assumption that the source set S is randomly sampled from V uniformly rather than picked arbitrarily. The assumption is motivated in practice that roundtrip distances from different parts of the network (*e.g.*, transportation networks) need to be approximated. Our algorithm is an adaption of [DW20] from preserving all-pairs roundtrip distances to only the sourcewise roundtrip distances.

Our algorithm starts by sampling a uniform random sample Q of $100\sqrt{s}\log s$ size from the source vertices S. This is one of the major changes in the sourcewise case: sampling only from S instead of all vertices V. Then it computes the shortest path trees $T^{in}(q), T^{out}(q)$ into and out of each $q \in Q$ and adds all the edges in these trees to a subgraph H of G. Let V' be vertices that are sufficiently close to vertices in Q in both directions, *i.e.*, $V' = \{v \in V \mid \exists q \in Q, d(v, q) \le d$ and $d(q, v) \le d\}$. Then if the shortest path between two vertices u and v passes through at least one vertex in V', we have $d_H(u, v) \le d_G(u, v) + 2d$.

Lemma 5. *Let $Q \subseteq S$ be a random sample of $100\sqrt{s}\log s$ vertices from S. In $\tilde{O}(m\sqrt{s})$ time we can compute the shortest path trees $T^{in}(q), T^{out}(q)$ into and out of each $q \in Q$. Let H be the subgraph of G consisting of the union of the edges in these trees. H has at most $\tilde{O}(ns^{0.5})$ edges. Let $V' = \{v \in V \mid \exists q \in Q, d(v, q) \le d$ and $d(q, v) \le d\}$. For any two vertices u, v with their shortest path passing through at least one vertex in V', we have $d_{H_1}(u, v) \le d_G(u, v) + 2d$.*

Proof. It is easy to see that the number of edges in H is $\tilde{O}(ns^{0.5})$ since we add the edges in the shortest path trees rooted at each of the $100\sqrt{s}\log s$ vertices in Q. Then let us focus on the (additive) distance approximation. Suppose that there is some vertex $x \in V'$ on the shortest path $u \to v$. Let $q \in Q$ be such that $d(x,q), d(q,x) \le d$. Then

$$
\begin{aligned}
d_H(u,v) &\le d_H(u,q) + d_H(q,v) = d_G(u,q) + d_G(q,v) \\
&\le d_G(u,x) + d_G(x,q) + d_G(q,x) + d_G(x,v) \\
&\le d_G(u,x) + d_G(x,v) + 2d \le d_G(u,v) + 2d
\end{aligned}
$$

The first and second inequalities are due to the triangle inequality while the last one is due to that x lies on the shortest path $u \to v$.   □

Following typical techniques in roundtrip-spanners [RTZ08, ZL17], we now separate the roundtrip distances into logarithmic intervals and consider each interval $[(1+\epsilon)^i, (1+\epsilon)^{i+1})$ for $i \in [0, \log_{1+\epsilon}(Mn))$. We define V_i' in terms of i,

$$
V_i' = \{v \in V \mid \exists q \in Q : d(v,q) \le (1+\epsilon)^{i+2} \text{ and } d(q,v) \le (1+\epsilon)^{i+2}\}.
$$

For the shortest path from or to a source that contains at least one vertex in V_i' we can simply turn to Lemma 5 for the distance approximation. In the following, we will concentrate on those shortest paths that do not go through any vertex in V_i' and show that they can be exactly preserved in the computed spanner. Let $Z_i = V \setminus V_i$. We can focus on the subgraph induced by Z_i. For each source $u \in S$ in Z_i, we would like to further separate vertices in Z_i into different parts according to their distance from u. For $j \le i$, let $B^j(u) = \{x \in V \mid (1+\epsilon)^j \le d(u,x) < (1+\epsilon)^{j+1}\}$ and $\overline{B}^j(u) = \{x \in V \mid d(u,x) < (1+\epsilon)^{j+1}\}$. Also, let $Z_i^j(u) = Z_i \cap B^j(u)$ and $\overline{Z}_i^j(u) = Z_i \cap \overline{B}^j(u)$. To include the boundary case, let $Z_i^0(u) = Z_i \cap B^0(u) = Z_i \cap \{x \in V \mid d(u,x) = 0\}$. When we say $j \le i$, it means $j = \{0, 1, ..., i\}$.

We show that for all $j \le i$, if the size of $Z_i^j(u)$ is larger than $O(\sqrt{n})$, we are able to find a subset $Z_i'^j(u) \subset Z_i^j(u)$ of size $O(\sqrt{n})$ while guaranteeing that for any vertex v with roundtrip distance from u, $d(u \leftrightarrows v) \in [(1+\epsilon)^i, (1+\epsilon)^{i+1})$, every vertex of the shortest path from u to v that is in $Z_i^j(u)$ must also be in $Z_i'^j(u)$. Then we can run a modified Dijkstra algorithm over $Z_i'^j(u)$ for all $j \le i$ to get the shortest path from u to v where $u \in S$ and $d(u \leftrightarrows v) \in [(1+\epsilon)^i, (1+\epsilon)^{i+1})$. Similarly for the backward direction, the shortest path from v to $u \in S$ can be computed efficiently by a modified Dijkstra search over some reduced subsets. In the followings, we prove the forward case and the backward case is analogous.

We prove Lemma 7 for the computation of the reduced sets $Z_i'^j(u)$. We need Lemma 2.3 of [DW20], restated in Lemma 6, in the proof.

Lemma 6 (Lemma 2.3 in [DW20]). *Let $G = (V, E)$ be a digraph with integer edge weights in $\{1, ..., M\}$. Let $Z \subseteq V$ with $|Z| > c\log n$ (for $c \ge 100/\log(10/9)$) and let d be a positive integer. Let R be a random sample of $c\log n$ vertices of Z and define $Z' = \{z \in Z \mid d(z,r) \le d, \forall r \in R\}$. If for every $z \in Z$ there are at most $0.2|Z|$ vertices $v \in V$ so that $d(z,v), d(v,z) \le d$, then $|Z'| \le 0.8|Z|$.*

Lemma 7. *Fix $i \in [0, \log_{1+\epsilon}(Mn))$, $\beta > 1$, and $\alpha \in (0,1)$. Let Q be a random sample of $\tilde{O}(s^\alpha)$ vertices from S, $d = \beta(1+\epsilon)^{i+1}$, and $V_i' = \{v \in V \mid \exists q \in Q : d(v,q) \leq d$ and $d(q,v) \leq d\}$. Then for each $u \in S$ and $j \in [0, \log_{1+\epsilon}(Mn))$, we can compute a sample set $R_i^j(u)$ of size $O(\log^2 n)$ from $\overline{Z}_i^j(u) = Z_i \cap \overline{B}^j(u)$, where the number of vertices in $\overline{Z}_i^j(u)$ of distance at most d from all vertices in $R_i^j(u)$ is $O(n^{1-\alpha})$, in $\tilde{O}(mn^\alpha)$ time whp.*

Proof. We first assume that one can get a random sample $R_i^j(u)$ of size $c \log n$ from $\overline{Z}_i^j(u)$. Consider any vertex w with at least $0.2|\overline{Z}_i^j(u)|$ vertices $v \in V$ such that $d(w,v), d(v,w) \leq d$. If the size of $\overline{Z}_i^j(u)$ is no larger than $10n^{1-\alpha}$, that set is of small enough size to use without modification. If $|\overline{Z}_i^j(u)| \geq 10n^{1-\alpha} \geq 10s^{1-\alpha}$, we have $0.2\overline{Z}_i^j(u) \geq 2s^{1-\alpha}$. Then according to the assumption that S is randomly sampled from V, whp, Q contains some q such that $d(w,q), d(q,w) \leq d$ and thus $w \in V_i'$. Therefore, for every $w \notin V_i'$ (or $w \in Z_i$), there are at most $0.2|\overline{Z}_i^j(u)|$ vertices $v \in V$ such that $d(w,v), d(v,w) \leq d$. Using Lemma 6, we can perform sampling recursively until getting a subset of size $O(n^{1-\alpha})$ from $\overline{Z}_i^j(u)$ that contains all the vertices in $Z_i^j(u)$ with distance at most d to all sampled vertices.

Specifically, we start with $Z_{i,0}^j = \overline{Z}_i^j(u)$. For every $k \in [0, 2\log n]$, we get a random sample $R_{i,k}^j$ of size $O(\log n)$ from $Z_{i,k}^j$ and let $Z_{i,k+1}^j = \{z \in \overline{Z}_i^j(u) \mid d(z,r) \leq d, \forall r \in \cup_{l=0}^k R_{i,l}^j\}$. Then according to Lemma 6, we have $|Z_{i,k+1}^j| \leq 0.8|Z_{i,k}^j|$ and thus $|Z_{i,k+1}^j| \leq 0.8^k|\overline{Z}_i^j(u)|$. After the last iteration, $|Z_{i,2\log n}^j| \leq 10n^{1-\alpha}$. $R_i^j(u) = \cup_{l=0}^k R_{i,l}^j$ is of size $O(\log^2 n)$.

Finally, the method of randomly sampling $R_i^j(u)$ from the unknown $\overline{Z}_i^j(u)$ can be adapted from [DW20] (See Procedure *RandomSamples* in Algorithm 2). For each possible i, j, k, get a random sample $P_{i,j,k}$ of Z_i by sampling each vertex with probability $p = O(\log n * n^{\alpha-1})$. For each of the $\tilde{O}(n^\alpha)$ vertices in $P_{i,j,k}$, run Dijkstra search in the forward and backward directions to get the distances from and to every vertex in V. Then formulate the set $T_{i,k}^j = \{w \in P_{i,j,k} \mid w \in \overline{Z}_i^j(u)$ and $d(w,r) \leq d, \forall r \in \cup_{l<k} R_{i,l}^j\}$ in polylog time by making use of the distances. Since the sampling of $P_{i,j,k}$ is independent, $T_{i,k}^j$ is a random sample of $Z_{i,k}^j$ by picking each vertex with probability p. If $|T_{i,k}^j| > 10\log n$, pick a random sample of $10\log n$ vertices as $R_{i,k}^j$, which is also a random sample of $Z_{i,k}^j$; otherwise, use $T_{i,k}^j$ directly. The running time of this procedure is $\tilde{O}(mn^\alpha)$. $\qquad \square$

By applying Lemma 7 with $\alpha = 0.5$ and $\beta = 1+\epsilon$, we, in $\tilde{O}(m\sqrt{n})$ time, whp get the sets $R_i^j(u)$ of size $O(\log^2 n)$ from $\overline{Z}_i^j(u)$ where the number of vertices in $\overline{Z}_i^j(u)$ of distance at most $(1+\epsilon)^{i+2}$ from all vertices in R_i^j is $O(n^{0.5})$. Then we compute $Z_i'^j(u) = \{v \in \overline{Z}_i^j(u) \mid d(v,r) \leq (1+\epsilon)^{i+2}, \forall r \in R_i^j(u)\}$ and it has size $O(n^{0.5})$. The fact that for each $x \in Z_i^j(u)$ in a roundtrip shortest path $u \leftrightarrows v$ with $(1+\epsilon)^i \leq d(u \leftrightarrows v) \leq (1+\epsilon)^{i+1}$, $x \in \{w \in \overline{Z}_i^j(u) \mid d(w,y) \leq (1+\epsilon)^{i+2}, \forall y \in$

$\overline{Z}_i^j(u)\}$ has been revealed by [DW20]. To see this, $d(x,y) \leq d(x,u) + d(u,y) = d(u \leftrightarrows v) - d(u,x) + d(u,y) \leq (1+\epsilon)^{i+1} - (1+\epsilon)^j + (1+\epsilon)^{j+1} \leq (1+\epsilon)^{i+2}$. Therefore, all vertices $x \in Z_i^j(u)$ that are in a roundtrip shortest path $u \leftrightarrows v$ with $(1+\epsilon)^i \leq d(u \leftrightarrows v) \leq (1+\epsilon)^{i+1}$ must also be in $Z_i'^j(u)$.

Algorithm 2. $(5+\epsilon)$-Sourcewise-Roundtrip-Spanner

Input: $G(V,E), \epsilon \in (0,1)$, and sources $S \subseteq V$
Output: Roundtrip-spanner H
1: $H \leftarrow \emptyset$
2: Get a uniform random sample set Q of size $100\sqrt{s}\log s$ from the sources S
3: **for** $u \in Q$ **do**
4:   Perform Dijkstra searches from and to u in G and add the edges in the computed shortest path trees to H
5: **for** each $i \in [0, \log_{1+\epsilon}(Mn))$ **do**
6:   $R^1(.), ..., R^i(.), d(.) \leftarrow RandomSamples(G, i, \epsilon, S)$
7:   **for** each $u \in S$ **do**
8:     $H \leftarrow H \cup ModDijkstra(G, u, i, \epsilon, R^1(u), ..., R^i(u), d(.))$
9: **return** H;

RandomSamples

Input: G, i, ϵ, and sources $S \subseteq V$
Output: $R^j(u)$ for all $u \in S$, $j \leq i$, and $d(p,v), d(v,p)$ for all $p \in \cup_{j,k} P_{j,k}$ and $v \in V$
1: **for** each $j \in \{1,...,i\}$ **do**
2:   **for** each $k \in \{1,...,2\log n\}$ **do**
3:     Let $P_{j,k} \subseteq V$ be a uniform random sample of $100\sqrt{n}\log n$ vertices.
4:     **for** each $p \in P_{j,k}$ **do**
5:       Run Dijkstra's to and from p to get for all v, $d(p,v)$ and $d(v,p)$
6: **for** each $u \in S$ **do**
7:   **for** each $j \in \{0,...,i\}$ **do**
8:     $R^j(u) \leftarrow \emptyset$
9:     **for** each $k \in \{1,...,2\log n\}$ **do**
10:       $T_k^j(u) \leftarrow \{p \in P_{j,k} \mid d(u,p) < (1+\epsilon)^{j+1}$ and $\forall y \in R^j(u) : d(p,y) \leq (1+\epsilon)^{i+2}\}$
11:       **if** $|T_k^j(u)| < 10\log n$ **then**
12:         $R^j(u) \leftarrow R^j(u) \cup T_k^j(u)$
13:         Exit this loop (over k).
14:       **else**
15:         Let $R_k^j(u)$ be a uniform random sample of $10\log n$ vertices from $T_k^j(u)$.
16:         $R^j(u) \leftarrow R^j(u) \cup R_k^j(u)$
17: **return** $R^j(u)$ for all $j \leq i$, $u \in S$, and $d(p,v), d(v,p)$ for all $p \in \cup_{j,k} P_{j,k}$ and $v \in V$

Theorem 3. *Let $u \in S$ and i be fixed. Suppose that for every $j \in \{\emptyset\} \cup \{1,...,i\}$ we have access to sets $Z_i'^j(u) \subseteq \overline{Z}_i^j(u)$ as defined above. Using these sets we can define a modified Dijsktra's algorithm which will, in $\tilde{O}(m\log(M)/(\epsilon\sqrt{n}))$ time,*

ModDijsktra

Input: $G, u, i, \epsilon, R^1(u), ..., R^i(u), d(.)$
Output: H_u (set of edges)
1: $F \leftarrow$ empty Fibonacci heap
2: *Extracted* $\leftarrow$ empty hash table
3: $F.insert(u, 0)$
4: $H_u \leftarrow \emptyset$
5: **while** F *is nonempty* **do**
6:      $(x, d[x]) \leftarrow F.extractmin$
7:      $Extracted.instert(x)$
8:      **if** $for\ every\ r \in R^j(u), d(x, r) \leq (1 + \epsilon)^{i+2}$ **then**
9:          **for all** $y\ s.t\ (x, y) \in E$ **do**
10:             **if** $y \notin Extracted$ **then**
11:                 **if** $y\ is\ in\ F$ **then**
12:                     $F.DecreaseKey(y, d[x] + w(x, y))$
13:                 **else**
14:                     $F.insert(y, d[x] + w(x, y))$
15:         $H_u \leftarrow H_u \cup \{e(x, y)\}$

---

find the shortest paths from u to each $v \in V$, where those paths do not contain a vertex of V_i'. The $\tilde{O}(\sqrt{n}\log(M)/\epsilon)$ edges returned by this algorithm contain all these shortest paths.

Proof. We define our modified Djikstra's algorithm on vertex u as follows (see Procedure *ModDijkstra* in the Appendix). We begin by placing u in the Fibonacci heap with $d[u] = 0$, and all others with $d[-] = \infty$. When we extract a vertex x with estimate $d[x]$, we determine the j for which $(1 + \epsilon)^j \leq d[x] \leq (1 + \epsilon)^{j+1}$. If $d[x] = 0$ we use our boundary case $j = \emptyset$. Then we determine if $x \in Z_i'^j(u)$. If it is not, we extract a new vertex. Otherwise, we go through all its out-edges (x, y), and if $d[y] \geq d[x] + w(x, y)$, we update $d[y] = d[x] + w(x, y)$. It is not difficult to see that this search can find the shortest path from u to $v \in V$ with $(1 + \epsilon)^i \leq d(u \leftrightarrows v) \leq (1 + \epsilon)^{i+1}$ that does not contain a vertex of V_i'. It is because for any v such that $(1 + \epsilon)^i \leq d(u \leftrightarrows v) \leq (1 + \epsilon)^{i+1}$, and every $j \leq i$, every vertex on the shortest path u to v that is in $Z_i^j(u)$ is also in $Z_i'^j(u)$.

Checking whether $x \in Z_i'^j(u) = \{v \in \overline{Z}_i^j(u) \mid d(v, r) \leq (1 + \epsilon)^{i+2}, \forall r \in R_i^j(u)\}$ takes only polylog time: we know that x is always in $\overline{B}^j(u)$, as $d(u, x) \leq d[x] < (1 + \epsilon)^{j+1}$. Then we only need to check if $x \in Z_i$ (which is easy since the distances from Q are known) and whether $d(x, r) \leq (1 + \epsilon)^{i+2}, \forall r \in R_i^j(u)$, which takes $O(\log^2 n)$ time. Since we only go through the edges of at most $O(\sqrt{n}\log(Mn)/\epsilon)$ vertices (Lemma 7) and all vertex degrees are $O(m/n)$, the running time is $O(m\log(Mn)/(\epsilon\sqrt{n}))$. This algorithm produces at most $O(\sqrt{n}\log(Mn)/\epsilon)$ edges. □

Combining these theorems and lemmas yields the proof of Theorem 2.

Proof. (Theorem 2) Algorithm 2 summarizes the algorithms mentioned in the above. We first prove the stretch factor. For roundtrip shortest path between

$u \in S$ and $v \in V$ with $(1 + \epsilon)^i \leq d(u \leftrightarrows v) \leq (1 + \epsilon)^{i+1}$, if the one-way shortest path $u \to v$ (or shortest path $v \to u$) contains a vertex in V', then we apply Lemma 5 with $d = (1 + \epsilon)^{i+2}$ to get that $d_H(u, v) \leq d(u, v) + 2(1 + \epsilon)^{i+2}$ ($d_H(v, u) \leq d(v, u) + 2(1 + \epsilon)^{i+2}$, respectively). Otherwise (*i.e.*, the one-way shortest path does not contain a vertex of V_i'), Theorem 3 provides guarantee that $d_H(u, v) = d(u, v)$ (or $d_H(v, u) = d(v, u)$). In either case, we have

$$d_H(u \leftrightarrows v) \leq d(u \leftrightarrows v) + 4(1 + \epsilon)^{i+2}$$
$$\leq d(u \leftrightarrows v)(1 + 4(1 + 3\epsilon)) = d(u \leftrightarrows v)(5 + 12\epsilon).$$

As we run $ModDijkstra$ (Theorem 3) for each $u \in S$ and $i \in [0, \log_{1+\epsilon}(Mn))$, the final number of edges to be unioned with graph H is $\tilde{O}(s\sqrt{n}\log^2(M)/\epsilon^2)$ and the running time is $\tilde{O}(ms\,\log^2(M)/(\epsilon^2\sqrt{n}))$. Combining with Lemma 5, the total number of edges in the spanner H is $\tilde{O}(n\sqrt{s} + s\sqrt{n}\log^2(M)/\epsilon^2)$. Since $s = \Omega(\sqrt{n})$, we have $n\sqrt{s} \geq s\sqrt{n}$ and thus the size of H is $\tilde{O}(n\sqrt{s}\log^2(M)/\epsilon^2)$. By adding the $\tilde{O}(m\sqrt{n}\log(M)/\epsilon)$ time to compute R_i^j ($\log_{1+\epsilon}(Mn)$ calls of $RandomSamples$ in Lemma 7) and $\tilde{O}(m\sqrt{s})$ time of Dijkstra's from Q (Lemma 5), the overall running time is $\tilde{O}(m\sqrt{n}\,\log^2(M)/\epsilon^2)$. □

5   Conclusion

We have developed two algorithms for the creation of sourcewise roundtrip-spanners with specific stretch. Our first algorithm improves upon the thought to be near optimal 3-roundtrip-spanner algorithm from [RTZ08], fully generalizing to the sourcewise case. Our second algorithm acts as an alternative to the first, create a spanner with worse stretch $5 + \epsilon$ in exchange for a faster runtime in the case that $\sqrt{n} \leq s$. In the future, we will look at reducing the runtime of our algorithm for the 3-sourcewise-roundtrip-spanner to $O(n\sqrt{s})$, as has been done for most other sourcewise roundtrip-spanner solutions. We can also look into improving further our $5 + \epsilon$ result by removing the requirement for S to be randomized.

References

[ABS+20] Ahmed, R., et al.: Graph spanners: a tutorial review. Comput. Sci. Rev. **37**, 100253 (2020)

[ADD+93] Althofer, I., Das, G., Dobkin, D.P., Joseph, D., Soares, J.: On sparse spanners of weighted graphs. Discrete Comput. Geom. **9**, 81–100 (1993)

[BKMP10] Baswana, S., Kavitha, T., Mehlhorn, K., Pettie, S.: Additive spanners and (α, β)-spanners. ACM Trans. Algorithms **7**(1), 1–26 (2010)

[Bod17] Bodwin, G.: Linear size distance preservers. In: Proceedings of SODA Conference, pp. 600–615 (2017)

[CDG20] Cen, R., Duan, R., Gu, Y.: Roundtrip spanners with $(2k - 1)$ stretch. In: Proceedings of ICALP Conference, pp. 24:1–24:11 (2020)

[CE06] Coppersmith, D., Elkin, M.: Sparse source-wise and pair-wise preservers. SIAM J. Discret. Math. **20**(2), 463–501 (2006)

[Che13] Chechik, S.: New additive spanners. In: Proceedings of SIAM SODA Conference, pp. 498–512 (2013)

[CL21] Chechik, S., Lifshitz, G.: Optimal girth approximation for dense directed graphs. In: Proceedings of the 2021 ACM-SIAM Symposium on Discrete Algorithms (SODA), pp. 290–300. SIAM (2021)

[CLRS20] Chechik, S., Liu, Y.P., Rotem, O., Sidford, A.: Constant girth approximation for directed graphs in subquadratic time. In: Proceedings of STOC Conference, pp. 1010–1023 (2020)

[DW20] Dalirrooyfard, M., Williams, V.V.: Conditionally optimal approximation algorithms for the girth of a directed graph. In: Proceedings of ICALP Conference, pp. 35:1–35:20 (2020)

[EP04] Elkin, M., Peleg, D.: $(1 + \epsilon, \beta)$-spanner constructions for general graph. SIAM J. Comput. **33**(3), 608–631 (2004)

[Pet09] Pettie, S.: Low distortion spanners. ACM Trans. Algorithms **6**(1), 1–22 (2009)

[PRS+18] Pachocki, J., Roditty, L., Sidford, A., Tov, R., Williams, V.: Approximating cycles in directed graphs: fast algorithms for girth and roundtrip spanners. In: Proceedings of SODA Conference, pp. 1374–1392 (2018)

[PS89] Peleg, D., Schaffer, A.A.: Graph spanners. J. Graph Theory **13**(1), 99–116 (1989)

[RTZ05] Roditty, L., Thorup, M., Zwick, U.: Deterministic constructions of approximate distance oracles and spanners. In: Caires, L., Italiano, G.F., Monteiro, L., Palamidessi, C., Yung, M. (eds.) ICALP 2005. LNCS, vol. 3580, pp. 261–272. Springer, Heidelberg (2005). https://doi.org/10.1007/11523468_22

[RTZ08] Roditty, I., Thorup, M., Zwick, U.: Roundtrip spanners and roundtrip routing in directed graphs. ACM Trans. Algorithms **4**(3), 1–17 (2008)

[SS10] Shpungin, H., Segal, M.: Near-optimal multicriteria spanner constructions in wireless ad hoc networks. IEEE/ACM Trans. Netw. **18**(6), 1963–1976 (2010)

[TZ05] Thorup, M., Zwick, U.: Approximate distance oracles. J. ACM **52**(1), 1–24 (2005)

[Woo10] Woodruff, D.P.: Additive spanners in nearly quadratic time. In: Proceedings of ICALP Conference, pp. 463–474 (2010)

[ZHL21] Zhu, C.J., Han, S., Lam, K.-Y.: A fast algorithm for source-wise round-trip spanners. Theor. Comput. Sci. **876**, 34–44 (2021)

[ZL17] Zhu, C., Lam, K.: Source-wise round-trip spanners. Inf. Process. Lett. **124**, 42–45 (2017)

[ZL18] Zhu, C., Lam, K.: Deterministic improved round-trip spanners. Inf. Process. Lett. **127**, 57–60 (2018)

Randomized Data Partitioning with Efficient Search, Retrieval and Privacy-Preservation

M. Oğuzhan Külekci$^{(\boxtimes)}$ (iD)

Department of Computer Science, Indiana University Bloomington,
Bloomington, IN, USA
okulekci@iu.edu

Abstract. We introduce a new data representation that serves mainly for privacy preserving data storage with efficient search and retrieval capabilities over the distributed systems. The cornerstone of the proposed scheme is based on a novel algorithm that splits an input bit sequence $B[1..n]$ into two as left and right partitions with well–control over the partition sizes, and the reconstruction of B in absence of either partition is hard to achieve. The algorithm processes the input bit stream in blocks of d–bits, where initially each block is replaced with another d–bit according to a randomly chosen permutation of the set $\{0, 1, ..2^d - 1\}$. Following the replacement, the leftmost bits of each block up until and including the qth set bit are appended to the left and the remaining bits to the right partition. We prove that the expected length of the left partition is $\ell \approx 2qn/d$ bits and the right partition becomes of length $|R| = n - \ell$ bits. Therefore, there is no overhead on the new representation with respect to original input. We also show that due to the randomization step, the input data B is not required to follow any special probability distribution to have the mentioned partitioning ratio $\rho = 2q/d$ and it is possible to tune the parameters d and q to support any desired ratio ρ on the input. We consider recursive application of that splitting algorithm on each partitions, which can be viewed as generating a full binary tree with k–leaves such that at each internal node the data is subject to the proposed splitting operation. Such a construction represents an input bit sequence $B[1..n]$ with k partitions as $P_1, P_2, \ldots, P_k$, where it is hard to reconstruct the original data in absence of any P_i.

Keywords: randomization · data representation · data partitioning · distributed data storage · cloud storage · privacy-preserving data representation

1 Introduction

We introduce a randomized data representation that preserves the privacy of the data, while still supporting efficient access and retrieval. The building block of

W. Wu and G. Tong (Eds.): COCOON 2023, LNCS 14422, pp. 310–323, 2024.
https://doi.org/10.1007/978-3-031-49190-0_22

the proposed scheme is a novel randomized algorithm that splits an input n–bit long sequence into two pieces, which we refer as the *left* and *right* partitions. The algorithm processes the input in blocks of a predetermined length d bits. The initial randomization step replaces each block on the input with its counterpart according to a randomly generated permutation [8] of d-bit integers. Following the replacement, the leftmost bits of each block up until and including the qth set bit are appended to the left, and the remaining bits to the right partition.

We prove that the randomization step guarantees the *expected* size of the left partition to be $\ell = n \cdot \rho$ bits, while the right partition becomes of length $(n - \ell)$ bits according to the input parameter $0 < \rho < 1$ such that the q value can be chosen to satisfy $\rho = 2q/d$. Access to any block on the original input can be achieved efficiently with the intrinsic addressing in the construction. We show that it is also possible to search for a queried pattern by filtering on one partition and verifying on the other as well.

The privacy of the data in the proposed representation depends on combination of several factors, which can be summarized as follows. First, the random permutation used in the construction is assumed to be secret, which can be maintained by using a secret initial seed in the random number generator [8]. It is known that the permutations are vulnerable to statistical attacks [1,16], but due to the splitting mechanism it becomes hard to make a frequency analysis on the left and right partitions. On the right partition, each block is represented with variable-length bits, where the code-word boundaries cannot be determined as these codes are not prefix-free, that lacks the possibility of frequency analysis. Actually, the left partition encodes those code-word boundary information. On the left partition, the code-word boundaries can be determined, however, since multiple randomly chosen symbols are represented with the same code, again running a frequency analysis becomes hard as 2^d symbols on the input are mapped to d symbols on the left partition, which introduces again an ambiguity. Therefore, in absence of either partition and the secret permutation it is hard to reconstruct the original input, that provides the privacy in the proposed scheme. It is noteworthy that even the prefix-free codes have been found to be hard to decode without proper information [9,10,18], where recent works on combining compression and security [4,7] considers the hardness of decoding asymetric numeral codes [6], which produce variable-length non-prefix-free codes.

By applying the proposed split operation recursively to each partition, we observe that an input sequence can be split into more than two partitions. This can be achieved by generating a full binary tree with k–leaves, where the root node is the original input data. Each internal node has two children and the data on internal nodes are subject to the proposed splitting operation. Notice that in such a multiple partition it becomes more and more difficult to analyze and reproduce the original in absence of any partition as deeper the tree the ambiguity in the partitions increases.

There might be several applications of the proposed scheme, where a basic scenario is that a user wants to store a large file on the cloud or on a decentralized network such as the IPFS [3], where keeping data secret from the owners of the storage medium is a serious concern. Splitting data into multiple pieces and encrypting those partitions before saving to the cloud or the distributed network provides the ultimate secrecy. However, search and retrieval, which are vital operations on any data management scheme, becomes difficult to achieve on such encrypted data. The encryption algorithms, mainly the homomorphic [14] and multi-party computation [5] schemes, support those operations, but their computational load might be prohibiting, particularly on distributed networks or IoT environments with limited resources [11,19]. On such a case, the proposed scheme would serve as an alternative solution.

Such a scenario have been previously considered and data splitting have been proposed by Li *et al.* [13], where the authors proposed to use a key and create two bit streams from the input by using this key. It becomes hard to decode the original again in absence of the partitions and the key. Each bit stream is of equal length with the input, and hence, total storage requirement increases to 2n bits for n bit input. Yet another point is although random access might be supported, the search operation is not possible. Therefore, when compared to [13], the randomized algorithm in this study becomes advantageous in two dimensions by not creating an overhead storage and supporting efficient search and access.

The outline of the paper is as follows. In Sect. 2 we introduce the proposed randomized data splitting algorithm and provide the proofs of well-control over the sizes of the partitions. The support mechanisms for efficient random access and search operations are explained in Sect. 3, which id=s followed by the analysis of the privacy aspects in Sect. 4. We describe how to partition a data into multiple splits according to desired proportions in Sect. 5. We conclude by a summary and discussions of possible further research directions.

2   The Split Coding and Its Properties

Definition 1 (Split–Coding). *The split coding of an input d-bit sequence $A = a_1 a_2 .. a_d$ according to the parameter q, $d > q > 0$, is denoted by $S(A, q) \rightarrow \langle L_A, R_A \rangle$ and produces the left partition L_A and right partition R_A as follows*

i) If A has at least q 1 bits, then $L_A = a_1 a_2 .. a_i = (0^*1)^q$ and $R_A = a_{i+1} ... a_d = (0|1)^{d-i}$
ii) Otherwise, $L_A = a_1 a_2 .. a_d = A$ and $R_A = \emptyset$ (empty).

where (0^*1) denotes any number of 0 bits followed by a 1 bit. $(0|1)$ means either 0 or 1 bit. The superscripts over them represent how many times they are repeated.

The split–coding that partitions an input d–bit integer into two according to a parameter q is given in Definition 1, where some examples are $S(01001101, 1) \rightarrow \langle 01, 001101 \rangle$, $S(01001101, 2) \rightarrow \langle 01001, 101 \rangle$, $S(01001101, 4) \rightarrow \langle 01001101, \emptyset \rangle$. Let the n-bits long input bit sequence is shown by $B[1..n] = b_1 b_2 ... b_n$ and $d \in O(\log n)$ is a chosen fixed block-length, which we assume $d|n$, as B can be padded with random bits otherwise. The input B can be shown by $B = D_1 D_2 ... D_m$, where $D_i = B[d(i-1)+1..i \cdot d]$ is a block for $1 \leq i \leq m$, and $m = n/d$. The

i	1	2	3	4	5	6	7	8
D_i	100	101	001	000	010	110	111	011
$D_i' = \pi(D_i)$	011	111	010	101	000	100	110	001
L_i	01	1	01	1	000	1	1	001
R_i	1	11	0	01		00	10	

Fig. 1. The split-coding of B = 10010100100010110111011 with d = 3, q = 1, and
$\pi = \langle 5, 2, 0, 1, 3, 7, 4, 6 \rangle$ as $P(B, d, \pi, q) \rightarrow (L = 01101100011001, R = 1110010010)$. The
decoding process to merge L_B and R_B partitions to restore B can also be traced on the
example strings. For example, scanning the initial bits from L returns 2 bits sequence
01 as we hit a set bit. Therefore, we need to append $1 = 3 - 2$ bits from R, which creates
$D_1' = 011$. Applying the inverse permutation $D_1 = \pi^{-1}(3) = 4$ restores the original bits
100.

d–bits long blocks are the binary encoding of the integers from the set
$U = \{0, 1, .., 2^d - 1\}$.

Let $\pi[0..2^d - 1] = \{\pi_0, \pi_1, \ldots, \pi_{2^d-1}\}$ be a *randomly generated* permutation
of the set U. We apply the permutation π on the input $B = D_1 D_2 \ldots D_m$ that
produces $B' = D_1' D_2' \ldots D_m'$ such that $D_i' = \pi[D_i]$. The random permutation π can
be constructed by using a pseudo-random number generator that is initialized
with a selected *seed*. Following the permutation step, we apply split–coding to
each d–bits long block $D_{1 \leq k \leq m}'$ that produces the left and right partitions as
$S(D_k', q) \rightarrow \langle L_k, R_k \rangle$ according to the chosen parameter q, 0<q<d. For each block,
the left partition L_k is the sequence of leftmost bits up until and including the
qth set bit, and the right partition R_k is the remaining bits succeeding the qth set
bit. As stated in Definition 1, if qth set bit appears on the last position or
there are less than qth set bits on the block, then the right partition will be
empty, and the block is copied into left partition and nothing is written to the
right partition. We concatenate the left partitions L_1 to L_m in order and generate
the L–partition as $L_B = L_1 L_2 .. L_m$. Similarly, $R_B = R_1 R_2 .. R_m$ denotes the R–partition.
This complete process is represented by $\texttt{SplitEncode}(B, d, \pi, q) \rightarrow \langle L, R \rangle$.

Given the L– and R–partitions, the permutation π, and the parameters q and
d, the decoding process reconstructs B by simply scanning the L- and R–partitions
from left-to-right. The decoding procedure starts by initializing the read pointers
on L_B and R_B to their first positions, and reads bits from L_B–partition until q set
bits are observed or total number of bits examined reaches d with less than q
set bits. Assuming t bits are read on L_B, the remaining (d−t) bits are read
from the R_B–partition in order. The concatenation of these bits creates a d–bits
long block, say T, from which the original block is restored via $\pi^{-1}[T]$. This
procedure is repeatedly applied until all m blocks are restored. A sample split
encoding and decoding is depicted on Fig. 1 with the parameters q = 1, d = 3,
and $\pi = \langle 5, 2, 0, 1, 3, 7, 4, 6 \rangle$ on a bitmap B = 10010100100010110111011.

We now analyze the expected bit lengths of the partitions created by split–
coding process on a n bits long input B and prove that the expected length of

the left partition is $n \cdot 2q/d$ bits long, which leaves the right partition to be $\approx n \cdot (d - 2q)/d$ bits long.

Proposition 1 (Length Invariance). *The sum of the lengths of the* L– *and* R– *partitions created by the split-coding* SplitEncode$(B, d, \pi, q) \to \langle L_B, R_B \rangle$ *is equal to the length of the input bit string* B.

Proof. Each d-bits long block D_i on B is split into two bit strings L_i and R_i without any insertion or deletion of bits. Therefore, per each block, the lengths of its corresponding bits on L_B and R_B partitions sum up to d as $L_i + R_i = d$ for all $i = 1$ to m. Thus, $\sum_{1 \le i \le m}(L_i + R_i) = m \cdot d = |B|$ holds.   □

Lemma 1 (Expected bit lengths of partitions for $q = 1$). *For* $q = 1$ *and* $d \ge 1$, *the* *expected* *bit lengths of the* L *and* R *partitions of a randomly chosen integer from* $U = \{0, 1, ..., 2^d - 1\}$ *are* $E(|L|) = 2 - \frac{1}{2^{d-1}}$ *and* $E(|R|) = d - 2 + \frac{1}{2^{d-1}}$.

Proof. A randomly selected d-bit integer i can take any value in $\{0, 1, ..., 2^d - 1\}$ with probability $1/2^d$. For all values of $i \ge 2^{d-1}$, the L partition will be equal to 1, and thus, $|L| = 1$. Since there are 2^{d-1} such i values, the probability that the length of the L partition will be one is $2^{d-1}/2^d = 1/2$. Similarly, if $2^{d-2} \le i < 2^{d-1}$, then L will be 01 that will dictate $|L| = 2$ with probability $2^{d-2}/2^d = 1/4$, which can be generalized to $|L| = \ell$ with probability $1/2^\ell$ for $1 \le \ell \le (d-1)$. When $i = 0$ or $i = 1$, the L partitions will completely include i, and thus, the corresponding length $|L|$ will be d with probability $2/2^d$. Therefore, the expected length $E(|L|)$ by the split coding can be calculated as follows.

$$E(|L|) = \frac{1}{2} \cdot 1 + \frac{1}{4} \cdot 2 + \cdots + \frac{1}{2^{d-1}} \cdot (d-1) + \frac{2}{2^d} \cdot d \; = \; \frac{2}{2^d} \cdot d + \sum_{\ell=1}^{d-1} \ell \cdot \frac{1}{2^\ell} \quad (1)$$

The algebraic identity $\sum_{i=1}^{n} i \cdot r^i = \frac{r - r^{n+1}(1 + n - nr)}{(1-r)^2}$ with $r = 1/2$ and $n = d - 1$, reduces Eq. 1 to

$$E(|L|) = \frac{2d}{2^d} + 4 \cdot \left(\frac{1}{2} - \frac{d+1}{2^{d+1}}\right) \; = \; \frac{2d}{2^d} + 2 - \frac{2d+2}{2^d} \; = \; 2 - \frac{1}{2^{d-1}} \quad (2)$$

For any integer i the sum of the lengths of L and R sum up to d as $|L| + |R| = d$ by definition and $E(|R|) = E(d - |L|) = d - E(|L|) = d - 2 + 1/2^{d-1}$ holds.   □

Lemma 2 (Expected bit lengths of the partitions on a d-bit block). *For any* $0 < q < d/2$, *the* *expected* *bit lengths of the* L– *and* R– *partitions of a randomly chosen d-bit integer from* $U = \{0, 1, ..., 2^d - 1\}$ *are*

$$q \le E(|L|) \le 2q - \frac{q}{2^{d-1}} \quad and \quad (d-q) > E(|R|) > d - 2q + \frac{q}{2^{d-1}}$$

Proof. Let $p_1, p_2, ..p_q$ denote the first q set bit positions on the randomly chosen integer $i \in \{0, 1, .., 2^d - 1\}$, and $x_1, x_2, ..x_q$ represent the differences in between the positions such that $x_1 = p_1$, and $x_j = p_j - p_{j-1}$. For instance, if the positions of the first four set bits are, say $2, 5, 6, 8$, then the $\langle x_1, x_2, x_3, x_4 \rangle = \langle 2, 3, 1, 2 \rangle$. The length of the L partition is then $|L| = x_1 + x_2 + .. + x_q$.

We observe that x_j is the length of the L–partition of a $d_j = d - \sum_{r=1}^{j-1} x_r$ bits long sequence. Due to Lemma 1, $E(x_j) = 2 - 1/2^{d_j-1}$. Since the lengths of the sequences decrease at each step as $d_1 > d_2 > .. > d_q$, the expected values decline accordingly $E(x_1) > E(x_2) > .. > E(x_q)$, e.g., for d=8, $E(|L|) = 2 - 1/2^7 = 1.992$, where for d=6, $E(|L|) = 2 - 1/2^7 = 1.984$. That defines $E(|L|)$ to be upper bounded by $q \cdot E(x_1)$ due to $E(x_1) > E(x_2) > .. > E(x_q)$ as shown in Eq. 5. Therefore, the expected value $E(|L|)$ can be represented as in Eq. 3 due to the linearity of expectations.

$$E(|L|) = E(x_1 + x_2 + .. + x_q) = \sum_{j=1}^{q} E(x_j) \tag{3}$$

$$= 2q - \left(\frac{1}{2^{d_1-1}} + \frac{1}{2^{d_2-1}} + .. + \frac{1}{2^{d_q-1}} \right) \tag{4}$$

$$\leq q \cdot E(x_1) = 2q - \frac{q}{2^{d-1}} \tag{5}$$

By Definition 1, the length of the L–partition is at least one bit, and thus, $x_i \geq 1$ holds for each i, which provides the lower bound for $E(|L|) \geq q$. Again by using $|L| + |R| = d$, $E(|R|) = d - E(|R|)$, the upper– and lower–bounds for $E(|R|)$ is obtained.   □

Theorem 1. *The split-coding of a given* n *bits long* B *with a randomly chosen permutation* π, *block length* d, *and parameter* $0 < q \leq d/2$ *generates left partition* L_B *with an expected length of* $E(|L_B|) \leq n \cdot (\frac{2q}{d} - \frac{q}{d \cdot 2^{d-1}})$. *Accordingly, the right partition is expected to be* $E(|R_B|) > n \cdot (1 - \frac{2q}{d} + \frac{q}{d \cdot 2^{d-1}})$ *bits long.*

Proof. Split-coding operates on the n-bits long input sequence by n/d blocks, each of which is a d-bit integer. The blocks are replaced by their corresponding integers in the set $\{0, 1, 2, \ldots, 2^d - 1\}$ with the randomly generated permutation π. Let $X_1, X_2, \ldots X_{2^d}$ denote the lengths of the L–partitions of randomly assigned integers to each distinct d-bit integers and $f_1, f_2, \ldots, f_{2^d}$ represent the frequencies of each integer. The expected length $E(|L(B)|)$ for the L–partition of input B can be expressed as

$$E(|L_B|) = E\left(\sum_{i=1}^{2^d} f_i \cdot X_i\right) = \sum_{i=1}^{2^d} f_i \cdot E(X_i) \qquad \text{by linearity of expectations}$$

$$\leq \sum_{i=1}^{2^d} f_i \cdot \left(2q - \frac{q}{2^{d-1}}\right) \qquad \text{by substituting } E(X_i) \text{ due to Lemma 2}$$

$$\leq \left(2q - \frac{q}{2^{d-1}}\right) \cdot \frac{n}{d} \qquad \text{since } \sum_{i=1}^{2^d} f_i = \frac{n}{d} \text{ by definition}$$

$$\leq n \cdot \left(\frac{2q}{d} - \frac{q}{d \cdot 2^{d-1}}\right)$$

Since $|L_B| + |R_B| = n$, $E(|R_B|) = n - E(|L_B|) > n \cdot \left(1 - \frac{2q}{d} + \frac{q}{d \cdot 2^{d-1}}\right)$. □

Corollary 1. *The lengths of the* L_B *and* R_B *partitions of an input* n*-bits long* B *approach respectively to* $2nq/d$ *and* $n(d - 2q)/d$ *bits as the term* $q/d \cdot 2^{d-1}$ *in Theorem 1 becomes insignificant especially on larger values of* d, *e.g., assuming* $d = 8$ *and* $q = 1$, $q/d \cdot 2^{d-1} = 1/(8 \cdot 2^7) = 1/2^{10} = 0.00097$, *which makes* $E(|L(B)|) = 0.24903n$ *and* $E(|R(B)|) = 0.75097n$ *bits that can be assumed to be* $0.25n$ *and* $0.75n$.

3 The Random Access and Search Operations

In this section we investigate the random access and search mechanisms of the split-coding scheme.

Lemma 3 (Random access support). *Let* $Q = \binom{d}{0} + \binom{d}{1} + .. + \binom{d}{q-1}$ *denote the number of distinct* d-bit *integers that has less than* q *set bits. Given the* L_B *and* R_B *partitions of an input sequence* B, *if the number of* **unique** d-bit *blocks on* B *is less than* $2^d - Q$, *then the random access to any block* D_i *of* $B = D_1 D_2 \ldots D_m$ *can be achieved in* $O(\log |L_B|)$*–time by using an extra space of* $o(|L_B|)$ *bits. Otherwise, it takes* $O(q \log(n/d))$*–time at the expense of* $O((n/d) \cdot (Q/2^d) \cdot \log n)$ *additional bits space.*

Proof. Once the L_B and R_B partitions are given, the retrieval of any d–bit block D_i' on $B' = D_1' D_2'..D_m'$ can be done by using the observation that $D_1' D_2'..D_{i-1}'$ is $w = d \cdot (i - 1)$ bits long, where ℓ of those bits have been deposited on L_B and the rest $(w - \ell)$ on R_B. Therefore, given ℓ, we know the starting bit position of the left and right partitions of queried D_i' on L_B and R_B, respectively.

We start reading the bits on L_B from the first position and keep scanning until we read either d bits or encounter q set bits, whichever happens first. Once, one of these cases occurs, then the bits are actually the left partition of the first block of the input. Re-initiating the same procedure starting from the next bit position will retrieve the left partition of the second block, and so on. After repeating this operation $(i - 1)$ times, we reach the first bit of the left partition L_i of the queried D_i'. If we have scanned ℓ bits until we reach the beginning of L_i, then it is immediate that the R_i starts just after $(w - \ell)$ bits on R_B. Therefore, once we retrieve L_i from L_B, we read R_i by depositing the remaining $(d - |L_i|)$ bits from the starting bit position $(w-\ell+1)$ of the R_B. Combining the extracted partitions will reconstruct $D'_i = L_i R_i$ and deploying the inverse permutation would return the original block $D_i = \pi^{-1}(D'_i)$.

The time complexity of that random access procedure depends on finding ℓ, since once ℓ is known, rest is simple constant time operation. Let's first assume each block in $D'_1 D'_2 .. D'_{i-1}$ includes at least q set bits. Then, the left partitions of these blocks have exactly q set bits and each end with a set bit on L_B, e.g., the initial bits of L_B up until and including the qth set bit are the left partition of D'_1. Therefore, $\ell+1$ is the position of the first bit after the $q \cdot (i - 1)$th set bit on L_B. Finding the position of the kth set bit on a static bitmap of n bits can be achieved in constant or logarithmic time by constructing a dictionary structure (the *rank/select*(R/S) dictionary) that occupies $o(n)$ bits space [15,17,20]. Therefore, the position of the $q \cdot (i - 1)$th set bit on L_B can be returned by creating that R/S data structure in expense of $o(|L_B|)$ additional space. In this case, detecting the ℓ value can be achieved in $O(\log |L_B|)$ in the worst case, and thus, the random access. On the other hand, possible occurrences of the blocks with less than q set bits, which are directly copied to the L_B, corrupts the above procedure. That requires maintaining the indices of such blocks separately. Assume S_z is the sorted list of block ids that have z set bits for all $z < q$. On list S_z, let the number of items less than the queried i is represented by c_z. Therefore, the total number of set bits of $D'_1 D'_2 .. D'_{i-1}$ on L_B is $\ell = q \cdot (i - \sum_{z=0}^{q-1} c_z) + \sum_{z=0}^{q-1}(z \cdot c_z)$. As a last step to find the initial position of D'_i, it is necessary to check whether the immediately preceding blocks are in S_0, and the correct position is computed accordingly. Algorithm 1 lists the complete pseudo-code to retrieve D_i.

The space overhead for the random access support by Algorithm 1 is the sum of the space occupied by the S_z lists and the additional space used by rank/select (R/S) data structure for efficient execution of line 9. The number of unique d–bit sequences that have less than q set bit is $Q = \binom{d}{0} + \binom{d}{1} + .. + \binom{d}{q-1}$. Assuming all distinct 2^d blocks are equally likely among observed (n/d) blocks of the input, the size of the S_z list is expected to be $\approx (n/d) \cdot (Q/2^d)$. Therefore, the space required to store the S_z lists becomes around $\approx (n/d) \cdot (Q/2^d) \cdot \log n$ bits, where the space overhead of the (R/S) dictionaries can be kept around $o(|L_b| \approx 2qn/d)$ with the state-of-the-art implementations. The time complexity of Algorithm 1 depends on detecting the blocks with less than q set bits via the `for` loop in between lines 3 and 7 and also the `select` query run on line 9. The `for` loop takes roughly $O(q \log(n/d))$–time where the select query can be executed in $O(1)$ or $O(\log(2qn/d))$ time according to the R/S data structure used. Thus, the total time complexity is upper bounded by $O(q \log(n/d))$. □

It is important to note that if the number of distinct d–bit symbols on the input is less than $2^d - Q$, then the random permutation over integers $\{Q, Q + 1, .., 2^d - 1\}$ can be used for the observed blocks, which guarantees to have at least q set bits per block on L_B, and thus, removes the necessity of maintaining S_z lists. In such a case, the time and space complexity of the random access is determined

---

Algorithm 1: RandomAccess($i, L_B, R_B, d, \pi, q, S_0, S_1, .., S_{q-1}$)

---

input : i is the queried block index, L_B and R_B are the partitions, d, q, π are the parameters of the coding, S_z is the list of block indices that has $z < q$ set bits.

output: The original d–bit block D_i.

1 $C \leftarrow 0$
2 $r \leftarrow 1$
3 **for** $z = 0$ **to** $(q-1)$ **do**
4 $\quad$ $t \leftarrow$ number of indices less than i in S_z
5 $\quad$ $C \leftarrow C + z \cdot t$
6 $\quad$ $r \leftarrow r + t$
7 **end**
8 $\ell \leftarrow q \cdot (i - r) + C$
9 $p \leftarrow 1 +$ The position of the ℓth set bit on L_B
10 $r \leftarrow i - 1$
11 **while** $(r \in S_0)$ **do** $p \leftarrow p + d, r \leftarrow r - 1$
12 detected $\leftarrow 0$, step $\leftarrow 0$
13 **while** $((\text{detected} < q) \wedge (\text{step} < d))$ **do**
14 $\quad$ **if** $L_B[p + \text{step}] == 1$ **then** detected++
15 $\quad$ step++
16 **end**
17 $L \leftarrow L_B[p..p + \text{step} - 1]$
18 $w \leftarrow d \cdot (i - 1) - p + 1$
19 $R \leftarrow R_B[w..w + (d - \text{step}) - 1]$
20 $D_i' \leftarrow L \frown R$
21 **return** $\pi^{-1}(D_i')$

---

solely by the integrated R/S dictionary data structure, where the state of the art solutions provide $O(\log |L_B|)$–time with $o(|L_B|)$ extra space.

Lemma 4 (Search on split encoded sequences). *When the number of distinct d bit blocks on input is less than $2^d - Q$, it is possible to efficiently detect all occurrences of a queried pattern P on a split encoded sequence B by maintaining a rank-support on left partition L_B.*

Proof. Let $P = p_1 p_2 \ldots p_m$ be a binary sequence that we would like to find its all occurrences on the split encoded B. We apply split encoding to P with the same parameters as $\texttt{SplitEncode}(P, d, \pi, q) \rightarrow (L_P, R_P)$. The basic idea of the search is to find the matching positions of L_P on L_B, and then verifying each match by comparing R_P with the corresponding position of the candidate on R_B. The first step is to search L_P on L_B. Notice that not all matching positions are valid left partitions as a match can appear bridging two consecutive blocks. Hence, while scanning for L_P, we should accept only completely matching left partitions on L_B. Since we assume the number of distinct d bit blocks on input is less than $2^d - Q$, the permutation π can be constructed over integers $\{Q, Q + 1, .., 2^d - 1\}$, which guarantees that each block has exactly q set bits on L_B. Therefore, the validity of a matching position can be confirmed by checking whether *i)* the bit preceding the match position on L_B is a set bit, and *ii)* total number of previous set bits is a multiple of q. For each such valid matching left partition on L_B, we can find the corresponding positions on R_B to extract its right partition via Algorithm 1. We then extract its corresponding right partition from R_B and verify whether the constructed pattern matches the query. The performance of the search process greatly depends the efficiency of the filtering phase, where short P sequences would result in larger set of verification points, and longer ones less.   □

4 The Privacy Aspects

Each d–bit block of input B is represented on R_B with variable number of bits whose lengths are ranging from 0 to $d - 1$. We observe that these variable-length blocks are not prefix-free, and thus, the code–word boundaries on R_B are uncertain, which can only be determined by using the L_B partition, assuming the parameters d and q are known. For instance, in Fig. 1, the code–word boundaries on $R_B = 1110010010$ can be detected by using the $L_B = 01101100011001$. One can simply use the random access function Algorithm 1 to detect the bits of a queried ith block on R_B. It is important to address that even after the construction the d–bit block, the permutation function π should be known to reach the original d–bit on B by finding the inverse permutation mapping of the block. Therefore, in absence of the L_B partition and the permutation π, it seems hard to reconstruct the original data B even we assume all other parameters d, q, n are known. We investigate this hardness by analyzing the information released by each partition about the other partition by counting the number of possible bit sequences of length n that produce the same left or right partition. Ideally, given the left partition, all bit sequences of length $n - |L_B|$ should be a valid right partition and decode a distinct B sequence. Therefore, the correct right partition is indistinguishable in the search space of size $2^{n-|L_B|}$. Lemma 5 proves that this is indeed the case.

Lemma 5 (Number of distinct sequences with the same L_B). *Given the left partition L_B of an input n bits long sequence B, there are exactly $2^{n-|L_B|}$ different bit sequences of length n that generates the same L_B.*

Proof. Let's assume we have chosen a random binary string B of length n and generated the left partition L_B and right-partition R_B. Since there is no restriction on the possible bits of the right partition, any binary string X of length $n - |L_B|$ will be valid and produce a different n–bit sequence by the decoding process. In other words, the L partition of all these $2^{n-|L_B|}$ sequences will be equal, where the correct right partition is only one of them.   □

However, we observe interestingly that the same does not hold for the right partition by Lemma 6, which means the knowledge of right partition narrows the possibilities of left partition. Although this exhibits an information leakage, it still provides a reasonable privacy with large enough search space.

Lemma 6 (Number of distinct sequences with the same R_B). *Given the right partition R_B of an input n bits long sequence B, there are less than $2^{n-|R_B|}$ different bit sequences of length n that generates the same R_B. It is also expected that the number of bit sequences sharing the same R_B is also more than $2^{2qn/d}/(2qn/d + 1)$.*

Proof. There are n/d blocks and the length of the L_B partition is $(n - |R_B|)$ bits. Each block has at least zero and at most q set bits on L_B, which means the left partition should include certain number of set bits. Therefore, the binary strings of length $(n - |R_B|)$ that include less number of set bits then dictated by the original input are not valid and can not be decoded properly, which means the number of possible n bit input strings generating the known R_B is less than $2^{n-|R_B|}$.

If we assume all blocks are represented by q set bits on L_B (in other words, the number of observed distinct blocks on the input is less than $2^d - Q$ and we used only blocks with at least q set bits in the permutation step), then the number of $|L_B|$ bits long binary strings with qn/d set bits are all valid left partitions for the given right partition and each of those strings spell a different B. According to Theorem 1, the length of L_B is expected to be $\approx 2qm$ bits with $m = n/d$. The number of sequences of length $2qm$ bits with qm set bits is greater than $2^{2qn/d}/(2qn/d + 1)$ due to the simple bounds of the central binomial coefficient $4^x = (1 + 1)^{2x} = \sum_{k=0}^{2x} \binom{2x}{k}$. □

Due to Lemma 5, given the left partition L_B of a split encoded bit sequence, there is no information released about the right partition R_B, since any bit sequence of length $n - |L_B|$ is a valid R_B partition and correctly decodes with the given L_B. On the other hand, due to Lemma 6, once the right partition R_B is given, not all bit sequences of length $n - |R_B|$ are valid left partitions, and thus, the knowledge of R_B narrows down the search space of L_B. However, this reduction is limited and on large input sequences, still it can provide some privacy. It is noteworthy that the above analyses assume the permutation π is known, where actually keeping that π secret improves security significantly.

5 Splitting Beyond Two Partitions

We observe that recursive application of the splitting mechanism on each partition will generate further splits. As long as the depth of this partitioning increases the partitions gets smaller and it gets harder to reconstruct the data in absence of the one piece. Such a cascaded splitting operation might make sense as in distributed storage systems or recent block-chain style storage like IPFS [3]. We generalize our observation and provide a mechanism such that an input bit stream is divided into k partitions, where any ratio on the size of the partitions can be supported, e.g., creating 5 partitions that will occupy approximately $4/16, 5/16, 1/16, 3/16, 3/16$ of the input size, respectively.

Lemma 7 (Splitting into multiple pieces). *Let $\{r_1, r_2, \dots, r_k\}$ represent the given ratios such that $\sum_{i=1}^{k} r_i = 1$ and $k > 2$. An input bit string of length n*

can be split into k *pieces* $p_1, p_2, .., p_k$ *by recursive application of the split coding,*
where $|p_i| \approx r_i \cdot n$.

Proof. Assume a *full* binary tree of $s = 2k - 1$ elements, which is shown by
an array $A[1..s]$. In this tree, the leaf nodes represent the final partitions we
aim to generate. We place the ratios to the last k elements of the array, i.e.,
$A[s - i] = r_{k-i}$ for $i = 0..(k - 1)$. We compute the values of the vacant positions
$A[t]$ in the array starting from $t = s - k$ down to the root $t = 1$ by summing
up their corresponding children, where the indices of the children of $A[t]$ are
$A[2t]$ and $A[2t + 1]$. Notice that the already filled positions are the leaf nodes
and thus, the internal nodes can be computed from them. As an example, for
$\langle r_1, r_2, r_3, r_4, r_5 \rangle = \langle 4/16, 5/16, 1/16, 3/16, 3/16 \rangle$, the array will end up with
$A[1..9] = [1, 10/16, 6/16, 6/16,\ 4/16,\ 5/16,\ 1/16,\ 3/16,\ 3/16]$.

Considering that the root node $A[1]$ corresponds to the input bit string, we
apply the split coding at each internal node with the most appropriate q and d
parameters to support the ratios mentioned in its children. For instance, in the
example case, since $A[2] = 10/16$ and $A[3] = 6/16$, choosing $d = 16$ and $q = 5$
provides us the left partition to be close to $n \cdot 10/16$ bits as desired and right
partition automatically scales to the remaining $n \cdot 6/16$. We keep applying the
split coding to the internal nodes until we produce all the leaves. In our sample
case, for instance, we need to split the $A[2]$ according to its childrens's ratios as
$A[4] = 6/16$ and $A[5] = 4/16$, which dictates the left partition $A[4]$ is desired to
be around $6/10$ of its parent $A[2] = 10/16$. This can be achived by setting $d = 10$
and $q = 3$. The bits deposited at the leaf nodes finalize the splitting construction,
where reconstruction of the original input requires traversing the tree from the
leaves to the root and applying split decode procedure for each internal node. $\square$

6   Conclusions and Further Studies

We have presented a randomized data splitting algorithm that provides well–
control on the sizes of the partitions with efficient search and access mechanisms.
Reconstruction of the original input is hard without having access to both partitions and the permutation used in the construction.

Such a privacy-preserving data splitting scheme may find application areas in
secure and searchable massive data storage on the cloud. For example, the user
may keep the permutation secret, and then store the left and right partitions on
different remote servers. Due to the privacy aspects, the admins of those remote
locations will not be able to extract the content, while the user will still be able
to achieve search and retrieval with the mechanisms described in this study. It
is also possible to maintain the left partition on-premise and save right partition
encrypted on a cloud to make it completely secure. The user can still perform
search operations on the encrypted cloud storage by filtering the candidates from
the data on-premise, and then verify the candidates by fetching and decrypting
them from the cloud. Obviously, such applications require further analysis of
some different possible architectures. Similar applications might be considered

for other distributed storage schemes including block-chain storage systems [12]. The recursive application of the proposed split coding may be used to privacy-preserving partitioning of the data into some desired proportions. This can also serve as an alternative in secret sharing schemes [2].

References

1. Bard, G.V., Ault, S.V., Courtois, N.T.: Statistics of random permutations and the cryptanalysis of periodic block ciphers. Cryptologia **36**(3), 240–262 (2012)
2. Beimel, A.: Secret-sharing schemes: a survey. In: Chee, Y.M., et al. (eds.) IWCC 2011. LNCS, vol. 6639, pp. 11–46. Springer, Heidelberg (2011). https://doi.org/10.1007/978-3-642-20901-7_2
3. Benet, J.: IPFS-content addressed, versioned, P2P file system. arXiv preprint arXiv:1407.3561 (2014)
4. Camtepe, S., et al.: Compcrypt-lightweight ANS-based compression and encryption. IEEE Trans. Inf. Forensics Secur. **16**, 3859–3873 (2021)
5. Du, W., Atallah, M.J.: Secure multi-party computation problems and their applications: a review and open problems. In: Proceedings of the 2001 Workshop on New Security Paradigms, pp. 13–22 (2001)
6. Duda, J.: Asymmetric numeral systems. arXiv preprint arXiv:0902.0271 (2009)
7. Duda, J., Niemiec, M.: Lightweight compression with encryption based on asymmetric numeral systems. arXiv preprint arXiv:1612.04662 (2016)
8. Durstenfeld, R.: Algorithm 235: random permutation. Commun. ACM **7**(7), 420 (1964). https://doi.org/10.1145/364520.364540
9. Fraenkel, A.S., Klein, S.T.: Complexity aspects of guessing prefix codes. Algorithmica **12**, 409–419 (1994)
10. Gillman, D.W., Mohtashemi, M., Rivest, R.L.: On breaking a huffman code. IEEE Trans. Inf. Theory **42**(3), 972–976 (1996)
11. Kaaniche, N., Laurent, M.: Data security and privacy preservation in cloud storage environments based on cryptographic mechanisms. Comput. Commun. **111**, 120–141 (2017)
12. Li, R., Song, T., Mei, B., Li, H., Cheng, X., Sun, L.: Blockchain for large-scale internet of things data storage and protection. IEEE Trans. Serv. Comput. **12**(5), 762–771 (2018)
13. Li, Y., Gai, K., Qiu, L., Qiu, M., Zhao, H.: Intelligent cryptography approach for secure distributed big data storage in cloud computing. Inf. Sci. **387**, 103–115 (2017). https://doi.org/10.1016/j.ins.2016.09.005. https://www.sciencedirect.com/science/article/pii/S0020025516307319
14. Martins, P., Sousa, L., Mariano, A.: A survey on fully homomorphic encryption: an engineering perspective. ACM Comput. Surv. (CSUR) **50**(6), 1–33 (2017)
15. Okanohara, D., Sadakane, K.: Practical entropy-compressed rank/select dictionary. In: 2007 Proceedings of the Ninth Workshop on Algorithm Engineering and Experiments (ALENEX), pp. 60–70. SIAM (2007)
16. Plackett, R.L.: The analysis of permutations. J. R. Stat. Soc.: Ser. C: Appl. Stat. **24**(2), 193–202 (1975)
17. Raman, R., Raman, V., Satti, S.R.: Succinct indexable dictionaries with applications to encoding k-ary trees, prefix sums and multisets. ACM Trans. Algorithms (TALG) **3**(4), 43-es (2007)

18. Rubin, F.: Cryptographic aspects of data compression codes. Cryptologia **3**(4), 202–205 (1979)

19. Sharma, P., Jindal, R., Borah, M.D.: Blockchain technology for cloud storage: a systematic literature review. ACM Comput. Surv. (CSUR) **53**(4), 1–32 (2020)

20. Vigna, S.: Broadword implementation of rank/select queries. In: McGeoch, C.C. (ed.) WEA 2008. LNCS, vol. 5038, pp. 154–168. Springer, Heidelberg (2008). https://doi.org/10.1007/978-3-540-68552-4_12

The k Edge-Vertex Domination Problem

Peng Li$^{(\boxtimes)}$, Xingli Zhou, and Zhiang Zhou

Chongqing University of Technology,
69 Hongguang Road, Chongqing 400054, China
lipengcqut@cqut.edu.cn

Abstract. Let $G = (V(G), E(G))$ be a simple n-vertex graph with m edges. Take any $e = uv \in E(G)$. We say e dominates a vertex $w \in V(G)$ provided w belongs to the closed neighborhood of u or v. Let $S \subseteq V(G)$, $D \subseteq E(G)$. Take a positive integer k. If w is edge-dominated by k edges of D, then D is called a k edge-vertex dominating set of G with respect to S. In this paper, we study the k edge-vertex domination problem and present $O(m \lg m + k|S| + n)$-time algorithms to find a minimum k edge-vertex dominating set of G with respect to any $S \subseteq V(G)$ on interval graphs. In addition, we design $O(n|S|)$-time algorithms to find a minimum k edge-vertex dominating set of T with respect to any $S \subseteq V(T)$ on trees.

Keywords: domination · edge-vertex domination · k edge-vertex domination · interval graphs · trees

1 Introduction

Domination is one of the most popular research directions over the last few years, which has attracted the interest of many mathematicians. It is widely used in many fields, such as RNA sequence [6], electric power networks [3], chemical materials which are used in drug chemistry [2], distribution centers in logistics [1] and computer science for investigation of the complexity problem [7,15]. The domination and its variations have attracted considerable attention and have been widely studied, see [4,5].

All the graphs are finite, simple and undirected in this paper. Let $G = (V(G), E(G))$ be any graph. For any $v \in V(G)$, the *open neighborhood* $N_G(v)$ is the set $\{u \in V(G) : uv \in E(G)\}$ and the *closed neighborhood* of v is the set $N_G[v] = N_G(v) \cup \{v\}$. Take any subset D of $V(G)$. If every vertex of G is either in D or adjacent to a vertex in D, then we say D is a *dominating set* of G. The *domination number* $\gamma(G)$ is the minimum cardinality of a dominating set in G.

Take a map $\mathcal{I}$ that assigns to every vertex $v \in V(G)$ a nonempty closed interval $\mathcal{I}(v) = [\ell_{\mathcal{I}}(v), r_{\mathcal{I}}(v)]$. If $vu \in E(G)$ if and only if $v \neq u$ and $\mathcal{I}(v) \cap \mathcal{I}(u) \neq \emptyset$ for all $v, u \in V(G)$, then we say $\mathcal{I}$ is an *interval representation* of G. We may assume that all the endpoints in $\{\mathcal{I}(x) : x \in V(G)\}$ are distinct in this article. If all intervals have the same length, then we refer to $\mathcal{I}$ as a *unit interval*

W. Wu and G. Tong (Eds.): COCOON 2023, LNCS 14422, pp. 324–334, 2024.
https://doi.org/10.1007/978-3-031-49190-0_23

representation. A graph is a *(unit) interval graph* if and only if it has a (unit) interval representation.

Take a positive integer k. For any $vw \in E(G)$ and $u \in V(G)$, we say vw is *vertex-dominated* by u, or u is *edge-dominated* by vw, provide $u \in N_G[v] \cup N_G[w]$ [10]. A set $T \subseteq V(G)$ is a k *vertex-edge* dominating set if each edge of $E(G)$ is vertex-dominated by k vertices of T. The k vertex-edge dominating number $\gamma_{kve}(G)$ of G is the minimum cardinality of a k vertex-edge dominating set of G. Take any $S \subseteq V(G)$. A set $D \subseteq E(G)$ is k *edge-vertex* dominating set of G w.r.t. S if each vertex of S is edge-dominated by k edges of D. The k edge-vertex dominating number $\gamma_{kev}(G)$ of G is the minimum cardinality of k edge-vertex dominating set of G w.r.t. $V(G)$. The k vertex-edge domination problem is to search a minimum k vertex-edge dominating set of G. Similarly, the k edge-vertex domination problem is to find a minimum k edge-vertex dominating set of G.

In 1986, Peters [10] introduced the vertex-edge domination problem. In 2007, Lewis [8] showed that the vertex-edge domination problem is NP-hard for bipartite, chordal, planar and circle graphs, and the independent vertex-edge domination problem is NP-hard even on bipartite and chordal graphs. Recently, Li and Wang [9] designed linear time algorithms for the vertex-edge domination and double vertex-edge domination problems on interval graphs, and $O(nm)$ time algorithms for k vertex-edge domination problem for any positive integer k on interval graphs.

Next, we introduce the research progress of edge-vertex domination. In 1986, Peters [10] introduced the edge-vertex domination problem. In 2018, Venkatakrishnan and Krishnakumari [14] obtained an improved upper bound of edge-vertex domination number of a tree. In 2021, Şahin [13] et al. found some relations between double edge-vertex domination and other domination parameters, investigated the relation between $\gamma_{2ev}(G)$ and $\gamma_{ev}(G)$, and determined $\gamma_{2ev}(G)$ on paths and cycles. There are few research results on the algorithm of this problem.

In this article, we study the minimum k edge-vertex domination problems on interval graphs and trees. Given some n-vertex interval graph G with m edges, a positive integer k and $S \subseteq V(G)$, we present an $O(m \lg m + k|S| + n)$-time algorithm to obtain some minimum k edge-vertex dominating set of G w.r.t. S, which implies that the minimum k edge-vertex domination problem can be solved in $O(m \lg m + kn)$ time. In addition, we present an $O(n|S|)$-time algorithm to obtain a minimum k edge-vertex dominating set of any n-vertex tree T w.r.t. any $S \subseteq V(T)$, which implies the minimum k edge-vertex domination problem can be solved in $O(n^2)$ time.

The rest of this article is organized as follows. Firstly, we give the necessary terminologies in Sect. 2. Then we present an algorithm to obtain a minimum k edge-vertex dominating set of interval graph G w.r.t. any $S \subseteq V(G)$ for each positive integer k, prove the correctness of this algorithm and explain how to implement it in Sect. 3. In Sect. 4, we study the minimum k edge-vertex dominating problem on trees. We design an algorithm to obtain a minimum k edge-vertex dominating set of a tree T w.r.t. some $S \subseteq V(T)$ for every positive integer k and prove the correctness of this algorithm. Finally, we raise some open questions and end the paper in Sect. 5.

2   Preliminaries and Notation

For any positive integers i and j with $i \leq j$, let $[i, j]$ denote the integers k with $i \leq k \leq j$. If $i = 1$, we often abbreviate $[1, j]$ as $[j]$. Let $\pi = (\pi_1, \pi_2, \ldots, \pi_n)$ be an ordering of some set V. For all $1 \leq i \leq j \leq n$, let $\pi[i, j]$ denote the ordering $(\pi_i, \pi_{i+1}, \ldots, \pi_j)$. Let $\pi = (\pi_1, \pi_2, \ldots, \pi_n)$ be an ordering of the vertex set of graph G. We say π is an *I-ordering* provided $\pi_i \pi_k \in E(G)$ implies $\pi_i \pi_j \in E(G)$ for every $1 \leq i < j < k \leq n$. Take some $v \in V(G)$. Let $r_\pi(v) = \max\{i : \pi_i \in N_G[v]\}$ and $\ell_\pi(v) = \min\{i : \pi_i \in N_G[v]\}$.

Lemma 2.1. *[11, 12] A graph G is an interval graph if and only if there is an interval ordering of $V(G)$.*

In the following, we always assume that G is an interval graph, $V(G) = \{v_1, v_2, \ldots, v_n\}$, and $\mathcal{I}$ is an interval representation of G where $r_\mathcal{I}(v_1) < r_\mathcal{I}(v_2) < \cdots < r_\mathcal{I}(v_n)$. Let's take an interval graph as an example, and the following is its interval representation.

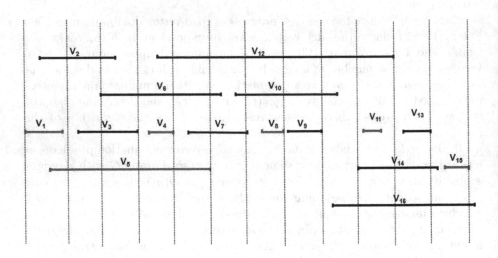

Fig. 1. An interval graph G and its interval representation.

3   The M-IG-EVD(G, S, k) Algorithm for Interval Graphs

For each $ab \in E(G)$, let $N_G[ab]$ denote the set $N_G[a] \cup N_G[b]$. Let $xy, zw \in E(G)$. Assume that $r_\mathcal{I}(x) < r_\mathcal{I}(y)$ and $r_\mathcal{I}(z) < r_\mathcal{I}(w)$. We say $xy <_\mathcal{I} zw$, if it provided

$r_{\mathcal{I}}(y) < r_{\mathcal{I}}(w)$, or $y = w$ and $r_{\mathcal{I}}(x) < r_{\mathcal{I}}(z)$. For each $v \in V(G)$, let $F_G[v]$ denote the set $\{ab \in E(G) : v \in N_G[ab]\}$. Let $W = \{u_1 w_1, u_2 w_2, \ldots, u_p w_p\} \subseteq E(G)$ be a set with $u_1 w_1 >_{\mathcal{I}} u_2 w_2 >_{\mathcal{I}} \cdots >_{\mathcal{I}} u_p w_p$. For any positive integer q with $q \le p$, denote the edge set $\{u_1 w_1, u_2 w_2, \ldots, u_q w_q\}$ by $M_q(W)$.

Now let's provide the algorithm to obtain a minimum k edge-vertex dominating set of G w.r.t. a subset $S \subseteq V(G)$ for each positive integer k.

M-IG-EVD(G, S, k)
1   ▷ **Input** interval graph G, $S \subseteq V(G)$, positive integer k;
2   ▷ **Output** a minimum k edge-vertex dominating set D of G w.r.t. S;
3   If there is any $x \in S$ with $|F_G[x]| < k$, output "there is not any
4   k edge-vertex dominating set of G w.r.t. S", exit.
5   Else, take an interval representation $\mathcal{I}$ of G where $V(G) = \{v_1, v_2, \ldots, v_n\}$ with
6    $r_{\mathcal{I}}(v_1) < r_{\mathcal{I}}(v_2) < \cdots < r_{\mathcal{I}}(v_n)$, let $S = \{v_{i_1}, v_{i_2}, \ldots, v_{i_h}\}$ with $i_1 < i_2 \cdots < i_h$.
7    $D \leftarrow \emptyset$;
8   **for** $j \leftarrow 1$ **to** h
9   **do if** $|D \cap (F_G[v_{i_j}])| < k$
10   **then** Let $q = k - |D \cap (F_G[v_{i_j}])|$.
11   Do $D \leftarrow D \cup M_q(F_G[v_{i_j}] \setminus D)$.
12   Output D.
13   ▷ **Exit**

Before proving the correctness of this algorithm, let's give an example to illustrate how it runs.

Example 3.1. *Let G be the interval graph as depicted in Fig. 1, $S = \{v_{i_1}, v_{i_2}, \ldots, v_{i_h}\} = \{v_1, v_4, v_5, v_8, v_{11}, v_{15}\}$. $h = 6, i_1 = 1, i_2 = 4, i_3 = 5, i_4 = 8, i_5 = 11, i_6 = 15$. Consider the algorithm M-IG-EVD$(G, S, 3)$.*

$F_G(v_1) = \{v_5 v_{12}, v_5 v_7, v_5 v_6, v_2 v_6, v_4 v_5, v_3 v_5, v_2 v_5, v_1 v_5, v_2 v_3, v_1 v_2\}$,

$F_G(v_4) = \{v_{12} v_{16}, v_{12} v_{14}, v_{11} v_{12}, v_{10} v_{12}, v_9 v_{12}, v_8 v_{12}, v_7 v_{12}, v_6 v_{12}, v_5 v_{12}, v_4 v_{12},$
$v_6 v_7,$

$v_5 v_7, v_5 v_6, v_4 v_6, v_3 v_6, v_2 v_6, v_4 v_5, v_3 v_5, v_2 v_5, v_1 v_5\}$,

$F_G(v_5) = \{v_{12} v_{16}, v_{12} v_{14}, v_{11} v_{12}, v_{10} v_{12}, v_9 v_{12}, v_8 v_{12}, v_7 v_{12}, v_6 v_{12}, v_5 v_{12}, v_4 v_{12},$
$v_7 v_{10},$

$v_6 v_7, v_5 v_7, v_5 v_6, v_4 v_6, v_3 v_6, v_2 v_6, v_4 v_5, v_3 v_5, v_2 v_5, v_1 v_5, v_2 v_3, v_1 v_2\}$,

$F_G(v_8) = \{v_{12} v_{16}, v_{12} v_{14}, v_{11} v_{12}, v_{10} v_{12}, v_9 v_{12}, v_8 v_{12}, v_7 v_{12}, v_6 v_{12}, v_5 v_{12}, v_4 v_{12},$
$v_9 v_{10},$

$v_8 v_{10}, v_7 v_{10}\}$,

$F_G(v_{11}) = \{v_{15} v_{16}, v_{14} v_{16}, v_{13} v_{16}, v_{12} v_{16}, v_{11} v_{16}, v_{10} v_{16}, v_{13} v_{14}, v_{12} v_{14}, v_{11} v_{14},$
$v_{11} v_{12},$

$v_{10} v_{12}, v_9 v_{12}, v_8 v_{12}, v_7 v_{12}, v_6 v_{12}, v_5 v_{12}, v_4 v_{12}\}$,

$F_G(v_{15}) = \{v_{15} v_{16}, v_{14} v_{16}, v_{13} v_{16}, v_{12} v_{16}, v_{11} v_{16}, v_{10} v_{16}\}$.

Iteration 0, $D_0 = \emptyset$;
Iteration 1, $|D_0 \cap (F_G[v_{i_1}])| = 0 < k = 3$, $q = k - |D_0 \cap (F_G[v_{i_1}])| = 3$,
$D_1 = D_0 \cup M_q(F_G[v_{i_1}] \setminus D_0) = \{v_5v_{12}, v_5v_7, v_5v_6\}$;
Iteration 2, $|D_1 \cap (F_G[v_{i_2}])| = 3, D_2 = D_1$;
Iteration 3, $|D_2 \cap (F_G[v_{i_3}])| = 3, D_3 = D_2$;
Iteration 4, $|D_3 \cap (F_G[v_{i_4}])| = 1 < k = 3$, $q = k - |D_3 \cap (F_G[v_{i_4}])| = 2$,
$D_4 = D_3 \cup M_q(F_G[v_{i_4}] \setminus D_3) = \{v_{12}v_{16}, v_{12}v_{14}, v_5v_{12}, v_5v_7, v_5v_6\}$;
Iteration 5, $|D_4 \cap (F_G[v_{i_5}])| = 3, D_5 = D_4$;
Iteration 6, $|D_5 \cap (F_G[v_{i_4}])| = 1 < k = 3$, $q = k - |D_5 \cap (F_G[v_{i_6}])| = 2$,
$D_6 = D_5 \cup M_q(F_G[v_{i_6}] \setminus D_5) = \{v_{15}v_{16}, v_{14}v_{16}, v_{12}v_{16}, v_{12}v_{14}, v_5v_{12}, v_5v_7, v_5v_6\}$;
output $D = D_6$, end.

Next we turn to show the correctness of the algorithm, then give an implementation of it. In the following of this section, let $S = \{v_{i_1}, v_{i_2}, \ldots, v_{i_h}\}$ with $i_1 < i_2 \cdots < i_h$. Take any positive integer k. Suppose that $|F_G[x]| \geq k$ holds for every $x \in S$. Recall that $\mathcal{I}$ is an interval representation of G and $V(G) = \{v_1, \ldots, v_n\}$ where $r_{\mathcal{I}}(v_1) < \cdots < r_{\mathcal{I}}(v_n)$.

Lemma 3.2. *Suppose ab and uw be edges of G with $ab >_{\mathcal{I}} uw$. Let v_p and v_q be two vertices of $V(G)$ with $p < q$. If $v_q \in N_G[uw] \setminus N_G[ab]$, then $v_p \notin N_G[ab]$.*

Proof. As $v_q \in N_G[uw] \setminus N_G[ab]$ and $ab >_{\mathcal{I}} uw$, it holds $\mathcal{I}(v_q) < \mathcal{I}(a)$ and $\mathcal{I}(v_q) < \mathcal{I}(b)$. Since $p < q$, we obtain $r_{\mathcal{I}}(v_p) < r_{\mathcal{I}}(v_q)$, hence $\mathcal{I}(v_p) < \mathcal{I}(a)$ and $\mathcal{I}(v_p) < \mathcal{I}(b)$, which means $v_p \notin N_G[ab]$. □

Lemma 3.3. *There is some minimum k edge-vertex dominating set X of G w.r.t. S which satisfies that $M_k(F_G[v_{i_1}]) \subseteq X$.*

Proof. Let $F_G[v_{i_1}] = \{u_1w_1, \ldots, u_pw_p\}$ where $u_1w_1 >_{\mathcal{I}} \cdots >_{\mathcal{I}} u_pw_p$. Notice that $M_k(F_G[v_{i_1}]) = \{u_1w_1, \ldots, u_kw_k\}$. Let X be a minimum k edge-vertex dominating set X of G w.r.t. S with maximum edges of $M_k(F_G[v_{i_1}])$. If $M_k(F_G[v_{i_1}]) \subseteq X$ were no true, take some edge u_jw_j of $M_k(F_G[v_{i_1}]) \setminus X$ with minimum j. Since X is a k edge-vertex dominating set of G w.r.t. S, it holds $|F_G[v_{i_1}] \cap X| \geq k$, hence there is some $u_sw_s \in (F_G[v_{i_1}] \cap X) \setminus M_k(F_G[v_{i_1}])$. Note that $j < s$.

Consider the set $X' = (X \setminus \{u_sw_s\}) \cup \{u_jw_j\}$. We want to show that X' is also a k edge-vertex dominating set of G w.r.t. S. If not, then there is $v_{i_t} \in S$ with $v_{i_t} \in N_G[u_sw_s] \setminus N_G[u_jw_j]$. Since $i_1 < i_t$, we deduce $v_{i_1} \notin N_G[u_jw_j]$, contradicting with $u_jw_j \in F_G[v_{i_1}]$, finishing the proof. □

Lemma 3.4. *Let $j \in [h]$ and D_{j-1} be some k edge-vertex dominating set of G w.r.t. $S_{j-1} = \{v_{i_1}, \ldots, v_{i_{j-1}}\}$. Assume that there is a minimum k edge-vertex dominating set D'_{j-1} of G w.r.t. S with $D_{j-1} \subseteq D'_{j-1}$. Suppose $|D \cap (F_G[v_{i_j}])| < k$. Let $q = k - |D \cap (F_G[v_{i_j}])|$, $W = M_q(F_G[v_{i_j}] \setminus D_{j-1})$ and $D_j = D_{j-1} \cup W$. Then D_j is a k edge-vertex dominating set of G w.r.t. $S_j = \{v_{i_1}, \ldots, v_{i_j}\}$. Furthermore, there is some minimum k edge-vertex dominating set D'_j of G w.r.t. S with $D_j \subseteq D'_j$.*

Proof. Notice that: (i) D_{j-1} is a k edge-vertex dominating set of G w.r.t. S_{j-1}; (ii) $q = k - |D \cap (F_G[v_{i_j}])|$, $W = M_q(F_G[v_{i_j}] \setminus D_{j-1})$ and $D_j = D_{j-1} \cup W$. It follows from (i) and (ii) that D_j is a k edge-vertex dominating set of G w.r.t. S_j.

Next we turn to show there is some minimum k edge-vertex dominating set D'_j of G w.r.t. S such that $D_j \subseteq D'_j$. Take some minimum k edge-vertex dominating set X of G w.r.t. S which contains D_{j-1} and with maximum edges of W. Notice that there is some minimum k edge-vertex dominating set D'_{j-1} of G w.r.t. S with $D_{j-1} \subseteq D'_{j-1}$, it holds $X \neq \emptyset$.

Suppose $W = \{u_1 w_1, \ldots, u_q w_q\}$ where $u_1 w_1 >_\mathcal{I} \cdots >_\mathcal{I} u_q w_q$. We shall show $W \subseteq X$. If this were not true, take any edge $u_s w_s \in W \setminus X$ with minimum s. As X is a k edge-vertex dominating set of G w.r.t. S, there is some $ab \in (F_G[v_{i_j}] \cap X) \setminus D_j$. By the definition of W and by the choice of s, we have $ab <_\mathcal{I} u_s w_s$. Consider the edge set $X' = (X \setminus \{ab\}) \cup \{u_s w_s\}$. Note that $|X'| = |X|$. By the choice of X and ab, X' is not a k edge-vertex dominating set of G w.r.t. S. Therefore, there is some $v_{i_t} \in S$ such that $|X' \cap (F_G[v_{i_t}])| < k$. Note that $D_{j-1} \subseteq D'_j$ and D_{j-1} is a k edge-vertex dominating set of G w.r.t. S_{j-1}, we get $t > j$. Because X is a k edge-vertex dominating set of G w.r.t. S, we deduce that $v_{i_t} \in N_G[ab] \setminus N_G[u_s w_s]$. By Lemma 3.2, $v_{i_j} \notin N_G[u_s w_s]$, a contradiction. So it holds $W \subseteq X$, hence there is some minimum k edge-vertex dominating set $D'_j = X$ of G w.r.t. S such that $D_j \subseteq D'_j$, finishing the proof. $\square$

Theorem 3.5. *Take a positive integer k. Let G be any n-vertex interval graph and $S \subseteq V(G)$. Suppose $|F_G[v]| \geq k$ holds for every $v \in S$. The output of M-IG-EVD(G, S, k), say D, is a minimum k edge-vertex dominating set of G w.r.t. S.*

Proof. Take an interval representation $\mathcal{I}$ of G so that $V(G) = \{v_1, \ldots, v_n\}$ and $r_\mathcal{I}(v_1) < \cdots < r_\mathcal{I}(v_n)$. Let $S = \{v_{i_1}, v_{i_2}, \ldots, v_{i_h}\}$ with $i_1 < i_2 \cdots < i_h$. For each $j \in [h]$, let D_j denote the set D right after step j. Denote the set $\{v_{i_1}, v_{i_2}, \ldots, v_{i_j}\}$ by S_j. Let $D_0 = S_0 = \emptyset$. It follows from Lemmas 3.3 and 3.4 that for each $j \in [h]$, D_j is a k edge-vertex dominating set of G w.r.t. S_j. In addition, there is a minimum k edge-vertex dominating set D'_j of G w.r.t. S such that $D_j \subseteq D'_j$. Now, we find that $D = D_h$ is a k edge-vertex dominating set of G w.r.t. $S_h = S$ and there is a minimum k edge-vertex dominating set D'_h of G w.r.t. S such that $D_h \subseteq D'_h$. Note that D itself is a k edge-vertex dominating set of G w.r.t. S, so it must hold that $D = D_h = D'_h$, as required. $\square$

Theorem 3.6. *Take a positive integer k. Let G be any n-vertex interval graph with m edges. For each $S \subseteq V(G)$, the algorithm M-IG-EVD(G, S, k) can be implemented in $O(m \lg m + k|S| + n)$ time.*

Proof. It is well known that constructing an interval representation $\mathcal{I}$ of G takes $O(n + m)$ time. Sorting the elements in set $E(G) = \{u_1 w_1, \ldots, u_m w_m\}$ so that $u_1 w_1 >_\mathcal{I} \cdots >_\mathcal{I} u_m w_m$ costs us $O(m \lg m)$ time. It needs $O(k|S|)$ time to determine whether there is any $x \in S$ with $|F_G[x]| < k$ or not. If $|D \cap (F_G[x])| < k$, it needs at most $O(k)$ time to do $D \leftarrow D \cup M_q(F_G[x] \setminus D)$ where $q = k - |D \cap (F_G[x])|$. So the whole algorithm can be implemented in $O(m \lg m + k|S| + n)$ time. $\square$

Corollary 3.7. *Let G be some n-vertex interval graph with m edges. The k edge-vertex domination problem can be solved in $O(m \lg m + kn)$ time for each positive integer k.*

4   The M-T-EVD(T, S, k) Algorithm for Trees

Let T be any n-vertex tree. Take some $v_0 \in V(T)$. For any nonnegative integer i, let denote the vertex set $\{u : d_T(u, v) = i\}$ by L_i. We also say $L^{-1}(u) = i$ if $u \in L_i$. For each $j \in [4]$ and each $u \in L_i$, let $E_j(u)$ denote the edge set $\{xy : d_T(u, x) = i - 3 + j, d_T(u, y) = i - 2 + j, (x \in N_T[u]) \vee (y \in N_T[u])\}$. For $u \in V(T)$, we sort the elements in set $F_T[u] = \{x_1 y_1, \ldots, x_p y_p\}$ so that $i < j$ provided $x_i y_i \in E_s(u)$ and $x_j y_j \in E_t(u)$ where $s < t$. The first q edges in $F_T[u]$ will be referred to as $M_q(F_T[u])$ for any positive integer q. After laying the groundwork above, let's provide the algorithm to find some minimum k edge-vertex dominating set of a tree T w.r.t. a subset $S \subseteq V(T)$ for any positive integer k.

M-T-EVD(T, S, k)

1   ▷ **Input** tree T rooted at v_0, $S \subseteq V(T)$, positive integer k;
2   ▷ **Output** a minimum k edge-vertex dominating set D of T w.r.t. S;
3   If there is any $x \in S$ where $|F_G[x]| < k$, then output "there is not any
4   k edge-vertex dominating set of T w.r.t. S", exit.
5   Else, compute $E_j(u)$ for each $j \in [4]$ and $u \in V(T)$. Sorting the elements in set
6   $F_T[u] = \{x_1 y_1, \ldots, x_p y_p\}$ where $i < j$ if $x_i y_i \in E_s(u)$, $x_j y_j \in E_t(u)$ and $s < t$.
7   Sorting the elements in $S = \{v_{q_1}, \ldots, v_{q_z}\}$ so that $L^{-1}(v_{q_1}) \geq \cdots \geq L^{-1}(v_{q_t})$.
8   $D \leftarrow \emptyset$;
9   **for** $i \leftarrow 1$ **to** z
10   **do if** $|D \cap (F_T[v_{q_i}])| < k$
11   **then** Let $r = k - |D \cap (F_T[v_{q_i}])|$
12   Do $D \leftarrow D \cup M_r(F_T[v_{q_i}] \setminus D)$.
13   Output D.
14   ▷ **Exit**

To help readers better understand the algorithm and the terminology involved in it, we will provide an example to describe how the algorithm runs.

Example 4.1. *Let T be the tree as depicted in Fig. 2, $S = \{v_1, v_2, v_5, v_7, v_8, v_{12}, v_{17}, v_{26}\}$. Consider the algorithm M-T-EVD$(T, S, 3)$.* $E_1(v_1) = \emptyset$, $E_2(v_1) = \{v_0 v_1, v_0 v_2, v_0 v_3\}$, $E_3(v_1) = \{v_1 v_4, v_1 v_5\}$, $E_4(v_1) = \{v_4 v_{11}, v_4 v_{12}, v_4 v_{13}, v_5 v_{14}\}$; $E_1(v_2) = \emptyset$, $E_2(v_2) = \{v_0 v_1, v_0 v_2, v_0 v_3\}$, $E_3(v_2) = \{v_2 v_6, v_2 v_7\}$, $E_4(v_2) = \{v_6 v_{15}, v_7 v_{16}, v_7 v_{17}\}$; $E_1(v_5) = v_0 v_1$, $E_2(v_5) = \{v_1 v_4, v_1 v_5\}$, $E_3(v_5) = \{v_5 v_{14}\}$, $E_4(v_5) = \{v_{14} v_{26}, v_{14} v_{27}\}$; $E_1(v_7) = v_0 v_2$, $E_2(v_7) = \{v_2 v_6, v_2 v_7\}$, $E_3(v_7) = \{v_7 v_{16}, v_7 v_{17}\}$, $E_4(v_7) = \{v_{17} v_{30}\}$; $E_1(v_8) = \{v_0 v_3\}$, $E_2(v_8) = \{v_3 v_8, v_3 v_9, v_3 v_{10}\}$, $E_3(v_8) = \{v_8 v_{18}\}$, $E_4(v_8) = \{v_{18} v_{31}, v_{18} v_{32}\}$; $E_1(v_{12}) = \{v_1 v_4\}$, $E_2(v_{12}) = \{v_4 v_{11}, v_4 v_{12}, v_4 v_{13}\}$, $E_3(v_{12}) = \{v_{12} v_{23}, v_{12} v_{24}, v_{12} v_{25}\}$, $E_4(v_{12}) = \{v_{23} v_{37}, v_{23} v_{38}\}$; $E_1(v_{17}) = \{v_2 v_7\}$, $E_2(v_{17}) = \{v_7 v_{16}, v_7 v_{17}\}$,

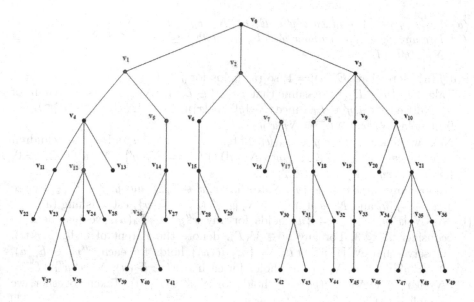

Fig. 2. A tree T which is rooted at v_0.

$E_3(v_{17}) = \{v_{17}v_{30}\}, E_4(v_{17}) = \{v_{30}v_{42}, v_{30}v_{43}\}; E_1(v_{26}) = \{v_5v_{14}\}, E_2(v_{26}) = \{v_{14}v_{26}, v_{14}v_{27}\}, E_3(v_{26}) = \{v_{26}v_{39}, v_{26}v_{40}, v_{26}v_{41}\}, E_4(v_{26}) = \emptyset.$

$S = \{v_{q_1}, \ldots, v_{q_z}\} = \{v_{26}, v_{17}, v_{12}, v_8, v_7, v_5, v_2, v_1\}, z = 8, q_1 = 26, q_2 = 17, q_3 = 12, q_4 = 8, q_5 = 7, q_6 = 5, q_7 = 2, q_8 = 1.$ $F_T[v_{q_i}] = \cup_{j\in[4]}(E_j(v_{q_i}))$ *for each $i \in [8]$.*

Iteration 0, $D_0 = \emptyset$.

Iteration 1, $|D_0 \cap (F_T[v_{q_1}])| = 0 < k = 3, r = k - |D \cap (F_T[v_{q_1}])|=3,$ $D_1 = D_0 \cup M_r(F_T[v_{q_1}] \setminus D_0) = \{v_5v_{14}, v_{14}v_{26}, v_{14}v_{27}\}.$

Iteration 2, $|D_1 \cap (F_T[v_{q_2}])| = 0 < k, r = k - |D \cap (F_T[v_{q_2}])|=3, D_2 = D_1 \cup M_r(F_T[v_{q_2}] \setminus D_1) = \{v_2v_7, v_5v_{14}, v_7v_{16}, v_7v_{17}, v_{14}v_{26}, v_{14}v_{27}\}.$

Iteration 3, $|D_2 \cap (F_T[v_{q_3}])| = 0 < k, r = k - |D \cap (F_T[v_{q_2}])|=3, D_3 = D_2 \cup M_r(F_T[v_{q_3}] \setminus D_2) = \{v_1v_4, v_2v_7, v_4v_{11}, v_4v_{12}, v_5v_{14}, v_7v_{16}, v_7v_{17}, v_{14}v_{26}, v_{14}v_{27}\}.$

Iteration 4, $|D_3 \cap (F_T[v_{q_4}])| = 0 < k, r = k - |D \cap (F_T[v_{q_3}])| = 3, D_4 = D_3 \cup M_r(F_T[v_{q_4}] \setminus D_3) = \{v_0v_3, v_1v_4, v_2v_7, v_3v_8, v_3v_9, v_4v_{11}, v_4v_{12}, v_5v_{14}, v_7v_{16}, v_7v_{17}, v_{14}v_{26}, v_{14}v_{27}\}.$

Iteration 5, $|D_4 \cap (F_T[v_{q_5}])| \geq 3, D_5 = D_4;$

Iteration 6, $|D_5 \cap (F_T[v_{q_6}])| \geq 3, D_6 = D_5;$

Iteration 7, $|D_6 \cap (F_T[v_{q_7}])| \geq 3, D_7 = D_6;$

Iteration 8, $|D_7 \cap (F_T[v_{q_8}])| \geq 3, D_8 = D_7.$

output $D = D_8$, end.

Now let's prove the correctness of this algorithm.

Lemma 4.2. *Let T be a tree rooted at v_0. Take any $u \in L_i$. Denote the set $\cup_{0\leq j\leq i}L_j$ by L_i^-. It holds that:*

(a) For any $j \in [4]$ and $xy, x'y' \in E_j(u)$, $N_T[xy] \cap L_i^- = N_T[x'y'] \cap L_i^-$.

(b) For any $xy \in E_j(u)$ and $x'y' \in E_{j'}(u)$ with $1 \le j < j' \le 4$, $N_T[xy] \cap L_i^- \supseteq N_T[x'y'] \cap L_i^-$.

Proof.(a) Note that $|E_1(u)| = 1$, so (a) holds for $j = 1$.

Take $xy, x'y' \in E_2(u)$. Assume that $x = x' \in L_{i+1} \cap N_T(u)$ (x is the parent of u). Notice that $y, y' \in L_i$. Since $(N_T[y] \backslash N_T[x]) \cap L_i^- = (N_T[y'] \backslash N_T[x]) \cap L_i^- = \emptyset$, it holds $N_T[xy] \cap L_i^- = N_T[x'y'] \cap L_i^-$.

Next we pick any $xy, x'y' \in E_3(u)$. Note that $x = x' = u$ and y, y' are children of u. Since $y, y' \in L_{i+1}$, $(N_T[y] \backslash N_T[u]) \cap L_i^- = (N_T[y'] \backslash N_T[u]) \cap L_i^- = \emptyset$, hence $N_T[xy] \cap L_i^- = N_T[x'y'] \cap L_i^-$.

Choose any $xy, x'y' \in E_4(u)$. Note that $x, x' \in L_{i+1}$ and $y, y' \in L_{i+2}$, we see $x, x' \in N_T[u]$ and $N_T[xy] \cap L_i^- = N_T[x'y'] \cap L_i^- = \{u\}$, establishing (a).

(b) As $N_T[x''y''] \cap L_i^- = \{u\}$ holds for any $x''y'' \in E_4(u)$, we just need to consider $j' \le 3$. For any $w \in V(T)$, denote the parent of w by $par(w)$. Observe that $N_T[x''y''] \cap L_i^- = \{u, par(u)\}$ holds for each $x''y'' \in E_3(u)$, $N_T[x''y''] \cap L_i^- = N_T[par(u)]$ holds for each $x''y'' \in E_2(u)$, $N_T[x''y''] \cap L_i^- = N_T[par(u)] \cup N_T[par(par(u))]$ holds for $x''y'' \in E_1(u)$. Taken together, we establish (b).   □

Theorem 4.3. *Take a tree T with n vertices and a positive integer k. The algorithm M-T-EVD(T, S, k) is correct for each $S \subseteq V(T)$.*

Proof. Suppose T is rooted at v_0, $S \subseteq V(T)$ and $|F_G[v]| \ge k$ holds for any $v \in S$. Assume that the elements in set $F_T[u] = \{x_1y_1, \ldots, x_py_p\}$ is sorted so that $i < j$ if $x_iy_i \in E_s(u), x_jy_j \in E_t(u)$ and $s < t$, and the elements in $S = \{v_{q_1}, \ldots, v_{q_z}\}$ is sorted with $L^{-1}(v_{q_1}) \ge \cdots \ge L^{-1}(v_{q_z})$. Let D_i denote the edge set D obtained just after step i. For each $0 \le i \le z$, we will prove the following statements:

(a) D_i is k edge-vertex dominating set of T w.r.t. $S_i = \{v_{q_1}, \ldots, v_{q_i}\}$.

(b) There is a minimum k edge-vertex dominating set of T w.r.t. S which contains D_i.

We shall proceed by induction on i. Note that (a) and (b) hold for $i = 0$. Assume that $i > 1$ and (a) and (b) hold for each $i' < i$. According to (b), there is some minimum k edge-vertex dominating set of T w.r.t. S which contains D_{i-1}, say W. Suppose $v_{q_i} \in L_s$. Notice that $L^{-1}(w) \le s$ holds for all $w \in S \backslash S_{i-1}$, we have $S \backslash S_{i-1} \subseteq L^{-1}(s)$. If v_{q_i} is already k edge-vertex dominated by D_{i-1}, then $D_i = D_{i-1}$ and statements (a) and (b) hold for i. If v_{q_i} is not k edge-vertex dominated by D_{i-1}, then $|D_{i-1} \cap (F_T[v_{q_i}])| < k$. Let $r = k - |D_{i-1} \cap (F_T[v_{q_i}])|$. Note that $D_i = D_{i-1} \cup M_r(F_T[v_{q_i}] \backslash D_{i-1})$. Obviously, D_i is k edge-vertex dominating set of T w.r.t. S_i, so (a) holds. By Lemma 4.2, we can replace a certain r elements in set W with set $M_r(F_T[v_{q_i}] \backslash D_{i-1})$, and the resulting set W' still satisfies the condition that W' is also a minimum k edge-vertex dominating set of T w.r.t. S. This establishes (b), as desired.

Theorem 4.4. *Take some n-vertex T be a tree and a positive integer k. Pick any $S \subseteq V(T)$, the algorithm M-T-EVD(T, S, k) can be implemented in $O(n|S|)$ time.*

Proof. Note that $|E(T)| = n - 1$. Suppose T is rooted at v_0. For every $u \in V(T)$, $F_G[u] = \cup_{1 \leq j \leq 4} E_j(u)$. To decide whether there is some $v \in S$ with $|F_G[v]| < k$ or not takes us $O(k|S|)$ time. Sorting the elements in $S = \{v_{q_1}, \ldots, v_{q_z}\}$ so that $L^{-1}(v_{q_1}) \geq \cdots \geq L^{-1}(v_{q_t})$ needs $O(n)$ time. Take any $u \in S$. To decide wether $|D \cap (F_T[v_{q_i}])| < k$ or not takes us $O(n)$ time for each $q_i \in S$. If $|D \cap (F_T[v_{q_i}])| < k$, it takes $O(n)$ time to get $M_r(F_T[v_{q_i}] \setminus D)$. So the entire algorithm can be implemented within $O(n|S|)$ time.

Corollary 4.5. *Let T be a tree with n vertices. The k edge-vertex domination problem can be solved in $O(n^2)$ time for each positive integer k.*

5   Conclusion

This article starts the research work on the k edge-vertex domination problems, mainly focusing on interval graphs and trees. We design $O(m \lg m + kn)$-time algorithms to solve these problems for any given interval graph G with n vertices and m edges, and $O(n^2)$-time algorithms to solve these problems for any tree with n vertices. Note that our algorithms are simple and natural, using only the interval representation of the interval graph and the tree representation of the tree graph, so their algorithm implementation is not difficult. We hope that the research method in this article can be extended to more general graph classes. The following questions are worth further investigation:

(1) Design polynomial time algorithms for the k edge-vertex domination problems on weighed interval graphs and trees;
(2) Design polynomial time algorithms for the k edge-vertex domination problems on other non-trivial graph classes;
(3) Find the connection between the k edge-vertex dominating number and the k vertex-edge dominating number for interval graphs and trees.

Acknowledgements. Thanks very much to the reviewers and editors for their hard work and valuable suggestions. This work is supported by the National Natural Science Foundation of China (Nos. 11701059, 12171061), Chongqing Natural Science Foundation Innovation and Development Joint Fund (Municipal Education Commission)(No. CSTB2022NSCQ-LZX0003) and Youth project of science and technology research program of Chongqing Education Commission of China (No. KJQN202101130).

References

1. Desormeaux, W.J., Haynes, T.W., Hedetniemi, S.T., Moore, C.: Distribution centers in graphs. Discret. Appl. Math. **243**, 286–293 (2018)
2. Ediz, S., Cancan, M.: On molecular topological properties of alkylating agents based anticancer drug candidates via some VE-degree topological indices. Curr. Comput. Aided Drug Des. **16**(2), 190–195 (2020)
3. Haynes, T.W., Hedetniemi, S.M., Hedetniemi, S.T., Henning, M.A.: Domination in graphs applied to electric power networks. SIAM J. Discret. Math. **15**(4), 519–529 (2002)

4. Haynes, T.W., Hedetniemi, S.T., Slater, P.J.: Domination in Graphs: Advanced Topics. Marcel Dekker Inc., New York (1998)
5. Haynes, T.W., Hedetniemi, S.T., Slater, P.J.: Fundamentals of Domination in Graphs. Marcel Dekker Inc., New York (1998)
6. Haynes, T.W., Knisley, D., Seier, E., Zou, Y.: A quantitive analysis of secondary RNA structure using domination based parameters on trees. BMC Bioinform. **7**, Article ID 108 (2006)
7. Henning, M.A., Pandey, A.: Alghoritmic aspects of semitotal domination in graphs. Theoret. Comput. Sci. **766**, 46–57 (2019)
8. Lewis, J.R.: Vertex-edge and edge-vertex parameters in graphs. Ph.D. thesis, Clemson, SC, USA (2007)
9. Li, P., Wang, A.: A polynomial time algorithm for k-vertex-edge dominating problem in interval graphs. J. Comb. Optim. **45**, 45 (2023). https://doi.org/10.1007/s10878-022-00982-8
10. Peters, J.K.W.: Theoretical and algorithmic results on domination and connectivity (Nordhaus-Gaddum, Gallai type results, max-min relationships, linear time, series-parallel). Ph.D. thesis, Clemson, SC, USA (1986)
11. Ramalingam, G., Rangan, C.P.: A uniform approach to domination problems on interval graphs. Inf. Process. Lett. **27**, 271–274 (1988)
12. Raychaudhuri, A.: On powers of interval and unit interval graphs. Congr. Numer. **59**, 235–242 (1987)
13. Şahin, B., Şahin, A.: Double edge–vertex domination. In: Kahraman, C., Cevik Onar, S., Oztaysi, B., Sari, I.U., Cebi, S., Tolga, A.C. (eds.) INFUS 2020. AISC, vol. 1197, pp. 1564–1572. Springer, Cham (2021). https://doi.org/10.1007/978-3-030-51156-2_182
14. Venkatakrishnan, Y.B., Krishnakumari, B.: An improved upper bound of edge-vertex domination of a tree. Inf. Process. Lett. **134**, 14–17 (2018)
15. Venkatakrishnan, Y.B., Naresh Kumar, H.: On the algorithmic complexity of double vertex-edge domination in graphs. In: Das, G.K., Mandal, P.S., Mukhopadhyaya, K., Nakano, S. (eds.) WALCOM 2019. LNCS, vol. 11355, pp. 188–198. Springer, Cham (2019). https://doi.org/10.1007/978-3-030-10564-8_15

Resource-Adaptive Newton's Method for Distributed Learning

Shuzhen Chen[1], Yuan Yuan[2(✉)], Youming Tao[1], Zhipeng Cai[3], and Dongxiao Yu[1]

[1] School of Computer Science and Technology, Shandong University, Qingdao 266237, China
{szchen,ym.tao99}@mail.sdu.edu.cn, dxyu@sdu.edu.cn
[2] School of Software & Joint SDU-NTU Centre for Artificial Intelligence Research (C-FAIR), Shandong University, Jinan 250100, China
yyuan@sdu.edu.cn
[3] Department of Computer Science, Georgia State University, Atlanta, GA 30303, USA
zcai@gsu.edu

Abstract. By leveraging curvature information for improved performance, Newton's method offers significant advantages over first-order methods for distributed learning problems. However, the practical applicability of Newton's method is hindered in large-scale and heterogeneous learning environments due to challenges such as high computation and communication costs associated with the Hessian matrix, sub-model diversity, staleness in training, and data heterogeneity. To address these challenges, this paper introduces a novel and efficient algorithm called Resource-Adaptive Newton Learning (RANL), which overcomes the limitations of Newton's method by employing a simple Hessian initialization and adaptive assignments of training regions. The algorithm demonstrates impressive convergence properties, which are rigorously analyzed under standard assumptions in stochastic optimization. The theoretical analysis establishes that RANL achieves a linear convergence rate while effectively adapting to available resources and maintaining high efficiency. Moreover, RANL exhibits remarkable independence from the condition number of the problem and eliminates the need for complex parameter tuning. These advantages make RANL a promising approach for distributed learning in practical scenarios.

Keywords: Distributed learning · Newton's method · resource-adaptation · sub-model diversity · data heterogeneity

---

Supported in part by National Natural Science Foundation of China (NSFC) under Grant 62122042 and 62302247, in part by Fundamental Research Funds for the Central Universities under Grant 2022JC016, in part by Shandong Natural Science Foundation, China under Grant ZR2022QF140.

W. Wu and G. Tong (Eds.): COCOON 2023, LNCS 14422, pp. 335–346, 2024.
https://doi.org/10.1007/978-3-031-49190-0_24

1   Introduction

In recent years, the field of machine learning has witnessed exponential growth
in both the volume and complexity of data. This surge in data has necessitated
the development of innovative approaches to process and analyze information
efficiently. Distributed machine learning has emerged as a pivotal solution, cap-
italizing on the parallelism and scalability of modern computing architectures.
By leveraging distributed systems, this approach enables the effective processing
of large datasets, thereby addressing the challenges posed by the ever-expanding
data landscape. Distributed machine learning encompasses a diverse range of
techniques that enable the collaborative training and inference of models across
multiple computing nodes. Among the key research problems in this field, distributed stochastic optimization has gained significant attention since it lies at
the heart of various distributed machine learning applications, including federated learning, distributed deep learning, and distributed reinforcement learning.

First-order methods, such as stochastic gradient descent (SGD), have become
popular for distributed stochastic optimization problems due to their simplicity
and scalability. However, first-order methods have limitations, especially when
the objective function is ill-conditioned or non-smooth. For example, they suffer
from slow convergence rates and are highly sensitive to the choice of step size
and the condition number[1] of the objective function [2,5,7,22]. Furthermore,
in distributed settings, first-order methods require frequent gradient exchanges
between machines, resulting in large number of communication rounds. In contrast, second-order methods, such as Newton's method, offer several advantages
over first-order methods. These methods leverage additional curvature information, leading to faster and more precise convergence. By utilizing second-order
derivatives, they can capture more intricate aspects of the objective function's
geometry, allowing for efficient optimization in complex landscapes. Moreover,
second-order methods can reduce the required communication rounds in distributed settings since they often require fewer iterations to converge compared
to first-order methods [1].

Although Newton's method offers distinct advantages over first-order methods, its application to distributed stochastic optimization is not trivial, especially
in practical large-scale and heterogeneous learning environments. The limited
computation, storage, and communication resources in such settings exacerbate
the convergence degradation. Computationally, inverting, storing, and transmitting the objective function's Hessian matrix can be prohibitive for large-scale
problems [4]. High-dimensional problems further intensify the issue, requiring
substantial storage and manipulation for the colossal and dense Hessian matrix.
Additionally, the communication overhead of transmitting the matrix across distributed workers and servers poses a significant challenge in distributed learning.
Resource limitations further impede the learning process, while the diverse constraints among workers, such as processing speeds, memory limitations, and battery life, hinder local model training. In practical scenarios, workers often access

---

[1] For L_g-smooth μ-strongly convex functions, the *condition number* is defined as L_g/μ.

or compute on only a subset of model parameters, forming sub-models. The heterogeneity and discrepancy among machines and data sources introduce noise, bias, and variance, impacting the convergence and accuracy of distributed learning [11]. Given the difficulties above, a natural question arises: *can Newton's method efficiently address distributed stochastic optimization in heterogeneous environments?* Motivated by this question, we aim to design an efficient and heterogeneous distributed learning algorithm for stochastic optimization based on Newton's method, capable of adapting to the dynamic and diverse resource constraints of different distributed workers.

Challenges. To achieve this, we have to tackle several non-trivial challenges.

(1). *Hessian computation and communication.* Developing a more streamlined computation and communication process for generating the Hessian matrix is necessary. Existing methods have explored strategies such as low-rank approximations [9], subsampling techniques [4], or Hessian-free methods [8] to leverage local curvature information without explicitly forming or storing the complete Hessian matrix. However, these methods may rely on heuristics or introduce additional approximation errors in practical implementations.

(2). *Sub-model diversity.* Sub-model diversity involves partitioning the model into different *regions* that can be trained independently by individual workers. This allows for adaptive resource allocation, as each worker can focus on training specific regions based on their available resources. This approach deviates from training all regions consistently in each round and promotes local optimization instead of global optimization. While sub-model diversity has been addressed in first-order methods of distributed stochastic optimization by researchers such as [20, 24], there is a lack of related research for second-order methods.

(3). *Staleness in training.* Since sub-models are constructed arbitrarily, not every region of the model can be trained by workers in each round. As a result, certain regions may remain untrained for extended periods. This lack of training for some regions can hinder the convergence of the entire distributed stochastic optimization process. This challenge remains unresolved in current research.

(4). *Data heterogeneity.* Data heterogeneity arises when different machines have varying data distributions or qualities due to their local contexts. Various techniques have been proposed to tackle this issue for first-order optimization methods, including data selection [13], transfer learning [25], regular optimization [12] and meta learning [10]. However, there is a dearth of efficient solutions for second-order methods to effectively handle data heterogeneity. Future research is needed to bridge this gap and develop effective strategies for addressing data heterogeneity in the context of second-order optimization methods.

Our Contributions. This paper presents an innovative algorithm called Resource-Adaptive Newton Learning (RANL) to address the above challenges. RANL employs simple Hessian initialization and adaptive assignments of training regions to tackle these challenges effectively. In RANL, during each optimization round, the server broadcasts the global model to all workers, who generate masks based on their own preferences without any constraints and train heterogeneous sub-models. The server then collects and aggregates the accumulated local updates for each region from the workers.

To address the challenge of Hessian computation and communication, a Hessian matrix is generated during the initialization phase, eliminating the need for repetitive computation and frequent communication. To tackle sub-model diversity, staleness in training, and data heterogeneity, RANL introduces a server aggregation mechanism that continuously monitors the latest updates for different regions of the global model in each worker and reuses them as approximations for the update of the current region. This approach efficiently addresses the challenges while minimizing memory requirements on the server side.

In summary, this paper introduces RANL, a novel distributed stochastic optimization algorithm based on Newton's method. A rigorous convergence analysis of RANL is provided, considering standard assumptions in stochastic optimization. The analysis demonstrates that RANL achieves a linear convergence rate and exhibits insensitivity to the condition number and complex parameter tuning compared to first-order methods.

2   Related Work

First-Order Methods. Various first-order methods have been proposed to tackle distributed stochastic optimization problems, including distributed SGD [14], variance reduction SGD [17], and accelerated SGD [19]. Pu *et al.* propose DSGD, achieving optimal network-independent convergence rates compared to centralized SGD [14]. Reddi *et al.* explore variance reduction algorithms in asynchronous parallel and distributed settings [17]. Shamir *et al.* systematically study distributed stochastic optimization, demonstrating the effectiveness of accelerated mini-batched SGD [19]. While these methods lower computation costs, they may escalate communication costs due to increased communication rounds. Moreover, they have limitations regarding the use of first-order information, potentially underutilizing curvature details in the objective function. Parameter tuning and dependence on the data's condition number can be intensive.

Second-Order Methods. Newton's method, utilizing the Hessian information, has demonstrated resilience in optimization problems, including federated learning with Newton-type methods [15,18], communication-efficient Newton-type methods [3,16], and quasi-Newton methods [21,23]. Qian *et al.* propose Basis Learn (BL) for federated learning, reducing communication costs through a change of basis in the matrix space [15]. Safaryan *et al.* introduce Federated Newton Learn (FedNL) methods, leveraging approximate Hessian information for improved solutions [18]. Bullins *et al.* present FEDSN, a communication-efficient stochastic Newton algorithm [3]. Qiu *et al.* develop St-SoPro, a stochastic second-order proximal method with decentralized second-order approximation [16]. Zhang *et al.* devise SpiderSQN, a faster quasi-Newton method that incorporates variance reduction techniques [23], and propose stochastic quasi-Newton methods with linear convergence guarantees [21].

However, there is limited work addressing heterogeneous methods for second-order optimization, particularly considering resource constraints and sub-model diversity. In this study, we investigate distributed learning problems concerning resource adaptation in stochastic Newton's method optimization.

3   Preliminaries

3.1   Distributed Learning Setup

We consider a distributed learning setup where N workers collaborate with a central server to solve the stochastic optimization problem. The learning process is synchronized and is divided into a series of communication rounds of $t = 1, 2, \cdots$ and thus resulting in a sequence of model parameters $x^1, x^2, \cdots, x^N$. The learning objective is to find an optimal model parameter vector x^* from the d-dimensional parameter space $\mathbb{R}^d$ that minimizes the global population risk $\tilde{f}$, which can be formulated as:

$$x^* = \arg\min_{x \in \mathbb{R}^d} \left\{ \tilde{f}(x) \triangleq \frac{1}{N} \sum_{i=1}^{N} \mathbb{E}_{\xi_i \sim \mathcal{D}_i}[F_i(x; \xi_i)] \right\}, \qquad (1)$$

where $\mathcal{D}_i$ is the local data distribution of worker i, ξ_i is a random data sample drawn from $\mathcal{D}_i$, and $F_i(x; \xi_i)$ is the per-sample loss of worker i incurred by the model parameter x. In each round t, we denote the group of data samples drawn by the workers as $(\xi_1^t, \xi_2^t, \cdots, \xi_N^t)$, and denote the model parameter at the beginning of round t as x^t. For any feasible $(\xi_1^t, \xi_2^t, \cdots, \xi_N^t)$ and x^t, we define

$$\tilde{F}(x^t; \xi_1^t, \xi_2^t, \cdots, \xi_N^t) \triangleq \frac{1}{N} \sum_{i=1}^{N} F_i(x^t; \xi_i^t), \qquad (2)$$

which represents the instant global loss incurred by the data samples generated in round t under model x^t. In the remainder of the paper, we use $\nabla \tilde{F}^t$ to denote the gradient of function F at x^t for simplicity, i.e., $\nabla \tilde{F}^t \triangleq \nabla_x F(x^t; \xi_1^t, \xi_2^t, \cdots, \xi_N^t)$.

3.2   Newton's Method

Typically, to apply Newton's method for solving the population risk minimization problem described in (1), one can modify the stochastic gradient descent (SGD) framework and leverage gradients preconditioned by the inverse Hessian therein. Let $\mathbf{H}^t \in \mathbb{R}^{d \times d}$ be the Hessian matrix of function F evaluated at x^t with data samples being $(\xi_1^t, \xi_2^t, \cdots, \xi_N^t)$, i.e., $\mathbf{H}^t \triangleq \nabla^2 F(x^t; \xi_1^t, \xi_2^t, \cdots, \xi_N^t)$. Then the model parameter updating step in round t can be written as:

$$x^{t+1} = x^t - [\mathbf{H}^t]^{-1} \nabla \tilde{F}^t, \qquad (3)$$

where $[\mathbf{H}^t]^{-1}$ is the inverse of $\mathbf{H}^t$.

3.3   Resource-Adaptive Learning via Online Model Pruning

One popular way to achieve resource-adaptive learning is online model pruning, which is first introduced in [24]. For each local loss $F_i(x, \xi_i)$ with parameters x and input data ξ_i. The pruning process takes F_i as the input and generates a new training model $F_i(x \odot m_i; \xi_i)$, where $m_i \in \{0,1\}^d$ is a binary mask for indicating whether each parameter is trainable by worker i, $\odot$ denotes element-wise multiplication, and $x \odot m_i$ denotes the pruned model at worker i, which has a reduced model size and is more efficient for communication and local training. The pruning mask m_i can be generated from various pruning policies in practice. In this work, we consider a generic pruning policy $\mathcal{P}$ such that (i) pruning masks are allowed to be time-varying, enabling online adjustment of pruned local models during the entire training process and (ii) pruning policies may vary for different clients, making it possible to optimize the pruned local models with respect to individual clients' heterogeneous computing resource and network conditions. We use m_i^t to denote the mask used by worker i in round t and define $x_i^t \triangleq x^t \odot m_i^t$ as the pruned local model that contains trainable parameters of worker i in round t. In cater to different available local resources among workers, the global model into Q disjoint regions with varying number of parameters before the training. During the learning process, different workers leverage adaptive online masks to train heterogeneous pruned models composed of varying regions according to local resources. Let $\mathcal{B}$ be the set of all regions. It is possible that (i) only part of the regions are trained by at least one worker in each round t due to the adaptive online pruning and (ii) certain region can stay untrained for several consecutive rounds. To characterize the regions covered by each round of training, we introduce the notation of $\mathcal{B}^t$ to denote the set of regions that get trained in round t. For each region $q \in \mathcal{B}^t$, we use $\mathcal{N}^{t,q}$ to denote the set of workers that train on region q in round t. With these notations, we define the minimum worker coverage number as follows:

$$\tau^* \triangleq \min_{t, q \in \mathcal{B}^t} |\mathcal{N}^{t,q}|. \tag{4}$$

To capture the delay between adjacent training rounds for each region, we introduce the notation of $\kappa_i^{t,q}$ to denote the maximum number of consecutive rounds where region q remains untrained by worker i till round t. Based on this, we define κ_t to denote the maximum number of consecutive rounds a region can remain untrained across all workers in round t:

$$\kappa_t \triangleq \max_{q \in \mathcal{B}, i \in [N]} \kappa_i^{t,q}. \tag{5}$$

3.4   Other Notations

We use $\| \cdot \|$ to denote the ℓ_2 norm for vectors or the spectral norm for matrices, and use $\| \cdot \|_F$ to denote the Frobenius norm for matrices. In accordance with (2), we define the following global empirical loss function F:

$$F(x_1^t, x_2^t, \cdots, x_N^t; \xi_1^t, \xi_2^t, \cdots, \xi_N^t) \triangleq \frac{1}{N} \sum_{i=1}^{N} F_i(x_i^t; \xi_i^t), \tag{6}$$

which measures the instant global empirical loss under the pruned models given the sampled data points in round t. For simplicity, we will just write $F(x)$ instead of $F(x_1^t, x_2^t, \cdots, x_N^t; \xi_1^t, \xi_2^t, \cdots, \xi_N^t)$ when it is clear from the context.

4   The Resource-Adaptive Newton Learning Algorithm

In this section, we provide a detailed description of our algorithm, namely *Resource-Adaptive Newton Learning* (RANL). RANL leverages only the initial second-order information, namely the Hessian $\mathbf{H}$ at the 0-th round. Moreover, each worker transmits a pruned gradient to the server with a pruned model, aiming to reduce communication costs. RANL can be roughly divided into two phases.

Phase I: Initialization. Each worker i calculates the local Hessian $\nabla^2 F_i(x^0, \xi_i^0)$ based on the initial global parameter x^0 and sends it to the server. The server then aggregates the Hessians, resulting in $\mathbf{H} = \frac{1}{N} \sum_{i=1}^{N} \nabla^2 F_i(x^0, \xi_i^0)$. Notably, this aggregation step is based on the initial second-order information and will be used throughout the subsequent learning rounds. Next, a projected Hessian estimate, denoted by $[\mathbf{H}]_\mu \in \{\mathbf{M} \in \mathcal{R}^{d \times d} : \mathbf{M}^\top = \mathbf{M}, \mu\mathbf{I} \preceq \mathbf{M}\}$, is computed, which is utilized for updating the global parameter in each updating step. The projection method satisfies the following definition.

Definition 1 (Projection [18]). *The projection of symmetric matrix $\mathbf{A}$ onto the cone of positive semi-definite matrices $\{\mathbf{M} \in \mathcal{R}^{d \times d} : \mathbf{M}^\top = \mathbf{M}, \mu\mathbf{I} \preceq \mathbf{M}\}$ is computed by*

$$[\mathbf{A}]_\mu := [\mathbf{A} - \mu\mathbf{I}]_0 + \mu\mathbf{I}, \quad [\mathbf{A}]_0 := \sum_{i=1}^{d} \max\{\lambda_i, 0\} u_i u_i^\top,$$

where $\sum_i \lambda_i u_i u_i^\top$ is an eigenvalue decomposition of the matrix $\mathbf{A}$.

Phase II: Resource-Adaptive Learning. In each round $t \in \{1, 2, \ldots, T\}$, all workers first receive the latest global parameter x^t from the server. Then each worker generates a local adaptive mask m_i^t, and compute a *pruned* local gradient $\nabla F_i^t \triangleq \nabla F_i(x^t \odot m_i^t, \xi_i^0) \odot m_i^t$ based on m_i^t. The pruned local gradients are used to generate the global gradient ∇F^t by the server. Now we expand on how to generate ∇F^t. ∇F^t can be partitioned into Q discontinuous fragments with each corresponding to a region of the model. Formally, we use $\nabla F^{t,q}$ to denote the gradient fragment corresponding to region q. Since only partial model regions can get trained by the workers in round t, to generate a reasonable global gradient, the server has to keep track of the latest gradient information for each model region. To this end, in RANL, the server maintains $C_i^{t,q}$ to store the most recent accessible local gradient fragment that is received from each worker i and corresponds to region q. $\nabla F^{t,q}$ is formed by aggregating all $C_i^{t,q}$'s for $i \in [N]$ and ∇F^t is generated by re-combining all fragments $\nabla F^{t,q}$'s for $q \in [Q]$. Finally, ∇F^t is used to update x^t via the preconditioned gradient descent step.

The overall scheme is summarized in Algorithm 1.

Algorithm 1: RANL: Resource-Adaptive Newton Learning

Input: Local datasets $\{\mathcal{D}_i\}_{i=1}^{N}$, pruning policy $\mathcal{P}$, initialized model x^0.

Output: x^T.

/* Phase I: Initialization   */

1 **Server** broadcasts x^0 to all workers;

2 **for every** *worker* i **in parallel do**

3   Compute local gradient $\nabla F_i(x^0, \xi_i^0)$ and local Hessian $\nabla^2 F_i(x^0, \xi_i^0)$;

4   Send $\nabla F_i(x^0, \xi_i^0)$ and $\nabla^2 F_i(x^0, \xi_i^0)$ to the server ;

5 **Server** aggregates the local Hessian matrices: $\mathbf{H} = \frac{1}{N} \sum_{i=1}^{N} \nabla^2 F_i(x^0, \xi_i^0)$;

6 **Server** initializes $C_i^{0,q}$ for each worker i and region q: $C_i^{0,q} = \nabla F_i^q(x^0, \xi_i^0)$;

7 **Server** updates the global model: $x^1 = x^0 - [\mathbf{H}]_\mu^{-1} \left(\sum_{i=1}^{N} \nabla F_i(x^0, \xi_i^0)/N \right)$;

8 **Server** broadcasts x^1 to all workers ;

/* Phase II: Resource-Adaptive Learning   */

9 **for round** $t = 1, \cdots, T$ **do**

10   **for every** *worker* i **in parallel do**

11   Generate the mask $m_i^t = \mathcal{P}(x^t, i)$;

12   Prune the model: $x_i^t = x^t \odot m_i^t$;

13   Compute the gradient after pruning: $\nabla F_i^t = \nabla F_i(x_i^t, \xi_i^t) \odot m_i^t$;

14   Send ∇F_i^t to the server ;

15   **for every region** $q = 1, \cdots, Q$ **do**

16   **Server** finds $\mathcal{N}^{t,q} = \{i : m_i^{t,q} = 1\}$;

17   **for** $i = 1$ *to* N **do**

18              $C_i^{t,q} = \begin{cases} \nabla F_i^{t,q} & if \quad i \in \mathcal{N}^{t,q} \\ C_i^{t-1,q} & if \quad i \notin \mathcal{N}^{t,q} \end{cases}$;

19   **Server** updates $\nabla F^{t,q} = \frac{1}{N} \sum_{i=1}^{N} C_i^{t,q}$;

20   **Server** updates the global parameter: $x^{t+1} = x^t - [\mathbf{H}]_\mu^{-1} \nabla F^t$;

21   **Server** broadcasts x^{t+1} to all workers ;

---

5   Convergence Analysis

In this section, we analyze the performance of Algorithm 1. Specifically, we aim to show that our proposed Algorithm 1 exhibits linear convergence rate. To achieve this, we first introduce some key concepts in optimization.

Definition 2 (Lipschitzness). *A function* $\ell : \mathbb{R}^d \to \mathbb{R}^d$ *is L-Lipschitz if for* $\forall x_1, x_2 \in \mathbb{R}^d$,

$$\|\ell(x_1) - \ell(x_2)\| \leq L\|x_1 - x_2\|.$$

Then, the function ℓ *has* L_g *Lipschitz continuous gradient and* L_h *Lipschitz continuous Hessian if for* $\forall x_1, x_2 \in \mathbb{R}^d$,

$$\|\nabla \ell(x_1) - \nabla \ell(x_2)\| \leq L_g\|x_1 - x_2\|.$$

$$\|\nabla^2\ell(x_1) - \nabla^2\ell(x_2)\| \le L_h\|x_1 - x_2\|.$$

Definition 3 (Bounded variance). *Define a function $L(x,\xi)$ and its unbiased expectation function $\ell(x)$. $\ell(x)$ is an unbiased estimator of $L(x,\xi)$ with a bounded variance if for $\forall x \in \mathbb{R}^d$,*

$$\mathbb{E}_{\xi \sim \mathcal{D}}\|L(x,\xi) - \ell(x)\|^2 \le \sigma^2.$$

Definition 4 (μ-strong convexity). *A differentiable function $\ell : \mathbb{R}^d \to \mathbb{R}^d$ is μ-strongly convex if for $\forall x_1, x_2 \in \mathbb{R}^d$,*

$$\ell(x_1) \ge \ell(x_2) + \langle \nabla\ell(x_2), x_1 - x_2 \rangle + \frac{\mu}{2}\|x_1 - x_2\|.$$

For twice differentiable functions, μ-strong convexity is equivalent to that $\lambda_{\min}\left(\nabla^2\ell(x)\right) \ge \mu$, where $\lambda_{\min}(\cdot)$ represents the minimum eigenvalue.

We adopt the following standard assumptions throughout the analysis.

Assumption 1. The average loss function $\widetilde{f}(x)$ is μ-strongly convex.

Assumption 2. For $\forall i \in [N]$, each function $f_i(x)$ is L-Lipschitz for any $x \in \mathbb{R}^d$ and twice continuously differentiable in respect of $x \in \mathbb{R}^d$. Each function $f_i(x)$ has a L_g-Lipschitz gradient and a L_h-Lipschitz Hessian for any $x \in \mathbb{R}^d$.

Assumption 3. There exists a constant $\Delta \ge 0$ and a constant $\sigma \ge 0$ such that: (i) for $\forall i \in [N]$ with any $x \in \mathcal{R}^d$, it holds that $\mathbb{E}_{\xi \sim \mathcal{D}_i}\|\nabla F_i(x,\xi) - \nabla f_i(x)\|^2 \le \Delta^2$; (ii) for $\forall i \in [N]$ with the $x^0 \in \mathcal{R}^d$, it holds that $\mathbb{E}_{\xi \sim \mathcal{D}_i}\|\nabla^2 F_i(x^0,\xi) - \nabla^2 f_i(x^0)\|_F^2 \le \sigma^2$.

Assumption 4. For $\forall i \in [N]$ and $\forall t \in [T]$, there exists a constant $\delta \ge 0$ such that $\mathbb{E}\|x^t - x_i^t\|^2 \le \delta^2$.

The above three assumptions are fairly standard and have been widely adopted in previous work, e.g., [18,23,24]. To establish the linear convergence of RANL, we need four Lemmas 1–4 as follows. All detailed proofs are included in the full version [6] due to space limitation.

Lemma 1. *The projected Hessian estimate $[\mathbf{H}]_\mu \in \{\mathbf{M} \in \mathcal{R}^{d\times d} : \mathbf{M}^\top = \mathbf{M}, \mu\mathbf{I} \preceq \mathbf{M}\}$ is computed by Definition 1. We have the following inequality*

$$\|\mathbf{H}_\mu - \mathbf{H}^*\|_F \le \|\mathbf{H} - \mathbf{H}^*\|_F, \quad \mathbf{H}^* = \nabla^2\widetilde{f}(x^*). \tag{7}$$

Lemma 1 quantifies the proximity of the projected Hessian matrix $[\mathbf{H}]_\mu$ to the optimal Hessian matrix $\mathbf{H}^*$.

Lemma 2. *Under Assumption 3(ii), we have*

$$\mathbb{E}\left\|\mathbf{H} - \nabla^2\widetilde{f}(x^*)\right\|^2 \le 2\mathbb{E}\left\|\nabla^2\widetilde{f}(x^0) - \nabla^2\widetilde{f}(x^*)\right\|_F^2 + 2\sigma^2. \tag{8}$$

Lemma 2 measures the stochastic error induced by the local Hessian aggregation based on workers' random local data in the initial stage.

Lemma 3. *Under Assumption 3(i), we have*

$$
\mathbb{E}\left\|\nabla^2 \widetilde{f}(x^*)(x^t - x^*) - \nabla F^t + \nabla \widetilde{f}(x^*)\right\|^2
$$
$$
\leq 2\mathbb{E}\left\|\nabla^2 \widetilde{f}(x^*)(x^t - x^*) - \nabla f^t + \nabla \widetilde{f}(x^*)\right\|^2 + \frac{2N}{\tau^*}\Delta^2 + 2\Delta^2. \tag{9}
$$

Lemma 3 quantifies the stochastic error of the local gradient aggregation during the learning process. The error consists of two parts based on whether regions are trained or not. The aggregated gradient information for the trained regions may not include all the N workers' gradients. Therefore, the error for the trained regions is limited by the defined minimum worker coverage number τ^*. For the untrained regions, the gradients of all the N workers are used, which reduces the stochastic error but introduces the delay error.

Lemma 4. *Under Assumption 2 and 4, we have*

$$
\mathbb{E}\left\|\nabla f^t - \nabla \widetilde{f}(x^*) - \nabla^2 \widetilde{f}(x^*)(x^t - x^*)\right\|^2
$$
$$
\leq \frac{N}{\tau^*}\left(2L_g^2\delta^2\right) + \kappa_t^2 \frac{8L^2 L_g^2}{\mu^2} + 4L_g^2\delta^2 + \frac{L_h^2 N}{2\tau^*}\mathbb{E}\left\|x^t - x^*\right\|^4. \tag{10}
$$

Lemma 4 quantifies the pruning error and the delay accumulation error by establishing the relationship between the gradient and the parameter.

Before presenting Theorem 1, we define some variables: $a = \frac{L_h^2 N}{2\tau^*}$, $b = \frac{\mu^2}{16} - \sigma^2$, $c = \frac{N}{\tau^*}\left(2L_g^2\delta^2 + \Delta^2\right) + \kappa_t^2 \frac{8L^2 L_g^2}{\mu^2} + 4L_g^2\delta^2 + \Delta^2$ and $\rho = b^2 - 4ac$.

Theorem 1. *Under Assumption 1, 2, 3 and 4, assume there exists a constant $\alpha \geq 0$ satisfying $c \leq \alpha\mathbb{E}\|x^0 - x^*\|^2$, $\mathbb{E}\|x^0 - x^*\|^2 \leq \frac{b-\alpha}{a}$ and $\mathbb{E}\left\|\nabla^2 \widetilde{f}(x^0) - \nabla^2 \widetilde{f}(x^*)\right\|_F^2 \leq \frac{\mu^2}{16}$ with $\rho \geq 0$. Then, Algorithm 1 converges linearly with the rate*

$$
\mathbb{E}\|x^{t+1} - x^*\|^2 \leq \frac{1}{2}\mathbb{E}\left\|x^t - x^*\right\|^2. \tag{11}
$$

Corollary 1. *Let all assumptions hold. Supposing $\alpha = \frac{b-\sqrt{\rho}}{2}$ in Theorem 1, $\mathbb{E}\|x^0 - x^*\|^2 \leq \frac{b+\sqrt{\rho}}{2a}$ and $\mathbb{E}\left\|\nabla^2 \widetilde{f}(x^0) - \nabla^2 \widetilde{f}(x^*)\right\|_F^2 \leq \frac{\mu^2}{16}$. Then, Algorithm 1 converges linearly with the rate*

$$
\mathbb{E}\|x^{t+1} - x^*\|^2 \leq \frac{1}{2}\mathbb{E}\left\|x^t - x^*\right\|^2. \tag{12}
$$

One can observe from (12) that the local linear convergence rate of the iterates does not rely on any specific constant. This constant is universal, meaning that it is independent of the problem's condition number, the training data's size, or the problem's dimension. In fact, the squared distance to the optimal solution is reduced by half at every iteration.

6   Conclusion

In this paper, we have presented a novel approach for addressing distributed stochastic optimization problems in resource-constrained and heterogeneous settings using Newton's method. Our proposed algorithm, RANL, combines an efficient Hessian initialization with adaptive region partitioning. We have demonstrated that RANL is an effective and adaptable distributed learning algorithm that can handle the dynamic and diverse resource limitations of distributed workers. Our work also opens up several promising avenues for future research. One potential direction is to extend RANL to asynchronous or decentralized scenarios and incorporating more sophisticated Hessian approximation methods. Overall, our contributions shed light on the challenges and opportunities of distributed stochastic optimization in resource-constrained and heterogeneous environments.

References

1. Agarwal, N., Bullins, B., Hazan, E.: Second-order stochastic optimization for machine learning in linear time. J. Mach. Learn. Res. **18**, 116:1–116:40 (2017)
2. Beck, A.: Introduction to Nonlinear Optimization - Theory, Algorithms, and Applications with MATLAB, MOS-SIAM Series on Optimization, vol. 19. SIAM (2014)
3. Bullins, B., Patel, K.K., Shamir, O., Srebro, N., Woodworth, B.E.: A stochastic newton algorithm for distributed convex optimization. In: Annual Conference on Neural Information Processing Systems, NeurIPS, pp. 26818–26830 (2021)
4. Byrd, R.H., Hansen, S.L., Nocedal, J., Singer, Y.: A stochastic quasi-newton method for large-scale optimization. SIAM J. Optim. **26**(2), 1008–1031 (2016)
5. Chen, J., Yuan, R., Garrigos, G., Gower, R.M.: SAN: stochastic average newton algorithm for minimizing finite sums. In: International Conference on Artificial Intelligence and Statistics, AISTATS. Proceedings of Machine Learning Research, vol. 151, pp. 279–318. PMLR (2022)
6. Chen, S., Yuan, Y., Tao, Y., Cai, Z., Yu, D.: Resource-adaptive newton's method for distributed learning. arXiv preprint arXiv:2308.10154 (2023)
7. Islamov, R., Qian, X., Richtárik, P.: Distributed second order methods with fast rates and compressed communication. In: Proceedings of the 38th International Conference on Machine Learning, ICML. Proceedings of Machine Learning Research, vol. 139, pp. 4617–4628. PMLR (2021)
8. James, M.: Deep learning via hessian-free optimization. In: Proceedings of the International Conference on Machine Learning (ICML), vol. 27, pp. 735–742 (2010)
9. James, M., Roger, G.: Optimizing neural networks with kronecker-factored approximate curvature. In: International Conference on Machine Learning, pp. 2408–2417. PMLR (2015)
10. Jiang, Y., Konečný, J., Rush, K., Kannan, S.: Improving federated learning personalization via model agnostic meta learning. CoRR abs/1909.12488 (2019)
11. Kairouz, P., McMahan, H.B., Avent, B., Bellet, A., Bennis, M., Bhagoji, A.N., et al.: Advances and open problems in federated learning. Found. Trends Mach. Learn. **14**(1–2), 1–210 (2021)
12. Li, T., Sahu, A.K., Zaheer, M., Sanjabi, M., Talwalkar, A., Smith, V.: Federated optimization in heterogeneous networks. In: Dhillon, I.S., Papailiopoulos, D.S., Sze, V. (eds.) Proceedings of Machine Learning and Systems, MLSys. mlsys.org (2020)

13. de Luca, A.B., Zhang, G., Chen, X., Yu, Y.: Mitigating data heterogeneity in federated learning with data augmentation. CoRR abs/2206.09979 (2022)
14. Pu, S., Olshevsky, A., Paschalidis, I.C.: A sharp estimate on the transient time of distributed stochastic gradient descent. IEEE Trans. Autom. Control **67**, 5900–5915 (2021)
15. Qian, X., Islamov, R., Safaryan, M., Richtárik, P.: Basis matters: better communication-efficient second order methods for federated learning. In: International Conference on Artificial Intelligence and Statistics, AISTATS. Proceedings of Machine Learning Research, vol. 151, pp. 680–720. PMLR (2022)
16. Qiu, C., Zhu, S., Ou, Z., Lu, J.: A stochastic second-order proximal method for distributed optimization. IEEE Control. Syst. Lett. **7**, 1405–1410 (2023)
17. Reddi, S.J., Hefny, A., Sra, S., Póczos, B., Smola, A.J.: On variance reduction in stochastic gradient descent and its asynchronous variants. In: Conference on Neural Information Processing Systems, pp. 2647–2655 (2015)
18. Safaryan, M., Islamov, R., Qian, X., Richtárik, P.: FEDNL: making newton-type methods applicable to federated learning. In: International Conference on Machine Learning, vol. 162, pp. 18959–19010. PMLR (2022)
19. Shamir, O., Srebro, N.: Distributed stochastic optimization and learning. In: 52nd Annual Allerton Conference on Communication, Control, and Computing, pp. 850–857. IEEE (2014)
20. Yuan, B., Wolfe, C.R., Dun, C., Tang, Y., Kyrillidis, A., Jermaine, C.: Distributed learning of fully connected neural networks using independent subnet training. Proc. VLDB Endow. **15**(8), 1581–1590 (2022)
21. Zhang, J., Liu, H., So, A.M., Ling, Q.: Variance-reduced stochastic quasi-newton methods for decentralized learning. IEEE Trans. Sig. Process. **71**, 311–326 (2023)
22. Zhang, J., You, K., Basar, T.: Achieving globally superlinear convergence for distributed optimization with adaptive newton method. In: 59th IEEE Conference on Decision and Control, CDC, pp. 2329–2334. IEEE (2020)
23. Zhang, Q., Huang, F., Deng, C., Huang, H.: Faster stochastic quasi-newton methods. IEEE Trans. Neural Netw. Learn. Syst. **33**(9), 4388–4397 (2022)
24. Zhou, H., Lan, T., Venkataramani, G., Ding, W.: On the convergence of heterogeneous federated learning with arbitrary adaptive online model pruning. CoRR abs/2201.11803 (2022)
25. Zhuang, F., et al.: A comprehensive survey on transfer learning. Proc. IEEE **109**(1), 43–76 (2021)

DR-Submodular Function Maximization with Adaptive Stepsize

Yanfei Li, Min Li, Qian Liu, and Yang Zhou(✉)

School of Mathematics and Statistics, Shandong Normal University, Jinan 250014,
China
zhouyang@sdnu.edu.cn

Abstract. The DR-submodular function maximization problem has
been gaining increasing attention due to its important applications in
many fields. In [1], a framework was proposed to describe algorithms
using differential dynamical systems and discretization to obtain implementable algorithms. In this framework, the time domain is discretized
with equal stepsizes, which also determined the computational complexity of the algorithm. In this paper, we propose an adaptive approach to determine the stepsize, which is applicable for various scenarios. With the guarantee of achieving the same approximation ratio as
the state-of-art results, the iteration complexity of our stepsize selec-
tion strategy is $O(\frac{\|\nabla F(\mathbf{0})\|_1}{\epsilon})$ when the objective function is monotone
and $O(n + \frac{\|\nabla F(\mathbf{0})\|_1}{\epsilon})$ when it is non-monotone, where F denotes the
objective function, and ϵ represents the approximation loss by discretization process. This strategy has been shown to have lower computational
complexity in some of the most common application scenarios for DR-
submodular function maximization.

Keywords: DR-submodular functions · adaptive stepsize · binary
search · computational complexity

1 Introduction

Diminishing-returns (DR) submodular functions have a wide range of applications in economics, engineering, machine learning, etc. The submodularity,
which describes a diminishing return property of a function, has a profound theoretical influence on optimization. It can achieve effective minimization [2] and
approximate maximization [3,4] in polynomial time. In [5], it is proved that the
continuous submodularity is equivalent to a weak form of diminishing returns
(DR) property.

This paper is supported by National Science Foundation of China (No. 12371099)
and Natural Science Foundation of Shandong Province (Nos. ZR2020MA029,
ZR2021MA100)) of China.

Maximizing a DR-submodular function is an NP-hard problem. For the unconstrained case, in [6] Niazadeh et al. presents a tight $\frac{1}{2}$-approximation algorithm. Specifically, quasi-linear time algorithm with $\frac{1}{2}$-approximation for DR-submodular maximization is designed that improves previous work in [7–9]. For the convex-constrained maximization of DR-submodular functions, there exists a $\left(1 - \frac{1}{e}\right)$-approximation algorithm when the objective function is monotone. If the objective function is non-monotone and the constraint is down-closed, there exists a $\frac{1}{e}$-approximation algorithm [7], otherwise, only a $\frac{1}{4}$-approximation algorithm is known [1,10].

In [1] a framework using Lyapunov function and exponential integrator method to design and analysing algorithms for DR-submodular function maximization problem is proposed. According to the given framework, some commonly used algorithms for DR-submodular function minimization problem, e.g. continuous greedy algorithm, Frank-Wolfe algorithm, can be regarded as a discretized form of initial-valued ordinary differential equations defined in time domain $[0, 1]$. Base on this point of view, the approximation ratio of the solution output by the discrete-time system at the end differs from that of the continuous-time system at the final moment in the approximation ratio only by a residual term, which will tend to 0 as the iteration stepsizes decreases to 0.

One possible approach for improving the computational complexity of the existing algorithms in above framework is to adaptively select stepsizes. However, a uniform result for this idea has not been established yet. Despite this, there have been some achievements in improving algorithm complexity by adopting an adaptive stepsize strategy under specific settings. In [11] Chen et al. propose a $\frac{1}{2}$-approximation algorithm to solve box-constrained DR-submodular function maximization problems, which needs only $\mathcal{O}(\frac{1}{\epsilon})$ adaptivity rounds. The stepsizes in each iteration of the algorithm is obtained by enumeration in a candidate set of size $O(\log \frac{1}{\epsilon})$ according to certain rules. In [12] Ene et al. consider maximizing the multilinear extension of submodular set functions with packing constraints. Their algorithms can achieve $1/e$-approximation guarantee with $\mathcal{O}(\log(\frac{n}{\epsilon}) \log(\frac{1}{\epsilon}) \log \frac{2n+m}{\epsilon^2})$ parallel rounds for the non-monotone case, and $1 - 1/e$ approximation guarantee with $\mathcal{O}(\log(\frac{n}{\epsilon}) \log \frac{m+n}{\epsilon^2})$ parallel rounds for the monotone case, where m denotes the number of the packing constraints.

1.1   Contributions

In this paper, we present an adaptive approach for determining the time stepsize in various scenarios. Our approach guarantees the same approximation ratio as the state-of-the-art results. The iteration complexity of our stepsize selection strategy is $O(\frac{\|\nabla F(0)\|_1}{\epsilon})$ for monotone objective functions and $O(n + \frac{\|\nabla F(0)\|_1}{\epsilon})$ for non-monotone objective functions, where F represents the objective function and ϵ accounts for the approximation loss due to discretization process. To evaluate functions, we employ the binary search method, which multiplies the complexity by $O(\log \frac{nL}{\epsilon})$ in addition to the computational complexity of the iterative steps. Moreover, our strategy exhibits lower computational complexity in common application scenarios of DR-submodular function maximization.

1.2   Organizations

The rest of this paper will be presented in the following order. In Sect. 2, we will provide detailed definitions of concepts and previous results used in this paper. In Sect. 3, we present the design ideas of our strategy and provide theoretical results for the case where the objective function is monotone. In Sect. 4 we propose algorithms and theoretical analysis of our strategy for the non-monotone case. In Sect. 5, we will discuss the computational complexity of applying our strategy to solve three common DR-submodular functions: multi-linear extension of set submodular functions, DR-submodular quadratic function, and softmax extension for DPP MAP problem.

2   Preliminaries

The problem that will be considered in this paper can be written as

$$\max_{x \in P} F(x), \tag{1}$$

where the feasibility region $P \subseteq [0, 1]^n$ is convex, and $F : [0, 1]^n \mapsto \mathbb{R}_+$ is a DR-submodular function, whose definition is given in Definition 1.

Definition 1. *A function* $F : [0, 1]^n \mapsto \mathbb{R}_+$ *is DR-submodular if for* $\forall x \leq y \in [0, 1]^n$ *and* $\forall a \geq 0$ *such that* $x + ae_i, y + ae_i \in [0, 1]^n$, *there holds*

$$F(x + ae_i) - F(x) \geq F(y + ae_i) - F(y) \qquad \forall i = 1, ..., n$$

Given two vectors $x, y \in \mathbb{R}^n$, the notation $x \leq y$ means that $x_i \leq y_i$ for $i \in [n]$. The notion $x \vee y$ denotes the coordinate-wise maximum of x and y, and $x \wedge y$ denotes the coordinate-wise minimum of x and y. When F is differentiable, the DR-submodularity is equivalent to the monotonicity of its gradient, that is, F is DR-submodular if and only if $\nabla F(x) \geq \nabla F(y)$ for $\forall x \leq y \in [0, 1]^n$. The concavity of a differentiable DR-submodular function in the non-negative direction has an important consequence as follows.

Proposition 1. *[13] If* F *is continuously differentiable and DR-submodular, then the following inequality holds*

$$\langle \nabla F(x), y - x \rangle \geq F(x \vee y) + F(x \wedge y) - 2F(x),$$

for $\forall x, y \in [0, 1]^n$.

To analyze the computational complexity of the algorithm, we need to introduce the concept of gradient Lipschitz continuity.

Definition 2. *A function* F *is L-smooth if for all* $x, y \in P$ *it holds that*

$$\| \nabla F(x) - \nabla F(y) \| \leq L \| x - y \|, \tag{2}$$

where $\| \cdot \|$ *denotes* $\| \cdot \|_2$ *for simplicity.*

A necessary but not sufficient condition of L-smoothness of function F is that

$$F(y) - F(x) \geq \langle \nabla F(x), y - x \rangle - \frac{L}{2} \| y - x \|^2. \tag{3}$$

All the algorithms and theoretical discussions regarding problem (1) in this paper are based on the following assumption.

Assumption 1. *Problem (1) that will be discussed in this paper is assumed to satisfy the following conditions.*

1. *F is DR-submodular and L-smooth on $[0, 1]^n$.*
2. *$F(\mathbf{0}) = 0$ and $F(x) \geq 0$ for $\forall x \in [0, 1]^n$.*
3. *P is convex and $\mathbf{0} \in P$.*
4. *There exists a Linear-Objective Optimization (LOO) oracle which returns an optimal solution of the optimization problem*

$$\max_{x \in P} c^T x,$$

for $\forall c \in \mathbb{R}^n$.

By Assumption 1.2 it is noteworthy to conclude that $\nabla F(\mathbf{0}) \geq 0$. This fact is natural since by the DR-submodularity we have $\nabla F(x) \leq \nabla F(\mathbf{0})$ for $\forall x \in P$. If for some $i \in [n]$ we have $\frac{\partial F}{\partial x_i}(\mathbf{0}) < 0$, then this entry will make no positive gain onto the objective function over the whole feasible region and thus it can be omitted throughout the algorithm. In the following text, unless otherwise specified, all the results regarding problem (1) are given based on Assumption 1.

From [1], the ideal algorithms described via differential systems for maximizing DR-submodular can be given as

$$v(x(t)) \in \arg\max_{v \in P} \langle \nabla F(x(t)), v \rangle$$
$$\dot{x}(t) = v(x(t))$$

for monotone case with $F(x(1)) \geq (1 - \frac{1}{e})F(x^*)$ where x^* denotes the optimal solution,

$$v(x(t)) \in \arg\max_{v \in P \cap \{v : v \leq 1 - x(t)\}} \langle \nabla F(x(t)), v \rangle$$
$$\dot{x}(t) = \alpha_t v(x(t))$$

for F is non-monotone, P is down-closed with $F(x(1)) \geq \frac{1}{e}F(x^*)$ and

$$v(x(t)) \in \arg\max_{v \in P} \langle \nabla F(x(t)), v \rangle$$
$$\dot{x}(t) = \alpha_t v(x(t))$$

for F is non-monotone, P is only convex with $F(x(1)) \geq \frac{1}{4}F(x^*)$. The initial condition of all the three systems above is $x(0) = \mathbf{0}$ due to the assumption that $\mathbf{0} \in P$.

3 Maximizing a Monotone DR-Submodular Function with Convex Constraint

In this section, we will discuss the adaptive stepsize algorithm for solving the problem of maximizing a monotonically increasing objective function and its approximation guarantee. In order to better illustrate the strategy of how to choose the stepsize, we first give an ideal algorithm who needs an oracle which can return a solution of a uni-variate continuous monotone equation and then propose the actual implementable algorithm.

3.1 Algorithm with Uni-Variate Equation Oracle

The ideal version mentioned above is shown as Algorithm 1. In this algorithm, the stepsize denoted as δ^j is determined by solving Eq. (4) whose LHS is monotonically decreasing with respect to the unique variable δ^j.

Algorithm 1: Continuous greedy via line-search (CGLS)

Input: feasible region P, function F which is **monotone**

$x^0 \leftarrow 0$, $j \leftarrow 0$, $t_0 \leftarrow 0$.

while $t_j < 1$ **do**

> Find $v^j \in \arg\max_{v \in P} \left\langle \nabla F(x^j), v \right\rangle$
>
> Find $\delta^j \in [0, 1 - t_j]$ ($\delta^j = 1 - t_j$ if not exists) such that
>
> $$\left\langle \nabla F(x^j + \delta^j v^j), v^j \right\rangle = \left\langle \nabla F(x^j), v^j \right\rangle - \epsilon \qquad (4)$$
>
> $x^{j+1} \leftarrow x^j + \delta^j v^j$;
>
> $j \leftarrow j + 1$.

end

Return: $x \leftarrow x^j$.

In contrast to the fixed stepsize strategy that $\delta^j \equiv \frac{1}{K}$ in [1] where K denotes the iteration number, in this algorithm we obtain a stepsize through solving Eq. (4). From a high level perspective, this stepsize is chosen to guarantee that the directional derivatives along the direction v^j between two adjacent iteration points differ exactly ϵ.

We must ensure that the output of Algorithm 1 is feasible before analyzing its complexity and approximation ratio. The following lemma establishes its feasibility.

Lemma 1. *The output of Algorithm 1 satisfies $x \in P$.*

Let K denote the total number of steps taken by the "while" loop. We will now prove the following lemma.

Lemma 2. *For Algorithm 1 we have $K \leq \frac{\min\{\|\nabla F(0)\|_1, nL\}}{\epsilon} + 2$.*

Proof. For $j = 0, \ldots, K - 2$ we have

$$\langle \nabla F(x^j) - \nabla F(x^{j+1}), \mathbf{1} \rangle \geq \langle \nabla F(x^j) - \nabla F(x^{j+1}), v^j \rangle = \epsilon.$$

Adding up the above equation for j from 0 to $K - 2$ yields

$$(K - 2)\epsilon \leq \sum_{j=0}^{K-3} \langle \nabla F(x^j) - \nabla F(x^{j+1}), \mathbf{1} \rangle$$
$$= \langle \nabla F(x^0) - \nabla F(x^{K-2}), \mathbf{1} \rangle.$$

Further noting that $\|\nabla F(0)\|_1 \geq \langle \nabla F(x^0) - \nabla F(x^{K-2}), \mathbf{1} \rangle$ due to the monotonicity of F, we can conclude that $K \leq \frac{\|\nabla F(0)\|_1}{\epsilon} + 2$.

On the other hand, by the L-smoothness and DR-submodularity of F, we also have

$$\langle \nabla F(x^0) - \nabla F(x^{K-2}), \mathbf{1} \rangle \leq \langle \nabla F(0) - \nabla F(1), \mathbf{1} \rangle \leq nL.$$

We complete the proof.   □

It is worth noting that the iteration complexity of this algorithm is upper-bounded by the iteration complexity presented in [1]. Additionally, even when the original function F is not L-smooth, the algorithm can still achieve its complexity by assuming the uni-variate equation oracle hypothesis.

Theorem 1. *When F is monotone, the solution x output by Algorithm 1 satisfies*

$$F(x) \geq (1 - e^{-1})F(x^*) - \epsilon.$$

3.2   Algorithm with Binary Search

Due to the general intractability of obtaining an exact solution for Eq. (4) within polynomial time, we plan to utilize a binary search approach to obtain an approximate solution without sacrificing the approximation ratio achieved by Algorithm 1. This will give rise to Algorithm 2. The only difference in the descriptions of the two algorithms lies in the requirement for the stepsize δ^j in each iteration.

Analogue with previous analysis for Algorithm 1, the iteration number K of the 'while' loop in Algorithm 2 can also be bounded.

Corollary 1. *For Algorithm 2 we have $K \leq \frac{2\min\{\|\nabla F(0)\|_1, nL\}}{\epsilon} + 2$.*

To obtain the complexity of the gradient evaluation of F, we also need to discuss the number of steps of the binary search process in each iteration.

Lemma 3. *For each iteration $j \in \{0, \ldots, K - 1\}$, the stepsize δ^j can be found after at most $M = 1 + \log_2 \frac{nL}{\epsilon}$ steps of binary search if it exists.*

Algorithm 2: Continuous greedy with binary search (CGBS)

Input: feasible region $P \subseteq [0, 1]^n$, function F which is monotone.
Initialization: $x^0 \leftarrow \mathbf{0}$, $j \leftarrow 0$, $t_0 \leftarrow 0$.
while $t_j < 1$ **do**

> Find $v^j \in \arg\max_{v \in P} \langle \nabla F(x^j), v \rangle$.
> Binary search a $\delta^j \in [0, 1 - t_j]$ such that ($\delta^j = 1 - t_j$ if not exists)
>
> $$\left\langle \nabla F(x^j), v^j \right\rangle - \left\langle \nabla F(x^j + \delta^j v^j), v^j \right\rangle \in [\tfrac{\epsilon}{2}, \epsilon]. \tag{5}$$
>
> $t_{j+1} \leftarrow t^j + \delta^j$;
> $x^{j+1} \leftarrow x^j + \delta^j v^j$;
> $j \leftarrow j + 1$.

end
Return $x \leftarrow x^j$.

Proof. Denote δ^* as the exact solution of equation

$$\phi(\delta^j) := \left\langle \nabla F(x^j), v^j \right\rangle - \left\langle \nabla F(x^j + \delta^j v^j), v^j \right\rangle - \frac{\epsilon}{2} = 0. \tag{6}$$

By the monotonicity and continuity of the uni-variate function $\phi(\delta^j)$, an interval that contains δ^* with length 2^{-M} can be found using the binary search. Let the right endpoint of the interval be δ^j and we can conclude that

$$0 \leq \delta^j - \delta^* \leq 2^{-M} \leq \frac{\epsilon}{2nL}.$$

By the L-smoothness of F, we then have

$$\left| \left\langle \nabla F(x^j + \delta^* v^j), v^j \right\rangle - \left\langle \nabla F(x^j + \delta^j v^j), v^j \right\rangle \right|$$
$$= \left| \left\langle \nabla F(x^j + \delta^* v^j) - \nabla F(x^j + \delta^j v^j), v^j \right\rangle \right|$$
$$\leq L(\delta^j - \delta^*)\|v^j\|^2 \leq \frac{\epsilon}{2},$$

and thus

$$\left\langle \nabla F(x^j), v^j \right\rangle - \left\langle \nabla F(x^j + \delta^j v^j), v^j \right\rangle$$
$$= \frac{\epsilon}{2} + \left\langle \nabla F(x^j + \delta^* v^j), v^j \right\rangle - \left\langle \nabla F(x^j + \delta^j v^j), v^j \right\rangle$$
$$\leq \epsilon.$$

On the other hand, by the monotonicity of ∇F, we have

$$\left\langle \nabla F(x^j + \delta^* v^j), v^j \right\rangle \geq \left\langle \nabla F(x^j + \delta^j v^j), v^j \right\rangle.$$

Together with the definition of δ^* the proof can be completed.   □

Theorem 2. *Assume that F is monotone. Then Algorithm 2 output a solution x satisfying*

$$F(x) \geq (1 - e^{-1})F(x^*) - \epsilon.$$

The LOO oracle complexity is at most $O(\frac{\|\nabla F(0)\|_1}{\epsilon})$. The gradient evaluation complexity is at most $O(\frac{\|\nabla F(0)\|_1}{\epsilon} \log \frac{nL}{\epsilon})$.

4 Non-monotone DR-Submodular Maximization

In this section, we will discuss how to address the problem of maximizing a non-monotone DR-submodular function. For this scenario, we will provide two different algorithms depending on whether the constraints satisfy the down-closed property.

4.1 Down-Closed Constraint

Algorithm 3 is designed for the down-closed case, which means that if there are $x \in P$ and $0 \leq y \leq x$ then there holds $y \in P$. Its main framework is originally from the measured continuous greedy algorithm in [14], which was first proposed for solving the maximization problem of the multilinear extension relaxation of submodular set functions, and later it was proven to be applicable to general DR-submodular functions maximization as well. In [1], this algorithm was proven to require $O(\frac{nL}{\epsilon})$ iterations to guarantee an approximation loss of ϵ. Similar to the previous section, we can also demonstrate that the use of adaptive stepsizes can improve the iteration complexity.

Algorithm 3: Measured continuous greedy via binary-search (MCGBS)

Input: feasible region $P \subseteq [0,1]^n$ which is down-closed, function F.
$x^0 \leftarrow \mathbf{0}$, $j \leftarrow 0$, $t_0 \leftarrow 0$.
while $t_j < 1$ **do**

> Find $v^j \in \arg\max_{v \in P \cap \{v:v \leq 1-x^j\}} \langle \nabla F(x^j), v \rangle$.
> Binary search a $\delta^j \in [0, 1 - t_j]$ such that ($\delta^j = 1 - t_j$ if not exists)
>
> $$\left\langle \nabla F(x^j), v^j \right\rangle - \left\langle \nabla F(x^j + \frac{\delta^j}{e^{\delta^j}}v^j), v^j \right\rangle \in [\frac{\epsilon}{2}, \epsilon]. \qquad (7)$$
>
> $t_{j+1} \leftarrow t^j + \delta^j$;
> $x^{j+1} \leftarrow x^j + \frac{\delta^j}{e^{\delta^j}}v^j$;
> $j \leftarrow j + 1$.

end
Return: $x \leftarrow x^j$.

The feasibility of the output of Algorithm 3 can be guaranteed by the down-closeness of P.

Lemma 4. *The output of Algorithm 3 satisfies $x \in P$.*

Due to the lack of monotonicity assumption of F, the analysis of the iteration complexity of Algorithm 3 is different with that of Algorithm 2.

Lemma 5. *For Algorithm 3, there holds $K \leq \min\{n + \frac{2\|\nabla F(0)\|_1}{\epsilon}, \frac{2nL}{\epsilon}\} + 2$.*

Theorem 3. *Let the feasible region $P \subseteq [0,1]^n$ be down-closed. Then Algorithm 3 outputs a solution x satisfying*

$$F(x) \geq e^{-1} F(x^*) - \epsilon.$$

The LOO oracle complexity is at most $O(n + \frac{\|\nabla F(0)\|_1}{\epsilon})$. The gradient evaluation complexity is at most $O((n + \frac{\|\nabla F(0)\|_1}{\epsilon}) \log \frac{nL}{\epsilon})$.

4.2 General Convex Constraint

In this section, a Frank-Wolfe type algorithm is presented for non-monotone DR-submodular function maximization with convex constraint via binary search. In the algorithm, the stepsize is determined by solving (8). Unlike the algorithms presented in [1] and the previous sections of this paper, this algorithm introduces a noteworthy variation. In this approach, it is necessary to keep track of the value and corresponding function value of each iteration point. Upon completion of the iteration process, the algorithm returns the iteration point with the highest function value, rather than solely using the last obtained point.

Algorithm 4: Frank-Wolfe with binary search (FWBS)

Input: feasible region $P \subseteq [0,1]^n$, function F.
Initialization: $x^0 \leftarrow 0$, $j \leftarrow 0$, $t_0 \leftarrow 0$.
while $t_j < 1$ **do**

Find $v^j \in \arg\max_{v \in P} \langle \nabla F(x^j), v \rangle$.
Binary search a $\delta^j \in [0, 1 - t_j]$ such that ($\delta^j = 1 - t_j$ if not exists)

$$\left\langle \nabla F(x^j), v^j \right\rangle - \left\langle \nabla F(x^j + \left(1 - 2^{-\delta^j}\right)(v^j - x^j)), v^j \right\rangle \in [\tfrac{\epsilon}{2}, \epsilon]; \quad (8)$$

$t_{j+1} \leftarrow t^j + \delta^j$;
$x^{j+1} \leftarrow x^j + \left(1 - 2^{-\delta^j}\right)(v^j - x^j)$;
$j \leftarrow j + 1$.
end
Return $x \leftarrow \arg\max_{x^j} F(x^j)$.

The feasibility of x output by Algorithm 4 is shown as follows.

Lemma 6. *Algorithm 4 outputs a solution satisfying $x \in P$.*

Theorem 4. *Algorithm 4 outputs a solution x satisfying*

$$F(x) \geq \frac{1}{4}F(x^*) - \frac{\epsilon}{2}.$$

The LOO oracle complexity is at most $O(n + \frac{\|\nabla F(0)\|_1}{\epsilon})$. The gradient evaluation complexity is at most $O((n + \frac{\|\nabla F(0)\|_1}{\epsilon}) \log \frac{nL}{\epsilon})$.

5   Examples

Our focus in this article lies in enhancing the complexity of maximizing DR-submodular functions. To verify that whether the adaptive stepsize strategy can be faster, we present three examples in this section.

Multilinear Extension (MLE) for Submodular Set Functions. Given a finite ground set V, and a function $f : 2^V \mapsto \mathbb{R}$, its multilinear extension is defined as follows:

$$F(x) = \mathbb{E}\left[f(R(x))\right] = \sum_{S \subseteq V} f(S) \prod_{i \in S} x_i \prod_{i \notin S}(1 - x_i),$$

where $x \in [0,1]^{|V|}$ and $R(x)$ denote a random subset of V that each element i belongs to $R(x)$ with probability x_i independently. It can be proven that function f is submodular (i.e., $f(X \cap Y) + f(X \cup Y) \leq f(X) + f(Y)$ for $\forall X, Y \subseteq V$) if and only if its multilinear extension F is DR-submodular. Continuous greedy algorithms designed using the concept of multilinear extension can be used to solve the maximization problem of submodular set functions under various constraints, and have been shown to have high approximation guarantees [15,16].

Denote the vector with the i-th entry equal to 1 and the others equal to 0 by e_i. The upper bound of $\|\nabla F(0)\|_1$ can be bounded as follows.

Lemma 7. *Assume that F is a multilinear extension of a submodular set function f and the feasible region $P \supseteq \{e_i\}_{i=1}^n$. Then we have*

$$\|\nabla F(0)\|_1 \leq nF(x^*).$$

The Lipschitz constant L of the multilinear extension for submodular set function is $L = O(n^2)F(x^*)$ [14].

Softmax Extension for DPP MAP Problems. Determinantal point processes are probabilistic models of repulsion, that have been used to model diversity in machine learning. Let A be the positive semi-definite kernel matrix of a DPP. The softmax extension of the DPP MAP problem is

$$F(x) = \log \det(diag(x)(A - I) + I), \ x \in [0,1]^n,$$

where I is the identity matrix, $diag(x)$ denotes the diagonal matrix induced by x. By Corollary 2 in [17], the gradient of the softmax extension $f(x)$ is

$$\nabla_i F(x) = ((diag(x)(A - I) + I)^{-1}(A - I))_{ii}, \ \text{for } \forall i \in [n],$$

and thus $\|\nabla_i F(\mathbf{0})\|_1 = \sum_{i=1}^{n} |A_{ii} - 1|$. By the application of DPP problem, the matrix A is usually a Gram matrix and A_{ii} is universally bounded. Hence we can conclude that the asymptotic bound of $\|\nabla_i F(\mathbf{0})\|_1$ is at most $O(n)$. However, the asymptotic bound of the gradient Lipshcitz constant with respect to n of the softmax extension remains unfounded.

DR-Submodular Quadratic Functions. Let $F(x)$ be a quadratic function with the form

$$F(x) = \frac{1}{2}x^T H x + h^T x + c.$$

In this scenario, $F(x)$ is DR-submodular when $H \in \mathbb{R}_{-}^{n \times n}$. It is easy to observe that $\|\nabla F(\mathbf{0})\|_1 = \|h\|_1 \leq n$ and the gradient Lipschitz constant $L = \|H\|_2$.

Table 1. Comparisons of complexities between adaptive stepsizes and constant stepsizes for three examples. (grad.eval: the complexity of gradient evaluation)

Examples	Adaptive stepsize		Constant stepsize
	LOO	grad.eval	LOO (grad.eval)
MLE	$O\left(\frac{n}{\epsilon}\right)$	$O(\frac{n}{\epsilon} \log \frac{n}{\epsilon})$	$O(\frac{n^3}{\epsilon})$
Softmax	$O\left(\frac{n}{\epsilon}\right)$	$O(\frac{n}{\epsilon} \log \frac{nL}{\epsilon})$	$O(\frac{nL}{\epsilon})$
quadratic	$O\left(\frac{\|h\|_1}{\epsilon}\right)$	$O(\frac{\|h\|_1}{\epsilon} \log \frac{n\|H\|_2}{\epsilon})$	$O(\frac{n\|H\|_2}{\epsilon})$

The complexities of algorithms to solve the above three constrained DR-submodular function maximization problems are shown as in Table 1. From the table, it can be observed that the adaptive stepsize strategy proposed in this paper has lower complexity compared to the constant stepsize strategy in both MLE and softmax scenarios. However, in the quadratic scenario, it is not possible to definitively compare their complexities as they depend on two different parameters of the original problem.

References

1. Du, D.: Lyapunov function approach for approximation algorithm design and analysis: with applications in submodular maximization (2022). arXiv preprint arXiv:2205.12442
2. Iwata, S., Fleischer, L., Fujishige, S.: A combinatorial strongly polynomial algorithm for minimizing submodular functions. J. ACM (JACM) **48**(4), 761–777 (2001)
3. Nemhauser, G.L., Wolsey, L.A., Fisher, M.L.: An analysis of approximations for maximizing submodular set functions–i. Math. Program. **14**, 265–294 (1978)
4. Krause, A., Golovin, D.: Submodular function maximization. Tractability **3**, 71–104 (2014)

5. Bian, Y., Buhmann, J.M., Krause, A.: Continuous submodular function maximization (2020). arXiv preprint arXiv:2006.13474
6. Niazadeh, R., Roughgarden, T., Wang, J.R.: Optimal algorithms for continuous non-monotone submodular and dr-submodular maximization. J. Mach. Learn. Res. **21**, 4937–4967 (2020)
7. Bian, A., Levy, K., Krause, A., Buhmann, J.M.: Continuous DR-submodular maximization: structure and algorithms. In: Advances in Neural Information Processing Systems, vol. 30 (2017)
8. Bian, A.A., Mirzasoleiman, B., Buhmann, J., Krause, A.: Guaranteed non-convex optimization: Submodular maximization over continuous domains. In: Artificial Intelligence and Statistics, PMLR, pp. 111–120 (2017)
9. Soma, T., Yoshida, Y.: Non-monotone DR-submodular function maximization. In: Proceedings of the AAAI Conference on Artificial Intelligence, vol. 31 (2017)
10. Mualem, L., Feldman, M.: Resolving the approximability of offline and online non-monotone dr-submodular maximization over general convex sets. In: International Conference on Artificial Intelligence and Statistics, PMLR, pp. 2542–2564 (2023)
11. Chen, L., Feldman, M., Karbasi, A.: Unconstrained submodular maximization with constant adaptive complexity. In: Proceedings of the 51st Annual ACM SIGACT Symposium on Theory of Computing, pp. 102–113 (2019)
12. Ene, A., Nguyen, H.: Parallel algorithm for non-monotone DR-submodular maximization. In: International Conference on Machine Learning, PMLR, pp. 2902–2911 (2020)
13. Hassani, H., Soltanolkotabi, M., Karbasi, A.: Gradient methods for submodular maximization. In: Guyon, I., et al. (eds.) Advances in Neural Information Processing Systems, vol. 30, Curran Associates, Inc. (2017)
14. Feldman, M., Naor, J., Schwartz, R., A unified continuous greedy algorithm for submodular maximization. In: IEEE 52nd Annual Symposium on Foundations of Computer Science. IEEE 2011, pp. 570–579 (2011)
15. Calinescu, G., Chekuri, C., Pál, M., Vondrák, J.: Maximizing a monotone submodular function subject to a matroid constraint. SIAM J. Comput. **40**(6), 1740–1766 (2011)
16. Chekuri, C., Vondrák, J., Zenklusen, R.: Submodular function maximization via the multilinear relaxation and contention resolution schemes. SIAM J. Comput. **43**(6), 1831–1879 (2014)
17. Gillenwater, J., Kulesza, A., Taskar, B.: Near-optimal map inference for determinantal point processes. In: NIPS 2012, pp. 2735–2743. Curran Associates Inc., Red Hook (2012)

On the Routing Problems in Graphs with Ordered Forbidden Transitions

Kota Kumakura, Akira Suzuki(iD), Yuma Tamura$^{(\boxtimes)}$, and Xiao Zhou

Graduate School of Information Sciences, Tohoku University, Sendai, Japan
kota.kumakura.s8@dc.tohoku.ac.jp, {akira,tamura,zhou}@tohoku.ac.jp

Abstract. Finding a path between two vertices of a given graph is one of the most classic problems in graph theory. Recently, problems of finding a route avoiding forbidden transitions, that is, two edges that cannot be passed through consecutively, have been studied. In this paper, we introduce the ordered variants of these problems, namely the PATH AVOID-ING ORDERED FORBIDDEN TRANSITIONS problem (PAOFT for short) and the TRAIL AVOIDING ORDERED FORBIDDEN TRANSITIONS problem (TAOFT for short). We show that both the problems are NP-complete even for bipartite planar graphs with maximum degree three. Since the problems are solvable for graphs with maximum degree two, the NP-completeness results are tight with respect to the maximum degree of a graph. Furthermore, we show that TAOFT remains NP-complete for cactus graphs. As positive results of PAOFT, we give a polynomial-time algorithm for bounded treewidth graphs and a linear-time algorithm for cactus graphs.

Keywords: Graph algorithms · NP-completenss · Forbidden transitions

1 Introduction

Nowadays, it is quite common to use car navigation systems and map applications to find a route from a current location to a destination. Road networks are typically represented as graphs. Finding routes on graphs has been studied since the earliest days in the fields of graph theory and graph algorithms. In particular, a path between two vertices of a graph can be obtained in linear time by using classic algorithms such as breadth-first search and depth-first search. However, the actual road networks have various constraints that cannot be fully captured by conventional graph modeling alone. For instance, to prohibit travel in a specified direction, such as "no left turn" or "no entry," additional constraints are required on the graph.

A. Suzuki—Partially supported by JSPS KAKENHI Grant Number JP20K11666, Japan.
Y. Tamura—Partially supported by JSPS KAKENHI Grant Number JP21K21278, Japan.
X. Zhou—Partially supported by JSPS KAKENHI Grant Number JP19K11813, Japan.

W. Wu and G. Tong (Eds.): COCOON 2023, LNCS 14422, pp. 359–370, 2024.
https://doi.org/10.1007/978-3-031-49190-0_26

One way to represent this constraint is to provide *forbidden transitions* on the graph [7]. A transition $\{e_1, e_2\}$ is a pair of adjacent edges e_1, e_2 in a graph. By making it impossible to consecutively pass through the two edges in the transition, possible routes between two vertices of the graph are restricted. Given a graph $G = (V, E)$, two vertices $s, t \in V$, and a set $\mathcal{F}$ of forbidden transitions in G, the PATH AVOIDING FORBIDDEN TRANSITIONS problem (PAFT for short) asks whether there exists an s-t path of G that avoids forbidden transitions in $\mathcal{F}$. While this problem is known to be NP-complete on general graphs [9] and even on grid graphs [5], it has been studied from viewpoints of fixed-parameter tractability [1,5] and an exact algorithm [4]. On the other hand, the problem of finding an s-t trail that avoids forbidden transitions, namely the TRAIL AVOID-ING FORBIDDEN TRANSITIONS problem (TAFT for short), is known to be solv-able in polynomial time [8]. Recall that a trail can pass through the same vertex more than once.

However, when focusing on the real road network again, the road network cannot be completely modeled with forbidden transitions. If "no left turn" (from a point a to a point b) at an intersection is represented by a forbidden transition on a graph, then "no right turn" (from b to a) is also provided. This is because a forbidden transition consists of an *unordered* pair of adjacent edges, not an *ordered* pair of adjacent edges.

1.1   Our Contributions

In this paper, motivated by the above situation, we introduce the PATH AVOID-ING ORDERED FORBIDDEN TRANSITIONS problem (PAOFT for short) and the TRAIL AVOIDING ORDERED FORBIDDEN TRANSITIONS problem (TAOFT for short). For a graph $G = (V, E)$, an *ordered transition* (e_1, e_2) in G is an ordered pair of adjacent edges $e_1, e_2 \in E$, that is, transitions (e_1, e_2) and (e_2, e_1) are distinguished. Let $\mathcal{F} \subseteq E \times E$ be a set of ordered forbidden transitions in G. A path or a trail $\langle e_1, e_2, \ldots, e_\ell \rangle$ consisting of consecutive edges in G is *compatible with* $\mathcal{F}$ or $\mathcal{F}$-*compatible* if $(e_i, e_{i+1}) \notin \mathcal{F}$ for every $i \in \{1, \ldots, \ell - 1\}$. The PAOFT problem (resp. the TAOFT problem) asks whether given a graph $G = (V, E)$, two vertices $s, t \in V$, and a set $\mathcal{F}$ of ordered forbidden transitions in G, there exists an $\mathcal{F}$-compatible s-t path (resp. an $\mathcal{F}$-compatible s-t trail) of G.

Since PAOFT is a generalization of PAFT, all intractable results of PAFT are inherited by PAOFT. In this paper, we show that the intractability of the ordered variants is more severe: PAOFT and TAOFT are NP-complete for bipartite planar graphs with maximum degree three. It is clear that PAOFT and TAOFT are solvable for graphs with maximum degree at most two, that is, paths or cycles. Thus, our NP-completeness results for PAOFT and TAOFT are tight with respect to maximum degree of a graph. Furthermore, we show that TAOFT remains NP-complete for cactus graphs.

In contrast to the intractability, we first observe that PAFT on graphs with maximum degree at most three is solvable in polynomial time. Combined with the result of Kanté et al. [5], this gives a complexity dichotomy of PAFT with respect to maximum degree of a graph. We then provide a polynomial-time

Table 1. The complexity of PAOFT, PAFT, TAOFT, and TAFT

	PAOFT	PAFT	TAOFT	TAFT
General graphs	NP-comp. [9]			Poly.time [8]
Grid graphs	NP-comp. [5]			
Graphs with max. degree 3	NP-comp. [Thm. 1]	Linear time [Thm. 3]	NP-comp. [Thm. 1]	
Bipartite planar graphs with max. degree 3				
Bounded treewidth graphs	Poly. time [Thm. 4]	Poly. time [5]	NP-comp. [Thm. 2]	
Cactus graphs	Linear time [Thm. 5]			

algorithm for PAOFT on bounded treewidth graphs, whereas TAOFT remains NP-complete for cactus graphs, which have bounded treewidth. Our algorithm for bounded treewidth graphs also improves the polynomial factor of the existing algorithm of PAFT for bounded treewidth graphs [5]. Finally, we design a linear-time algorithm of PAOFT for cactus graphs. Since all cactus graphs are planar graphs, the tractability of PAOFT on cactus graphs gives a nice contrast to the hardness result for planar graphs. Moreover, the linear-time algorithm for cactus graphs runs polynomially faster than the algorithm for bounded treewidth graphs.

We summarize our results in Table 1, which highlights interesting differences between the problems involving unordered or ordered forbidden transitions. In the unordered variants, TAFT is more tractable than PAFT because TAFT is solvable in polynomial time for general graphs [8]. In the ordered variants, however, a reversal of the tractability occurs in some sense: TAOFT is intractable even for cactus graphs, whereas PAOFT is tractable for bounded treewidth graphs.

2 Preliminaries

Let $G = (V, E)$ be a graph: we denote by $V(G)$ and $E(G)$ the vertex set and the edge set of G, respectively. We assume that all the graphs in this paper are simple, undirected and unweighted. For a vertex v of G, we denote by $N_G(v)$ the *neighborhood* of v in G, that is, $N_G(v) = \{w \in V \mid vw \in E\}$. The *degree* of a vertex v in G is defined as the size of $N_G(v)$. For a positive integer n, let K_n and P_n denote the complete graph and the path with n vertices, respectively. For two graphs $G_1 = (V_1, E_1)$ and $G_2 = (V_2, E_2)$, we denote by $G_1 + G_2$ the disjoint union of G_1 and G_2, that is, $G_1 + G_2 = (V_1 \cup V_2, E_1 \cup E_2)$. For positive integers i, we write $[i]$ as the shorthand for the set $\{1, 2, \ldots, i\}$ of integers. A graph G is said to be *connected* if there is a path between any two vertices of G. A maximal connected subgraph of G is called a *connected component* of G.

362   K. Kumakura et al.

A graph G is *planar* if G can be embedded in a plane without crossing edges. A graph G is a *cactus* if any two cycles share at most one vertex. Every cactus graph is planar.

The *endpoints* of an edge e in a graph G are vertices incident to e. Edges e and e' of G are said to be *adjacent* if e shares an endpoint with e'. A sequence $\langle e_1, e_2, \ldots, e_\ell \rangle$ of edges in G is called a *walk* W of G if there is a sequence $\langle v_1, v_2, \ldots, v_{\ell+1} \rangle$ of vertices in G such that $e_i = v_i v_{i+1}$ for each $i \in [\ell]$, where ℓ is the *length* of W, that is, the number of edges in W. For an edge e of G, we say that a walk $W = \langle e_1, e_2, \ldots, e_\ell \rangle$ *passes through* e if $e = e_i$ for some $i \in [\ell]$. Similarly, for a vertex v of G, we say that the walk *passes through* v if v is an endpoint of e_i for some $i \in [\ell]$. A walk is a *path* (resp. a *trail*) of G if the walk passes through each vertex (resp. edge) at most once.

The problems PAFT, TAFT, PAOFT, and TAOFT are defined as in Introduction. Instead of a set $\mathcal{F}$ of ordered forbidden transitions in a graph $G = (V, E)$, we sometimes use a set $\mathcal{A}$ of *ordered allowed transitions* in G defined as $\mathcal{A} = E \times E \setminus \mathcal{F}$. For adjacent two edges e_1 and e_2 of G, we write $\{e_1, e_2\}$ as a shorthand for two ordered transitions (e_1, e_2) and (e_2, e_1).

3 NP-Completeness

3.1 PAOFT and TAOFT for Graphs with Maximum Degree Three

Recall that PAFT remains NP-complete even for grid graphs [5]. This directly means that PAFT and PAOFT are NP-complete for bipartite planar graphs with maximum degree four. In this section, inspired by the proof in [5], we show that PAOFT and TAOFT remain NP-complete for more restricted graphs.

Theorem 1. PAOFT *and* TAOFT *are NP-complete for bipartite planar graphs with maximum degree three.*

Due to space limitation, we here show the NP-completeness of PAOFT on graphs with maximum degree three. Clearly, PAOFT is in NP. To show the NP-hardness of PAOFT, we reduce the SATISFIABILITY problem (SAT for short) to PAOFT on graphs with maximum degree three. Recall that SAT asks whether there exists a satisfying truth assignment of a CNF formula ϕ. SAT is a well-known NP-complete problem [6].

We transform a CNF formula ϕ for SAT to a graph for PAOFT. We first give an overview of the transformation. Suppose that ϕ has n variables and m clauses. For simplicity of the reduction, we assume that m is even. (If m is odd, add the clause $(x_1 \vee \neg x_1)$ into ϕ.) For each clause C_i and each variable x_j, where $i \in [m]$ and $j \in [n]$, we prepare a variable gadget $G_{i,j}$ whose edges are assigned to red or blue. Then, we arrange them in a grid pattern as shown in Fig. 1, and connect adjacent graphs. In the grid, rows and columns correspond to clauses and variables of ϕ, respectively. We will construct a set $\mathcal{F}$ of ordered forbidden transitions in G such that, there exists an $\mathcal{F}$-compatible s-t path in G if and only if the path changes the color of passing edges once during traversing of

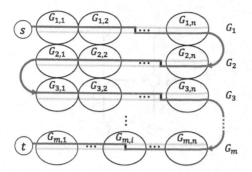

Fig. 1. We arrange variable gadgets in a grid pattern such that ϕ is satisfiable if and only if there exists an $\mathcal{F}$-compatible s-t path.

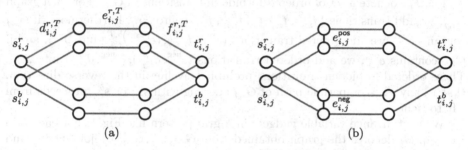

Fig. 2. (a) A variable gadget $G_{i,j}$ when C_i contains no variable x_j, and (b) a variable gadget $G_{i,j}$ when C_i contains both a positive literal x_j and a negative literal $\neg x_j$. Some labels are omitted for visibility.

$G_{i,1}, \ldots, G_{i,n}$ for each $i \in [m]$. The changes of the colors indicate that there is a satisfying truth assignment of ϕ.

We now explain the details of the reduction. A literal in ϕ is said to be *negative* if it has a negation; otherwise, *positive*. For each $i \in [m]$ and each $j \in [n]$, a *variable gadget* $G_{i,j}$ consists of a red cycle and a blue cycle as shown in Fig. 2(a). We denote by $s^r_{i,j}$ and $t^r_{i,j}$ the left-most vertex and the right-most vertex of the red cycle, respectively. Similarly, $s^b_{i,j}$ and $t^b_{i,j}$ are the left-most vertex and the right-most vertex of the blue cycle, respectively. The upper $s^r_{i,j}$-$t^r_{i,j}$ path in red and the upper $s^b_{i,j}$-$t^b_{i,j}$ path in blue of $G_{i,j}$ are called *positive paths*, and the lower $s^r_{i,j}$-$t^r_{i,j}$ path in red and the lower $s^b_{i,j}$-$t^b_{i,j}$ path in blue of $G_{i,j}$ are called *negative paths*. For $c \in \{r, b\}$, let $\langle d^{c,T}_{i,j}, e^{c,T}_{i,j}, f^{c,T}_{i,j} \rangle$ denote the edge sequence corresponding to the $s^c_{i,j}$-$t^c_{i,j}$ positive path, and $\langle d^{c,F}_{i,j} e^{c,F}_{i,j}, f^{c,F}_{i,j} \rangle$ the edge sequence corresponding to the $s^c_{i,j}$-$t^c_{i,j}$ negative path. As shown in Fig. 2(b), if a clause C_i contains a positive literal x_j, then add an edge $e^{pos}_{i,j}$ between positive paths, and if C_i contains a negative literal $\neg x_j$, then add an edge $e^{neg}_{i,j}$ between negative paths,

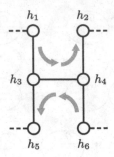

Fig. 3. A gate gadget. The gray arrows represent ordered allowed transitions.

Let $\mathcal{F}_{i,j}$ denote a set of ordered forbidden transitions in $G_{i,j}$. For each graph $G_{i,j}$, we add transitions $\{f_{i,j}^{r,T}, f_{i,j}^{r,F}\}, \{f_{i,j}^{b,T}, f_{i,j}^{b,F}\}$ into $\mathcal{F}_{i,j}$. Furthermore, if $G_{i,j}$ contains $e_{i,j}^{pos}$, we add ordered transitions $(e_{i,j}^{pos}, d_{i,j}^{r,T}), (e_{i,j}^{pos}, d_{i,j}^{b,T})$ into $\mathcal{F}_{i,j}$, and if $G_{i,j}$ contains $e_{i,j}^{neg}$, we add ordered transitions $(e_{i,j}^{neg}, d_{i,j}^{r,F}), (e_{i,j}^{neg}, d_{i,j}^{b,F})$ into $\mathcal{F}_{i,j}$. These ordered forbidden transitions prohibit traveling in the reverse direction, that is, any $\mathcal{F}_{i,j}$-compatible path of $G_{i,j}$ starting with $s_{i,j}^r$ or $s_{i,j}^b$ proceeds from left to right.

We next arrange variable gadgets in a grid pattern like Fig. 1. For each odd $i \in [m]$, G_i denotes the graph obtained from $G_{i,1}, \dots, G_{i,n}$ by joining $t_{i,j}^r$ and $s_{i,j+1}^r$ by a red edge, and joining $t_{i,j}^b$ and $s_{i,j+1}^b$ by a blue edge for each $j \in [n-1]$. On the other hand, for each even $i \in [m]$, G_i denotes the graph obtained from $G_{i,1}, \dots, G_{i,n}$ by join $s_{i,j}^r$ and $t_{i,j+1}^r$ by a red edge, and join $s_{i,j}^b$ and $t_{i,j+1}^b$ by a blue edge for each $j \in [n-1]$. In addition, for each $i \in [m-1]$, we join $t_{i,n}^b$ in G_i and $s_{i+1,n}^b$ in G_{i+1} by a blue edge if i is odd; otherwise, join $t_{i,1}^r$ in G_i and $s_{i+1,1}^r$ in G_{i+1} by a red edge. Let $s = s_{1,1}^r$ and $t = t_{m,1}^r$ (under the assumption that m is even). We denote by G' the graph constructed above. We also define a set $\mathcal{F}'$ of ordered forbidden transitions of G' as a union of $\mathcal{F}_{i,j}$ for $i \in [m]$ and $j \in [n]$. Note that every $\mathcal{F}'$-compatible s-t path of G' traverses $G_{i,1}, G_{i,2}, \dots, G_{i,n}$ in this order if i is odd; otherwise, it traverses $G_{i,n}, G_{i,n-1}, \dots, G_{i,1}$ in this order. Moreover, for each $i \in [m]$, it must pass through an edge $e_{i,j}^{pos}$ or $e_{i,j}^{neg}$ for some $j \in [n]$; otherwise, the path cannot reach G_{i+1} if $i < m$ and t if $i = m$.

As previously mentioned, our goal is to construct a graph G so that there exists an $\mathcal{F}$-compatible s-t path P in G if and only if P changes the color of passing edges during traversing of G_i. The graph G' already satisfies the condition, but P still cannot correspond to a satisfying truth assignment of ϕ. For distinct $i, i' \in [m]$, the path P may pass through edges of positive paths in $G_{i,j}$ and may pass through edges of negative paths in $G_{i',j}$. We desire that P passes through edges of positive paths in $G_{1,j}$ if and only if P passes through edges of positive paths in $G_{i,j}$ for every $i \in [m]$.

To this end, we modify G' by using a *gate gadget* H illustrated in Fig. 3. Let $\mathcal{A}_H = \{(h_1 h_3, h_3 h_4), (h_3 h_4, h_4 h_2), (h_6 h_4, h_4 h_3), (h_4 h_3, h_3 h_5)\}$ be a set of ordered allowed transitions and let $\mathcal{F}_H = E(H) \times E(H) \setminus \mathcal{A}_H$ be a set of ordered forbidden

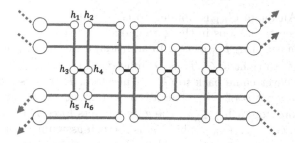

Fig. 4. Bonding the negative paths of $G_{i,j}$ and the positive paths of $G_{i',j}$ by four gate gadgets, where $i' = i+1$ if $i < m$ and $i' = 1$ if $i = m$. The center vertices correspond to h_3 and h_4 of the gate gadgets.

transitions of H. If an $\mathcal{F}_H$-compatible path P starts from h_1, then it must pass through h_3, h_4 and reach h_2. If P starts from h_6, then it must pass through h_4, h_3 and reach h_5. Thus, for any gate gadget H in G, an $\mathcal{F}_H$-compatible s-t path of G cannot visit both h_1 and h_6 of H. This property of H allows us to restrict positive or negative paths that can be traversed by an s-t path avoiding ordered forbidden transitions. For each $i \in [m]$ and each $j \in [n]$, we subdivide $f_{i,j}^{c,B}$ into four vertices for each pair of $c \in \{r, b\}$ and $B \in \{T, F\}$, and bond the negative paths of $G_{i,j}$ and the positive paths of $G_{i',j}$ as shown in Fig. 4, where $i' = i+1$ if $i < m$ and $i' = 1$ if $i = m$. Notice that, if i is even, each gate gadget is flipped horizontally because P proceeds from right to left on G_i and from left to right on $G_{i'}$. By these gate gadgets, P traverses a positive path of $G_{i,j}$ if and only if P also traverses a positive path of $G_{i',j}$. Thus, for each $j \in [n]$, P traverses the same type of paths in $G_{1,j}, \ldots, G_{m,j}$. This completes the construction of a graph G. We define $\mathcal{F}$ as a union of $\mathcal{F}'$ and a set of all ordered forbidden transitions added by gate gadgets. It is not hard to see that G has maximum degree three and our construction of G takes polynomial time. The following lemma gives the correctness of our reduction.

Lemma 1. *Let ϕ be a CNF-formula and G be the graph obtained from ϕ by the above construction. Then, there exists a satisfying truth assignment of ϕ if and only if G has an $\mathcal{F}$-compatible s-t path.*

Proof. Suppose that there exists a satisfying truth assignment of ϕ. We show that it is possible to depart from s and arrive at t without passing through ordered forbidden transitions in $\mathcal{F}$. For each $i \in [m]$ and each $j \in [n]$, traverse positive paths of $G_{i,j}$ if $x_j = T$; otherwise, traverse negative paths of $G_{i,j}$. In particular, if a clause C_i contains a literal x_j (resp. $\neg x_j$) evaluated as True, pass through $e_{i,j}^{pos}$ (resp. $c_{i,j}^{neg}$) so that the color of passing edges changes exactly once during traversing G_i. Since ϕ has a satisfying truth assignment, we can change the color for each G_i, and hence successfully arrive at t. Our walking corresponds to an $\mathcal{F}$-compatible s-t path of G.

Conversely, suppose that P is an $\mathcal{F}$-compatible s-t path of G. We claim that for every $G_{i,j}$, P traverses either positive paths or negative paths. The path P

starts from s. Although $G_{1,1}$ is bonded to $G_{2,1}$ and $G_{m,1}$ by gate gadgets, the ordered forbidden transitions in the gate gadgets prevent P from reaching $G_{2,1}$ and $G_{m,1}$ without visiting $t_{1,1}^r$ or $t_{1,1}^b$. In addition, a subset $\mathcal{F}_{1,1}$ of $\mathcal{F}$ forces P to proceed from left to right in G. Thus, P must visit $t_{1,1}^r$ or $t_{1,1}^b$, reach vertices of $G_{1,2}$, and then never come back to $G_{1,1}$. Therefore, P traverses either positive paths or negative paths of $G_{1,1}$. The above argument can be applied to every $G_{i,j}$. Furthermore, for each $j \in [n]$, the gate gadgets force P to travel the same type of paths in $G_{1,j}, \ldots, G_{m,j}$. This yields a truth assignment of ϕ. Since there exists $j \in [n]$ for each $i \in [m]$ such that P passes through $e_{i,j}^{\text{pos}}$ or $e_{i,j}^{\text{neg}}$, which corresponds to a literal evaluated as True, the truth assignment indeed satisfies ϕ. This completes the proof of lemma.   □

3.2   TAOFT for Cactus Graphs

As we will see in Sect. 4.2, PAOFT is solvable in polynomial time for bounded treewidth graphs. In this section, on the contrary, we show that TAOFT remains NP-complete even for cactus graphs, whose treewidth is at most two.

Theorem 2. TAOFT *remains NP-complete even for cactus graphs.*

Clearly, TAOFT is in NP. We reduce TAOFT on general graphs to TAOFT on cactus graphs. Let $(G, s, t, \mathcal{F})$ be an instance of TAOFT. We denote $V(G) = \{v_1, v_2, \ldots, v_n\}$ and $E(G) = \{e_1, e_2, \ldots, e_m\}$, where $s = v_1, t = v_n$. For each $e_i \in E(G)$, we construct a cycle C_i of length three. Select an arbitrary vertex of C_i for each $i \in [m]$ and then identify them with a single vertex c. For $e_i = v_{k_1}v_{k_2} \in E(G)$, let $e_i^{k_1}$ and $e_i^{k_2}$ denote edges of C_i incident to c, and let e_i' denote the other edge of C_i. We also join a vertex s' (resp. t') and c by an edge e_s (resp. e_t). Then, G' denotes the graph constructed above. We next define a set $\mathcal{A}'$ of ordered allowed transitions in G' as follows:

$$\mathcal{A}' = \{\{e_i^{k_1}, e_i'\}, \{e_i', e_i^{k_2}\} : i \in [m], k_1, k_2 \in [n], \text{and}$$
$$v_{k_1}, v_{k_2} \text{ are distinct endpoints of } e_i\}$$
$$\cup \{(e_i^k, e_j^k) : i, j \in [m], k \in [n], (e_i, e_j) \notin \mathcal{F}, \text{and}$$
$$e_i \text{ shares an endpoint } v_k \text{ with } e_j\}$$
$$\cup \{(e_s, e_i^1) : i \in [m] \text{ and } e_i \text{ is incident to } s\}$$
$$\cup \{(e_i^n, e_t) : i \in [m] \text{ and } e_i \text{ is incident to } t\}$$

Finally, let $\mathcal{F}' = E(G') \times E(G') \setminus \mathcal{A}'$ be a set of ordered forbidden transitions of G'. This completes the construction of a new instance $(G', s', t', \mathcal{F}')$ of TAOFT. Clearly, the construction can be done in polynomial time and G' is a cactus graph. The correctness of this reduction is given by the following lemma, whose proof is omitted due to space limitation.

Lemma 2. G *has an* $\mathcal{F}$-*compatible* s-t *trail if and only if* G' *has an* $\mathcal{F}'$-*compatible* s'-t' *trail.*

4 Polynomial-Time Algorithms

4.1 PAFT for Graphs with Maximum Degree Three

We first revisit PAFT and give the following theorem in order to contrast with Theorem 1. The result in [5] that PAFT is NP-complete for grid graphs also indicates that Theorem 3 is tight with respect to maximum degree. We here omit its proof because it follows from the result of Szeider [9].

Theorem 3. PAFT *is solvable in linear time for graphs with maximum degree at most three.*

4.2 PAOFT for Bounded Treewidth Graphs

Kanté et al. designed a polynomial-time algorithm[1] of PAFT for bounded treewidth graphs [5]. We extend the algorithm so that it can be applied to PAOFT.

Theorem 4. *Let G be an n-vertex graph with maximum degree Δ. Given a tree decomposition of width k, there exists an algorithm that solves PAOFT on G in $O(h(k)\Delta^{2k+3}n)$ time, where h is some computable function.*

Proof Sketch. Our algorithm solves PAOFT by means of dynamic programming over a nice tree-decomposition $\mathcal{T} = \langle T, \{X_u \subseteq V(G) : u \in V(T)\}\rangle$ of G. (See [2] for its definition.) For a node $u \in V(T)$, let T_u be a subtree of T rooted at u and let $V_u = \bigcup_{v \in V(T_u)} X_v$. We denote by G_u the subgraph of G induced by V_u. For an $\mathcal{F}$-compatible s-t path P of G, the intersection of P with G_u forms $\mathcal{F}$-compatible subpaths of G_u such that the subpaths are pairwise vertex-disjoint and both endpoints of each subpath are in X_u except for s and t. Based on the subpaths, Kanté et al. considered a 5-tuple $(X_u^0, X_u^1, X_u^2, M_u, S_u)$ consisting of the following sets: for $d \in \{0, 1, 2\}$, X_u^d is a set of vertices $a \in X_u$ such that the subpaths pass through exactly d edges in E_u^a, where E_u^a is a set of edges incident to a in G_u; M is a matching of X_u^1 whose element $\{a, b\}$ indicates that a and b are endpoints of the same subpath; and S is a set of edges e such that e is incident to a vertex in X_u^1 and one of the subpaths passes through e. In PAOFT, however, the direction of the subpaths must also be taken into account. This is accomplished by giving orders to pairs in M. For each node u of T and each possible 5-tuple $(X_u^0, X_u^1, X_u^2, M_u, S_u)$, we determine whether there exist subpaths of G_u that satisfy the conditions above. For a join node u of T, we compute the existence of such subpaths by combining solutions of two 5-tuples for children of u. We omit the details, but the number of possible 5-tuples is bounded by $O(h'(k)\Delta^{k+1})$, where h' is some function, and the computation takes $O(k\Delta)$ time for each pair of 5-tuples. Since T has $O(kn)$ nodes, we conclude that there exists a computable function h such that our algorithm runs in $O(h(k)\Delta^{2k+3}n)$, as claimed in Theorem 4.

$\square$

---

[1] In [5], only a sketch of the algorithm was given. For more details, see a full version from the following URL: https://inria.hal.science/hal-01115395/file/PAFT.pdf.

4.3   PAOFT for Cactus Graphs

Since every cactus graph has treewidth at most two, we have already designed the algorithm that solves PAOFT on cactus graphs in polynomial time by applying Theorem 4. In this section, we give a faster algorithm of PAOFT for cactus graphs.

Theorem 5. PAOFT *is solvable in* $O(|\mathcal{F}| + n)$ *time for cactus graphs.*

We first remove redundant vertices from a given graph G. A *cut vertex* of G is a vertex whose removal from G increases the number of connected components. A graph G is called *2-connected* if G has no cut vertex. A *block* of a graph G is a maximal 2-connected subgraph of G. For a connected cactus graph G with at least two vertices, every block of G is either a cycle or a path with one edge.

One can verify that the following proposition is true.

Proposition 1. *For a connected graph G and two vertices $s, t \in V(G)$, suppose that G is divided into two graphs G_1 and G_2 by a cut vertex c of G such that $s \in V(G_1)$. Then, one of the following two properties holds:*

(1) when $t \in V(G_1)$, any s-t path of G is contained in G_1; and
(2) when $t \in V(G_2)$, any s-t path of G passes through c.

Proposition 1 ensures that if $s, t \in V(G_1)$, then $V(G_2) \setminus \{c\}$ can be removed from G. Let G' be an induced subgraph obtained by removing all redundant vertices from G based on Proposition 1. It is not hard to see that G' consists of a sequence of blocks in G. Moreover, if G is a cactus graph, since each block of G forms either a cycle or a path with one edge, then G' has maximum degree at most four. Applying Theorem 4 to G', we can also obtain an $O(|\mathcal{F}| + n)$-time[2] algorithm for cactus graphs. However, a very large constant is hidden in the computation time of the algorithm. Moreover, a simpler algorithm can be helpful. For these reasons, we here design a faster and simpler algorithm for cactus graphs.

We label the blocks of G' as $B_1, B_2, \ldots, B_\ell$ so that an s-t path P of G traverses B_i prior to B_j for every integers i, j with $1 \leq i < j \leq \ell$. Note that $s \in V(B_1)$ and $t \in V(B_\ell)$. We denote by c_i the cut vertex of G' that contains in both B_i and B_{i+1} for each $i \in [\ell - 1]$. For algorithmic convenience, we also set $c_0 = s$ and $c_\ell = t$. For $i \in [\ell]$, let E_i^h and E_i^t be sets of edges in B_i that are incident to c_{i-1} and c_i, respectively. Observe that $1 \leq |E_i^h| = |E_i^t| \leq 2$ for each $i \in [\ell]$. We denote by G'_i the subgraph induced by $V(B_1) \cup V(B_2) \cup \cdots \cup V(B_i)$. For each $i \in [\ell]$ and each $e \in E_i^t$, we define a Boolean function $g_i(e)$ as follows:

$$
g_i(e) = \begin{cases} 1 & \text{if there is an } \mathcal{F}\text{-compatible } s\text{-}c_i \text{ path } P \text{ of } G'_i \\ & \text{that passes through } e, \\ 0 & \text{otherwise.} \end{cases}
$$

---

[2] A set of ordered forbidden transitions for G' can be constructed by scanning $\mathcal{F}$ and removing redundant transitions from $\mathcal{F}$ in $O(|\mathcal{F}|)$ time. Then, PAOFT on G' can be solved in $O(n)$ time.

Since $c_\ell = t$ and $G'_\ell = G'$, G' has an $\mathcal{F}$-compatible s-t path if and only if $g_\ell(e) = 1$ for some $e \in E_\ell$.

We compute $g_i(e)$ for each $i \in [\ell]$ and each $e \in E_i^t$ by means of dynamic programming. If $i = 1$, then an s-c_1 path that passes through e is uniquely determined because G_1 is a cycle or a path with one edge. Thus, $g_1(e)$ can be computed. For $i > 1$, assume that $g_{i-1}(e')$ for each $e' \in E_{i-1}^t$ has already obtained. If G'_i has an $\mathcal{F}$-compatible s-c_i path P that passes through e, then it holds that

1. G'_{i-1} has an $\mathcal{F}$-compatible s-c_{i-1} path P' that passes through some edge $e' \in E_{i-1}^t$;
2. B_i has an $\mathcal{F}$-compatible c_{i-1}-c_i path P'' that passes through e and some edge $e'' \in E_i^h$; and
3. $(e', e'') \notin \mathcal{F}$.

One can see that the converse is also true. We define a Boolean function $b_i(e'', e)$ such that $b_i(e'', e) = 1$ if and only if B_i has an $\mathcal{F}$-compatible c_{i-1}-c_i path P'' such that $e'', e \in E(P'')$. Then, we have

$$g_i(e) = \bigvee_{e' \in E_{i-1}^t, e'' \in E_i^h} g_{i-1}(e') \wedge b_i(e'', e) \wedge (e', e'') \notin \mathcal{F}.$$

We bound the running time of our algorithm. For a graph $G = (V, E)$, we denote $n = |V|$ and $m = |E|$. We first find all blocks in $O(n + m)$ time [3], and then obtain a reduced graph G' of G. We also obtain a value of $b_i(e'', e)$ in advance for each $i \in [\ell]$, $e'' \in E_i^h$, and $e \in E_i^t$. Since $1 \le |E_i^h| = |E_i^t| \le 2$ and B_i is a path or a cycle, it can be done in $O(n)$ time. Then, compute $g_i(e)$ for each $i \in [\ell]$ and each $e \in E_i^t$ by the above formula. By a naive implementation such that the whole of $\mathcal{F}$ is scanned to decide whether $(e', e'') \notin \mathcal{F}$, the computation runs in $O(|\mathcal{F}|)$ time for each $i \in [\ell]$. This yields an $O(|\mathcal{F}|n)$-time algorithm for PAOFT on cactus graphs. To improve the running time, we prepare $m' = |E(G')|$ buckets in advance, each of which corresponds to an edge of G' and is accessible in $O(1)$ time. If there is an ordered forbidden transition (e_1, e_2) in $\mathcal{F}$, then we store it into the bucket corresponding to e_1. Notice that, after completing the preprocessing in $O(|\mathcal{F}|)$ time, each bucket has $O(1)$ ordered forbidden transitions because G' has maximum degree at most four. Thus, we decide whether $(e', e'') \notin \mathcal{F}$ in $O(1)$ time. As a conclusion, since it is known that $m = O(n)$ holds for cactus graphs, PAOFT is solvable in $O(|\mathcal{F}| + n)$ time for cactus graphs, completing the proof. $\square$

5   Conclusion

In this paper, we introduced PATH AVOIDING ORDERED FORBIDDEN TRANSITIONS and TRAIL AVOIDING ORDERED FORBIDDEN TRANSITIONS to generalize the unordered variants PATH AVOIDING FORBIDDEN TRANSITIONS and TRAIL AVOIDING FORBIDDEN TRANSITIONS, and investigated the complexity of the ordered variants. Combined some known results for the unordered variants, we provided interesting contrasts to these problems as summarized in Table 1. In particular, for the problems of finding a path avoiding forbidden transitions, we gave the tight hardness results with respect to maximum degree of a given graph.

We finally discuss future work. For an n-vertex graph G and a set $\mathcal{F}$ of forbidden transitions in G, Kanté et al. pointed out that, if $|\mathcal{F}| = O(\log n)$, there is an algorithm that solves PATH AVOIDING FORBIDDEN TRANSITIONS in polynomial time [4]. Indeed, on the actual road network, the number of forbidden transitions is expected to be considerably smaller than the total number of transitions. It would be interesting to design such algorithms for the ordered variants.

Acknowledgements. We thank the referees for their valuable comments and suggestions which greatly helped to improve the presentation of this paper.

References

1. Bellitto, T., Li, S., Okrasa, K., Pilipczuk, M., Sorge, M.: The complexity of routing problems in forbidden-transition graphs and edge-colored graphs. Algorithmica **85**(5), 1202–1250 (2023). https://doi.org/10.1007/s00453-022-01064-1
2. Bodlaender, H.L., Kloks, T.: Efficient and constructive algorithms for the pathwidth and treewidth of graphs. J. Algorithms **21**(2), 358–402 (1996). https://doi.org/10.1006/jagm.1996.0049
3. Hopcroft, J., Tarjan, R.: Algorithm 447: efficient algorithms for graph manipulation. Commun. ACM **16**(6), 372–378 (1973). https://doi.org/10.1145/362248.362272
4. Kanté, M.M., Laforest, C., Momège, B.: An exact algorithm to check the existence of (elementary) paths and a generalisation of the cut problem in graphs with forbidden transitions. In: van Emde Boas, P., Groen, F.C.A., Italiano, G.F., Nawrocki, J., Sack, H. (eds.) SOFSEM 2013. LNCS, vol. 7741, pp. 257–267. Springer, Heidelberg (2013). https://doi.org/10.1007/978-3-642-35843-2_23
5. Kanté, M.M., Moataz, F.Z., Momège, B., Nisse, N.: Finding paths in grids with forbidden transitions. In: Mayr, E.W. (ed.) WG 2015. LNCS, vol. 9224, pp. 154–168. Springer, Heidelberg (2016). https://doi.org/10.1007/978-3-662-53174-7_12
6. Karp, R.M.: Reducibility among combinatorial problems. In: Miller, R.E., Thatcher, J.W., Bohlinger, J.D. (eds.) Complexity of Computer Computations, pp. 85–103. Springer (1972). https://doi.org/10.1007/978-1-4684-2001-2
7. Kotzig, A.: Moves without forbidden transitions in a graph. Matematický časopis **18**(1), 76–80 (1968). https://eudml.org/doc/33972
8. Nguyên, L.T.D.: Unique perfect matchings, forbidden transitions and proof nets for linear logic with Mix. Logical Methods in Computer Science 16(1) (2020). https://doi.org/10.23638/LMCS-16(1:27)2020
9. Szeider, S.: Finding paths in graphs avoiding forbidden transitions. Discret. Appl. Math. **126**(2), 261–273 (2003). https://doi.org/10.1016/S0166-218X(02)00251-2

Delaying Decisions and Reservation Costs

Elisabet Burjons[1], Fabian Frei[2]([✉]), Matthias Gehnen[3], Henri Lotze[3],
Daniel Mock[3], and Peter Rossmanith[3]

[1] York University, Toronto, Canada
burjons@yorku.ca
[2] ETH Zürich, Zürich, Switzerland
fabian.frei@inf.ethz.ch
[3] RWTH Aachen University, Aachen, Germany
{gehnen,lotze,mock,rossmani}@cs.rwth-aachen.de

Abstract. We study the FEEDBACK VERTEX SET and the VERTEX
COVER problem in a natural variant of the classical online model that
allows for *delayed decisions* and *reservations*. Both problems can be characterized by an obstruction set of subgraphs that the online graph needs
to avoid. In the case of the VERTEX COVER problem, the obstruction
set consists of an edge (i.e., the graph of two adjacent vertices), while
for the Feedback Vertex Set problem, the obstruction set contains all
cycles. In the *delayed-decision* model, an algorithm needs to maintain a
valid partial solution after every request, thus allowing it to postpone
decisions until the current partial solution is no longer valid for the current request. The *reservation* model grants an online algorithm the new
and additional option to pay a so-called reservation cost for any given
element in order to delay the decision of adding or rejecting it until the
end of the instance. For the FEEDBACK VERTEX SET problem, we first
analyze the variant with only delayed decisions, proving a lower bound
of 4 and an upper bound of 5 on the competitive ratio. Then we look at
the variant with both delayed decisions and reservation. We show that
given bounds on the competitive ratio of a problem with delayed decisions imply lower and upper bounds for the same problem when adding
the option of reservations. This observation allows us to give a lower
bound of $\min\{1 + 3\alpha, 4\}$ and an upper bound of $\min\{1 + 5\alpha, 5\}$ for the
FEEDBACK VERTEX SET problem, where $\alpha \in \mathbf{R}_{\geq 0}$ is the reservation
cost per reserved vertex. Finally, we show that the online Vertex Cover
problem, when both delayed decisions and reservations are allowed, is
$\min\{1 + 2\alpha, 2\}$-competitive.

1 Introduction

In contrast to classical offline problems, where an algorithm is given the entire
instance it must then solve, an online algorithms has no advance knowledge
about the instance it needs to solve. Whenever a new element of the instance is
given, some irrevocable decision must be taken before the next piece is revealed.

An online algorithm tries to optimize an objective function that is dependent
on the solution set formed by its decisions. The *strict competitive ratio* of an

© The Author(s), under exclusive license to Springer Nature Switzerland AG 2024
W. Wu and G. Tong (Eds.): COCOON 2023, LNCS 14422, pp. 371–383, 2024.
https://doi.org/10.1007/978-3-031-49190-0_27

algorithm, as defined by Sleator and Tarjan [12], is the worst-case ratio of the performance of an algorithm compared to that of an optimal solution computed by an offline algorithm for the given instance, over all instances. The competitive ratio of an online *problem* is then the best competitive ratio over all online algorithms. For a general introduction to online problems, we refer to the books of Borodin and Ran El-Yaniv [4] and of Komm [11].

Not all online problems admit a competitive algorithm (i.e., one whose competitive ratio is bounded by a constant) under the classical model. In particular, this is the case for the problems VERTEX COVER and FEEDBACK VERTEX SET discussed in this paper.

The goal in the general VERTEX COVER problem is, given a graph $G = (V, E)$, to find a minimum set of vertices $S \subseteq V$ such that $G[V \setminus S]$ contains no edges, i.e., the obstruction set is a path of length 1. In the classical online version of VERTEX COVER, the graph is revealed vertex by vertex, including all induced edges, and an online algorithm must immediately and irrevocably decide for each vertex whether to add it to the proposed vertex cover or not.

The goal of the FEEDBACK VERTEX SET problem is, given a graph $G = (V, E)$, to find a minimum set of vertices $S \subseteq V$ such that $G[V \setminus S]$ contains no cycles. In this case, the obstruction set contains all cycles.

In both problems, the non-competitiveness is easy to see: If the first vertex is added to the solution set, the instance stops and thus leaving a single-vertex instance with an optimal solution size of zero. On the other hand, not selecting the first vertex will lead to an instance where this vertex becomes a central vertex, either of a star at VERTEX COVER, or of a friendship graph at FEEDBACK VERTEX SET.

These adversarial strategies are arguably pathological and unnatural, as decisions are enforced that are not based in the properties of the very problem to be solved: We need to start constructing a vertex cover before any edge is presented or a feedback vertex set without it being clear if there are even any cycles in the instance. To address this issue in general online Vertex- and Edge-Deletion problems, Boyar et al. [5,6] introduced the *late accept* online model, which was re-introduced by Chen et al. [9] as the *delayed-decision* model. This model allows an online algorithm to remain idle until a "need to act" occurs, which in our case means waiting until a graph from the obstruction set appears in the online graph. The online algorithm may then choose to delete any vertices in the current online graph. The main remaining restriction is that an online algorithm may not undo any of these deletions.

Definition 1. *Let G be an online graph induced by its vertices $V(G) = v_1, \ldots, v_n$, ordered by their occurrence in an online instance. The DELAYED VERTEX COVER problem is to select, for every i, a subset of vertices $S_i \subseteq \{v_1, \ldots, v_i\}$ with $S_1 \subseteq \ldots \subseteq S_n$ such that the induced subgraph $G[\{v_1, \ldots, v_i\} \setminus S_i]$ contains no edge. The goal is to minimize $|S_n|$.*

The definition of the DELAYED FEEDBACK VERTEX SET problem is identical, except that "contains no edge" is replaced by "is cycle-free."

A constant competitive ratio of 2 for the DELAYED VERTEX COVER problem is simple to prove and given in the introduction of the paper by Chen et al. [9]. The DELAYED FEEDBACK VERTEX SET problem, in contrast, is more involved. We show that no algorithm can admit a competitive ratio better than 4 and adapt results by Bar-Yehuda et al. [1] to give an algorithm that is strictly 5-competitive as an upper bound.

We also consider the model where decisions can be delayed even further by allowing an algorithm to *reserve* vertices (or edges) of an instance. If removing the reserved vertices from the instance would mean that a valid solution is maintained, the instance continues. Once an instance has ended, the algorithm can freely select the vertices to be included in the final solution (in addition to the already irrevocably chosen ones) among all presented vertices, regardless of their reservation status. This reservation is not free: When computing the final competitive ratio, the algorithm has to pay a constant $\alpha \in \mathbf{R}_{\geq 0}$ for each reserved vertex; these costs are then added to the size of the chosen solution set.

Definition 2. *Let $\alpha \in \mathbf{R}_{\geq 0}$ be a constant and G an online graph induced by its vertices $V(G) = v_1, \ldots, v_n$, ordered by their occurrence in an online instance. The* DELAYED VERTEX COVER *problem with reservations is to select, for every i, vertex subsets $S_i, R_i \subseteq \{v_1, \ldots, v_i\}$ with $S_1 \subseteq \ldots \subseteq S_n$ and $R_1 \subseteq \ldots \subseteq R_n$ such that $G[\{v_1, \ldots, v_i\} \setminus (S_i \cup R_i)]$ contains no edge. The goal is to minimize the sum $|S_n| + |T| + \alpha|R_n|$, where $T \subseteq V(G)$ is a minimal vertex subset such that $G - (S_n \cup T)$ contains no edge.*

Again, the definition for the DELAYED FEEDBACK VERTEX SET *problem with reservations* is identical, except for replacing "contains no edge" with "is cycle-free."

For reservation costs of $\alpha = 0$, the problem becomes equivalent to the offline version, whereas for $\alpha \geq 1$ taking an element directly into the solution set becomes strictly better than reserving it, rendering this reservation option useless. The results for DELAYED VERTEX COVER and DELAYED FEEDBACK VERTEX SET, each with reservations, are depicted in Fig. 1.

The reservation model is still relatively new and has been applied to the simple knapsack problem [3] and the secretary problem [8]. We note that the two cited papers consider relative reservation costs, while for the two problems in the present paper the cost per item are fixed.

The online VERTEX COVER problem has not received a lot of attention in the past years. Demange and Paschos [10] analyzed the online VERTEX COVER problem with two variations of how the online graph is revealed: either vertex by vertex or in clusters, per induced subgraphs of the final graph. The proven competitive ratios are functions on the maximum degree of the graph. Zhang et al. [15] looked at a variant called the Online 3-Path VERTEX COVER problem, where every induced path on three vertices needs to be covered. In this setting, the competitive ratio is again dominated by the maximum degree of the graph. Buchbinder and Naor [7] considered online integral and fractional covering problems formulated as linear programs where the covering constraints arrive online.

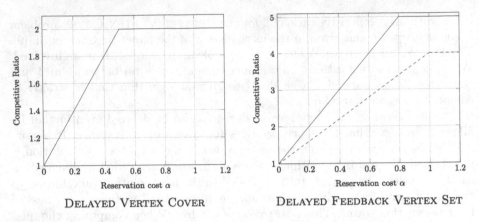

Fig. 1. Upper and lower bounds on the competitive ratios of DELAYED VERTEX COVER (left) and DELAYED FEEDBACK VERTEX SET (right), each with reservations.

As these are a strong generalization of the online VERTEX COVER problem, they achieve only logarithmic and not constant competitive ratios.

There has been some work on improving upon the bound of 2 for some special cases of VERTEX COVER in the model with delayed decisions (under different names). For the VERTEX COVER problem on *bipartite* graphs where one side is offline, Wang and Wong [14] give an algorithm achieving a competitive ratio of $\frac{1}{1-1/e} \approx 1.582$. Using the same techniques they achieve a competitive ratio of 1.901 for the full online VERTEX COVER problem for bipartite graphs and for the online fractional VERTEX COVER problem on general graphs.

To the best of our knowledge, the FEEDBACK VERTEX SET problem has received no attention in the online setting so far, most likely due to the fact that there is no competitive online algorithm for this problem in the classical setting. The offline FEEDBACK VERTEX SET problem, however, has been extensively studied, especially in the approximation setting. One notable algorithm is the one in the paper of Bar-Yehuda et al. [1], yielding an approximation ratio of $4 - 2/n$ on an undirected, unweighted graph. We adapt their notation in Sect. 2, and our delayed-decision algorithm with a competitive ratio of 5 is based on their aforementioned approximation algorithm. The currently best known approximation ratio of 2 by an (offline) polynomial-time algorithm was given by Becker and Geiger [2].

The paper is organized as follows. We first look at the DELAYED FEEDBACK VERTEX SET problem, giving a lower bound of 4 and an upper bound of 5 on the competitive ratio. Then, we discuss how bounds on obstruction set problems without reservation imply bounds on the equivalent problems with reservations and vice versa, and how this applies to the DELAYED FEEDBACK VERTEX SET problem. Finally, we consider the DELAYED VERTEX COVER *problem with reservations*, giving tight bounds dependent on the reservation costs.

2 Feedback Vertex Set with Delayed Decisions

In this section, we consider the DELAYED FEEDBACK VERTEX SET problem, which is concerned with finding the smallest subset of the vertices of a graph such that their removal yields a cycle-free graph. We give almost matching bounds on the competitive ratio in the delayed decision model.

Theorem 1 (Lower Bound). *Given an $\varepsilon > 0$, there is no algorithm for* DELAYED FEEDBACK VERTEX SET *achieving a competitive ratio of* $4 - \varepsilon$.

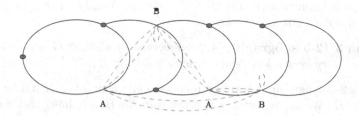

Fig. 2. Sketch of the lower bound graph, revealed from left to right. The marked vertices are deleted by an algorithm. Then dashed edges (gray) and finally the self-loops (blue) force more deletions. The competitive ratio tends to 4 with increasing instance size. (Color figure online)

Proof. The adversarial strategy depicted in Fig. 2 provides the lower bound.

First, a cycle is presented, which forces any algorithm to delete a single vertex. The adversary now repeats the following scheme n times: From the cycle that was presented in the previous step, select two different vertices. Then connect these vertices by a new path, thus forming a new cycle (the large, "layered" cycles in the figure).

Every time such a semicycle is added, any algorithm has to delete a vertex from it. This is sketched in Fig. 2 by the black cycles with red dots for exemplary deletions of some algorithm. We assume w.l.o.g. that an algorithm chooses one of the vertices of degree 3 between two cycles for deletion, we call these vertices of degree 3 branchpoints. If an algorithm deletes vertices of degree 2 instead, the adversary can force the algorithm to delete even more vertices.

After following this scheme, we split the remaining branchpoints (one for each new cycle) into two sets A and B alternatingly. We now take each (a, b)-pair with $a \in A$ and $b \in B$ and connect the two vertices with two paths, forming a new cycle between each pair of branching vertices. In order to remove all cycles with the lowest amount of deletions, any algorithm has to delete either all vertices from the set A or all vertices from the set B. Once such a set is chosen, say A, the adversary adds "self-loops", i.e., new cycles each only connected to a vertex of B, forcing the deletion of all (branchpoint) vertices of B.

The optimal solution consists only of the vertices of B and possibly, and depending on whether n is even and on the choice of A or B, another vertex from the first and the last cycle.

Thus, any online algorithm has to delete at least $2n - 1$ vertices, while an optimal algorithm only deletes at most $n/2+1$ vertices. The resulting competitive ratio is hence worse than $4 - \varepsilon$ for large enough values of n.   □

Next, we present Algorithm 1, which guarantees a competitive ratio of 5. It is a modified version of the 4-approximative algorithm for FEEDBACK VERTEX SET by Bar-Yehuda et al. [1]. Algorithm 1 maintains a maximal so-called 2-3-subgraph H in the presented graph G and selects a feedback vertex set F for G within H. It is important to note that H is not necessarily an *induced* subgraph of G and also does not have to be connected.

Definition 3 (2-3-subgraph). *A 2-3-subgraph of a graph G is a subgraph H of G where every vertex has degree exactly 2 or 3 in H.*

Given a 2-3-subgraph H, a vertex is called a *branchpoint of H* if it has exactly degree 3 in H. We say a vertex v is a *linkpoint in H* if it has exactly degree 2 in H and there is a cycle in G whose only intersection with H is v. A cycle is called *isolated in H* if every vertex of this cycle (in G) is contained in H and has exactly degree 2 in H.

Algorithm 1 adds every branchpoint, linkpoint and one vertex per isolated cycle without linkpoints in H to the solution set F.

Note that it is possible that vertices, who were added to F as linkpoints or due to isolated cycles, can still become branchpoints with degree 3, if H is expanded in later steps. For the sake of the analysis, we consider them as branchpoints. The extension of H is not necessarily unique, as often there are several ways of extending and the algorithm just chooses an arbitrary way.

Algorithm 1. Online SubG-2-3

$H \leftarrow \emptyset, F \leftarrow \emptyset$
for every new edge in G **do**
    if H is not a maximal 2-3-subgraph **then**
        Extend H to a maximal 2-3-subgraph, converting linkpoints to branchpoints whenever possible.
    $F \leftarrow F \cup$ All new branchpoints
    $F \leftarrow F \cup$ All new linkpoints
    $F \leftarrow F \cup$ A set of one vertex per new isolated cycle without linkpoints in H
return F

Lemma 1 (Correctness of Online SubG-2-3). *Algorithm 1 returns a feedback vertex set for the given input graph G.*

Proof. By contradiction, assume there is a cycle C in G without a vertex in F.

If C contains no point of H, then the complete cycle C can be added to the subgraph H as an isolated cycle, thus H is not maximal. This contradicts the procedure of Algorithm 1.

Therefore, we can assume that there is at least one vertex of C in H. If the cycle contains no branchpoints, three situations are possible: First, C is an isolated cycle, where all vertices are in H and one vertex is added to F. Second, C intersects with H at just a single vertex which will be a linkpoint and part of F as well. The third case is that C intersects H on two or more vertices and none of them are branchpoints (thus of degree 2). Then H would have been extended from one of those points along C towards another one, making the two vertices branchpoints.

Thus, every cycle in G has to contain at least one vertex of F, which is a feedback vertex set at the end. □

Proving that Algorithm 1 is 5-competitive is more tricky. First, note that an optimal feedback vertex set does not change if we consider a *reduction graph G'* instead of a graph G.

Definition 4 (Reduction Graph). *A reduction graph G' of a graph G is obtained by deleting vertices of degree 1 and their incident edges, and by deleting vertices of degree 2 without a self-loop and connecting the two neighbors directly.*

The following lemma bounds the size of a reduced graph by its maximum degree and the size of any feedback vertex set. This will be used in the analysis of Algorithm 1. The lemma is due to Voss [13] and is used in the proof of 4-competitiveness by Bar-Yehuda et al. [1].

Lemma 2. *Let G be a graph where no vertex has degree less than 2. Then for every feedback vertex set F, that contains all vertices of degree 2,*

$$|V(G)| \leq (\Delta(G) + 1)|F| - 2$$

holds, where $\Delta(G)$ is the maximum degree of G. In particular, if every vertex has a degree of at most 3, $|V(G)| \leq 4|F| - 2$ holds.

Theorem 2. *Algorithm 1 achieves a strict competitive ratio of $5 - 2/|V(G)|$ for the online* FEEDBACK VERTEX SET *with delayed decisions.*

Proof. At the end of the instance, after running Algorithm 1, call the set of branchpoints $B \subseteq F$, the set of linkpoints $L \subseteq F$, and the set of vertices added due to isolated cycles I. Again note that vertices can become branchpoints in later steps, even if they were added to F as a linkpoint or for an isolated cycle. Let μ be the size of an optimal feedback vertex set for the graph G.

The vertices in I are part of pairwise independent cycles. This follows from the fact that every isolated cycle was added completely to H, thus new isolated cycles cannot contain a vertex of one already added to H. Therefore, $|I| \leq \mu$, since an optimal solution must contain at least one vertex for each of the pairwise independent cycles.

Moreover, the set of cycles which caused adding vertices as linkpoints to F are also pairwise independent, if no linkpoint was relabeled to a branchpoint later:

For a contradiction, assume that two cycles overlap at a vertex v and intersect with H at the linkpoints ℓ_1 and ℓ_2. Now H could easily be extended by a path from ℓ_1 through v to ℓ_2, if none of those vertices are added to H before. This would make ℓ_1 and ℓ_2 branchpoints, contradicting the assumption.

The algorithm would also convert either ℓ_1 or ℓ_2 to a branchpoint, if—at one point—it was possible to extend H by connecting one of the linkpoints along the mentioned path to one vertex in H of degree 2 (with respect to H).

The only remaining case is when the first vertex of H along this path was a branchpoint immediately at the time it was added to H, thus making it impossible to connect it with one of the linkpoints.

Note that w.l.o.g. ℓ_1 must have been considered as a linkpoint at this time, since—by definition—no cycle would have caused any vertex to become a linkpoint if some vertices of the cycle were already considered as branchpoints. In particular, x cannot have been added immediately to H when it was presented.

Since the adversary presents the instance vertex-wise, there must be a vertex s such that there was no possibility to add x to H before s was presented, but such that x immediately could become a branchpoint when s was added to G.

Therefore, there must be at least two independent paths in $G \setminus H$ from x to s. Therefore, in this situation it would also be possible to extend H by adding the two independent paths from x to s towards H, and also one path from x to ℓ_1 along the cycle. Since the algorithm prioritizes extensions to H which convert linkpoints to branchpoints over others that do not, it has to add the path from x to ℓ_1 first. Therefore ℓ_1 becomes a branchpoint, which finally contradicts our assumption. Note that priority is unambiguous, since there are no paths in $G \setminus H$ from x to some other linkpoints: Then ℓ_1 and the other linkpoint would already be connected within H, thus making them branchpoints.

Thus we have $|L| \leq \mu$. This also shows that there cannot be a branchpoint inside a cycle that was used to delete a linkpoint, without also converting the linkpoint to a branchpoint.

It follows that $|L| + |I| \leq 2\mu$. If $|B| \leq 2|L|$, then we have $|F| = |I| + |L| + |B| \leq 3|L| + |I| \leq 4\mu$, which proves the statement.

In every other case, assume $|B| > 2|L|$. We now consider a reduction graph H' of the graph $H \setminus L$, and delete every component consisting of only a single vertex. Every vertex in the resulting graph has degree 3. In the graph H we can have up to $2|L|$ more branchpoints than in the resulting graph here.

By Lemma 2, $|B| - 2|L|$ is less than $4\mu(H') - 2$, where $\mu(H')$ is the optimal size of a feedback vertex set for the graph H'.

The size of an optimal feedback vertex set of the original graph G must be at least $\mu(H') + |L|$ since every linkpoint in G is part of a cycle that is not intersecting H'.

The inequality chain $|F| = |I| + |L| + |B| = |I| + 3|L| + |B| - 2|L| \leq |I| + 4|L| + 4\mu(H') - 2 \leq |I| + 4\mu(G) - 2 \leq 5\mu(G) - 2$ concludes the proof. $\square$

One could think that the reason Algorithm 1 does not match the competitive ratio of 4 is because the algorithm deletes vertices even in cases where it is not necessary. However, this is not the case.

Lemma 3. *The competitive ratio of Algorithm 1 is at least* $5 - 2/|V(G)|$, *even if vertices are only deleted whenever necessary.*

3 Adding Reservations to the Delayed-Decision Model

We now extend the previous results for the delayed-decision model without reservations by presenting two general theorems that translate both upper and lower bounds to the model with reservation.

The delayed-decision model allows us to delay decisions free of cost as long as a valid solution is maintained. Combining delayed decisions with reservations, we have to distinguish minimization and maximization: For a minimization problem the default is that the union of selected vertices and reserved ones constitutes a valid solution at any point. For a maximization problem, in contrast, it is that the selected vertices without the reserved ones always constitute a valid solution.

Whether the goal is minimizing or maximizing, any algorithm for a problem without reservations is also an algorithm for the problem with reservations, it just never uses this third option. We present a slightly smarter approach for any *minimization* problem such as the DELAYED FEEDBACK VERTEX SET problem.

Theorem 3. *In the delayed-decision model, a c-competitive algorithm for a minimization problem without reservation yields a* $\min\{c, 1 + c\alpha\}$-*competitive algorithm for the variant with reservation.*

Proof. Modify the c-competitive algorithm such that it reserves whatever piece of the input it would usually have immediately selected until the instance ends – incurring an additional cost $c \cdot \alpha \cdot Opt$ – and then pick an optimal solution for a cost of Opt. This already provides an upper bound of $1 + c\alpha$ on the competitive ratio. But running the algorithm without modification still yields an upper bound of c of course. The algorithm can now choose the better of these two options based on the given α, yielding an algorithm that is $\max\{c, 1 + c\alpha\}$-competitive. $\square$

Corollary 1. *There is a* $\min\{5, 1 + 5\alpha\}$-*competitive algorithm for the* DELAYED FEEDBACK VERTEX SET *problem with reservations.*

Theorem 4. *In the delayed-decision model, a lower bound of c on the competitive ratio for a minimization problem without reservations yields a lower bound of* $\min\{1 + (c - 1)\alpha, c\}$ *on the competitive ratio for the problem with reservation.*

Proof. The statement is trivial for $\alpha \geq 1$; we thus consider now the case of $\alpha < 1$. Assume that we have an algorithm with reservations with a competitive ratio better than $1 + (c - 1)\alpha$. Even though this algorithm has the option of reserving, it must select a definitive solution, at the latest when the instance ends. This definitive solution is of course at least as expensive as the optimal solution. Achieving a competitive ratio better than $1 + (c - 1)\alpha$ is thus possible only if less the algorithm is guaranteed to reserve fewer than $c - 1$ input pieces in total. But in this case, we can modify the algorithm with reservation such that it immediately accepts whatever it would have only reserved otherwise. This increases the incurred costs by $1 - \alpha$ for each formerly reserved input piece, yielding an algorithm without reservations that achieves a competitive ratio better than $1 + (c - 1)\alpha + (c - 1)(1 - \alpha) = c$. ☐

Corollary 2. *There is no algorithm solving the* DELAYED FEEDBACK VERTEX SET *problem with reservations that can achieve a lower bound better than* $\min\{1 + 3\alpha, 4\}$.

4   Vertex Cover

As already mentioned, the DELAYED VERTEX COVER has a competitive ratio of 2 without reservation. We now present tight bounds for all reservation-cost values α, beginning with the upper bound.

Theorem 5. *There is an algorithm for the* DELAYED VERTEX COVER *problem with reservations that achieves a competitive ratio of* $\min\{1 + 2\alpha, 2\}$ *for any reservation value.*

These upper bounds, depicted in Fig. 1, have matching lower bounds. We start by giving a lower bound for $\alpha \leq \frac{1}{2}$.

Theorem 6. *Given an* $\varepsilon > 0$, *there is no algorithm for the* DELAYED VERTEX COVER *problem with reservations achieving a competitive ratio of* $1 + 2\alpha - \varepsilon$ *for any* $\alpha \leq \frac{1}{2}$.

Proof. We present the following exhaustive set of adversarial instances as depicted in Fig. 3. First an adversary presents two vertices u_1 and v_1 connected to each other. Any algorithm for online VERTEX COVER with reservations is forced to cover this edge either by irrevocably choosing one vertex for the cover or by placing one of the vertices in the temporary cover (i.e., reserving it). In the first case, assume w.l.o.g. that v_1 is the chosen vertex for the cover. The adversary then presents a vertex v_2 connected only to u_1 and ends the instance. Such an algorithm would have a competitive ratio of 2 as it must then cover the edge (u_1, v_2) by placing one of its endpoints in the cover. Choosing vertex u_1 alone would have been optimal, however. This is the same lower bound as given for the model without reservations.

If, again w.l.o.g., the vertex v_1 is temporarily covered instead, the adversary still presents a vertex v_2 connected to u_1. Now an algorithm has four options to

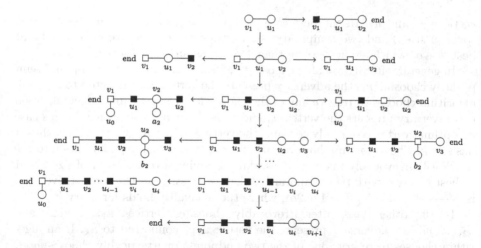

Fig. 3. Illustration for the proof of Theorem 5. Adversarial strategy for $\alpha \leq \frac{1}{2}$. Black squares are vertices irrevocably included into the vertex cover, white square vertices are reserved to be temporarily in the vertex cover, and round vertices are not yet chosen.

cover the edge (u_1, v_2): Each of the two vertex u_1 or v_2 can be either irrevocably chosen or temporarily reserved. If u_1 or v_2 are temporarily covered, the instance will end here and the reservation costs of the algorithm will be 2α. Both the algorithm and the optimal solution will end up choosing only vertex u_1, which implies a final competitive ratio of $1 + 2\alpha$ in this case. If vertex v_2 is chosen, the instance will also end, yielding a competitive ratio worse than 2.

Thus, the only option remaining is to irrevocably cover the vertex u_1. In this case, the adversary presents a vertex u_2 connected to v_2. An algorithm can then irrevocably or temporarily take v_2 or u_2 respectively. If an algorithm temporarily takes v_2 or u_2, the adversary will present one more vertex u_0 connected to v_1 and end the instance. This results in a graph that can be minimally covered by the vertices v_1 and v_2. The algorithm, however, will have 3 vertices in the cover and additional reservation costs of 2α for the temporarily chosen vertices. Thus it will have a competitive ratio of $\frac{3}{2} + \alpha$, which is larger than $1 + 2\alpha$ for the considered values of α. If an algorithm irrevocably takes u_2, the same vertex u_0 will be presented and then the instance will end with another auxiliary vertex a_2 connected to v_2. An optimal vertex cover would take vertices v_1 and v_2. Any algorithm that has already irrevocably chosen u_1 and u_2, however, will have to choose two more vertices in order to cover the edges $\{v_1, u_0\}$ and $\{v_2, a_2\}$; thus, its competitive ratio will be worse than 2.

Again, the only remaining option is to irrevocably choose the vertex v_2, after which an adversary presents a vertex v_3 connected to u_2. An algorithm may choose to irrevocably or temporarily take the vertex u_2 or v_3. If an algorithm decides to temporarily take any vertex or irrevocably choose v_3, then the adversary presents an auxiliary vertex b_2 connected to u_2 and ends the request sequence. An optimal vertex cover in this case has size two, containing only the vertices u_1 and u_2. In the best case, however, such an algorithm has a vertex cover of size 3 and two temporary covers, thus its competitive ratio will be at best $\frac{3}{2} + \alpha$, which again is worse than $1 + 2\alpha$, as already observed.

In general, after irrevocably choosing $u_1, \ldots, u_{k-1}$ and $v_2, \ldots, v_{k-1}$, and temporarily choosing v_1, the adversary presents the vertex u_k connected to v_k. If an algorithm chooses to reserve any of the endpoints or irrevocably selects u_i, then the adversary presents the vertex u_0 and ends the request sequence. In this case an optimal vertex cover only contains the vertices v_i for every $i = 1, \ldots k$, thus it has size k. The algorithm, however, will have to take v_1 and v_k in addition to the previously irrevocably taken vertices, thus obtaining a vertex cover of size $2k - 1$ at best together with two temporarily taken vertices. Thus the competitive ratio is $\frac{2k-1+2\alpha}{k} = 2 - \frac{1-2\alpha}{k} \geq 1 + 2\alpha$, where the inequality holds for every $k \geq 1$.

In the other case, after irrevocably choosing vertices $u_1, \ldots, u_{k-1}$ and $v_2, \ldots, v_k$, the adversary presents the vertex v_{k+1} connected to u_k. If an algorithm chooses to reserve one of the two endpoints or irrevocably chooses v_{k+1}, then the adversary stops the request sequence. An optimal vertex cover of such a graph consists of the vertices u_i for every $i = 1, \ldots, k$ and it has size k. The algorithm will have to choose u_i in order to obtain a vertex cover at all, obtaining a vertex cover of size $2k - 1$ at best together with at least one reservation, thus achieving a competitive ratio of $\frac{2k-1+\alpha}{k} \geq 1 + 2\alpha - \varepsilon$ for any $k \geq \frac{1}{\varepsilon}$.   □

For larger values of α the same adversarial strategy holds, but it gives us the following lower bound.

Theorem 7. *For $\alpha > 1/2$, no algorithm for the* DELAYED VERTEX COVER *problem with reservations is better than 2-competitive.*

Proof. The lower bound of Theorem 6 for $\alpha = 1/2$ is $2 - \varepsilon$. For larger values of α the same adversarial strategy will give us a lower bound of 2. This is because, at all points during the analysis, either the value of the competitive ratio for each strategy was at least 2, or it had a positive correlation with the value of α, meaning that for larger values of α any algorithm following that strategy obtains strictly worse competitive ratios.   □

5   Conclusion

We have shown that some problems that are non-competitive in the classical model become competitive in modified, but natural variations of the classical online model. Some questions remain open, such as the best competitive ratio for the DELAYED FEEDBACK VERTEX SET problem, which we believe to be 4.

It may be worthwhile to investigate which results can be found for restricted graph classes. For example, it is easy to see that the online version of DELAYED FEEDBACK VERTEX SET is 2-competitive on graphs with maximum degree three.

In addition we also introduced the reservation model on graphs, providing an upper and a lower bound for general graph problems. It would be interesting try to find matching bounds, also on specific graph problems.

References

1. Bar-Yehuda, R., Geiger, D., Naor, J., Roth, R.M.: Approximation algorithms for the feedback vertex set problem with applications to constraint satisfaction and bayesian inference. SIAM J. Comput. **27**(4), 942–959 (1998)
2. Becker, A., Geiger, D.: Optimization of pearl's method of conditioning and greedy-like approximation algorithms for the vertex feedback set problem. Artif. Intell. **83**(1), 167–188 (1996)
3. Böckenhauer, H., Burjons, E., Hromkovič, J., Lotze, H., Rossmanith, P.: Online simple knapsack with reservation costs. In: STACS 2021. LIPIcs, vol. 187, pp. 16:1–16:18 (2021)
4. Borodin, A., El-Yaniv, R.: Online Computation and Competitive Analysis. Cambridge University Press, Cambridge (1998)
5. Boyar, J., Eidenbenz, S.J., Favrholdt, L.M., Kotrbčík, M., Larsen, K.S.: Online dominating set. In: Pagh, R. (ed.) 15th Scandinavian Symposium and Workshops on Algorithm Theory, SWAT 2016, June 22–24, 2016, Reykjavik, Iceland. LIPIcs, vol. 53, pp. 21:1–21:15. Schloss Dagstuhl - Leibniz-Zentrum für Informatik (2016). https://doi.org/10.4230/LIPIcs.SWAT.2016.21
6. Boyar, J., Favrholdt, L.M., Kotrbčík, M., Larsen, K.S.: Relaxing the irrevocability requirement for online graph algorithms. In: WADS 2017. LNCS, vol. 10389, pp. 217–228. Springer, Cham (2017). https://doi.org/10.1007/978-3-319-62127-2_19
7. Buchbinder, N., Naor, J.: Online primal-dual algorithms for covering and packing. Math. Oper. Res. **34**(2), 270–286 (2009)
8. Burjons, E., Gehnen, M., Lotze, H., Mock, D., Rossmanith, P.: The secretary problem with reservation costs. In: Chen, C.-Y., Hon, W.-K., Hung, L.-J., Lee, C.-W. (eds.) COCOON 2021. LNCS, vol. 13025, pp. 553–564. Springer, Cham (2021). https://doi.org/10.1007/978-3-030-89543-3_46
9. Chen, L., Hung, L., Lotze, H., Rossmanith, P.: Online node- and edge-deletion problems with advice. Algorithmica **83**(9), 2719–2753 (2021)
10. Demange, M., Paschos, V.T.: On-line vertex-covering. Theoret. Comput. Sci. **332**(1), 83–108 (2005)
11. Komm, D.: An Introduction to Online Computation - Determinism, Randomization, Advice. Texts in Theoretical Computer Science, Springer, Cham (2016). https://doi.org/10.1007/978-3-319-42749-2
12. Sleator, D.D., Tarjan, R.E.: Amortized efficiency of list update and paging rules. Commun. ACM **28**(2), 202–208 (1985)
13. Voss, H.J.: Some properties of graphs containing k independent circuits. Theory of Graphs. In: Proceedings of Colloquium Tihany, pp. 321–334 (1968)
14. Wang, Y., Wong, S.C.: Two-sided online bipartite matching and vertex cover: beating the greedy algorithm. In: Halldórsson, M.M., Iwama, K., Kobayashi, N., Speckmann, B. (eds.) ICALP 2015. LNCS, vol. 9134, pp. 1070–1081. Springer, Heidelberg (2015). https://doi.org/10.1007/978-3-662-47672-7_87
15. Zhang, Y., Zhang, Z., Shi, Y., Li, X.: Algorithm for online 3-path vertex cover. Theory Comput. Syst. **64**(2), 327–338 (2020)

A PTAS Framework for Clustering Problems in Doubling Metrics

Di Wu[1,2], Jinhui Xu[3], and Jianxin Wang[1,2,4(✉)]

[1] School of Computer Science and Engineering, Central South University,
Changsha, China
[2] Xiangjiang Laboratory, Changsha 410205, China
[3] Department of Computer Science and Engineering, State University of New York
at Buffalo, Buffalo, NY, USA
[4] The Hunan Provincial Key Lab of Bioinformatics, Central South University,
Changsha, China
`jxwang@mail.csu.edu.cn`

Abstract. Constrained clustering problems have been studied extensively in recent years. In this paper, we focus on a class of constrained k-median problems with general constraints on facilities, denoted as GCF k-CMedian problems. We present a randomized polynomial-time approximation scheme (PTAS) framework based on the split-tree decomposition and dynamic programming process for GCF k-CMedian problems, such as k-median with service installation costs, k-facility location and priority k-median problems in doubling metrics.

Keywords: Approximation algorithm · k-median

1 Introduction

Clustering algorithms provide a fundamental tool in many machine learning libraries. The goal of clustering is to partition the given dataset into several clusters according to the similarity. Among different clustering objectives, the k-median is widely used in applications such as operations research, data summation and data analysis [10, 14, 26]. It was known that the k-median problem is NP-hard [18]. The current best-known approximation ratio in polynomial time for the k-median problem is $2.67059 + \epsilon$ given by Cohen-Addad et al. [12], which was obtained based on the LP rounding technique.

A crucial property used in the approximation algorithms for the standard k-median problem is that all clients of a cluster in an optimal solution lie fully in the Voronoi cell of the corresponding facility. However, in many applications involving constrained clustering, the clusters are no longer obtained from the

This work was supported by National Natural Science Foundation of China (62172446, 62350004), Open Project of Xiangjiang Laboratory (22XJ02002), and Central South University Research Programme of Advanced Interdisciplinary Studies (2023QYJC023).

Voronoi cell of the opened facilities, which means that the partition of clients and locations of the opened facilities in an optimal solution might be quite arbitrary. Moreover, the certain constraints can be imposed at either client or facility level [15]. Client-level constraints are mainly imposed on clients inside each cluster, such as fair constraints [5,6]. Facility-level constraints are restrictions on the facility, mainly referring to the constrained k-median problems with general constraints on facilities, such as service installation constraint [28], and facility-opening constraints [22,31]. In recent years, much attention has been paid to various types of constrained clustering problems with client-level [1,19,21]. The constant factor approximation for constrained clustering problems with client-level in metrics was given by Zhang et al. [17], who showed that the reduced search space technique yields $O(1)$-approximation with FPT(k)-time. Moreover, several sampling-based FPT(k)-time algorithms yield $(1 + \epsilon)$-approximation for constrained clustering problems with client-level in Euclidean space [15,16,23]. However, how to design a $(1 + \epsilon)$-approximation algorithm for the constrained clustering problems with facility-level is still an open problem.

The current results for the constrained k-median problems with general constraints on facilities are $O(1)$-approximation algorithms with polynomial time in metric space [8,25,30], where the approximation ratio is still large compared to the optimal solution. It was pointed out in [13] that identifying a near-optimal solution within $n^{O(1)}g(k)$ time (where g is a positive function) is unlikely for the constrained k-median problems with general constraints on facilities in general metrics. In [13], Cohen-Addad et al. showed that if the Gap-Exponential Time Hypothesis [7,27] holds, then any algorithm for the GCF k-CMedian problem with an approximation ratio better than $1 + e/2 - \epsilon$ must run in time at least $n^{k^{g(\epsilon)}}$. However, this negative result does not reject the possibility of obtaining a near-optimal solution within bounded time for the problem in doubling metrics. The doubling metric space is a metric space with fixed doubling dimension, where the doubling dimension of the metric space is the smallest d such that any ball of radius $2r$ can be covered by 2^d balls of radius r. Gupta et al. [20] first proposed the concept of doubling metrics. This concept for metrics abstraction has been proposed as a natural analog to Euclidean space. Following [20], Euclidean metrics of dimension d have doubling dimension $\Theta(d)$. Moreover, the submetrics of a given metric exhibit reduced dimensions, and this concept generalizes the growth restriction assumption outlined in [20]. According to [32], metric embeddings have proved to be extremely useful tools for algorithm design. Such low dimension restrictions are natural in several practical applications such as data analysis [33] and peer-to-peer networks [29]. Hence, it is urgent to design polynomial-time approximation schemes with $(1+\epsilon)$-approximation for the constrained k-median problems with general constraints on facilities in doubling metrics (Fig. 1).

In this work, we consider the constrained k-median problems with general constraints on facilities (GCF k-CMedian) such as k-median with service installation costs [30], k-facility location [8], and priority k-median [34] problems in

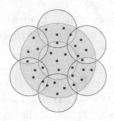

Fig. 1. In 2-dimensional Euclidean space, seven balls of radius $r/2$ can cover any ball of radius r, so 2-dimensional Euclidean space is a doubling metric space with doubling dimension $\log_2 7$.

doubling metrics. The following is the definition of the constrained k-median with general constraints on facilities considered in this paper.

Definition 1 (GCF k-CMedian). *Given a set P of n points in metric space with fixed doubling dimension d, a positive integer k, some constraint $\mathbb{C}$ on the opened facilities and each opened facility is associated with a facility cost, the constrained k-median problem with general constraints on facilities (GCF k-CMedian) is to partition P into k clusters satisfying the constraint $\mathbb{C}$ and minimize the sum of client-connection cost and opened facility costs.*

1.1 Our Results and Techniques

In this paper, we develop a randomized polynomial-time approximation framework with $(1+\epsilon)$-approximation for the k-median with service installation costs, k-facility location, and priority k-median problems in metric space with fixed doubling dimension d.

Our framework for the GCF k-CMedian problems is based on portal approach and dynamic programming method. Intuitively speaking, portals are the bridges to get the pairwise distances between the points in the tree embedding. Given a dataset in doubling metrics, we first construct a split-tree decomposition of the input metrics. The split-tree decomposition is a hierarchical decomposition that partitions the metrics into several subsets, which are called *blocks*. For each block, a subset of points, called *portals*, is constructed as bridges to get the pairwise distances between the points in the split-tree. The distance between any two points at different blocks on the split-tree can be defined as the length of the path that is constructed by the subpath through the portals of the blocks. In the following, we briefly remark on the commonly used techniques for the clustering problems to show the obstacles in obtaining PTAS for the GCF k-CMedian problems in doubling metrics.

- As far as we know, many results have been proposed by applying portal approaches and dynamic programming methods (e.g., [2,24,32]). In [2], they gave a quasi-polynomial time approximation scheme (QPTAS) for the k-median problem in Euclidean space. It should be noted that they showed the running time of the dynamic programming process depends exponentially on the number of portals. Kolliopoulos and Rao [24] improved the above result in [2] and obtained a PTAS for the k-median problem and facility location problem in d-dimensional Euclidean space. In [24], based on an adaptive decomposition scheme in Euclidean space, they proposed a new structural theorem to characterize the structure of the near-optimal solutions such that the number of portals can be reduced to a constant number. Unfortunately, Kolliopoulos and Rao's algorithm is not directly applicable to doubling metrics as it relies on the specific structure of Euclidean space.

- For applying a dynamic programming process to solve the k-median problem, the enumeration of the pairwise distances between portals and opened facilities is inevitable for using the standard dynamic programming method [2,9,11,24,32]. The difficulty in extending this approach to the GCF k-CMedian problems is due to the constraints associated with the facilities. In the GCF k-CMedian problems, the enumeration of the pairwise distances between portals and opened facilities leads to an $O(m \log n)$ loss on the number of parameters of the dynamic programming table entry where m is the size of portal set in a block, which increases the time complexity of the dynamic programming process.

To bypass the above obstacles in solving the GCF k-CMedian problems, we propose a PTAS framework for the GCF k-CMedian problems in doubling metrics. In this work, we present a new dynamic programming framework based on the split-tree decomposition to solve the GCF k-CMedian problems, which is based on the relation between the subproblem and the subproblems of its children. Furthermore, it can reduce the enumeration loss from $O(m \log n)$ to $O(m)$ based on newly designed multiple auxiliary bipartite graphs and minimum weighted perfect matching. Our portal approach is based on the badly-cut technique. Intuitively speaking, the general idea behind this technique is to optimize the solution by estimating the distance between the client and the facility of the optimal solution serving it, such that the exponential dependence of the dynamic programming process on $O(\log n)$ factor is avoided. The idea underlying this technique is that when the distance between a client and the optimal facility serving it is large, using the dynamic programming process to determine the best facility for this client becomes time-consuming. Then, these clients are denoted as special badly-cut points to deal with. It is difficult to directly extend this approach to the GCF k-CMedian problems due to the constraints associated with the facilities. In such settings, it cannot be proved that there exists a near-optimal solution with the opened badly-cut facilities by directly embedding the badly-cut facilities into the optimal solution since the reassignment of the clients may violate the additional constraints. To overcome the difficulties caused by the additional constraints in the proof, we construct a new combination model based on the reassignment of the clients.

Theorem 1. *Given an instance for the k-median with service installation costs problem in metrics of fixed doubling dimension d, with probability at least $1 - \epsilon$, there exists a randomized $(1 + \epsilon)$-approximation algorithm with running time*

$\tilde{O}(n^{O(\frac{1}{\epsilon})^{O(d)}}|\mathcal{S}|)^1$, where the size of X is n and $|\mathcal{S}|$ is an integer that denotes the number of the services in the metrics.

Theorem 2. *Given an instance for the k-facility location problem in metrics of fixed doubling dimension d, with probability at least $1-\epsilon$, there exists a randomized $(1+\epsilon)$-approximation algorithm with running time $\tilde{O}(n^{O(\frac{1}{\epsilon})^{O(d)}})$, where the size of X is n.*

Theorem 3. *Given an instance for the priority k-median problem in metrics of fixed doubling dimension d, with probability at least $1-\epsilon$, there exists a randomized $(1+\epsilon)$-approximation algorithm with running time $\tilde{O}(n^{O(\frac{1}{\epsilon})^{O(d)}}|\mathcal{P}|)$, where the size of X is n and $|\mathcal{P}|$ is an integer that denotes the number of the priorities of the points in the metrics.*

2   Preliminaries

Definition 2 (metric space). *Given a space $\mathcal{X} = (X, \text{dist})$ where X is a set of points associated with a distance function $\text{dist}: X \times X \to \mathbb{R}$, space $\mathcal{X}$ is called a metric space if for any point x, y, z in X, the following properties are satisfied: (1) $\text{dist}(x, y) = 0$ iff $x = y$; (2) $\text{dist}(x, y) = \text{dist}(y, x)$; (3) $\text{dist}(x, y) + \text{dist}(y, z) \geq \text{dist}(x, z)$.*

Definition 3 (doubling dimension). *Given a metric space (X, dist) and an arbitrary non-negative real number r, for any point $x \in X$, let $B(x, r) = \{y \in X \mid \text{dist}(x, y) \leq r\}$ denote the set of points in the ball around x with radius r. The doubling dimension of the metric space (X, dist) is the smallest integer d such that any ball $B(x, 2r)$ can be covered by at most 2^d balls $B(y_i, r)$. A metric space with bounded doubling dimension is called a doubling metric space.*

Given a metric space $\mathcal{X} = (X, \text{dist})$, let Δ be the ratio between the largest and the smallest inter-point distance in $\mathcal{X}$, which is called the *aspect ratio* of $\mathcal{X}$. For any subset $Y \subseteq X$, the *aspect ratio* of Y is defined as $\frac{\max_{y,y' \in Y} \text{dist}(y,y')}{\min_{y \neq y' \in Y} \text{dist}(y,y')}$. For a positive integer l, let $[l] = \{1, 2, ..., l\}$.

For an instance $(X, \text{dist}, \mathcal{C}, \mathcal{F}, k, \mathbb{C})$ of the GCF k-CMedian problem, given a solution (F, μ) of instance $(X, \text{dist}, \mathcal{C}, \mathcal{F}, k, \mathbb{C})$, the facilities in F are called candidate facilities. For a facility $f \in F$ and a client $c \in C$, if $\mu(c) = f$, then it is called that f *serves* client c. Let $X = \mathcal{C} \cup \mathcal{F}$, and assume that the size of X is n.

3   Decomposition of Metrics

In this section, we introduce a split-tree decomposition of the metrics. Our dynamic programming process is based on the split-tree decomposition. (All proofs are in the complete version.)

---
[1] We use $\tilde{O}(\cdot)$ notation to hide polylogarithmic factors.

Given an instance $(X, \text{dist}, \mathcal{C}, \mathcal{F}, k, \mathbb{C})$ of the GCF k-CMedian problem where $\mathbb{C}$ denotes the specific constraint of the k-median problem, following the assumptions in [4,11], the aspect ratio Δ of the input metrics should be at most $O(n^4/\epsilon)$ (where n is the size of $\mathcal{C} \cup \mathcal{F}$, and ϵ is a constant). In order to satisfy the aspect ratio assumption, we first deal with the points in X by the following preprocessing step and we use OPT to denote an optimal solution to the GCF k-CMedian problem. We now show that the aspect ratio of the metrics can be polynomially bounded by the following lemma, which induces an additional $\epsilon \text{cost}(\text{OPT})$ loss in the approximate guarantee.

Lemma 1. *Given an instance $\mathcal{I} = (X, \text{dist}, \mathcal{C}, \mathcal{F}, k, \mathbb{C})$ of the GCF k-CMedian problem and a real number $\epsilon > 0$, there exists an instance $\mathcal{I}' = (X, \text{dist}', \mathcal{C}, \mathcal{F}, k, \mathbb{C})$ of the GCF k-CMedian problem with metric distance function, such that the aspect ratio of the metrics is at most $O(n^4/\epsilon)$, and any λ-approximation solution to $\mathcal{I}'$ is a $(1+\epsilon)\lambda$-approximation to $\mathcal{I}$ for any constant $\lambda > 1$. A set of $|\mathcal{C}||\mathcal{F}|$ instances containing such an instance $\mathcal{I}'$ can be constructed in polynomial time.*

For ease of presentation, we reset the metric distance function dist$'$ as dist in the following.

3.1 Split-Tree Decomposition

For any set Y of points, a ρ-*covering* of Y is a subset $S \subseteq Y$ if for any point $y \in Y$, there is a point $s \in S$ such that $\text{dist}(y, s) \leq \rho$. A ρ-*packing* of Y is a subset $S \subseteq Y$ if for any $s, s' \in S$, $\text{dist}(s, s') \geq \rho$. A subset $S \subseteq Y$ is called a ρ-*net* in Y if it is both a ρ-*covering* of Y and a ρ-*packing* of Y. The size of a net in metrics with fixed doubling dimension d is bounded by the following lemma.

Lemma 2 ([20]). *Let (X, dist) be a metric space with fixed doubling dimension d and aspect ratio Δ, and let $S \subseteq X$ be a ρ-net. Then $|S| \leq (\frac{\Delta}{\rho})^d$.*

We use the randomized *split-tree* decomposition of [32] for doubling metrics. A *decomposition* of the metric (X, dist) is a partition of X, and a partition of X is a disjoint collection of subsets, called *block*. A *hierarchical decomposition* consists of a sequence of $\ell+1$ decompositions $D_0, D_1, ..., D_\ell$ such that every block of D_{i+1} is the union of blocks in D_i with $D_\ell = \{X\}$ and $D_0 = \{\{x\} \mid x \in X\}$. D_i is called the i-th *level* of the hierarchical decomposition. Given a hierarchical decomposition $D_0, D_1, ..., D_\ell$, the corresponding split-tree is as follows: the root node is the level ℓ block $D_\ell = \{X\}$, and the leaves are the singletons such that $D_0 = \{\{x\} \mid x \in X\}$. We now construct a split-tree with the above properties. Given a metric space (X, dist), let $\Delta = O(2^\ell)$ be the diameter of the metrics. We start by constructing a sequence of sets $X = Y_0 \supseteq Y_1 \supseteq Y_2 \dots \supseteq Y_{\ell-1}$ such that Y_i is a 2^{i-2}-net of Y_{i-1}. By the triangle inequality, Y_i is a 2^{i-1}-covering of X, and a 2^{i-2}-packing of X. Choose ϱ uniform randomly from $[\frac{1}{2}, 1)$, and let π be a random ordering of the points in X. For each point $x \in X$, let $\pi(x)$ be the order of x in π. For the set Y_i, let $\{y_1, \dots, y_h\}$ ($h \leq 2^{O(d)}$) be the set of points in Y_i with decreasing order in π. Starting from $D_\ell = \{X\}$, we recursively partition the metrics until each block only contains one point. For $0 \leq i \leq \ell - 1$, the blocks at level i can be constructed as follows. Let $r_i = 2^i \varrho$. For each block $B_l \in D_{i+1}$ and for each point $y_j \in Y_i$ $(1 \leq j \leq h)$, a new block $B_l^{y_j}$ can be obtained at level i by the following way. Let $B_l^{y_j} = \{x \in B_l \mid x \in B(y_j, r_i)$ and $(\forall j' > j : \pi(y_{j'}) < \pi(y_j))[x \notin B(y_{j'}, r_i)]\}$. That is, for each point $x \in B_l$, if x is contained in $B_l^{y_j}$, then x is not contained in any block $\{B_l^{y_{j+1}}, \ldots, B_l^{y_h}\}$. In brief, each level $i + 1$ block is the union of at most $2^{O(d)}$ level i blocks. The followings are the specific properties of the split-tree decomposition.

Lemma 3 ([32]). *Given a split-tree of metric space (X, dist) with a sequence of decompositions $D_0, D_1, \ldots, D_\ell$, the split-tree decomposition has the following properties:*

1. *The total number of levels ℓ is $O(\log n)$ (since $\Delta = O(n^4/\epsilon)$).*
2. *Each block of level i has diameter at most 2^{i+1}, namely the maximum distance between any pair of points in a block of level i is at most 2^{i+1}.*
3. *Each level i block is the union of at most $2^{O(d)}$ level $i - 1$ blocks.*
4. *For any pair of points $u, v \in X$, the probability that they are in different sets corresponding to the blocks at level i of the split-tree is at most $O(d) \cdot \frac{\mathrm{dist}(u,v)}{2^i}$.*

Lemma 4. *For any metric (X, dist) with fixed doubling dimension d, given a randomized split-tree T, for any $x \in X$ and radius r, the probability that the points of $B(x, r)$ are in different sets corresponding to the blocks at level i of split-tree is at most $4d \cdot r/2^i$.*

Given an instance $(X, \mathrm{dist}, \mathcal{C}, \mathcal{F}, k, \mathbb{C})$ of the GCF k-CMedian problem, let T be the split-tree obtained by using the methods in [32]. For each block B at level i in T, we compute a $\rho 2^{i+1}$-net P of block B. Each point in P is called a portal, and P is also called a portal set. By Lemma 2, the number of portals at a given block is at most $\rho^{-O(d)}$. Moreover, the split-tree with the portal sets can be found in time $(1/\rho)^{O(d)} n \log \Delta$ [3,11]. For the portal set P of a block B at level i in T, there is an important property as follows.

Lemma 5 ([32]). *For the portal set P of a block B of level i in T, assume that the children of B at level $i - 1$ are $B_1, B_2, \ldots, B_u$ and each block B_j has the portal set $P_j \subseteq B_j$. Then, P is a subset of the portal sets computed for the descendant blocks of B, i.e. $P \subseteq P_1 \cup P_2 \cup \ldots P_u$.*

Given a split-tree T and two blocks B_1 and B_2 at level i of T, and for two points $u \in B_1$, and $v \in B_2$, the distance between u and v on the split-tree is defined as the length of the path which is constructed by the subpath from u to a portal p_i of B_1, the subpath from p_i to a portal p_j of B_2, and the subpath from p_i to v. Given a block B at level i of split-tree T and a portal set $P \subseteq B$, if there is a client c_1 outside B that is assigned to a facility f_1 inside B crossing at a portal $p_1 \in P$, then we say that client c_1 enters B through portal p_1. Similarly, if there is a client c_2 inside B that is assigned to a facility f_2 outside B crossing at a portal $p_2 \in P$, then we say that client c_2 leaves B through portal p_2. In the following, all the distances considered are the distances on the split-tree T.

Lemma 6. *For any metric* (X, dist) *with fixed doubling dimension* d *and any* $\rho > 0$, *given a randomized split-tree* T, *a pair of blocks* B_1 *and* B_2 *of level* i, *the portal set* P_1 *of* B_1 *where* P_1 *is a* $\rho 2^{i+1}$-net *of block* B_1, *the portal set* P_2 *of* B_2 *where* P_2 *is a* $\rho 2^{i+1}$-net *of block* B_2, *and any two points* $u \in B_1$ *and* $v \in B_2$, *for the distance between* u *and* v *on the split-tree* T *and the distance* $\text{dist}(u, v)$ *in metric space, we have:* $\min_{p_i \in P_1, p_j \in P_2} \{\text{dist}(u, p_i) + \text{dist}(p_i, p_j) + \text{dist}(p_j, v)\} \leq$ $\text{dist}(u, v) + 4 \cdot \rho 2^{i+1}$.

3.2 Definition of Badly-Cut Points

In this section, we give the formal definition of the badly-cut points as follows, which is similar to the definition in [11]. For any pair of points $u, v \in X$, if they are in different sets corresponding to the blocks at level i of the split-tree, we say that u, v are cut at level i. For any point $x \in X$ and radius r, if the points of $B(x, r)$ are in different sets corresponding to the blocks at level i of the split-tree, we say that $B(x, r)$ is cut at level i. We start by finding an $O(1)$-approximate feasible solution L to the specific GCF k-CMedian problem. For each client $c \in C$, we denote $L(c)$ as the facility of L that serves c. Let L_c denote the distance between c and $L(c)$. For each facility f of L, we denote $\text{OPT}(f)$ as the nearest facility to f in OPT. Let OPT_f denote the distance between f and $\text{OPT}(f)$. We now give the formal definition of the badly-cut points as follows.

Definition 4. ([11]) *Let* (X, dist) *be a metric space with fixed doubling dimension* d, *given a randomized split-tree* T *and a constant* $\epsilon > 0$, *a client* $c \in C$ *is called badly-cut if there exists an integer* i *such that* $B(c, 2^i)$ *is cut at level* j *greater than* $i + \log d + 3 + \log \log(1/\epsilon) + \log(1/(\frac{\epsilon^2}{2}))$ *(where* $2^i \in [\epsilon L_c, L_c/\epsilon]$).

A facility f *of* L *is called badly-cut if there exists an integer* i *such that* $B(f, 2^i)$ *is cut at level* j *greater than* $i + \log d + 3 + \log \log(1/\epsilon) + \log(1/(\frac{\epsilon^2}{2}))$ *(where* $2^i \in [\epsilon \text{OPT}_f, \text{OPT}_f/\epsilon]$).

Let $\zeta(\epsilon, d) = \log d + 3 + \log \log(1/\epsilon) + \log(1/(\frac{\epsilon^2}{2}))$ for short in the following. With a slight modification of the definition in [11], we show the probability of each point being badly-cut in the following lemma.

Lemma 7. *For a metric space* (X, dist) *with fixed doubling dimension* d, *given a randomized split-tree* T, *for a point* $x \in X$, *the probability that* x *is badly-cut is at most* $\frac{\epsilon^2}{2}$.

4 A PTAS for k-median with Service Installation Costs Problem

In the k-median with service installation costs problem [35], we are given a set C of clients and a set $\mathcal{F}$ of facilities, a non-negative integer k, and a set $\mathcal{S}$ of services. Moreover, each client $c \in C$ is associated with a service $M(c) \in \mathcal{S}$, and each service $s \in \mathcal{S}$ is associated with a cost $f(s) > 0$ for installing it at a facility.

The goal of the k-median with service installation costs problem is to find a subset $F \subseteq \mathcal{F}$ of size at most k and an assignment function $\mu : \mathcal{C} \to F$, such that: (1) for any $f \in F$, a set $S(f) \subseteq \mathcal{S}$ of services is installed at f, and for each client $c \in \mathcal{C}$, c is assigned to an opened facility $f \in F$ satisfying $M(c) \in S(f)$; (2) the total cost (including the sum of service installation and client-connection costs) is minimized.

In this section, we show how to use the portal approach and the dynamic programming process based on the badly-cut technique to obtain a PTAS for the k-median with service installation costs (k-MSIC) problem in doubling metrics. Our portal approach is based on the framework of the badly-cut technique outlined in [11]. For a given instance, the precondition of applying a dynamic programming process in this instance is that there exists an optimal solution in the instance. To reduce the time complexity of the dynamic programming process, we modify the given instance to reduce the assignment cost of the badly-cut clients to 0. Hence, since the given instance of the k-MSIC problem has been modified, it is necessary to prove that there exists a near-optimal solution to the k-MSIC problem in the new instance. In the following, we first give our main proofs for solving the k-MSIC problem.

4.1   Modification of Instance for k-MSIC Problem

In this section, we show how to modify the instance for the k-MSIC problem by relocating the badly-cut clients.

We now show that the modified instance induces an arbitrarily small loss in the approximate guarantee with probability at least $1 - \epsilon$.

Lemma 8. *Given an instance $\mathcal{I}' = (X, \mathrm{dist}, \mathcal{C}, \mathcal{F}, k, \mathcal{S}, M, f)$ of the k-MSIC problem and a real number $\epsilon > 0$, let T be a randomized split-tree, we can convert it to into a new instance $\mathcal{I}_T = (X, \mathrm{dist}^*, \mathcal{C}, \mathcal{F}, k, \mathcal{S}, M, f)$ by moving each badly-cut client c to the position of $L(c)$, such that with probability at least $1 - \epsilon$, the difference between the cost of any solution S to $\mathcal{I}_T$ and the solution S to $\mathcal{I}'$ can be bounded by at most $O(\epsilon \mathrm{cost}(L))$. That is, $\max_{S} \{\mathrm{cost}(\mathcal{I}(S)) - \mathrm{cost}(\mathcal{I}_T(S)), \mathrm{cost}(\mathcal{I}_T(S)) - \mathrm{cost}(\mathcal{I}(S))\} \leq O(\epsilon \mathrm{cost}(L))$.*

4.2   Construction of Near-Optimal Solution for k-MSIC Problem

In this section, we prove that there exists a near-optimal solution to the k-MSIC problem with the badly-cut facilities of L based on the modified instance. We now give the general idea of the proof.

Given an instance $\mathcal{I}_T = (X, \mathrm{dist}^*, \mathcal{C}, \mathcal{F}, k, \mathcal{S}, M, f)$ obtained by modifying $\mathcal{I}'$, for the facilities of L in the instance $\mathcal{I}_T$, if a facility of L is badly-cut, open this facility. Thus, based on the probability that a point is badly-cut, it holds trivially that there are at most ϵk badly-cut facilities of L opened. To deal with the case that the obtained solution with more than k facilities, for each badly-cut facility f of L, we open f and close some facility f' of OPT, and reassign the clients assigned to f' to f. This guarantees that the resulting solution is of cost at most $(1+O(\epsilon))\mathrm{cost}(\mathrm{OPT})+O(\epsilon)\mathrm{cost}(L)$. The following lemma shows that there exists a near-optimal solution S' containing the opened badly-cut facilities of L in $\mathcal{I}_T$.

Lemma 9. *Given an instance* $\mathcal{I}_T = (X, \mathrm{dist}^*, \mathcal{C}, \mathcal{F}, k, \mathcal{S}, M, f)$ *of the k-MSIC problem and a real number $\epsilon > 0$, with probability at least $1 - \epsilon$, there exists a near-optimal solution S' containing the opened badly-cut facilities of L with cost at most $(1 + O(\epsilon))\mathrm{cost}(\mathrm{OPT}) + O(\epsilon)\mathrm{cost}(L)$.*

4.3 Dynamic Programming Process for k-MSIC Problem

In this section, we show how to design the dynamic programming process based on the split-tree. Before the dynamic programming process, we should open each badly-cut facility of L based on Sect. 4.2.

Given an instance $\mathcal{I}_T = (X, \mathrm{dist}^*, \mathcal{C}, \mathcal{F}, k, \mathcal{S}, M, f)$ of the k-MSIC problem, let T be a randomized split-tree, the dynamic programming process proceeds on T from the leaves to the root. For each block B of the split-tree T, let T_B denote the subtree rooted at block B, and each subtree is a subproblem in our dynamic programming process, which includes the information that how many clients with each service entering or leaving the subtree through the portals of the subtree. A table entry in the dynamic programming process is a tuple $M[B, k_B, I, O]$, where the parameters are defined as follows: (1) m is the size of portal set in B and $m = \rho^{-O(d)}$; (2) B is the root node of the subtree T_B; (3) k_B ($0 \le k_B \le k$) is the number of opened facilities in B; (4) I is a tuple with $m|\mathcal{S}|$ elements where $I = [N_1^I, N_2^I, \ldots, N_m^I]$, N_i^I ($i \in [m]$) in I is an $|\mathcal{S}|$-tuple where $N_i^I = [n_1^{I(i)}, n_2^{I(i)}, \ldots, n_{\mathcal{S}}^{I(i)}]$, and for $t \in [|\mathcal{S}|]$, $n_t^{I(i)}$ ($0 \le n_t^{I(i)} \le n$) in N_i^I denotes the number of clients with the t-th service entering B through the i-th portal; (5) O is a tuple with $m|\mathcal{S}|$ elements where $O = [N_1^O, N_2^O, \ldots, N_m^O]$, N_i^O ($i \in [m]$) in O is an $|\mathcal{S}|$-tuple where $N_i^O = [n_1^{O(i)}, n_2^{O(i)}, \ldots, n_{\mathcal{S}}^{O(i)}]$, and for $t \in [|\mathcal{S}|]$, $n_t^{O(i)}$ ($-n \le n_t^{O(i)} \le 0$) in N_i^O is a non-positive number such that $n_t^{O(i)} = -n_t^{O(i)'}$, where $n_t^{O(i)'}$ denotes the number of clients with the t-th service leaving B through the i-th portal.

The cost of table entry $M[B, k_B, I, O]$ consists of the following three parts: (1) The cost of assigning clients inside of B to facilities inside of B; (2) For the clients inside of B assigned to facilities outside of B through portals of B, the cost from the clients inside of B to portals of B; (3) For the clients outside of B assigned to facilities inside of B through portals of B, the cost from portals of B to the assigned facilities inside of B.

The solutions at the root R of T are in the table entry $M[R, k, \phi, \phi]$, where ϕ is a tuple with $m|\mathcal{S}|$ zero components. Among all these solutions in $M[R, k, \phi, \phi]$, the algorithm outputs the one with the minimum cost.

The base case of the dynamic programming process is located at the leaves (which are singletons) of the split-tree. For a leaf of the split-tree, the corresponding block has only one portal and we only need to set N_1^I and N_1^O, while the parameters in the remaining N_i^I's and N_i^O's can be set as 0. Since there is only one facility or client in each leaf of the split-tree, we consider the following three cases for the table entry of leaf node B: (1) If there is only one facility in the block and this facility is opened, then the table value of this block is $M[B, 1, I, \phi] = \sum_{t=1}^{|S|} x_t^{I(1)} \cdot f^t(s)$, where $I = [N_1^I, \phi', \ldots, \phi']$ and ϕ' is a tuple with $|S|$ zero components. In addition, the t-th value of N_1^I is denoted as n_t, i.e., n_t is the number of clients with the t-th service outside of B that are served by the facility inside B. $x_t^{I(1)}$ denotes whether there are clients with the t-th service entering B through the first portal, and $f^t(s)$ denotes the cost of installing the t-th service at the opened facility. If $n_t \neq 0$, $x_t^{I(1)} = 1$. Otherwise, $x_t^{I(1)} = 0$; (2) If there is only one facility in the block but this facility is not opened, then the table value of this block is $M[B, 0, \phi, \phi] = 0$; (3) If there is only one client in the block, then the table value of this block is $M[B, 0, \phi, O] = 0$. If the client is associated with the t-th service ($t \in [|S|]$), then the t-th value in N_1^O is -1 and the other values in O are 0.

Now we consider the subproblem on the subtree T_B, where block B is at level i of the split-tree T. We formalize the children of B as the following parameters. We define the children of B as $B_1, B_2, \ldots, B_u$, where u is at most $2^{O(d)}$. Let $\{B_1, B_2, \ldots, B_u\}$ be the set of children of block B. Assume that there is an ordering from left to right in $\{B_1, B_2, \ldots, B_u\}$. Let (B, k_B, I, O) denote the configuration of B and (B_j, k_{B_j}, I_j, O_j) be the configuration of B_j, where B_j ($j \in [u]$) denotes the j-th child of B, k_{B_j} is the number of opened facilities in B_j, I_j is a tuple with $m|S|$ elements and the i-th element $N_i^{I(j)}$ in I_j ($i \in [m], j \in [u]$) is an $|S|$-tuple, and each value $n_t^{I(ij)}$ in $N_i^{I(j)}$ ($0 \leq n_t^{I(ij)} \leq n, t \in [|S|]$) denotes the number of clients with the t-th service entering B_j through the i-th portal of B_j, O_j is a tuple with $m|S|$ elements and the i-th element $N_i^{O(j)}$ in O_j ($i \in [m], j \in [u]$) is an $|S|$-tuple, and each value $n_t^{O(ij)}$ in $N_i^{O(j)}$ ($-n \leq n_t^{O(ij)} \leq 0, t \in [|S|]$) is a non-positive number such that $n_t^{O(ij)} = -n_t^{O(ij)'}$, where $n_t^{O(ij)'}$ denotes the number of clients with the t-th service leaving B_j through the i-th portal of B_j. The values of k_{B_j}, I_j and O_j have the following constraints: (1) $k_{B_1} + k_{B_2} + \ldots + k_{B_u} \leq k_B$; (2) For each $t \in [|S|]$, $\sum_{i=1}^{m} \sum_{j=1}^{u} n_t^{I(ij)} + \sum_{i=1}^{m} \sum_{j=1}^{u} n_t^{O(ij)} = \sum_{i=1}^{m} n_t^{I(i)} + \sum_{i=1}^{m} n_t^{O(i)}$.

Lemma 10. *Given a subproblem B at level i of the split-tree T ($i \neq 0$) and its children $B_1, B_2, \ldots, B_u$, where u is at most $2^{O(d)}$, let (B, k_B, I, O) be the configuration of B and (B_j, k_{B_j}, I_j, O_j) be the configuration of each children B_j ($1 \leq j \leq u$), we have $M[B, k_B, I, O] = \min_{j \in [u]: k_{B_j}, I_j, O_j} \sum_{j=1}^{u} \{M[B_j, k_{B_j}, I_j, O_j] + \omega\}$ with (1) $k_{B_1} + k_{B_2} + \ldots + k_{B_u} \leq k_B$; (2) $\forall t \in [|S|], \sum_{i=1}^{m} \sum_{j=1}^{u} n_t^{I(ij)} + \sum_{i=1}^{m} \sum_{j=1}^{u} n_t^{O(ij)} = \sum_{i=1}^{m} n_t^{I(i)} + \sum_{i=1}^{m} n_t^{O(i)}$, where ω is the cost of moving clients between portals of $B_1, B_2, \ldots, B_u$ and the ones in B, and can be calculated in polynomial time.*

4.4 Analysis and Running Time

In this section, we analyze the running time of our algorithm for solving the k-MSIC problem in doubling metrics.

Given an instance $(X, \text{dist}, \mathcal{C}, \mathcal{F}, k, \mathcal{S}, M, f)$ of the k-MSIC problem in doubling metrics, with probability at least $1 - \epsilon$, there exists a randomized algorithm such that a solution of cost at most $(1 + O(\epsilon))\text{cost}(\text{OPT}) + O(\epsilon\text{cost}(L))$ can be obtained, and the running time is at most $\tilde{O}(n^{O(\frac{1}{\epsilon})^{O(d)}}|\mathcal{S}|)$.

5 Extension to k-facility Location Problem and Priority k-median Problem in Doubling Metrics

Our techniques can be generalized to variants of the clustering problems. We consider two of them: k-facility location problem and priority k-median problem. We analyze the running time of the algorithms for the k-facility location and priority k-median problems by the same approach of the k-MSIC problem.

In the k-facility location problem, we are given a set $\mathcal{C}$ of clients and a set $\mathcal{F}$ of facilities, a non-negative integer k, and a facility-opening cost function $o : \mathcal{F} \to \mathbb{R}$. The goal of the k-facility location problem is to find a subset $F \subseteq \mathcal{F}$ of size at most k and an assignment function $\mu : \mathcal{C} \to F$ such that the sum of facility-opening and client-connection costs is minimized. Given an instance $(X, \text{dist}, \mathcal{C}, \mathcal{F}, k, o)$ of the k-facility location problem in doubling metrics, with probability at least $1-\epsilon$, there exists a randomized algorithm such that a solution of cost at most $(1 + O(\epsilon))\text{cost}(\text{OPT}) + O(\epsilon\text{cost}(L))$ can be obtained, and the running time is at most $\tilde{O}(n^{O(\frac{1}{\epsilon})^{O(d)}})$.

In the priority k-median problem, we are given a set $\mathcal{C}$ of clients and a set $\mathcal{F}$ of facilities, a non-negative integer k, and a set $\mathcal{P} = \{1, 2, \ldots, |\mathcal{P}|\}$ of priorities. Moreover, each client $c \in \mathcal{C}$ is associated with a priority $p(c) \in \mathcal{P}$, and each priority $p \in \mathcal{P}$ is associated with a cost $f(p) > 0$ for opening any facility at the priority, where $f(p_1) \geq f(p_2)$ for each $p_1, p_2 \in \mathcal{P}$ with $p_1 > p_2$. The goal of the priority k-median problem is to find a subset $F \subseteq \mathcal{F}$ of size at most k and an assignment function $\mu : \mathcal{C} \to F$, such that each client is assigned to a facility opened at the same or higher priority and the sum of facility-opening and client-connection costs is minimized. Given an instance $(X, \text{dist}, \mathcal{C}, \mathcal{F}, k, \mathcal{P}, p, f)$ of the priority k-median problem in doubling metrics, with probability at least $1 - \epsilon$, there exists a randomized algorithm such that a solution of cost at most $(1 + O(\epsilon))\text{cost}(\text{OPT}) + O(\epsilon\text{cost}(L))$ can be obtained, and the running time is at most $\tilde{O}(n^{O(\frac{1}{\epsilon})^{O(d)}}|\mathcal{P}|)$.

References

1. Adamczyk, M., Byrka, J., Marcinkowski, J., Meesum, S.M., Wlodarczyk, M.: Constant-factor FPT approximation for capacitated k-median. In: Proceedings of 27th Annual European Symposium on Algorithms, p. 1 (2019)
2. Arora, S., Raghavan, P., Rao, S.: Approximation schemes for Euclidean k-medians and related problems. In: Proceedings of 30th Annual ACM Symposium on Theory of Computing, pp. 106–113 (1998)
3. Bartal, Y., Gottlieb, L.A.: A linear time approximation scheme for Euclidean TSP. In: Proceedings of 54th Annual Symposium on Foundations of Computer Science, pp. 698–706 (2013)
4. Behsaz, B., Friggstad, Z., Salavatipour, M.R., Sivakumar, R.: Approximation algorithms for min-sum k-clustering and balanced k-median. Algorithmica 81(3), 1006–1030 (2019)
5. Bera, S., Chakrabarty, D., Flores, N., Negahbani, M.: Fair algorithms for clustering. In: Proceedings of 33rd Advances in Neural Information Processing Systems, pp. 4954–4965 (2019)
6. Bercea, I.O., et al.: On the cost of essentially fair clusterings. In: Proceedings of 22nd International Conference on Approximation Algorithms for Combinatorial Optimization Problems and 23rd International Conference on Randomization and Computation, p. 18 (2019)
7. Chalermsook, P., et al.: From gap-ETH to FPT-inapproximability: clique, dominating set, and more. In: Proceedings of 58th IEEE Annual Symposium on Foundations of Computer Science, pp. 743–754 (2017)
8. Charikar, M., Guha, S., Tardos, É., Shmoys, D.B.: A constant-factor approximation algorithm for the k-median problem. In: Proceedings of 31st Annual ACM Symposium on Theory of Computing, pp. 1–10 (1999)
9. Cohen-Addad, V.: Approximation schemes for capacitated clustering in doubling metrics. In: Proceedings of 14th Annual ACM-SIAM Symposium on Discrete Algorithms, pp. 2241–2259 (2020)
10. Cohen-Addad, V., Esfandiari, H., Mirrokni, V., Narayanan, S.: Improved approximations for Euclidean k-means and k-median, via nested quasi-independent sets. In: Proceedings of 54th Annual ACM SIGACT Symposium on Theory of Computing, pp. 1621–1628 (2022)
11. Cohen-Addad, V., Feldmann, A.E., Saulpic, D.: Near-linear time approximations schemes for clustering in doubling metrics. In: Proceedings of 60th Annual Symposium on Foundations of Computer Science, pp. 540–559 (2019)
12. Cohen-Addad, V., Grandoni, F., Lee, E., Schwiegelshohn, C.: Breaching the 2 LMP approximation barrier for facility location with applications to k-median. In: Proceedings of 34th Annual ACM-SIAM Symposium on Discrete Algorithms, pp. 940–986 (2023)
13. Cohen-Addad, V., Gupta, A., Kumar, A., Lee, E., Li, J.: Tight FPT approximations for k-median and k-means. In: Proceedings of 46th International Colloquium on Automata, Languages, and Programming, pp. 42-1 (2019)
14. Current, J., Daskin, M., Schilling, D., et al.: Discrete network location models. In: Facility Location: Applications and Theory, vol. 1, pp. 81–118 (2002)
15. Ding, H., Xu, J.: A unified framework for clustering constrained data without locality property. Algorithmica 82(4), 808–852 (2020)
16. Feng, Q., Hu, J., Huang, N., Wang, J.: Improved PTAS for the constrained k-means problem. J. Comb. Optim. 37(4), 1091–1110 (2019)

17. Feng, Q., Zhang, Z., Huang, Z., Xu, J., Wang, J.: A unified framework of FPT approximation algorithms for clustering problems. In: Proceedings of 31st International Symposium on Algorithms and Computation, pp. 51–517 (2020)
18. Guha, S., Khuller, S.: Greedy strikes back: improved facility location algorithms. J. Algorithms **31**(1), 228–248 (1999)
19. Guo, Y., Huang, J., Zhang, Z.: A constant factor approximation for lower-bounded k-median. In: Proceedings of 16th International Conference Theory and Applications of Models of Computation, pp. 119–131 (2020)
20. Gupta, A., Krauthgamer, R., Lee, J.R.: Bounded geometries, fractals, and low-distortion embeddings. In: Proceedings of 44th Annual IEEE Symposium on Foundations of Computer Science, pp. 534–543 (2003)
21. Hajiaghayi, M., Hu, W., Li, J., Li, S., Saha, B.: A constant factor approximation algorithm for fault-tolerant k-median. ACM Trans. Algorithms **12**(3), 1–19 (2016)
22. Han, L., Xu, D., Du, D., Zhang, D.: A local search approximation algorithm for the uniform capacitated k-facility location problem. J. Comb. Optim. **35**(2), 409–423 (2018)
23. Jaiswal, R., Kumar, A., Sen, S.: A simple D^2-sampling based PTAS for k-means and other clustering problems. Algorithmica **70**(1), 22–46 (2014)
24. Kolliopoulos, S.G., Rao, S.: A nearly linear-time approximation scheme for the Euclidean k-median problem. SIAM J. Comput. **37**(3), 757–782 (2007)
25. Kumar, A., Sabharwal, Y.: The priority k-median problem. In: Proceedings of 27th Foundations of Software Technology and Theoretical Computer Science, pp. 71–83 (2007)
26. Love, R., Morris, J., Wesolowsky, G.: Facilities location: models and methods. Operations Research Series, vol. 7 (1988)
27. Manurangsi, P., Raghavendra, P.: A birthday repetition theorem and complexity of approximating dense CSPs. In: Proceedings of 44th International Colloquium on Automata, Languages, and Programming, p. 78 (2017)
28. Markarian, C.: Online non-metric facility location with service installation costs. In: ICEIS, pp. 737–743 (2021)
29. Ng, T.E., Zhang, H.: Predicting internet network distance with coordinates-based approaches. In: Proceedings of 21st Annual Joint Conference of the IEEE Computer and Communications Societies, pp. 170–179 (2002)
30. Shmoys, D.B., Swamy, C., Levi, R.: Facility location with service installation costs. In: Proceedings of 15th Annual ACM-SIAM Symposium on Discrete Algorithms, pp. 1088–1097 (2004)
31. Swamy, C., Kumar, A.: Primal-dual algorithms for connected facility location problems. Algorithmica **40**(4), 245–269 (2004)
32. Talwar, K.: Bypassing the embedding: algorithms for low dimensional metrics. In: Proceedings of 36th Annual ACM Symposium on Theory of Computing, pp. 281–290 (2004)
33. Tenenbaum, J.B., Silva, V.D., Langford, J.C.: A global geometric framework for nonlinear dimensionality reduction. Science **290**(5500), 2319–2323 (2000)
34. Zhang, Z., Feng, Q., Xu, J., Wang, J.: An approximation algorithm for k-median with priorities. Sci. China Inf. Sci. **64**(5), 150104 (2021)
35. Zhang, Z., Zhou, Y., Yu, S.: Better guarantees for k-median with service installation costs. Theoret. Comput. Sci. **923**, 292–303 (2022)

A Physical Zero-Knowledge Proof for Sumplete, a Puzzle Generated by ChatGPT

Kyosuke Hatsugai[1]($\boxtimes$), Kyoichi Asano[1], and Yoshiki Abe[1,2]

[1] The University of Electro-Communications, 1-5-1 Chofugaoka, Chofu,
Tokyo 182-8585, Japan
{hatsugai,k.asano,yoshiki}@uec.ac.jp
[2] National Institute of Advanced Industrial Science and Technology, 2-3-26 Aomi,
Koto-ku, Tokyo 135-0064, Japan

Abstract. In March 2023, ChatGPT generated a new puzzle, Sumplete. Sumplete consists of an $n \times n$ grid, each whose cell has an integer. In addition, each row and column of the grid has an integer, which we call a *target value*. The goal of Sumplete is to make the sum of integers in each row and column equal to the target value by deleting some integers of the cells. In this paper, we prove that Sumplete is NP-complete and propose a physical zero-knowledge proof for Sumplete. To show the NP-completeness, we give a polynomial reduction from the subset sum problem to Sumplete. In our physical zero-knowledge proof protocol, we use a card protocol that realizes the addition of negative and positive integers using cyclic permutation on a sequence of cards. To keep the solution secret, we use a technique named *decoy technique*.

Keywords: Physical Zero-knowledge Proof · Card-based Cryptographic Protocol · Sumplete

1 Introduction

1.1 Background

Chat Generative Pre-trained Transformer (ChatGPT) is an artificial intelligence chatbot developed by OpenAI [18]. ChatGPT can response to questions more naturally than usual artificial intelligence. Moreover, it can work on generative tasks such as writing sentences, programming, drawing pictures and making puzzles.

In March 2023, ChatGPT generated a puzzle named *Sumplete*[1] [19]. Sumplete is a puzzle consisting of a grid with $n \times n$ cells. Each cell in the grid has an integer. In addition, each row and column in the grid also has an integer, and we call it the *target value* hereafter. The goal of Sumplete is to delete some

---

[1] This name was also named by ChatGPT [19].

-3	3	2	-7	9	-8
5	-1	-9	8	-7	-5
1	8	6	3	5	14
9	9	-6	8	-8	3
-7	3	-9	5	8	16
11	10	-7	6	0	

-3	✗	2	-7	✗	-8
5	-1	-9	✗	✗	-5
✗	8	6	✗	✗	14
9	✗	-6	8	-8	3
✗	3	✗	5	8	16
11	10	-7	6	0	

Fig. 1. Example of a problem (left) and its solution (right) of Sumplete.

integers in the cells so that the sum of the non-deleted integers in every row (resp., column) is equal to the target value of the row (resp., column). Figure 1 shows an example of the Sumplete problem and its solution: deleted integers' locations.

Since there are 2^{n^2} possible combinations of locations to delete integers, it seems hard to judge whether the given Sumplete problem has a solution if n becomes large. Indeed, as described later, Sumpelte is NP-complete. It is interesting that AI generates an NP-complete puzzle.

This paper considers a situation where a contestant wants to convince a challenger the existence of a solution to a given Sumplete problem while the contestant does not want to tell the solution to the challenger. Although these requirements may seem contradictory at first glance, they can be satisfied simultaneously using a cryptographic technique called Zero-knowledge proof (ZKP).

ZKP is an interactive proof proposed by Goldwasser, Micali, and Rackoff in 1989 [7]. ZKP allows a prover P, who knows the solution to a problem, to convince a verifier V of the existence of the problem's solution without revealing the solution itself. Usually, ZKP protocols are implemented using computers. However, ZKP protocols implemented by physical tools like cards instead of computers are also studied, called *physical zero-knowledge proof* (physical ZKP). The first physical ZKP protocol is that for a pencil puzzle, Sudoku, proposed by Gradwohl, Naor, Pinkas, and Rothblum in 2007 [8]. Since then, physical ZKP protocols are proposed for various puzzles such as Sudoku [8,20,28,29], Nonogram [4,21], Slitherlink [11,12], Akari [2], Numberlink [23,24], Norinori [6], Makaro [3,27], Takuzu [2,14], Kakuro [2,15], Shikaku [26], and Pancake sorting [10].

Most card operations used in physical ZKP protocols come from card-based cryptography, secure multiparty computation (MPC) using cards. The study of card-based cryptographic protocol began with the protocol for computing the two-input logical AND function by den Boar in 1989 [1] and that for computing the two-input logical XOR function by Crepeau and Kilian in 1993 [5]. After their work, to realize MPC with cards for more complex functions, card shuffle operations called pile-shifting shuffle [30,31] and pile-scramble shuffle [9] are proposed. We also use these operations in this paper.

1.2 Our Contributions

Neither NP-completeness of Sumplete nor zero-knowledge proof protocol for Sumplete has been proven. In this paper, we prove that Sumplete is NP-complete. Our proof is based on the reduction from the Subset Sum Problem (SSP), known as an NP-complete problem.

In addition, we propose a physical zero-knowledge proof protocol for Sumplete. Our card-based ZKP protocol allows a prover P to convince a verifier V that integers in a Sumplete instance's cells can be deleted to satisfy the sum conditions in each row and column. On the other hand, during the protocol, the verifier V cannot get any information about which integers in the cells are deleted. Because of the NP-completeness of Sumplete, the larger problem's size n becomes, the more difficult it is to obtain the solution. Therefore, it is worth proposing a zero-knowledge proof protocol for Sumplete.

1.3 Organization

The rest of this paper is organized as follows. In Sect. 2, we introduce zero-knowledge proof and card operations. Section 3 is devoted to show the NP-completeness of Sumplete. In Sect. 4, we propose a physical ZKP protocol for Sumplete. Then, in Sect. 5, we show our proposed protocol satisfies the conditions required for ZKP. Finally, Sect. 6 concludes this paper.

2 Preliminaries

2.1 Notation

(Multi)sets are denoted by uppercase letters of the italic font, e.g., $\mathcal{A} = \{a_1, a_2, a_3\}$. Vectors are denoted by lowercase letters of the bold font and the i-th element of a vector $\mathbf{v}$ is denoted by v_i, e.g., $\mathbf{v} = (v_1, v_2, v_3)$. Matrices are denoted by uppercase letters of the bold font and the element of the i-th row and j-th column of a matrix $\mathbf{M}$ is denoted by $m_{i,j}$, e.g.,

$$\mathbf{M} = \begin{pmatrix} m_{1,1} & m_{1,2} \\ m_{2,1} & m_{2,2} \end{pmatrix}.$$

2.2 Zero-Knowledge Proof

Zero-knowledge proof is an interactive proof between a prover P and a verifier V. From now, we assume that P is a probabilistic Turing machine with unbounded computational ability and V is a probabilistic polynomial-time Turing machine. Let x be an instance of a NP language. For a given x which has the witness, we suppose that P can calculate the witness from x, however V cannot. We note that x has the witness if and only if P knows the witness since the computational ability of P is unbound. In zero-knowledge proof, P interacts with V and finally convince V that the problem x has the witness without revealing the witness. Zero-knowledge proof protocols must achieve the following three conditions.

Completeness. If P knows the witness, V is always convinced.

Soundness. If P does not know the witness, V is not convinced with more than negligible probability. The probability that V is convinced when P does not know the witness is called *soundness error*. If soundness error is less than 1, it approaches asymptotically to 0 by executing proof many times. However, repeating physical protocols by human hands is hard. Therefore, it is desirable that soundness error of physical zero-knowledge proof is 0.

Zero Knowledge. V cannot obtain any information of the witness. Formally, for every V, there exists a probabilistic polynomial-time algorithm S that does not know the witness. If the output of S is indistinguishable from the output of the interaction of P and V, any information about the solution is not leaked during the interaction.

2.3   Card Operations

In this paper, we use cards whose front side is either ♣ or ♡ and whose back side is ?. We assume that the front sides of cards with the same suit are identical, i.e., we cannot distinguish them. In addition, the back sides of all cards are also assumed to be indistinguishable. For understanding, we denote a face-down card ♣ (resp., ♡) by ?♣ (resp., ?♡).

Cyclic Shift. Let a $\mathbf{c} := (c_1, c_2, \ldots, c_k)$ be a card sequence of k cards. *Left cyclic shift* over $\mathbf{c}$ is defined as a operation that outputs

$$(c_{\rho^{-1}(1)}, c_{\rho^{-1}(2)}, \ldots, c_{\rho^{-1}(k)}),$$

where ρ is a cyclic permutation $\rho := (1\ k\ k-1\ \ldots\ 3\ 2)$.

Similarly, *right cyclic shift* over $\mathbf{c}$ is defined as a operation that outputs

$$(c_{\sigma^{-1}(1)}, c_{\sigma^{-1}(2)}, \ldots, c_{\sigma^{-1}(k)}),$$

where σ is a cyclic permutation $\sigma := (1\ 2\ 3\ \ldots\ k-1\ k)$.

Pile-Scramble Shuffle. Let a $\mathbf{p} := (p_1, p_2, \ldots, p_k)$ be a sequence of k piles of cards. Note that each pile has the same number of cards. *Pile-scramble shuffle* over $\mathbf{p}$ is a operation that outputs

$$(p_{\pi^{-1}(1)}, p_{\pi^{-1}(2)}, \ldots, p_{\pi^{-1}(k)}),$$

where a permutation π is an element of S_k, the symmetric group of degree k.

2.4   Representation of an Integer

Here, we show a representation of an integer using cards. In physical ZKP protocols and card-based cryptography protocols, e.g., [10,13,22,25,32], integers from

1 to n are represented by a sequence of n cards, which consists of a $\boxed{\heartsuit}$ card and $n-1$ $\boxed{\clubsuit}$ cards. Specifically, an integer i $(1 \leq i \leq n)$ is represented by the position of $\boxed{\heartsuit}$ in the sequence: if $\boxed{\heartsuit}$ is at the i-th leftmost position, the sequence represents the integer i. For example, 1 and 3 are represented as follows.

$$1 = \boxed{\heartsuit}\,\boxed{\clubsuit}\,\boxed{\clubsuit}\,\boxed{\clubsuit}\,\boxed{\clubsuit}$$
$$3 = \boxed{\clubsuit}\,\boxed{\clubsuit}\,\boxed{\heartsuit}\,\boxed{\clubsuit}\,\boxed{\clubsuit}$$

In this paper, we apply this method to represent integers including less than 1.

Definition 1 (Integer Counter). Let α and β be positive integers. An integer i $(-\alpha \leq i \leq \beta)$ is represented by a sequence of $\alpha + \beta + 1$ cards, consisting of a $\boxed{\heartsuit}$ card and $\alpha + \beta$ $\boxed{\clubsuit}$ cards. Specifically, if $\boxed{\heartsuit}$ is at the i-th leftmost position of the sequence, the sequence represents the integer $i - \alpha - 1$. We call the card sequence to represent an integer an *integer counter*.

For example, 0 and -2 are represented as follows when $\alpha = 3$ and $\beta = 2$.

$$0 = \boxed{\clubsuit}\,\boxed{\clubsuit}\,\boxed{\clubsuit}\,\boxed{\heartsuit}\,\boxed{\clubsuit}\,\boxed{\clubsuit}$$
$$-2 = \boxed{\clubsuit}\,\boxed{\heartsuit}\,\boxed{\clubsuit}\,\boxed{\clubsuit}\,\boxed{\clubsuit}\,\boxed{\clubsuit}$$

We execute the addition by shifting the counter. This method is used by Shinagawa et al. [31] and Ruangwises and Itoh [26]. Suppose there is an integer counter representing $x \in \mathbb{Z}$, where the number of cards (i.e., the values α and β for the counter) is large enough so that $\boxed{\heartsuit}$ does not overflow. Then, we can obtain the counter representing $x + y$ $(y \in \mathbb{Z})$ as follows: if $y < 0$, we execute the *left* cyclic shift over the card sequence $|y|$ times; otherwise (i.e., if $y \geq 0$), we execute the *right* cyclic shift over the card sequence $|y|$ times. Since these cyclic shift operations can be performed even if the cards of the counter are face-down, the addition of y can be performed without revealing the value of x.

3 NP-Completeness of Sumplete

In this section, we prove that Sumplete is NP-complete. To prove the NP-completeness, we show a polynomial-time reduction from an NP-complete problem SSP to Sumplete.

3.1 Formal Definition of Problems

Before proving Sumplete's NP-completeness, we formally define the decisional version of Sumplete and SSP.

Definition 2 (Sumplete). An instance of Sumplete consists of three ingredients: an $n \times n$ matrix $\mathbf{G} \in \mathbb{Z}^{n \times n}$ that represents each integer of the corresponding cell, a vector $\mathbf{R} \in \mathbb{Z}^n$ that represents each row's target value, and a vector

$\mathbf{C} \in \mathbb{Z}^n$ that represents each column's target value. The answer of the instance $S = (\mathbf{G}, \mathbf{R}, \mathbf{C})$ is *Yes* if there exists an $n \times n$ matrix $\hat{\mathbf{G}} \in \{0,1\}^{n \times n}$ that satisfies following equations:

$$r_i = \sum_{j=1}^{n} g_{i,j}\hat{g}_{i,j} \qquad \text{for all } i \in \{1,\ldots,n\},$$

$$c_j = \sum_{i=1}^{n} g_{i,j}\hat{g}_{i,j} \qquad \text{for all } j \in \{1,\ldots,n\}.$$

If there does not exist such $\hat{\mathbf{G}}$, the answer is *No*.

For instance, the example of Fig. 2 can be represented as follows:

$$\mathbf{G} = \begin{pmatrix} -3 & 3 & 2 & -7 & 9 \\ 5 & -1 & -9 & 8 & -7 \\ 1 & 8 & 6 & 3 & 5 \\ 9 & 9 & -6 & 8 & -8 \\ -7 & 3 & -9 & 5 & 8 \end{pmatrix},$$

$$\mathbf{R} = (-8 \ -5 \ 14 \ 3 \ 16),$$

$$\mathbf{C} = (11 \ 10 \ -7 \ 6 \ 0).$$

The answer of above instance is Yes since there exists the following solution $\hat{\mathbf{G}}$:

$$\hat{\mathbf{G}} = \begin{pmatrix} 1 & 0 & 1 & 1 & 0 \\ 1 & 1 & 1 & 0 & 0 \\ 0 & 1 & 1 & 0 & 0 \\ 1 & 0 & 1 & 1 & 1 \\ 0 & 1 & 0 & 1 & 1 \end{pmatrix}.$$

Definition 3 (Subset Sum Problem). The Subset Sum Problem (SSP) consists of a multiset $\mathcal{A} \subset \mathbb{Z}^n$ and an integer $N \in \mathbb{Z}$. The answer of the instance SSP $= (\mathcal{A}, N)$ is *Yes* if there exists a subset $\mathcal{A}' \subseteq \mathcal{A}$ that satisfies $\sum_{a \in \mathcal{A}'} a = N$. If there does not exist such $\mathcal{A}'$, the answer is *No*.

For example, let us consider the instance SSP $= (\mathcal{A}, N)$ where $\mathcal{A} := \{-3, 3, 2, -7, 9\}$ and $N := -8$. The answer of this instance is Yes since there exists the solution $\mathcal{A}' = \{-3, 2, -7\} \subset \mathcal{A}$.

3.2 Proof of NP-Completeness

To show that Sumplete is NP-complete, we prove that the following holds.

(1) Sumplete is in NP.
(2) Sumplete is polynomial-time reductive from SSP.

Proof of (1). We prove the existence of a non-deterministic polynomial-time algorithm which can decide whether Yes or No for a given instance of Sumplete. Let us consider the non-deterministic algorithm M that works as follows.

1. M non-deterministically chooses some cells.
2. M deletes integers in chosen cells.
3. For every row and column, M calculates the sum of non-deleted integers and compares it with the target value. M rejects if there are rows or columns whose sum does not equal the target value. Otherwise, M accepts.

Since each operation ends in polynomial time, M halts in polynomial time. Thus, (1) holds.

Proof of (2). We prove (2) by showing the following three conditions hold.

  (i) There exists a polynomial-time reduction f from SSP to Sumplete.
 (ii) For arbitrary SSP's instance SSP, if the answer of SSP is Yes, then the answer of the reduced Sumplete's instance $S':=f(\text{SSP})$ is Yes.
(iii) For arbitrary SSP's instance SSP, if the answer of $S':=f(\text{SSP})$ is Yes, the answer of SSP is Yes.

First, we show (i). Let us consider the following polynomial-time reduction f (See also Fig. 2).

- f receives an instance of SSP, denoted by $\text{SSP}:=(\mathcal{A}, N)$ where $\mathcal{A}:=\{a_1, a_2, \ldots, a_n\} \in \mathbb{Z}^n$ and $N \in \mathbb{Z}$.
- f outputs an instance of Sumplete, denoted by $S':=(\mathbf{G}', \mathbf{R}', \mathbf{C}')$, where $\mathbf{G}'$, $\mathbf{R}'$, and $\mathbf{C}'$ are defined as follows:

$$\mathbf{G}':= \begin{pmatrix} a_1 & a_2 & \cdots & a_n \\ a_{\rho^{-1}(1)} & a_{\rho^{-1}(2)} & \cdots & a_{\rho^{-1}(n)} \\ \vdots & \vdots & \ddots & \vdots \\ a_{\rho^{-(n-1)}(1)} & a_{\rho^{-(n-1)}(2)} & \cdots & a_{\rho^{-(n-1)}(n)} \end{pmatrix},$$

$$\mathbf{R}':= \begin{pmatrix} N & N & \cdots & N \end{pmatrix},$$
$$\mathbf{C}':= \begin{pmatrix} N & N & \cdots & N \end{pmatrix},$$

where $\rho:=(1\ n\ n-1\ \cdots\ 3\ 2)$ is a cyclic permutation.

We note that all the target values in S' are N. In addition, each element of $\mathcal{A}$ appears once for each row and column. In reduction f, $n^2 + 2n$ times writing of integers and n times shifting are executed. Therefore, the running time of f is polynomial in n.

Next, we show (ii). Since the answer of SSP is Yes, there exists the solution $\mathcal{A}' \subseteq \mathcal{A}$ such that $\sum_{a' \in \mathcal{A}'} a' = N$. Here, let us consider to delete all integers in $\mathbf{G}'$ excluding all integers in $\mathcal{A}'$. Then, the set of the non-deleted integers for each row and column in $\mathbf{G}'$ equals $\mathcal{A}'$ since each row and column in $\mathbf{G}'$ equals

a_1	a_2	$\cdots$	a_n	N
a_2	a_3	$\cdots$	a_1	N
$\vdots$	$\vdots$	$\ddots$	$\vdots$	$\vdots$
a_n	a_1	$\cdots$	a_{n-1}	N
N	N	$\cdots$	N	

Fig. 2. Instance of Sumplete constructed form an instance of the subset sum problem.

$\mathcal{A}$. Thus, the sum of the non-deleted integers for each row and column is N. Therefore, the answer of S′ is Yes.

Finally, we show (iii). For each row and column in S′ after deletion, the set of non-deleted integers equals the solution of SSP. Therefore, if the answer of S′ is Yes, then the answer of SSP is Yes.

From (i), (ii), and (iii), (2) holds.   □

Hence, from (1) and (2), Sumplete is NP-complete.

4   Physical Zero-Knowledge Proof Protocol for Sumplete

In our proposed protocol, the prover represents the solution of a Sumplete's instance using cards, and the verifier checks whether the sum of non-deleted integers equals the target value for each row and column.

4.1   Idea of Proposed Protocol

The outline of our proposed protocol is as follows. The prover calculates the sum of non-deleted integers for each row and column of the grid using an integer counter. Then, the verifier checks that the sum equals the target value. If the sum equals the target value for all rows and columns, the verifier can be convinced of the solution's existence.

Decoy Technique. To realize the above operation without revealing the solution, i.e., the locations of cells whose integer is deleted, we prepare two integer counters called a *true counter* and a *false counter* for each row and column. The true (resp., false) counter is used to calculate the sum of the non-deleted (resp., deleted) integers in a row or column. In our protocol, for each integer in a row or column, the prover add the integer to the true or false counter depending on the solution: if it is a non-deleted (resp., deleted) integer, it is added to the true (resp., false) counter. By using a technique we call *decoy technique*, we can add integers while hiding the solution, i.e., which counter they were added to. Similar technique is widely seen in physical cryptography, e.g., [16,17].

4.2 Proposed Protocol

Proposed protocol proceeds as follows.

1. For each cell, the prover places a pair of face-down cards 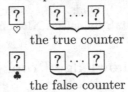 on non-deleted integer's cells and ? ? on deleted integer's cells.

2. For each row and column, the prover and the verifier execute the following operations.
   (a) The prover calculates α (resp., β), an absolute value of the sum of negative (resp., positive) integers. The prover also makes a true counter and a false counter which can represent an integer i ($-\alpha \le i \le \beta$). Then, the prover places the false counter below the true counter. In addition, the verifier checks that both counters indicates 0.
   (b) The prover places $\heartsuit$ on the left of the true counter and $\clubsuit$ on the left of the false counter. After placing them, the prover makes all the cards face-down. We call these two sequences of cards a *card matrix*.

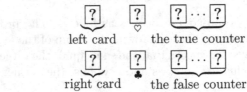

   (c) For every cell in the row (or column), the prover and the verifier execute the following operations.
      i. The prover picks a pair of cards on the cell and places left (resp., right) card of the pair to leftmost position of the upper (resp., lower) sequence of the card matrix made in Step 2(b).

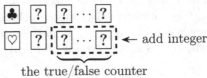

      ii. The prover regards each of the upper row and lower row in the matrix made in Step 2(c)i as a pile and applies pile-scramble shuffle to these two piles.
      iii. The prover opens the leftmost card of each row and adds the integer of the cell to the counter in the row whose opened leftmost card is $\heartsuit$. Note that this addition must be executed keeping the cards of the counter face-down.

$$\begin{array}{ccccc} \clubsuit & ? & ? & \cdots & ? \\ \heartsuit & ? & \boxed{? & \cdots & ?} \end{array} \leftarrow \text{add integer}$$

the true/false counter

iv. The prover makes all cards face-down. The prover regards upper and lower row as two piles and applies pile-scramble shuffle in same way as Step 2(c)ii.

v. The verifier opens the second leftmost card of each row. The verifier regards the upper and lower row as two piles and replaces these rows so that the row in which $\boxed{\heartsuit}$ is opened becomes the upper row.

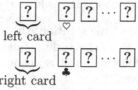

vi. The prover makes all the cards face-down and returns the pair of leftmost cards of the rows to the cell.

$$\underbrace{\boxed{?}}_{\text{left card}}_{\heartsuit} \quad \boxed{?}\,\boxed{?}\cdots\boxed{?}$$

$$\underbrace{\boxed{?}}_{\text{right card}}_{\clubsuit} \quad \boxed{?}\,\boxed{?}\cdots\boxed{?}$$

(d) The verifier opens the leftmost card of each row and opens the counter in the row whose opened leftmost card is $\boxed{\heartsuit}$ card. If the value indicated by opened counter does not equal the target value, the verifier rejects.

3. If the verifier did not reject fot all rows and columns, the verifier accepts.

4.3 Decoy Technique

The decoy techniqueis used in Step 2(c)ii and Step 2(c)iii. In Step 2(c)iii, the counter in the same row with $\boxed{\heartsuit}$ is the true (resp., false) counter if the integers' cell is non-deleted (resp., deleted). Thanks to the pile-scramble shuffle in Step 2(c)ii, the location of $\boxed{\heartsuit}$ in the leftmost column is independent from the card matrix made in Step (c)i. Therefore, the verifier cannot know which counter appears on the right of $\boxed{\heartsuit}$.

4.4 The Numbers of Cards and Shuffles

At first, we consider the number of cards used in our protocol. In Step 1, the prover uses n^2 pairs of $\boxed{\heartsuit}$ and $\boxed{\clubsuit}$ cards. In Step 2, the prover and verifier check n rows and n columns. Let M be the max value of $\alpha + \beta$ in each row and column, where α and β are those in Step 2(a). Then, Step 2(a) needs a $\boxed{\heartsuit}$ card and at most M $\boxed{\clubsuit}$ cards for each counter. In addition, Step 2(b) needs a $\boxed{\heartsuit}$ card and a $\boxed{\clubsuit}$ card. Since we can reuse the cards required for Steps 2(a) and 2(b), it is sufficient to consider the maximum number of cards required for the counter within the $2n$ times check. Thus, the prover and the verifier need 3 $\boxed{\heartsuit}$ cards and $2M + 1$ $\boxed{\clubsuit}$ cards in Step 2 Therefore, our protocol needs $2n^2 + 2M + 4$ cards (specifically, $n^2 + 3$ $\boxed{\heartsuit}$ cards and $n^2 + 2M + 1$ $\boxed{\clubsuit}$ cards) in total.

Next, we consider the number of shuffle in our protocol. Since pile-scramble shuffle is executed in Step 2(c)ii and Step 2(c)iv, $2n$ shuffles are executed in Step 2(c). In addition, Step 2(c) is executed $2n$ times in Step 2 Thus, our protocol needs $2n \times 2n = 4n^2$ shuffles.

5   Proof of Security

Here, we show our protocol satisfies the three conditions for ZKP.

5.1   Completeness

Lemma 1. If the prover knows the solution, the verifier always accepts.

Proof. If the prover knows the solution, the sum of the integers in cells, where two cards $\boxed{\heartsuit}\boxed{\clubsuit}$ are placed in this order, equals the target value for all rows and columns. Thus, the verifier does not reject for all rows and columns. Therefore, the verifier always accepts.   □

5.2   Soundness

Lemma 2. The soundness error is 0, that is, if the prover does not know the solution, the verifier always rejects.

Proof. We prove a contraposition of this lemma: if the verifier accepts, the prover knows the solution of the given instance of Sumplete. When the verifier accepts, the sum of cells where $\boxed{\heartsuit}\boxed{\clubsuit}$ are placed equals the target value for every row and column. Thus, we can see that the positions of the integers in the cells where $\boxed{\clubsuit}\boxed{\heartsuit}$ are placed are the solution. This fact implies that the prover knows the solution. Since the above argument holds with probability 1, the contraposition of Lemma 2 holds.   □

5.3   Zero Knowledge

Lemma 3. During the protocol, the verifier learns nothing about the solution of the given instance.

Proof. To prove zero-knowledge, it is sufficient that all distributions of opened cards can be simulated without the solution. In order to prove the zero-knowledge property, it is sufficient to show that all distributions of cards opened during the protocol execution can be simulated without the solution.

- In Step 2(c)iii, we open the leftmost column in the card matrix made in Step 2(c)i. Before the opening operation, we apply the pile-scramble shuffle to the two piles; one consists of the upper row, and the other consists of the lower row of the card matrix. Thus, $\boxed{\heartsuit}$ appears at each row with the same probability, i.e., probability 1/2. Therefore, the distribution of cards opened in Step 2(c)iii can be simulated without the solution.

- In Step 2(c)v, we open the second leftmost column in the card matrix made in Step 2(c)i. Before the opening operation, we apply the pile-scramble shuffle to the two piles of the upper row and the lower row of the card matrix. Thus, ♡ appears at each low with the probability 1/2. Therefore, the distribution of cards opened in Step 2(c)v can be simulated without the solution.
- In Step 2(d), we open the true counter after adding integers in the row or column for every row and column. If the prover knows the solution, the value represented by the true counter equals the target value of the row (or column). Therefore, the distribution of the true counter' cards opened in Step 2(d) can be simulated without the solution. Specifically, the true counter represents the target value with the probability of 1.

Therefore, the verifier learns nothing about the solution.   □

6   Conclusion

In this paper, we proved that Sumplete, a puzzle generated by ChatGPT, is NP-complete and proposed a physical zero-knowledge proof protocol. In our zero-knowledge proof protocol, we realized the addition of not only positive integers but negative integers by expansion the usual technique. Moreover, we use the decoy technique to conceal the solution from the verifier.

Acknowledgments. This work was supported by JSPS KAKENHI Grant Numbers JP22KJ1362 and JP23KJ0968.

References

1. Boer, B.: More efficient match-making and satisfiability *the five card trick*. In: Quisquater, J.-J., Vandewalle, J. (eds.) EUROCRYPT 1989. LNCS, vol. 434, pp. 208–217. Springer, Heidelberg (1990). https://doi.org/10.1007/3-540-46885-4_23
2. Bultel, X., Dreier, J., Dumas, J., Lafourcade, P.: Physical zero-knowledge proofs for Akari, Takuzu, Kakuro and KenKen. In: FUN, vol. 49, pp. 8:1–8:20 (2016)
3. Bultel, X., et al.: Physical zero-knowledge proof for Makaro. In: SSS, pp. 111–125 (2018)
4. Chien, Y., Hon, W.: Cryptographic and physical zero-knowledge proof: from Sudoku to Nonogram. In: FUN, pp. 102–112 (2010)
5. Crépeau, C., Kilian, J.: Discreet solitary games. In: Stinson, D.R. (ed.) CRYPTO 1993. LNCS, vol. 773, pp. 319–330. Springer, Heidelberg (1994). https://doi.org/10.1007/3-540-48329-2_27
6. Dumas, J., Lafourcade, P., Miyahara, D., Mizuki, T., Sasaki, T., Sone, H.: Interactive physical zero-knowledge proof for Norinori. In: COCOON, pp. 166–177 (2019)
7. Goldwasser, S., Micali, S., Rackoff, C.: The knowledge complexity of interactive proof systems. SIAM J. Comput. **18**(1), 186–208 (1989)
8. Gradwohl, R., Naor, M., Pinkas, B., Rothblum, G.N.: Cryptographic and physical zero-knowledge proof systems for solutions of Sudoku puzzles. Theory Comput. Syst. **44**(2), 245–268 (2009)

9. Ishikawa, R., Chida, E., Mizuki, T.: Efficient card-based protocols for generating a hidden random permutation without fixed points. In: UCNC, vol. 9252, pp. 215–226 (2015)
10. Komano, Y., Mizuki, T.: Card-based zero-knowledge proof protocol for pancake sorting. In: SecITC, pp. 222–239 (2022)
11. Lafourcade, P., et al.: How to construct physical zero-knowledge proofs for puzzles with a "single loop condition." Theor. Comput. Sci. **888**, 41–55 (2021)
12. Lafourcade, P., Miyahara, D., Mizuki, T., Sasaki, T., Sone, H.: A physical ZKP for slitherlink: how to perform physical topology-preserving computation. In: ISPEC, pp. 135–151 (2019)
13. Miyahara, D., Hayashi, Y., Mizuki, T., Sone, H.: Practical card-based implementations of yao's millionaire protocol. Theor. Comput. Sci. **803**, 207–221 (2020)
14. Miyahara, D., et al.: Card-based ZKP protocols for Takuzu and Juosan. In: FUN, pp. 20:1–20:21 (2021)
15. Miyahara, D., Sasaki, T., Mizuki, T., Sone, H.: Card-based physical zero-knowledge proof for Kakuro. IEICE Trans. Fundam. Electron. Commun. Comput. Sci. **102-A**(9), 1072–1078 (2019)
16. Mizuki, T., Sone, H.: Six-card secure AND and four-card secure XOR. In: FAW, pp. 358–369 (2009)
17. Nakai, T., Tokushige, Y., Misawa, Y., Iwamoto, M., Ohta, K.: Efficient card-based cryptographic protocols for millionaires' problem utilizing private permutations. In: CANS, pp. 500–517 (2016)
18. OpenAI: GPT-4 technical report (2023)
19. Penguin, P.: ChatGPT invented its own puzzle game (2023). https://puzzledpenguin.substack.com/p/chatgpt-invented-its-own-puzzle-game
20. Ruangwises, S.: Two standard decks of playing cards are sufficient for a ZKP for sudoku. New Gener. Comput. **40**(1), 49–65 (2022)
21. Ruangwises, S.: An improved physical ZKP for nonogram and nonogram color. J. Comb. Optim. **45**(5), 122 (2023)
22. Ruangwises, S., Itoh, T.: Securely computing the n-variable equality function with 2n cards. In: TAMC, pp. 25–36 (2020)
23. Ruangwises, S., Itoh, T.: Physical zero-knowledge proof for Numberlink. In: FUN, vol. 157, pp. 22:1–22:11 (2021)
24. Ruangwises, S., Itoh, T.: Physical zero-knowledge proof for numberlink puzzle and k vertex-disjoint paths problem. New Gener. Comput. **39**(1), 3–17 (2021)
25. Ruangwises, S., Itoh, T.: Physical zero-knowledge proof for Ripple Effect. In: WAL-COM, vol. 12635, pp. 296–307 (2021)
26. Ruangwises, S., Itoh, T.: How to physically verify a rectangle in a grid: a physical ZKP for shikaku. In: FUN, pp. 24:1–24:12 (2022)
27. Ruangwises, S., Itoh, T.: Physical ZKP for makaro using a standard deck of cards. In: TAMC, vol. 13571, pp. 43–54 (2022)
28. Sasaki, T., Miyahara, D., Mizuki, T., Sone, H.: Efficient card-based zero-knowledge proof for sudoku. Theor. Comput. Sci. **839**, 135–142 (2020)
29. Sasaki, T., Mizuki, T., Sone, H.: Card-based zero-knowledge proof for sudoku. In: FUN, vol. 100, pp. 29:1–29:10 (2018)
30. Shinagawa, K., et al.: Multi-party computation with small shuffle complexity using regular polygon cards. In: ProvSec, pp. 127–146 (2015)
31. Shinagawa, K., et al.: Card-based protocols using regular polygon cards. IEICE Trans. Fundam. Electron. Commun. Comput. Sci. **100-A**(9), 1900–1909 (2017)
32. Takashima, K., et al.: Card-based protocols for secure ranking computations. Theor. Comput. Sci. **845**, 122–135 (2020)

Author Index

W. Wu and G. Tong (Eds.): COCOON 2023, LNCS 14422, pp. 411–413, 2024.
https://doi.org/10.1007/978-3-031-49190-0

Printed in the United States
by Baker & Taylor Publisher Services

Printed in the United States
by Baker & Taylor Publisher Services